Groundwater Hydraulics and Pollutant Transport

Randall J. Charbeneau

Environmental and Water Resources Engineering Program
Department of Civil Engineering
The University of Texas at Austin

Prentice
Hall

PRENTICE HALL
Upper Saddle River, NJ 07458

Library of Congress Cataloging-in-Publication Data

Charbeneau, Randall J.
    Groundwater hydraulics and pollutant transport / Randall J. Charbeneau.
       p.   cm.
    Includes bibliographical references and index.
    ISBN 0-13-975616-7
    1. Groundwater flow.   2. Water—Pollution—Mathematical models.
  I. Title.
  TC176.C43   2000
  628.1′68—dc21

                                                     99-38307
                                                       CIP

Editor-in-chief: *Marcia Horton*
Acquistions editor: *Marcia Horton*
Production supervision/composition: *D&G Limited, LLC.*
Assistant managing editor: *Eileen Clark*
Executive managing editor: *Vince O'Brien*
Art director: *Jayne Conte*
Cover design: *Bruce Kenselaar*
Manufacturing buyer: *Beth Sturla*
Assistant vice president of production and manufacturing: *David W. Riccardi*

©2000 by Prentice-Hall, Inc.
Upper Saddle River, New Jersey 07458

The author and publisher of this book have used their best efforts in preparing this book. These efforts include the development, research, and testing of the theories to determine their effectiveness.

Printed in the United States of America

10 9 8 7 6 5 4 3 2 1

ISBN 0-13-975616-7

Prentice-Hall International (UK) Limited, *London*
Prentice-Hall of Australia Pty. Limited, *Sydney*
Prentice-Hall Canada Inc., *Toronto*
Prentice-Hall Hispanoamericana, S. A., *Mexico*
Prentice-Hall of India Private Limited, *New Delhi*
Prentice-Hall of Japan, Inc., *Tokyo*
Simon & Schuster Asia Pte. Ltd., *Singapore*
Editora Prentice-Hall do Brasil, Ltda., *Rio de Janeiro*

This textbook is dedicated to
Gerald T. Charbeneau
Who inspired me to be a teacher,
And showed me how it should be done.

Contents

Chapter Four

The Vadose Zone and Groundwater Recharge   179

Chapter Five

Groundwater Contamination   247

Chapter Six

Solute Transport by Advection   293

Chapter Nine

Multiphase Flow and Hydrocarbon Recovery   458

About the Author

Randall J. Charbeneau is Professor of Civil Engineering at the University of Texas at Austin where he has been on the faculty since 1978. He obtained his bachelor's degree from the University of Michigan (1973), his M.S. from Oregon State University (1975) and his Ph.D. from Stanford University (1978). Dr. Charbeneau, who is a licensed professional engineer, served as Director of the University of Texas Center for Research in Water Resources (CRWR) from 1989-1997. He currently serves as Associate Dean for Research in the College of Engineering. In addition to his research in the areas of groundwater hydrology, groundwater hydraulics, and subsurface mass transport, Charbeneau has maintained an active research program in stormwater management and radiological assessments. He has also served on numerous advisory boards, review panels and committees of the U.S. Environmental Protection Agency, the U.S. Department of Energy, the National Research Council, the National Science Foundation, and others.

Preface

Groundwater Hydraulics and Pollutant Transport is a textbook for upper level undergraduate and graduate courses in groundwater hydrology, groundwater hydraulics, and mass transport of subsurface contaminants. It is also a reference for practicing hydrologists, hydrogeologists, and environmental engineers. The subsurface flow of water and chemical mass transport are the focus of the book.

My primary objective in writing this book was to produce a textbook on subsurface flow and mass transport that is sufficiently rigorous and comprehensive, with problems and example problems with each chapter, to address the needs of upper level undergraduate and graduate students. This is a textbook for classroom use and self-study. The fundamentals that are presented are essential for students in natural sciences and engineering; however, applications focus on engineered systems for control of groundwater flow and contaminant transport. I have tried to write in a style that is readily accessible to students in engineering, drawing on their previous course work and developing quantitative material from basic principles founded in fluid mechanics.

The subject of subsurface fluid flow is covered in Chapters 1 to 4. The textbook includes discussion of basic concepts of groundwater hydrology. Darcy's law and continuity relations are presented in detail and are tied to their basis in fluid mechanics. In addition, applications of numerical models for groundwater flow are introduced. Special attention is given to the hydraulics of groundwater wells. Aquifer and well testing with pumping wells is presented, and applications of well hydraulics in dewatering and groundwater control are discussed. Chapter 4 provides an introduction to the assessment of unsaturated zone hydrologic processes, including discussion of many significant properties and their measurement.

Subsurface contamination and pollutant transport are covered in Chapters 5 to 9. Problems associated with groundwater pollution and its regulation are described in Chapter 5. This chapter also introduces methods for calculation of solute partitioning and degradation within the multiphase subsurface environment. The transport of solutes by groundwater flow is presented and discussed through application of potential and residence time distribution theory, including problems of pollutant transport and control using multiple well systems. Mass transport by molecular diffusion in porous media is discussed with applications including the volatilization loss of chemicals. Special attention is also given to advection-dispersion mass transport including its characterization through laboratory and field experiments and representation through the mechanical dispersion tensor. Development and application of analytical and numerical models for subsurface mass transport are presented. Chapter 9 introduces principles of multiphase flow in porous media and their application to problems of groundwater contamination by nonaqueous phase liquids such as solvents and petroleum hydrocarbon liquids.

This book is used for two courses at the University of Texas at Austin: an undergraduate course in groundwater hydraulics that follows courses in fluid mechanics and hydraulics, and a graduate course in groundwater contamination and transport. The undergraduate, senior-level course uses most of the material in Chapters 1 to 4, and includes a major assignment on application of numerical simulation modeling to a problem of controlling or assessing changes in a groundwater flow system. The graduate course uses most of the material in Chapters 5 to 9. It also includes a numerical simulation modeling assignment on the design of a groundwater remediation system that uses pumping wells. Generally, undergraduate curriculum in Civil Engineering and graduate curriculum in Environmental and Water Resources Engineering have courses on environmental chemistry, so I have not included standard material on geochemistry in this textbook. The included material on solute partitioning and dissolution within multiphase systems is essential for characterization of the source term of many groundwater contamination problems.

I view the application of groundwater hydraulics and pollutant transport as a quantitative field. Our predictions and assessments have numerical values and uncertainties associated with them. The presentation in *Groundwater Hydraulics and Pollutant Transport* reflects this viewpoint. Our quantitative language is vector calculus and our findings are often stated in the form of algebraic equations containing "special" functions. After one becomes familiar with this framework, it is easy to loose track of the fact that our quantitative framework may be built on shaky grounds. While the calculus is exact, the parameter fields that appear in our basic equations are poorly known. Laboratory and field methods provide estimates for *ideal* fields that are usually homogeneous, though calibration methods allow for estimation of heterogeneous ideal fields. What impact does use of *ideal* (instead of *real*) fields in our basic equations have on our model predictions? Unfortunately, we do not have an adequate answer to this question, which remains a significant area of research. Experience with groundwater dewatering and control applications, however, shows that use of estimated fields could provide useful answers for engineering design and analysis. The question remains open for mass transport applications, and the relationship between *ideal* and *real* fields needs to be considered on a continuing basis.

Numerical simulation and spreadsheet models are presented for use with this textbook. The numerical models MODFLOW and MOC3D are introduced in Chapters 2 and 8, respectfully. Both of these models were developed by the U.S. Geological Survey, are public domain, have a user interface (MFI) that is also public domain, and may be downloaded from the World Wide Web along with user documentation (see Table 2.5.1). These characteristics, along with the fact that MODFLOW is the most widely used numerical simulation model for groundwater flow, dictated their selection for this book. Many of the calculations presented can be performed through use of computer spreadsheets, and a number of these have been developed and described in the text and appendices. In addition, a number of modules that may be downloaded and incorporated into user-developed spreadsheets are presented in the appendices.

I wish to express my thanks to the many individuals who have contributed to this effort. Foremost, my students over the past 20 years have done much to change the way that I view the subject and how it can be presented. Many of them will identify their contribution to this textbook. Wade Hathhorn provided many useful comments on parts of an earlier draft of the text. Graham Fogg and Ken Rainwater provided extensive comments and suggestions that have greatly contributed to the present form and content of the manuscript. Finally, I wish to thank my wife Nancy and children Cynthia and Robert for their support and extreme patience with my work habits. I cannot promise that these habits will change, but at least this effort has reached completion.

Randall J. Charbeneau
Austin, Texas

Chapter One
Introduction to Groundwater Hydrology

Hydrology is the scientific study of water. It is concerned with the waters of the earth, their occurrence, circulation and distribution, their chemical and physical properties, and their reaction with their environment, including their relationship to living things. The domain of hydrology embraces the full life history of water on the earth (National Research Council, 1992). This definition of hydrology closely follows the definition given by O. Meinzer 50 years earlier (Meinzer, 1942). Hydrology, and its engineering applications, generally considers separately water that occurs above and below the surface of the earth. The science and engineering applications of surface water hydrology are mainly concerned with the origins of rainfall, the relationships between rainfall and runoff from a watershed, the routing of water through rivers and surface water reservoirs, and the occurrence of floods and droughts. **Groundwater hydrology** is the science that considers the occurrence, distribution, and movement of water below the surface of the earth. It is concerned both with the quantity and quality aspects of this water. In this text, we are principally concerned with groundwater hydrology, including its natural recharge and discharge.

Beneath the surface of the earth, groundwater flows through **porous media**, **fractured media**, and **large passages** (**karst**). A porous medium contains relatively small openings (pore space or voids) within the solid matrix, and the medium is permeable, allowing the passage of fluids (Dullien, 1992). A fractured medium has small openings in one direction (the fracture aperture), and relatively large openings in two directions along the plane of the fracture. A karst medium, such as limestone with large **dissolution cavities** (caves), has regions with large openings in three dimensions. Within karst media and fractured media, the nature of subsurface flow depends upon the lateral extent and connectivity of the openings. These are site-specific conditions and cannot be generalized. In this text, we are primarily concerned with fluid flow through porous media, though fractured media are discussed in Chapter 2, "Darcy's Law and Continuity Relations," and in Chapter 9, "Multiphase Flow and Hydrocarbon Recovery."

1

In this introductory chapter, we review the characterization of porous media and the distribution of water contained within their pore space. Parameters that are used to describe the pore space are introduced, and their foundational aspects are discussed. Components of the hydrologic cycle are discussed, different types of hydrogeologic formations are defined, and representative configurations for groundwater-bearing geologic formations are described.

1.1   Porous Media

The porous media that are of interest include natural soils, unconsolidated sediments, and sedimentary rock. Natural soils differ from unconsolidated sediments in their degree of weathering (Brady, 1974). Both are granular materials containing amorphous mineral and organic matter. The behavior of these materials in terms of retaining soil water and allowing fluid flow are determined by the range of particle sizes and pore-space openings. Based on their particle-size distribution, natural soils and unconsolidated sediments are classified into various textures.

Particle-Size Distribution

The grain or particle-size distribution curve gives the relative percentage of grain sizes. For coarser material, the grain size distribution is determined by sieving a soil sample through a series of successively smaller screens and measuring the percent mass retained. The test material is shaken on a sieve with square openings of specified size, and the size of the fraction that passes through the sieve is based on the size of the opening. Before mechanical analysis, the organic matter—which is ordinarily small in quantity—is normally removed by oxidation. For fine-grained material (approximately less than 0.05 mm), the grain size distribution is determined by the rate of settlement in an aqueous suspension. Gee and Bauder (1986) provide more detailed method descriptions for analyzing grain size.

Figure 1.1.1 shows particle-size distribution curves for three soils. Particle size is plotted on the x-axis, and the percent mass retained (percent larger than the given size) is plotted on the y-axis. For example, for Soil A in Figure 1.1.1, 90 percent of the soil mass has particle-size greater than 0.002 mm, and 20 percent has particle-size greater than 0.05 mm.

Many different organizations have established soil classification standards for use in different disciplines. The *United States Department of Agriculture* (USDA) system is one of the most widely used within the groundwater and environmental industries. Table 1.1.1 shows the USDA particle-size classification system. Based on the USDA classification system, Soil A in Figure 1.1.1 has 10 percent mass of clay size, 70 percent mass of silt-size particles, and 20 percent mass classified as sand.

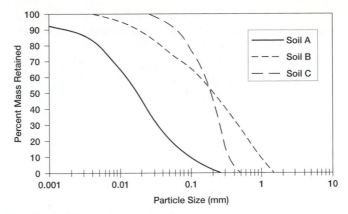

FIGURE 1.1.1 Example particle-size distribution curves

TABLE 1.1.1 USDA Classification of Soil Particles According to Size

Class	Diameter (mm)
Gravel	>2
Sand	0.05–2
very coarse	1–2
coarse	0.5–1
medium	0.25–0.5
fine	0.10–0.25
very fine	0.05–0.10
Silt	0.002–0.05
Clay	<0.002

Soil Texture

Natural soils and unconsolidated sediments may be composed of particles varying greatly in size and shape, and **soil textural class** names are used to convey some idea of their textural makeup and give some indication of their physical properties. **Soil texture** refers to the size range of particles in the soil. As such, the term carries both qualitative and quantitative connotations. Qualitatively, the term represents the "feel" of the soil material, whether coarse and gritty or fine and smooth. Quantitatively, soil texture denotes the measured distribution of particle sizes or proportions of the various size ranges of particles that occur in a given soil. It is a permanent, natural attribute of the soil that is often used to characterize its physical makeup (Hillel, 1982). Three broad, texture-based groups include sand, loam, and clay (Brady, 1974).

The grain size determines the particle size classification. The relative mass percentages of different particle sizes determine the soil texture for a soil sample. The USDA standard classification is provided by the **soil texture triangle**

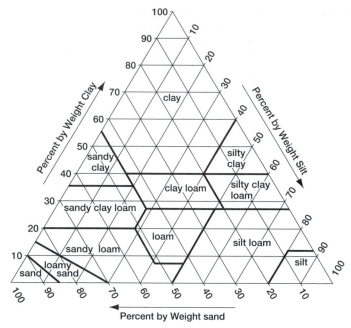

FIGURE 1.1.2 Textrue triangle, showing the fractions of clay, silt, and sand in the soil textural classes (courtesty U.S. Soil Conservation Service)

shown in Figure 1.1.2. Each point on the triangle corresponds to different percentages by mass (weight) of sand-, silt-, and clay-sized soil particles. For example, Soil A in Figure 1.1.1 consists of 20 percent sand, 70 percent silt, and 10 percent clay. This soil has a silt loam texture.

As might be expected, many of the physical and chemical characteristics of soils are related to their texture. Fine-grained soils have a much larger surface area than coarse-grained soils. In addition, their mineral structure is different, resulting in a much greater capacity for sorption of chemicals. On the other hand, well-sorted, coarse-grained soils have a much smaller sorption capacity and a much larger permeability. One of the significant issues in contaminant hydrogeology is to identify the permeability and sorption characteristics of different soils.

Cemented Sediments and Rock

Cemented sediments include sandstone and shale. Depending on the degree of cementation, sandstone can be very permeable and can serve as an excellent source for water supply. Shale, which is indurated clay, has very low permeability. Shale deposits are hydrogeologically significant when they act as confining beds, bounding more permeable strata. They can impede vertical flow over large areas. Carbonate rocks (limestone and dolomite) generally have relatively high amounts of primary (original) void space but have low permeability. Limestone, however, and to a lesser extent dolomite, may readily dissolve in flowing water. The solution openings provide significant permeability to these rocks, making them some of the most productive formations for water supply. Karst landscapes, which exhibit irregularities in surface form, are developed by significant rock dissolution.

Igneous and metamorphic rocks have small amounts of primary void space, commonly less than 1 percent of the rock mass. Furthermore, the few pores that are present are small and generally are not interconnected, resulting in permeabilities that may be regarded as zero for almost all practical problems (Davis and DeWiest, 1966). All rocks, however—**sedimentary**, **igneous**, and **metamorphic**—may be fractured by earth stresses. Fractured rock is quite different hydrogeologically from rock that has not been fractured. Water may move easily through the fractures, giving the rock mass a high permeability although not much capability for storing water within the pore space. Rock that is fractured by earth stresses may develop directional characteristics to its permeability, having a greater potential for allowing flow in certain directions.

1.2   Distribution of Subsurface Water

Porous media consist of solid material and void or pore space. Air and water and possibly non-aqueous phase liquids (oil, gasoline, etc.) occupy the pore space. Figure 1.2.1 provides a schematic view of the regions that make up the subsurface profile. The soil profile is shown in the center of the figure. On the left, the classification of these regions is shown, while the profile on the right shows the vertical distribution of water content.

The **vadose zone**[1], also called the unsaturated or partially saturated zone, constitutes all subsurface media above the water table. This zone contains three subzones. The first of these subzones is the **zone of soil moisture**. The zone of soil moisture is adjacent to the ground surface and extends through the base of the root zone. The water content and distribution within this zone depends on soil texture, vegetation, and atmospheric conditions. During periods of rainfall when water is entering the soil profile, the direction of flow is downward, and the soil may become saturated (except for entrapped air) —provided the infiltration rate is large enough. Following rainfall, the water content in this zone will decrease because of continued downward drainage as well as upward flow near the ground surface, with water loss occurring as a result of evaporation, and will also decrease because of water uptake by the roots of plants, with subsequent water loss to the atmosphere through transpiration.

---

[1]Todd (1959) notes that the term vadose is derived from the Latin 'vadosus', meaning shallow.

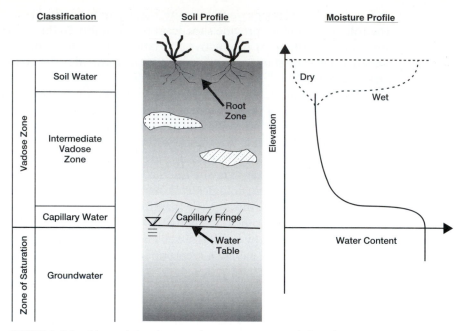

FIGURE 1.2.1   Vertical distribution of water content and classification system

After a period of time, the drainage rate slows considerably, and the remaining water content is called **field capacity**. This term refers to the water that is held by capillary forces and is largely unable to move because of gravitational forces alone. The water content of the soil water zone will decrease even after drainage has ceased because of combined evaporation and transpiration (collectively referred to as evapotranspiration). Eventually, the water content becomes so small that water ceases to be available to plants; therefore, the corresponding moisture content is known as the **wilting point**. For water contents below the wilting point, large "adhesive" forces hold the water, causing it to become immobile within the soil profile. The water content will remain near this low level until the next recharge event, though the soil immediately adjacent to the ground surface may continue to dry because of evaporation of soil moisture. Thus, the water content within the zone of soil moisture will vary from near saturation to nearly air-dry conditions.

The second subregion of the vadose zone is the **intermediate vadose zone**. The intermediate vadose zone extends from the base of the zone of soil moisture to the upper limit of the capillary zone. Water is held in this zone by capillary forces with water content at or near field capacity. Temporarily, the water content increases above field capacity as drainage from the zone of soil water moves downward to the capillary zone and the water table to become groundwater recharge. As the rate of water movement decreases, the water content again returns to field capacity. The intermediate vadose zone may not be present in areas with a shallow water table.

TABLE 1.2.1 Capillary rise in soils having similar porosity (0.41) after 72 days (from Lohman, 1972)

Material	Grain Size (mm)	Capillary Rise (cm)
Fine gravel	2–5	2.5
Very coarse sand	1–2	6.5
Coarse sand	0.5–1	13.5
Medium sand	0.2–0.5	24.6
Fine sand	0.1–0.2	42.8
Silt	0.05–0.1	105.5
Silt	0.02–0.05	200[1]

[1]Still rising after 72 days

The intermediate vadose zone terminates below with the **capillary zone** or capillary fringe. The transition from the intermediate zone to the capillary fringe is rather abrupt in coarse-grained soils but is very gradual and poorly defined in fine-grained soils such as silts and clays. Table 1.2.1 shows the capillary rise in soils of varying texture. The thickness of the capillary fringe varies from a few inches or centimeters in coarse-grained soils to a few feet or meters in fine-grained soils. The water content within the capillary zone is at saturation near its base, while at its upper extent it is near field capacity. This zone exists because water is pulled upward from the water table by surface tension, just as water rises in a capillary tube. The smaller the pore sizes, the greater the extent of capillary rise. The more uniform the pore size, the more abrupt the transition from the capillary zone to the intermediate vadose zone.

The vadose zone is separated from the zone of saturation by the **water table**, also known as the phreatic surface. The water table is defined as the surface or locus of points that are at atmospheric pressure. This area corresponds to the water elevation that is observed in a monitoring well. The diameter of a monitoring well is so large that capillary forces are unimportant and there is no capillary rise. Below the water table the gauge pressure is positive, while above the water table the water pressure is negative.

Water in the region that is below the water table is under positive pressure, and the porous matrix is saturated. This water is called **groundwater**, and the zone is called the **zone of saturation**. Both terms are somewhat misleading. Groundwater should refer to that water below the surface of the ground, while water in the lower part of the capillary fringe is saturated. This terminology is currently accepted, however.

1.3   Porosity and Related Properties of Soils

From a hydrologic point of view, the fundamental interests in a porous medium are its ability to hold and transmit water. There are a number of terms that relate to the water-holding potential of a medium. The most important of these is a medium's **porosity**, which is defined as the volume of voids per bulk volume (as shown in Equation 1.3.1):

$$n = \frac{\text{volume voids}}{\text{bulk volume}} = \frac{V_v}{V_t} \qquad (1.3.1)$$

Here, V_v is the volume of voids, and V_t is the bulk or total volume. Equation 1.3.1 may be written

$$n = \frac{V_t - V_s}{V_t} = 1 - \frac{M_s}{V_t} \times \frac{V_s}{M_s}$$

where V_s and M_s are the volume and mass of solids contained within the total volume. M_s/V_t, however, is the bulk density of the soil, ρ_b, while M_s/V_s is the soil particle density, ρ_s. Thus, the relationship between the porosity and bulk density (shown in Equation 1.3.2) is given by

$$n = 1 - \frac{\rho_b}{\rho_s} \qquad (1.3.2)$$

Equation 1.3.2 relates the porosity, bulk density, and particle density of a soil.

For most mineral soils, the particle density usually varies in the range 2.60 to 2.75 g/cm^3. This range occurs because quartz, feldspar, and the colloidal silicates with densities within this range usually make up the major portion of mineral soils. Organic matter weighs much less than an equal volume of mineral solids, having a particle density of 1.2 to 1.5 g/cm^3. Consequently, some mineral topsoil high in organic matter may have particle densities as low as 2.4 g/cm^3. Nevertheless, for general calculations, the average surface soil may be considered to have a particle density of about 2.65 g/cm^3. A soil with a porosity of 0.45 will have a bulk density of 1.45 g/cm^3, while if it is compacted to a porosity of 0.35, its bulk density will increase to 1.72 g/cm^3. Typical values of the porosity are shown in Table 1.3.1.

TABLE 1.3.1 Typical Range of Porosity Values

Material	n
Unconsolidated Material	
Gravel	0.20–0.40
Sand	0.25–0.55
Silt	0.35–0.60
Clay	0.35–0.65
Sedimentary Rock	
Sandstone	0.05–0.50
Limestone, dolomite	0–0.30
Karst limestone	0.05–0.50
Shale	0–0.10
Crystalline Rock	
Basalt	0.05–0.35
Fractured basalt	0.05–0.50
Dense crystalline rock	0–0.05
Fractured crystalline rock	0–0.10

The definition of porosity given by Equation 1.3.1 appears to be straightforward. There are a number of important and subtle questions that must be addressed, however, prior to proceeding with any scientific and engineering calculations. These questions concern the technical definition of parameters such as porosity. A basic objective in the scientific and engineering applications of groundwater hydrology is to develop a quantitative framework for analysis of processes associated with subsurface flow and mass transport. For engineering analysis, the foundational framework must be quantitative. That is, we must be able to apply the rules of calculus to problems of groundwater hydraulics. With this goal in mind, we ask the following questions: "What do we mean by porosity? Can we define porosity as a function, such as $n(\tilde{x})$?" Here, the meaning of $n(\tilde{x})$ is that the porosity is defined as a spatial field with its value at each point $\tilde{x}$ given by the value of the function $n(\tilde{x})$. (See Appendix A for a discussion of notation used in this textbook.) A moment's consideration shows that formally, at any given point, we must have $n = 0$ or $n = 1$, depending on whether the point lies within the solid matrix or within the void space, respectively. This function is rapidly discontinuous, and surely the rules of calculus will fail us. The question now concerns how to proceed to develop a quantitative framework for analysis of subsurface flow.

Two approaches have been developed to provide a fundamental basis for the quantitative analysis of subsurface flow and transport. The first of these is the notion of spatial averaging over a *representative elementary volume* (REV). Figure 1.3.1 shows a porous medium with different sizes of averaging volumes V^* centered at point $\tilde{x}$. We assign the point value of a property as that associated with an averaging volume centered at that point. For the porosity, the formal definition is

$$n(\tilde{x}) = \frac{1}{V^*} \int_{v^*(\tilde{x})} \chi(\tilde{x} - \tilde{y}) d\tilde{y} \qquad (1.3.3)$$

where $\chi(\tilde{x})$ is an indicator function for void space: $\chi(\tilde{x}) = 0$ if point $\tilde{x}$ is located in the solid phase $(\tilde{x} \in V_s)$, and $\chi(\tilde{x}) = 1$ if point $\tilde{x}$ is located in the void space $(x \in V_v)$. $V^*(\tilde{x})$ is the averaging volume centered on the point $\tilde{x}$. Figure 1.3.1 shows the estimated porosity value as a function of averaging volume size. If the

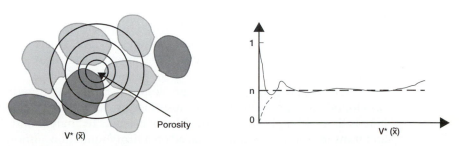

FIGURE 1.3.1 Estimation of porosity by averaging over a representative elementary volume (REV)

volume is too small, then the average value is not well defined and in the limit approaches either 0 or 1. If the averaging volume is too large, then soils with different textures may be included, and the resulting average depends on the nature of the heterogeneous soil surrounding the point. There may be a range of volumes surrounding the point for which the average is well defined, however. These volumes constitute representative elementary volumes for assigning a point value of the property being measured (porosity). The value assigned represents an extrapolated limit.

The idea of an extrapolated limit has already been used in fluid mechanics when we think of the density of a fluid at a point. In essence, the extrapolation is associated with the passage from the molecular scale to the fluid continuum scale. The same idea is brought forward for porous medium properties. An immediate question concerns the required size of the REV. For sandy soils, an averaging volume with a radius on the order of 10 to 20 grain diameters appears to be adequate for obtaining a well-defined average. Should the size of the averaging volume be the same for a fractured media or a karst media as for an alluvial soil? Obviously, the answer to this question is no if the average is to have any relevant physical meaning. The theory provides little guidance in this regard, however, except that the volume must be large enough so that the average is well behaved and small enough so that we do not average over large-scale trends in the variable. Bear (1972) provides a detailed discussion of REV concepts.

The second approach is based on the theory of ***random functions*** (RF). This approach has been used extensively by Dagan (1989), who provides a detailed discussion. The underlying idea is that because we cannot sample the medium in sufficient detail to know the spatial distribution of a variable with any great precision, the porous medium remains unknown to us. As such, the medium may be represented by any of a number of equivalent media, for example, based on the randomness of the depositional processes from which the sediments came. These equivalent random media form an ensemble of media. Using this definition, the value of a parameter is derived from a probabilistic average over this ensemble:

$$n(\tilde{x}) = E\{\chi(\tilde{x})\} = \langle\chi(\tilde{x})\rangle \qquad \textbf{(1.3.4)}$$

where $E\{\ \}$ represents the expectation operator (or average) from probability theory. Dagan (1989) and Marsily (1986) note that the RF approach is more powerful in that it allows one to deal with discontinuous media. More importantly, it allows one to look at other statistical properties of the "random" media, such as its variance and its spatial correlation structure. This latter characteristic in particular provides guidance as to the appropriate size of a REV for spatial averaging.

Actually, the two approaches are complementary and together provide a firm foundation for the analysis of subsurface processes. The main point to note here is that our application of the rules of calculus to analyze and discuss flow and transport processes can be placed on a firm foundation, although we leave the details to more advanced literature on the subject.

The porosity gives the total amount of a volume that can become filled with water or other fluids. The actual fraction of the volume that is filled with water is called the **water content** (or moisture content). This value is given by

$$\theta = \frac{\text{volume water}}{\text{bulk volume}} = \frac{V_w}{V_t} \qquad (1.3.5)$$

where θ is the water content and V_w is the volume of water contained in the total volume V_t. It is apparent that the water content must lie within the range

$$0 \le \theta \le n \qquad (1.3.6)$$

In applications where the subsurface void space is occupied by three or more fluids, it is convenient to introduce the phase saturation, which is the volume of the given phase per volume of void space. For example, the water saturation would be defined by

$$S_w = \frac{\text{volume water}}{\text{volume voids}} = \frac{\theta}{n} \qquad (1.3.7)$$

The saturation range must satisfy

$$0 \le S_w \le 1 \qquad (1.3.8)$$

In the case of multiphase flow, one fluid or another must occupy all of the pore space, resulting in the following relationship between the phase saturation values:

$$\sum S_i = 1 \qquad (1.3.9)$$

where S_i is the saturation of the ith phase and the summation is over all phases present. In the following discussion, we will use the volumetric water and phase contents in our quantitative analyses. We will often refer to the degree of saturation of the medium with regard to the various fluid phases present in general, qualitative terms, however.

Before ending this discussion, several additional terms should be introduced to the reader regarding terminology with respect to the porosity, namely the primary and secondary forms. The **primary porosity** refers to the original porosity of the medium upon deposition. **Secondary porosity** refers to that portion of the total porosity resulting from diagenetic processes such as dissolution. For example, limestone has a very low primary porosity upon deposition. If raised above sea level, however, dissolution processes can lead to the formation of caves with very large secondary porosity. A similar interpretation applies to the distinction between solid and fractured rock. The rock matrix has very low porosity. The presence of fractures increases the secondary porosity over the primary porosity.[2] The secondary porosity of a medium is not always

---

[2]The major role of fractures, however, is to greatly increase the permeability of the rock.

greater than its primary porosity. Consider a sand deposit that may become cemented over time to form sandstone. The chemically precipitated cementing agents occupy part of the pore space, therein reducing the overall porosity—sometimes significantly.

Finally, we mention the part of the total porosity that is important in problems of groundwater transport of pollutants. Some of the water in the pore space is not able to migrate under an induced hydraulic gradient because it might be isolated, adsorbed, or held in 'dead-end' pores, for example. In a saturated, fractured medium, only the water residing within the fractures is able to migrate at any appreciable rate. When one is concerned with the kinematics of flow (advection), then an appropriate definition of the pore water porosity is

$$n_e = \frac{\text{volume of water able to circulate}}{\text{total volume of rock}} \qquad \textbf{(1.3.10)}$$

This porosity may be called the kinematic porosity or the **effective porosity**. While the latter term is widely used, one must be aware that it is used by different authors to designate the specific yield (to be discussed in Chapter 2). Still, others do not distinguish between the two, leading to some confusion if those using the terminology are not careful in their interpretation.

1.4   Subsurface Hydrologic Cycle

The concept of the hydrologic cycle ties together the study of surface water with groundwater hydrology. These two subjects are linked through processes that occur within the vadose zone. From the standpoint of groundwater hydrology, the study of vadose zone processes allows one to estimate the potential recharge and evapotranspiration losses that may occur. A schematic view of the hydrologic cycle is shown in Figure 1.4.1.

If the hydrologic cycle has an origin, one might think of it beginning with **rainfall** or precipitation (which may include snow). Some of the rainfall or precipitation is intercepted, wetting vegetation and other surfaces. The remainder reaches the ground surface. At the beginning of a rainfall event, all of the precipitation that reaches the ground surface will enter the subsurface. The process of water entry into the soil profile, together with the associated downward flow away from the ground surface, is called **infiltration**. As the soil profile is wetted, the infiltration capacity of the soil decreases. If the rainfall event continues with a sufficient intensity, eventually a point is reached where all of the precipitation cannot infiltrate the profile, and **surface ponding** and the generation of **surface runoff** occurs. These are significant events for surface water hydrology studies. In rainfall-runoff studies where the water losses to infiltration are examined, interception and surface ponding are called **abstractions**.

After a rainfall event has ceased, the movement of subsurface water does not stop. **Redistribution** of subsurface water refers to the movement of soil water following infiltration. Here, the water spatially redistributes itself under various driving mechanisms, including gravity, capillarity, and temperature gradients.

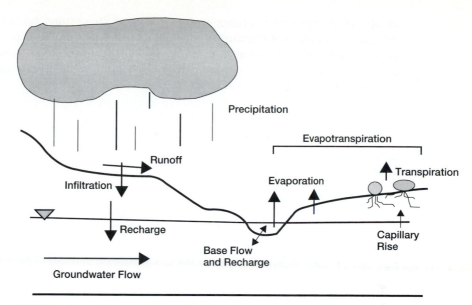

FIGURE 1.4.1   Schematic view of the hydrologic cycle

Initially, the primary direction of movement of water is downward under the influence of gravitational forces. The water immediately adjacent to the ground surface, however, will move into the atmosphere at a rate controlled by atmospheric conditions (solar radiation, wind speed, relative humidity, and other factors). This maximum rate is called the evaporation capacity of the atmosphere, and the process of soil water transfer is called **evaporation**. As water evaporates from the ground surface, it induces a gradient that pulls water from the soil profile to the surface for evaporative loss. When considered in conjunction with plant uptake and **transpiration** losses of soil water to the atmosphere, the combined process is called **evapotranspiration**. Eventually, as the soil profile dries out, the rate at which the soil profile can transmit soil water to the ground surface becomes less than the rate at which the atmosphere can receive the soil water. Under such conditions, the rate of evapotranspiration becomes dependent on the soil water distribution within the soil water zone. At this point, vadose zone processes control the evaporation rate.

Water that is able to drain from the zone of soil water and enter the intermediate vadose zone will continue to migrate downward to the water table. **Groundwater recharge** is the process by which water enters the saturated zone and moves away from the water table. It is important to note that not all infiltration becomes groundwater recharge. Within the zone of saturation, the water will migrate under the existing hydraulic energy gradients from the area of recharge to locations where it is lost again to the surface water components of the hydrologic cycle. **Groundwater discharge** refers to the removal of water from the saturated zone across the water table, together with the associated flow toward the water table within the saturated zone. This water may appear as surface springs, subterranean seepage into rivers (base flow), lakes or marine bodies, or as evapotranspiration losses associated with capillary rise from the zone of saturation. In regions where groundwater resources have been heavily developed, the major component of groundwater discharge is from production of water by pumping wells.

1.5   Hydrogeologic Formations

A geologist may study subsurface formations to obtain information on their lithology, stratigraphy, and structure. The **lithology** is the physical characteristics of the sediments or rocks of the geological system, including mineral composition, grain size, and grain packing. **Stratigraphy** describes the geometrical and age relations between the various lenses, beds, and formations in geological systems of sedimentary origin. **Structural** features including fractures, folds, and faults are produced by deformation after deposition and crystallization of the geologic system. In unconsolidated deposits, the lithology and stratigraphy constitute the most important controls (Freeze and Cherry, 1979).

A hydrogeologist (groundwater hydrologist) studies a subsurface formation, on the other hand, to determine its water holding capacity and its ability to transmit fluids. The terminology used by hydrogeologists is founded on these two properties. **Aquifers** are formations that are characterized by their ability to store and transmit water. These formations will yield significant quantities of water to wells and springs. Aquifers are bounded by **confining beds**, which are relatively impermeable material located stratigraphically adjacent to one or more aquifers. Three types of confining beds are distinguished. An **aquiclude** is a saturated but relatively impermeable material, such as clay. An **aquifuge** is a relatively impermeable material that does not contain water. An example is unfractured granite. Finally, an **aquitard** is a saturated material with low permeability that impedes water movement and will not yield appreciable water to wells. An aquitard may transmit appreciable water volumes between adjacent aquifers, thus serving as a medium for leakage from one aquifer to another. An example of an aquitard could be a sandy clay formation, depending on its permeability and thickness.[3] Aquifers and confining beds can be laterally extensive, or they can be discontinuous.

Based on the nature of the confining beds, we can distinguish three types of aquifers. The first is a **confined aquifer**, which is bounded above and below by an aquiclude or aquifuge. Confined groundwater is under pressure, and the

---

[3]Todd (1959) notes that the Latin origin of aquifer can be traced to *Aqui,* which is a combining form of *aqua,* meaning water, and *-ferre,* to bear. Hence, an aquifer is a water bearer. The suffix *-clude* is derived from the Latin *claudere,* to shut or close. The suffix *-fuge* comes from *fugere,* meaning to drive away.

water level in a well that is open to (screened) the aquifer will rise above the top of the aquifer. If the water level in a well rises above the ground surface, then the well will freely flow and the aquifer is called an **artesian aquifer**.

The second type of aquifer is one without a confining bed above so that the aquifer can be directly recharged by rainfall. Such an aquifer is called an **unconfined aquifer** or a **phreatic aquifer**.[4] As we shall see later, shallow, unconfined aquifers are much preferred for water resources development. The third type of aquifer is a **leaky aquifer** or **semi-confined aquifer**, in which one of the confining beds is an aquitard.

There exists a wide range of aquifer configurations in nature. One example is a regional **synclinal** system such as that found in the East Texas Basin or the London Basin. Such a system is shown in Figure 1.5.1. Water enters the system at its outcrop, where the aquifer is unconfined. The water then moves downgradient and deeper into the confined or semiconfined sections of the aquifer. Eventually, it crosses the upper confining bed to enter either an adjacent confined aquifer or a phreatic aquifer.

A second type of system is a **homoclinal** aquifer, such as that shown in Figure 1.5.2. Homoclinal aquifer systems are found along the entire Gulf of Mexico. The historical transgressions and regressions of sea level have resulted in a layered sequence of fluvial, littoral, and marine depositional systems. The fluvial and near-shore littoral deposits contain sufficient coarse-grained materials to form aquifers, at least locally. The offshore marine deposits form confining beds. These systems show greater dip and confinement than synclinal systems, along with a thinner fresh water layer. Thus, the "deep basin" saline system is located closer to the outcrop.

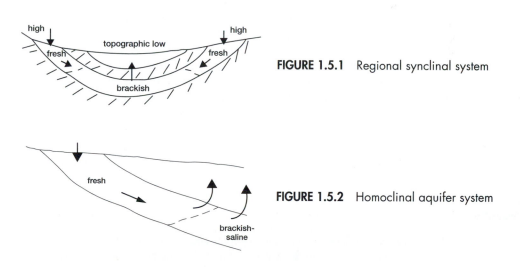

FIGURE 1.5.1 Regional synclinal system

FIGURE 1.5.2 Homoclinal aquifer system

---

[4]Todd (1959) notes that the term phreatic is derived from the Greek *phrearatos*, meaning a well.

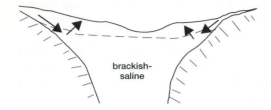

brackish-
saline

FIGURE 1.5.3   Basin and range aquifer system

The Desert Basin and Range Aquifer System, such as shown in Figure 1.5.3, typify a third type of aquifer system. Thick alluvial deposits are built up by coalescing alluvial fans, resulting in coarser material near mountain fronts and at shallow intervals. The permeability commonly decreases with depth within such systems. These systems show more local recharge, primarily along mountain fronts and ephemeral streams where freshwater resources may be found. Prominent discharge is associated with streamed flow and groundwater pumpage.

It is estimated that groundwater accounts for 30 percent of the world's freshwater (UNESCO, 1978). Considering that nearly 70 percent of the freshwater is locked within polar ice and other ice and snow, the importance of groundwater becomes apparent. When looking for a water supply system, surface water supplies are more economically feasible to develop. Some large municipalities, however, rely solely on groundwater (i.e., Tucson and San Antonio). In the western United States, where surface water supplies are scarce and expensive to develop, there is a much greater reliance on groundwater for water supply systems and agricultural use.

Groundwater resource evaluation and management is important for water use in public water supply, domestic and commercial self-supply, irrigation and livestock, industrial water use, mining, and other applications. Groundwater use in the United States over the period 1950-1995 is shown in Figure 1.5.4. Groundwater use more than doubled over the period 1950 to 1975. The peak groundwater use was in 1980, with nearly 315 million cubic meters per day (m^3/d) being utilized. Groundwater use in 1995 was approximately 290 million m^3/d (Solley et al., 1998). By far, irrigation water use made up the largest demand, with a daily use of 185 million m^3. Public water supply and domestic self-supply accounted for 70 million cubic meters per day in 1995 (Solley et al., 1998). Significantly, nearly 99 percent of domestic self-supply comes from groundwater.

Problems

1.1.1. For the soil particle-size distribution curves shown in Figure 1.1.1, what are the soil texture classifications for Soil B and Soil C using the USDA classification system?

1.1.2. A soil consists of 20 percent sand-sized particles, 60 percent silt, and 20 percent clay, by mass. What is its texture classification?

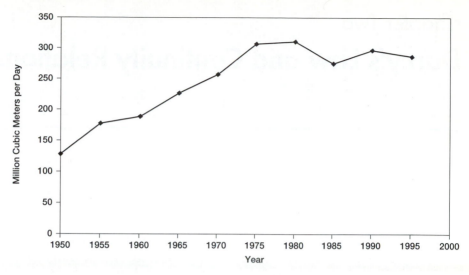

FIGURE 1.5.4   Groundwater use in the United States over the period 1950-1995 (from Solley et al., 1998)

1.3.1. A soil has a solids mass density of 2.60 g/cm^3 and a dry bulk density of 1.60 g/c^3. What is its porosity? If a moist sample of the same soil has a density of 1.90 g/cm^3, what is its volumetric water content? What is its water saturation?

1.3.2. The porosity of a porous medium is often measured by the gas expansion method. The apparatus consists of two chambers connected by a tube with a valve. The volumes of the chambers are V_1 and V_2. One of the chambers, V_2, is evacuated with a vacuum pump, and the valve is closed. A soil sample of measured volume V_t is placed in the other chamber, and the air pressure is increased to p_1. The valve is then opened, and the air pressure in both chambers is equilibrated at p_2 with the temperature remaining constant. With $V_2 = 0.5V_1$, $V_t = 0.2V_1$, and $p_2 = 0.627p_1$, use **Boyle's law** (for a fixed amount of gas, the product pV is constant at a constant temperature) to estimate the (connected) porosity for the rock sample.

1.3.3. Geotechnical engineers often use the **void ratio**, e, in lieu of the porosity, for quantifying the void space of a porous medium. The void ratio is the ratio of volume of voids to the volume of soil solids. What is the algebraic relationship between the void ratio and the porosity? If a soil has a porosity of 0.4, what is its void ratio?

1.3.4. The **gravimetric water content**, w, is the ratio of the weight (mass) of water contained in the soil to the dry weight (mass) of the soil sample. What is the algebraic relationship between the gravimetric water content and the volumetric water content? If a soil sample has a mass of 26.2 g when moist, 22.4 g when dry, and a bulk density of 1.65 g/cm^3, what is the gravimetric water content and the volumetric water content of the moist soil sample?

1.5.1. What is the origin of the term **aquitard**?

Chapter Two
Darcy's Law and Continuity Relations

In this chapter, we apply the principles of fluid mechanics to subsurface hydraulics. In an introductory fluid mechanics course, one learns how to apply the laws of conservation of mass, momentum, and energy to various problems. The usual approach is through an integrated or control volume analysis (see White, 1986, for example). As one must expect, the conservation of mass principle remains of utmost importance for all problems. For groundwater flow, inertial effects are small, and the momentum equation takes the simplified form called Darcy's Law. The energy equation plays a significant role for applications where temperature changes are important such as the thermal balance of the soil profile, analysis of high level radioactive waste disposal, or geothermal energy. We will not consider such applications.

For problems in subsurface hydraulics, the control volume approach has not been found to be as useful as it is in elementary fluid mechanics. Instead, we apply the fundamental equations at each point in space. The result of this process is that our fundamental equations will take the form of partial differential equations describing the flux of water and its conservation. In the first part of this chapter, we will look at the flux equations, while in later sections we look at how the conservation of mass principle is applied and how one deals with compressibility of water, as well as the porous matrix. The flow of air through porous media is discussed, and a look is taken at development and application of groundwater management models.

2.1   Principles from Fluid Mechanics

The basic equation of groundwater hydraulics is called **Darcy's Law**. The law expresses a linear relationship between the groundwater velocity and the head (energy) gradient. For many applications, this law may be written as follows:

$$\tilde{q} = K\tilde{I} \qquad\qquad (2.1.1)$$

where $\tilde{q}$ is the **Darcy velocity** vector, otherwise known as the filtration velocity, specific discharge, or volumetric flux. (Much of the mathematical formalism used in this text is summarized in Appendix A.) Inherent within this formulation is the assumptions that the flow field is steady (or that inertial effects associated with transient phenomena are not significant in the momentum balance within the fluid). In Equation 2.1.1, the parameter K is called the **hydraulic conductivity**. In earlier literature, this parameter is sometimes called the coefficient of permeability, although this term may be confused with the intrinsic permeability, to be discussed later. The vector $\tilde{I}$ is the **hydraulic gradient**,

$$\tilde{I} = -\nabla h \qquad (2.1.2)$$

where h is the hydraulic head. The objective of the following discussion is to gain physical insight into the law expressed by Equation 2.1.1 and its generalizations, through application of principles from fluid mechanics.

To start, we consider the condition of a fluid under hydrostatic conditions. The hydrostatic pressure equation quantifies how the fluid pressure changes with elevation under conditions of hydrostatic equilibrium; that is, when the fluid is not in motion. The general form of the hydrostatic pressure equation is

$$\nabla p + \rho g \hat{k} = 0 \qquad (2.1.3)$$

where p is the fluid pressure, ρ is the fluid density, g is the gravitational constant ($\rho g = \gamma$, the specific weight), and $\hat{k} = \nabla z$ is the upward unit vector. Equation 2.1.3 expresses the fact that the pressure does not change on a given level surface (the horizontal component of grad(p) vanishes) and that the pressure increases with depth at a rate corresponding to the local specific weight of the fluid. If the density is constant, Equation 2.1.3 may be written as

$$\nabla\left(\frac{p}{y} + z\right) = \nabla h = 0 \qquad (2.1.4)$$

where $h = \dfrac{p}{\gamma} + z$ is the **piezometric head**. Because the gradient of h vanishes h is constant. Accordingly, as long as h is constant, there will be no flow. If h varies from point to point, however, then there will be a head (energy) gradient present and a corresponding force to drive fluid flow. Thus, $-\nabla h$ may be interpreted as the driving force for flow (per unit volume). The minus sign corresponds to the fact that flow goes "down the energy hill".

Newton's Second Law equates the change in momentum for a fluid element to the sum of the forces acting on the element. If the fluid is incompressible, the momentum equation ($\tilde{F} = m\tilde{a}$) takes the following form:

$$\rho\frac{d\tilde{u}}{dt} = \tilde{f}_{\text{gravity}} + \tilde{f}_{\text{pressure}} + \tilde{f}_{\text{viscous}}$$

where $\tilde{u}$ is the local velocity at the point of interest, and $\tilde{f}_i$ is the force per unit volume due to mechanism i. For a Newtonian fluid such as water, the relationship between the viscous force and the velocity gradient is a linear one, giving the **Navier-Stokes equations** for flow:

$$\rho\frac{d\tilde{u}}{dt} = -\rho g\hat{k} - \nabla p + \mu\nabla^2\tilde{u}$$

In light of Equation 2.1.4, the above relation may be written as:

$$-\rho g\nabla h = \rho\frac{d\tilde{u}}{dt} - \mu\nabla^2\tilde{u} \tag{2.1.5}$$

Equation 2.1.5 states that the presence of a driving force results in acceleration of the fluid and development of a balancing viscous resistance force.

If the conditions driving the flow change abruptly, the time period of fluid acceleration is short (Harr, 1962), and the acceleration term of Equation 2.1.5 may generally be neglected. Equation 2.1.5 then simplifies to the **Stokes equation**.

$$-\rho g\nabla h = -\mu\nabla^2\tilde{u} \tag{2.1.6}$$

Equation 2.1.6 specifies a balance between the forces driving the flow and those resisting the fluid movement.

Equation 2.1.6 describes the flow of water and other Newtonian liquids under conditions where inertial forces are negligible, as is the case for flow in porous media under most conditions of interest. Equation 2.1.6 is not difficult to solve if the corresponding boundary conditions take a simple form. For a viscous fluid (such as water), the appropriate boundary conditions specify no-slip at the water-solid interface. Unfortunately, a quantitative description of the geometry of the water-solid interface for natural porous media is all but impossible. In general, because of this geometric difficulty, Equation 2.1.6 cannot be solved for lack of specificity with respect to the boundary conditions.

If the geometry is simple, Equation 2.1.6 may be readily solved. Such solutions are of interest because they show the basic character of the law of fluid flow that is retained in more geometrically complex porous media. One example of interest concerns flow in a circular capillary tube of radius R. For this problem, Equation 2.1.6 reduces to the following form:

$$-\frac{1}{r}\frac{d}{dr}\left(r\frac{du}{dr}\right) = \frac{\rho g I}{\mu}$$

where u is the axial component of the fluid velocity and I is the constant axial component of the hydraulic gradient. With the boundary condition $u = 0$ at $r = R$, this equation may integrated to find

$$u(r) = \frac{\rho g I}{4\mu}(R^2 - r^2) \tag{2.1.7}$$

which gives the parabolic velocity profile for laminar flow in a capillary tube. The average velocity is found by integration to be

$$v = \frac{\rho g R^2 I}{8\mu} = \frac{\rho g R_h^2 I}{2\mu} \tag{2.1.8}$$

where R_h is the hydraulic radius (cross-section area divided by wetted perimeter), which is equal to R/2 for a circular cross-section. The equations describing flow in circular capillary tubes are named after Hagan (1939) and Poiseuille (1840), who verified them experimentally.

A similar analysis may be applied for flow between two parallel plates separated by a distance H, such as that presented by localized flow in a smooth-walled fracture. (For fracture flow, H would be called the fracture aperture.) The resulting average velocity is given by

$$v = \frac{\rho g H^2 I}{12\mu} = \frac{\rho g R_h^2 I}{3\mu} \tag{2.1.9}$$

where the hydraulic radius is equal to H/2.

Given the widely differing geometry (circular capillary tube and infinite parallel plates), the similarity of Equations 2.1.8 and 2.1.9 is surprising. A comparison shows that they also have the same form as Equation 2.1.1. There are some significant differences, however. First, in Equations 2.1.8 and 2.1.9, the velocity v is the average velocity within the fluid phase, while in Equation 2.1.1, q is the "fictitious" velocity defined by the discharge per total cross-sectional area. If the capillaries or fractures formed a straight parallel system, then the two velocities would be related through

$$q = nv \tag{2.1.10}$$

where n may be interpreted as the area of flow per bulk unit area (area porosity). Equation 2.1.10 is fundamental to the study of solute transport in porous media. If Equation 2.1.10 is assumed true for natural porous media, then it is clear that v cannot be interpreted as a simple average of the velocity $\tilde{u}$ from Equation 2.1.6 over the pore space, because the flow paths are curved and not all parallel to the mean direction of flow. Nevertheless, comparison of Equation 2.1.1 with 2.1.8 or 2.1.9 and 2.1.10 suggests

$$K = \frac{\rho g n R_h^2}{c\mu}$$

where c is a parameter with a magnitude on the order of 2 or 3, depending on the pore structure. The important point to notice is that the hydraulic conductivity may be factored into (1) those properties depending on the characteristics of the fluid occupying the pore space and (2) those characteristics of the geometry and structure of the pore space of the medium. The appropriate generalization is

$$K = \frac{\rho g k}{\mu} \tag{2.1.11}$$

where k is called the **intrinsic permeability** and depends only on the characteristics of the porous matrix, independent of fluid properties. The intrinsic permeability has units of length-squared. (In petroleum engineering literature, k is often measured in the 'darcy' unit, where 1 darcy = 0.987×10^{-8} cm^2).

There have been many studies on the theoretical basis of Darcy's Law. Scheidegger (1974), Bear (1972), and Dullien (1992) provide comprehensive reviews. Bear (1972) derives Darcy's Law through spatial averaging over an REV. Dagan (1989) provides a derivation by way of the theory of random functions. Derivations such as those noted in these references are useful because they provide (1) physical insight into the content and limitations of the law and (2) a means of estimating various parameters from other physical information which may be more easily measured and interpreted. Here, we will examine a theory that is fairly straightforward and has found use in the theory of filtration and in flow through sand soils.

Kozeny-Carman Theory

There have been many attempts at deriving Darcy's Law from more fundamental equations in fluid mechanics. One such theory is the hydraulic radius theory first presented by Kozeny in 1927 and later modified by Carman (see Carman, 1937 and 1956). According to the hydraulic radius theory,

$$u = \frac{\rho g R_h^2}{c\mu} \frac{\Delta h}{\Delta L_e}$$

where u is the velocity within the capillary, ΔL_e is the length of the capillary, and c is a coefficient which varies from 2 (for a circular capillary, Equation 2.1.8) to 3 (for flow between flat plates, Equation 2.1.9). We cannot actually see the tortuous path through the media, and we envision it as a linear path of length L with an average linear velocity v along this path. This idea is shown schematically in Figure 2.1.1.

The time to travel between two points in our fictitious media must be the same as that through the actual media. Further, the average linear velocity is related to the Darcy velocity through $v = q/n_e$ where n_e is the effective porosity. Thus,

$$\frac{\Delta L}{v} = \frac{\Delta L_e}{u} = \frac{\Delta L n_e}{q}$$

$$q = \frac{\Delta L}{\Delta L_e} n_e u = \frac{\Delta L}{\Delta L_e} n_e \frac{\rho g R_h^2}{c\mu} \frac{\Delta h}{\Delta L_e}$$

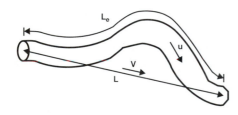

FIGURE 2.1.1 Actual capillary flow length compared with average linear distance

The head gradient is also a spatially averaged quantity over direct linear distances, however, so

$$-\operatorname{grad}(h) = \left(\frac{\Delta h}{\Delta L}\right)_{ave}$$

Thus,

$$q = -\left(\frac{\Delta L}{\Delta L_e}\right)^2_{ave} n_e \frac{\rho g R_h^2}{c\mu} \operatorname{grad}(h)$$

The parameter $\tau = \left(\dfrac{\Delta L}{\Delta L_e}\right)^2_{ave}$ is called the **tortuosity** of the medium. With ΔL measured in the direction of the mean flow and ΔL_e measured along the actual length of the capillary through which flow is occurring, then

$$\Delta L = \Delta L_e \cos(\theta)$$

as shown in Figure 2.1.2. If the pore orientation is random, then the tortuosity may be found by averaging $\cos^2\theta$ over the hemisphere with solid angle $d\omega = \sin\theta\, d\theta\, d\phi$. Thus,

$$\left(\frac{\Delta L}{\Delta L_e}\right)^2_{ave} = \frac{1}{2\pi}\int_0^{2\pi}\int_0^{\pi/2} \cos^2(\theta)\, \sin(\theta) d\phi = \frac{1}{3}$$

On the other hand, Carman (1937) noted from observation of flow of a colored tracer through a bed of glass beads that the average direction of the tracer movement through the pores made an angle of 45 degrees with respect to the mean direction of flow. This observation suggests that the average value of $\Delta L/\Delta L_e$ is equal to $1/\sqrt{2} \cong 0.707$ and that $\tau \cong 1/2$. Carman (1937) used this latter value

$$q = -\frac{\rho g n_e R_h^2}{2c\mu} \operatorname{grad}(h) \tag{2.1.12}$$

To proceed further, we need to evaluate the hydraulic radius. We have

$$R_h = \frac{A}{P} = \frac{An_e}{Pn_e} = \frac{\text{volume flow}}{\text{surface area of solids}} = \frac{n_e}{S_{sp}}$$

where A is the pore cross-section area, P is the wetted perimeter, and S_{sp} is the specific surface area (surface area per unit bulk volume). It is easier to calculate the surface area per volume of solids, S_o, which is related to S_{sp} through $S_o = \dfrac{S_{sp}}{1-n}$, so

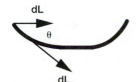

FIGURE 2.1.2 Linear distance and length along a capillary tube

$$R_h^2 = \frac{n_e^2}{(1 - n)^2 S_o^2}$$

and we finally obtain

$$q = -\left(\frac{n_e^3}{2c\,(1 - n)^2\,S_o^2}\right)\left(\frac{\rho g}{\mu}\right)\mathrm{grad}(h) \qquad \textbf{(h)}$$

The result has been factored into three terms. For an equivalent spherical particle,

$$S_o = \frac{4\pi r^2}{\frac{4}{3}\pi r^3} = \frac{3}{r} = \frac{6}{d_e}$$

where d_e is the effective particle diameter. With $c \cong 2.5$, we finally obtain

$$q = -\left(\frac{n_e^3 d_e^2}{180(1 - n)^2}\right)\left(\frac{\rho g}{\mu}\right)\mathrm{grad}(h) \qquad \textbf{(2.1.13)}$$

This equation is the *Kozeny-Carman equation*, and n_e is usually replaced by the total porosity, n. Comparison with Equation 2.1.1 suggests that the hydraulic conductivity depends on fluid properties and on the size and packing (through the porosity) of the porous medium.

Fair and Hatch (1933) independently arrived at a result that is similar to the Kozeny-Carman equation. They, however, express the hydraulic conductivity in the form

$$K = \frac{\rho g}{\mu}\frac{n^3}{(1 - n)^2}\left[m\left(\alpha\sum_i\frac{f_i}{d_m}\right)^2\right]^{-1} \qquad \textbf{(2.1.14)}$$

where m is a packing factor found experimentally to be about 5, α is a sand shape factor varying from 6.0 for spherical grains to 7.7 for angular grains, f_i is the fraction of sand held between adjacent sieves, and d_m is the geometric mean of the rated size of adjacent sieves. Notice that for uniform spherical grains, Equation 2.1.14 reduces to Equation 2.1.13. Neither Equations 2.1.13 nor 2.1.14 work well when fine-grained material is present within the porous medium.

2.2   Darcy's Law

Henry Philibert Gaspard Darcy (1803–1858) was a native of Dijon, France, was educated in Paris, and was a member of the French Corps. By popular demand, he was stationed in Dijon where his major work was the design and execution of a municipal water-supply system. Through this work, he carried out a series of studies on the flow of water in both pipes and permeable soils. In the study of fluid mechanics, we are most familiar with Darcy's name through its association with that of the Julius Weisbach (1806–1871) in the Darcy-Weisbach equation for the head loss in a closed conduit. It is his 1856 Paris treatise, however, "Les Fontaines Publiques de la Ville de Dijon," that is of most interest to us here. In a brief appendix to this work, he states the pertinent results from his filtration studies—that the loss of head through a filter bed was proportional to the rate of flow, rather than to its square root as popularly supposed (Rouse and Ince, 1957).

Darcy was engaged in modernizing and enlarging the public water works of Dijon, and in these efforts he needed to design a suitable filter for the system. Darcy needed to know how large a filter would be required for a given quantity of water per day and, unable to find the desired information in the published literature, he proceeded to obtain it experimentally (Hubbert, 1956). The apparatus that Darcy used is shown in Figure 2.2.1. This device consisted of a vertical iron pipe, 0.35 m in diameter and 3.50 m in length. At a height of 0.20 m above the base of the column was placed a horizontal screen supported by an iron grillwork. This structure supported a column of loose sand approximately 1 m in length. Water could enter the system by means of a pipe tapped into the column near its top, from the building water supply, and could be discharged through a faucet from the open chamber near its bottom. The faucet discharged into a measuring tank, and the flow rate could be controlled by means of adjustable valves in both the inlet pipe and the outlet faucet. Mercury manometers were used to measure the water pressure, one tapped into each of the open chambers above and below the sand column. As a unit of pressure, Darcy employed 'meters of water,' and all manometer readings were reported in meters of water measured above the bottom of the sand that was taken as an elevation datum (Hubbert, 1956).

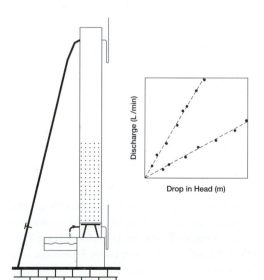

FIGURE 2.2.1 Experimental Apparatus Used by Darcy

Darcy's experiments comprised several series of observations made over a four-month period in 1855–1856. For each series, the system was charged with different sand and was completely filled with water. By adjustment of the inlet and outlet valves, the water was made to flow downward through the sand at a series of successively increasing rates. For each flow rate, the pressure difference between the upper and lower open chambers was read from the manometers and was recorded. The results from two series of measurements are sketched to the right on Figure 2.2.1. Through these experiments, Darcy arrived at the following empirical law relating the volume of water passing through the system in unit time (Q), the area of cross section (A), water elevation change across the filter sand measured from heights above a standard reference elevation in equivalent water manometers ($z_1 - z_2$), and the thickness of sand (L):

$$Q = KA\frac{z_1 - z_2}{L} \tag{2.2.1}$$

In Equation 2.2.1, K is a factor of proportionality. A look at the dimensions of all the terms shows that K has the same units as velocity. The factor K is called the hydraulic conductivity of the porous medium.

Applying the energy equation between the upper and lower open chambers, we have

$$z_1 = z_2 + h_{\text{sand}}$$

where h_{sand} is the head loss through the filter sand. Thus,

$$\frac{z_1 - z_2}{L} = \frac{h_{\text{sand}}}{L} = I$$

where I is the hydraulic gradient across the sand. The specific discharge or Darcy velocity is defined as the total discharge from the cross-section area:

$$q = \frac{Q}{A}$$

With these changes, we can write Darcy's Law in a more general form:

$$q = KI \tag{2.2.2}$$

which is Equation 2.1.1.

The actual implementation of Equation 2.2.1 will often times focus on the appropriate selection for K. Typical values of the hydraulic conductivity are shown in Table 2.2.1. The important point to notice from this table is that for natural porous media, K varies by more than 12 orders of magnitude. There are few physical parameters that show this great range of variation in magnitude. Variations of 3 orders of magnitude are common for a single site, and in modeling of regional hydrogeologic systems, variations of more than 6 orders of magnitude can be expected.

TABLE 2.2.1 Typical range of hydraulic conductivity values

Material	K (cm/s)	K (m/d)
Unconsolidated Material		
Gravel	10^{-1}–10^{1}	10^{2}–10^{4}
Sand	10^{-4}–10^{0}	10^{-1}–10^{3}
Silt	10^{-7}–10^{-3}	10^{-4}–10^{0}
Clay and glacial till	10^{-11}–10^{-6}	10^{-8}–10^{-3}
Sedimentary Rock		
Sandstone	10^{-8}–10^{-3}	10^{-5}–10^{0}
Limestone, dolomite	10^{-7}–10^{-1}	10^{-4}–10^{2}
Karst limestone	10^{-4}–10^{0}	10^{-1}–10^{3}
Shale	10^{-11}–10^{-6}	10^{-8}–10^{-3}
Crystalline Rock		
Basalt	10^{-9}–10^{-5}	10^{-6}–10^{-2}
Fractured basalt	10^{-5}–10^{0}	10^{-2}–10^{3}
Dense crystalline rock	10^{-12}–10^{-8}	10^{-9}–10^{-5}
Fractured crystalline rock	10^{-6}–10^{-2}	10^{-3}–10^{1}

Homogeneous and Isotropic Media

The hydraulic conductivity is perhaps the most important property of a porous medium. Because the medium is defined over a three-dimensional space, the assignment of physical parameters, such as hydraulic conductivity and porosity, to that space leads to the medium's representation as a parametric field. If the value of K is the same at every point, then the field is said to be **homogeneous**; otherwise, it is a **heterogeneous** field. If the magnitude of K is independent of direction, then the field is **isotropic**. However, under the most general conditions, the hydraulic conductivity field is both heterogeneous and **anisotropic**. We will see how to deal with this case later. First, however, we consider the simplest case where the hydraulic conductivity field is homogeneous and isotropic.

For a homogeneous and isotropic hydraulic conductivity field, Darcy's Law takes the form

$$\tilde{q} = -K\nabla h = K\tilde{I} \tag{2.2.7}$$

where K is a scalar constant. Equation 2.2.7 says that the vectors $\tilde{q}$ and $\tilde{I}$ are parallel with components $q_x = KI_x$, etc.

2.2.1 **Hydraulic Head and Gradient**

In fluid mechanics, the mechanical energy content of a fluid is specified by its hydraulic head, h, where

$$h = \frac{v^2}{2g} + \frac{p}{\gamma} + z \tag{2.2.3}$$

The hydraulic head, or simply the head, is the mechanical energy per unit weight and quantifies in units of length. Equation 2.2.3 shows that the head is the sum of the velocity head $\frac{v^2}{2g}$, which is the kinetic energy per unit weight, the pressure head $\frac{p}{\gamma}$, which is a measure of the ability of the fluid to do work, and the elevation head, z, which is the potential energy per unit weight. For problems of subsurface flow, the velocity head is negligible, so

$$h \cong \frac{p}{\gamma} + z \qquad (2.2.4)$$

The right-hand side of Equation 2.2.4 is known as the *piezometric head*, which specifies the level to which water will rise in a piezometer, such as a well penetrating an aquifer.

Piezometers

A **piezometer** is the basic field device for measurement of the piezometric head in saturated media. As shown in Figure 2.2.2, the device consists of a tube or pipe in which the water level can be measured. Water will enter or leave the screened section until the energy in the piezometer is equal to the energy within the porous medium. With reference to Equation 2.2.4, the elevation of the screened section of the piezometer corresponds to the potential energy measured with reference to a chosen datum, and the height of the column of water in the casing (pipe) is the pressure head.

If measurements from a number of different piezometers are available, then one can estimate the piezometric head distribution. Piezometers that are screened at the same elevation can be used to determine the horizontal direction of flow, while a set of piezometers that are screened at different elevations (but at the same location) may be used to determine the vertical gradient. It is important to note that the energy measured by the piezometer screened over some interval is the average energy over that sampled interval.

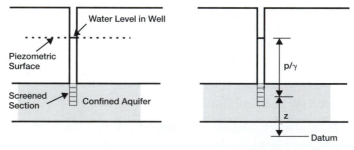

FIGURE 2.2.2   A field piezometer for measurement of the hydraulic head

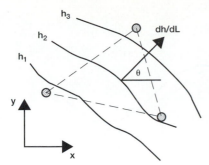

FIGURE 2.2.3 Piezometric surface map with $h_1 < h_2 < h_3$

If the flow is essentially **horizontal**, then the piezometric surface may be used to determine the direction and magnitude of the hydraulic gradient, as shown in Figure 2.2.3. In this figure, dh/dL is the slope of the piezometric surface in the direction of ∇h. The magnitude (or modulus) of the hydraulic gradient vector is given by

$$|I| = \sqrt{I_x^2 + I_y^2} \tag{2.2.5}$$

where I_x and I_y are the components of the vector. Each component is given by

$$I_i = -\frac{\partial h}{\partial x_i}$$

where x_i can represent either x or y. The direction in the plan view pointing up the energy hill is given by

$$\theta = \tan^{-1}\left(\frac{\partial h/\partial y}{\partial h/\partial x}\right) \tag{2.2.6}$$

With measurements of the piezometric head at three wells, we can fit a plane through the observed head field. The equation of this plane is given by

$$h = a + bx + cy$$

where the coefficients a, b, and c are found from the three piezometer observations:

$$h(x_1, y_1) = h_1$$
$$h(x_2, y_2) = h_2$$
$$h(x_3, y_3) = h_3$$

These observations may be used to solve for a, b, and c, and then within the triangle formed by the three wells, we have

$$I = \sqrt{b^2 + c^2}$$
$$\theta = \tan^{-1}(c/b)$$

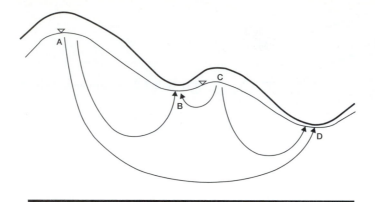

FIGURE 2.2.4   Cross-section of a groundwater flow system

Groundwater Flow Systems

On a regional scale and under undeveloped conditions, one generally finds that groundwater enters the subsurface at regional topographic highs and leaves the system at topographic lows, as shown in Figure 2.2.4. These high and low topographic elevations correspond to equivalent high and low points of potential energy. The distribution of permeability of the host medium controls the flow pattern between the recharge and discharge zones and the subsurface energy distribution. If water enters and leaves the subsurface at adjacent topographic highs and lows, then the flow system is considered to be a **local flow system**. Travel times for these systems vary from months to tens of years. If there are one or more intervening topographic highs and lows between the recharge and discharge points, then the system is an **intermediate flow system**. Travel times here vary from tens of years to hundreds of years. A flow system leading from the regional topographic high to the regional topographic low is a **regional flow system**. Travel times here may be on the order of thousands to hundreds of thousands of years. Important early quantitative investigations of regional flow systems include the work of Toth (1962, 1963). Freeze and Cherry (1979) and Domenico and Schwartz (1990) present reviews of subsequent work.

Pressure-depth analysis is sometimes used for determining the potential for vertical flow. A pressure-depth map shows the pressure head at a well screen versus the depth to the screen below the ground surface. If freshwater hydrostatic conditions exist or if the flow is horizontal, then the data would plot with a unit slope. Otherwise, the slope can differ from unity. For example, if a set of piezometers were located near points A or C in Figure 2.2.4, then the piezometers that were screened at greater depths below the ground surface would have smaller pressure heads for the corresponding depths than wells screened at shallower depths. On a plot of pressure head versus depth, these points would fall below the unit-slope line. Similarly, for a set of piezometers located near points B or D in Figure 2.2.4, the pressure head versus depth measurements

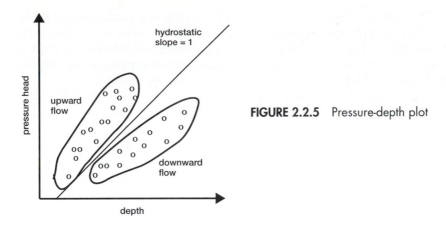

FIGURE 2.2.5   Pressure-depth plot

would fall above the unit-slope line. These measurements may be used to identify recharge and discharge locations. A schematic example of a pressure-depth map is shown in Figure 2.2.5.

Pressure-depth maps are not well suited to vertically nonaligned data where pressure changes are realized horizontally as well as vertically. Nevertheless, pressure changes are typically much less horizontally than vertically, and pressure-depth data within certain local neighborhoods can be used. A better approach is to construct 3-D potentiometric surfaces (or regress linear fields on sub-areas, $h = ax + by + cz + e$), although this process may be difficult with the sparse data typically available.

2.2.2   Factors that Influence K

From Equation 2.1.11 we have seen that the hydraulic conductivity of the porous medium is a function of both fluid and medium properties. In Equation 2.1.11, k is the intrinsic permeability that may vary from point to point in space, as well as with direction. There are other influences that may be of importance under certain circumstances. One such influence is the temperature of the groundwater. While temperature does not appear in Darcy's Law, it does influence water's properties, the most important of which is viscosity. At 20 degrees centigrade, the dynamic viscosity of water is $\mu = 1.002 \times 10^{-3}$ kg/(m − s), and its kinematic viscosity is $\nu = \mu/\rho = 1.004 \times 10^{-6}$ m²/s. A temperature change from 15 degrees C to 25 degrees C results in a kinematic viscosity decrease from 1.138 to 0.894 $\times 10^{-6}$ m²/s and an increase in hydraulic conductivity of about 24 percent.

The effect of temperature on the density of water is not as great. When the density of water varies from point to point, however, because of salt content for example, then the form of Darcy's Law must be modified, and we cannot work with the hydraulic conductivity of the medium. For an isotropic medium, the general form of Darcy's Law is

$$\tilde{q} = -\frac{k}{\mu}(\nabla p + \rho g \hat{k}) \tag{2.2.8}$$

This form is valid for both uniform and nonuniform fluids, and it must be used when we deal with problems such as salt-water intrusion (Section 3.10, "The Interface in the Coastal Aquifer"). It is also the standard form used by petroleum engineers who must deal with multiple fluids occupying the pore-space (see Chapter 9, "Multiphase Flow and Hydrocarbon Recovery").

Another factor that has an important influence on the hydraulic conductivity of a clay soil is the **cation composition**. Many clay soils can exhibit either a flocculated or dispersed structure, depending on the cations present and the salt concentration. The structure that exists is associated with the layer of cations surrounding the negatively charged clay particles. It takes fewer higher valence cations, such as Ca^{++} compared with Na^+, in this layer to balance the negative charge of the clays, and the resulting ion layer is thin. Such soils tend to flocculate and form aggregates, creating a larger permeability. On the other hand, low valence cations, such as Na^+, result in an expanded cation layer and dispersed clays. Such soils show poor structure with low permeabilities and poor drainage. This situation is complicated to some extent in that a soil solution with a high salt concentration can compress the ion layer and lead to good structure, even if the dominant cation is Na^+. A classic example is that of a soil invaded by seawater. Effects of ion composition on the structure and permeability of soils are discussed by Hillel (1980), Bresler et al. (1982), and Jury et al. (1991).

2.2.3   Heterogeneity and Anisotropy

Aquifers have heterogeneous hydraulic conductivity fields. This characteristic is due to geologic processes that do not yield uniform porous media characteristics over appreciable areas and to the layering of different sediment packages. The resulting heterogeneities occur on all length scales, from pore to pore on the scale of millimeters through facies (appearance and character of the porous media) changes that occur over distances of tens of kilometers. At the field scale, one can recognize at least three types of heterogeneities. The first and most common type is layering, resulting from successive deposition of beds with varying sediment sizes and textures. The second type of heterogeneity is trending, associated with lateral and vertical changes in soil texture, possibly due to variation of ambient energy in the depositional systems. Fluvial sediments often have fining-upward sequences. The third type of heterogeneity is contact, due to faults, alluvium-rock contacts, etc. In problems of groundwater hydraulics and pollutant transport for alluvial systems, we are most interested in layering heterogeneities on the scale of tens of centimeters to meters, and trending heterogeneities on the scale of tens to hundreds of meters.

In engineering analysis of groundwater flow problems, it is often useful to replace the true heterogeneous medium with an **equivalent** homogeneous medium. Here, the term 'equivalent' refers to a conceptually simpler system that yields the same calculated results as those obtained from the calculations using the actual medium. This concept is similar to the formal definition of the mean used in probability theory. According to Chisini (see De Finetti, 1974, p. 56), "x is said to be the mean of n numbers $x_1, x_2, \ldots x_n$, with respect to a problem in which a function of them $f(x_1, x_2, \ldots x_n)$ is of interest, if the function assumes the same value when all the x_i are replaced by the mean value x: $f(x_1, x_2, \ldots x_n) = f(x, x, \ldots x)$." This definition of the mean or equivalent value makes it clear that the same set of numbers, x_i, may yield different mean values depending on the problem of interest. This identification of an equivalent medium may formally be done through averaging the medium properties over an appropriate REV or through representation of the medium as a random function and considering ensemble averages. See Bear (1972), Marsily (1986), or Dagan (1989) for a discussion of these problems. Here, we are first interested in how anisotropy arises from consideration of effective or averaged media.

Layered Heterogeneity

Consider a porous medium consisting of uniform layers, each layer with a thickness b_i and hydraulic conductivity K_i. Figure 2.2.6 shows such a system with three layers. The total thickness of the system is b. In the following discussion, we compute an equivalent medium for either horizontal or vertical flow with respect to the actual medium's layering. We will first consider horizontal flow and calculate the equivalent horizontal hydraulic conductivity of the system, and then do the same for vertical flow.

For horizontal flow $(q_h \neq 0, q_v = 0)$ the hydraulic gradient must be the same in each layer:

$$I_1 = I_2 = I_3 = \ldots = I$$

This situation must be true so that no vertical hydraulic gradients are developed. The horizontal discharge through the layered system is

$$U_h = \Sigma K_i I_i b_i = I \Sigma K_i b_i$$

where U_h is the horizontal discharge per unit aquifer width (not thickness). By definition, the discharge through the equivalent homogeneous medium must be the same and is calculated from

$$U_h = K_h I b$$

where K_h is the equivalent horizontal hydraulic conductivity. The equivalence of these last two results gives the relation between K_h and the actual layer K's as

$$K_h = \frac{\Sigma K_i b_i}{b} \tag{2.2.9}$$

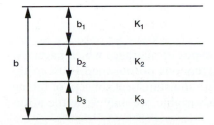

FIGURE 2.2.6 Porous medium consisting of three layers

Equation 2.2.9 shows that for flow along a system of **layers in parallel**, the equivalent hydraulic conductivity is an **arithmetic average** of the conductivities of the individual layers with each layer weighted by its thickness.

Now consider the case of **vertical flow** ($q_h = 0$; $q_v \neq 0$). The specific discharge through each layer must be the same (or else, where would the difference in flow go?):

$$q_1 = q_2 = \dots = q_v$$

For each layer, Darcy's Law gives

$$\Delta h_i = \frac{q_v b_i}{K_i}$$

where Δh_i is the head loss across the ith layer. The equivalent homogeneous medium must have the same discharge, and the corresponding head loss is

$$\Delta h = \frac{q_v b}{K_v}$$

where Δh is the total head loss across the system of layers. This value must be equal, however, to the sum of the individual layer head losses, so

$$\Delta h = \Sigma \Delta h_i = q_v \Sigma \frac{b_i}{K_i}$$

Comparing these two results, we find that the equivalent vertical hydraulic conductivity is given by

$$K_v = \frac{b}{\Sigma \dfrac{b_i}{K_i}} \qquad (2.2.10)$$

Equation 2.2.10 shows that the equivalent hydraulic conductivity for flow across a system of **layers in series** is given by the **harmonic average** of the conductivities of the individual layers, again weighted by their thickness. From statistics, we know that the arithmetic average is always greater than the harmonic average.

$$K_h \geq K_v \qquad (2.2.11)$$

The equivalent medium has a larger permeability parallel with the layers than transverse to them; it is an anisotropic medium, although the individual layers are homogeneous and isotropic.

Types of Averages

In Equations 2.2.9 and 2.2.10, we see two types of averages which naturally occur in describing the behavior of heterogeneous porous media: the arithmetic and harmonic means, respectively. With the statistical concept of taking the mathematical expectation of a random variable, we may write the **arithmetic** average as

$$\overline{K}_a = E\{K\} \qquad\qquad\qquad\text{(2.2.12)}$$

and the **harmonic** average as

$$\frac{1}{\overline{K}_h} = E\left\{\frac{1}{K}\right\} \qquad\qquad\qquad\text{(2.2.13)}$$

where $\overline{K}_a$ and $\overline{K}_h$ represent the mean values of the hydraulic conductivity and $E\{\ \}$ is the expectation operator. A third type of average that is important is the **geometric** average, defined by

$$ln(\overline{K}_g) = E\{ln(K)\} \qquad\qquad\qquad\text{(2.2.14)}$$

The geometric average is important because many naturally occurring hydraulic conductivity fields appear to be represented as pointwise lognormally distributed random variables. If K is **lognormally** distributed and the samples are uncorrelated, then the appropriate field "mean" or average is calculated from the geometric mean written as

$$\overline{K} = (\Pi K_i)^{1/N} \qquad\qquad\qquad\text{(2.2.15)}$$

where Π is the multiplicative product. In general, the order of averages is $\overline{K}_a \geq \overline{K}_g \geq \overline{K}_h$.

Anisotropic Porous Media

We have seen that for layered heterogeneity, the resulting equivalent homogeneous medium exhibits directional properties, with the permeability being greater along the layers than across them. These directional characteristics require the development of a generalized form of Darcy's Law for application to anisotropic media. For isotropic media, we have Darcy's Law written as

$$\tilde{q} = K\tilde{I} = -K\nabla h = -K\frac{\partial h}{\partial x_i} \qquad\qquad\qquad\text{(2.2.16)}$$

which states that the vectors $\tilde{q}$ and $\tilde{I}$ are parallel. Now, consider the layered system shown in Figure 2.2.7. Because the equivalent hydraulic conductivity is greater in the direction parallel with the layers than perpendicular to them, the components of the Darcy velocity must be greater and less respectively than their corresponding magnitudes for an isotropic medium. This idea means that the vectors $\tilde{q}$ and $\tilde{I}$ can no longer be parallel, although they are still linearly related (e.g., a 50 percent increase in $\tilde{I}$ will give a 50 percent increase in $\tilde{q}$). The question is how to generalize Equation 2.2.16?

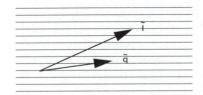

FIGURE 2.2.7 Anisotropy in a layered porous medium

The appropriate generalization of Darcy's Law is made through the introduction of tensors. Just like stress (force per area), the hydraulic conductivity system has two directions associated with its value at any point. The components of the hydraulic conductivity tensor give the component of the Darcy velocity in one direction caused by the hydraulic gradient in another direction. The two directions can be the same. A tensor is defined by its transformation rules under a rotation of coordinate system, though we may write and operate with a second order tensor as a 3×3 matrix, just as we can write a vector as a 1×3 matrix. The general form of Darcy's Law is

$$\begin{pmatrix} q_x \\ q_y \\ q_z \end{pmatrix} = \begin{bmatrix} K_{xx} & K_{xy} & K_{xz} \\ K_{yx} & K_{yy} & K_{yz} \\ K_{zx} & K_{zy} & K_{zz} \end{bmatrix} \cdot \begin{pmatrix} I_x \\ I_y \\ I_z \end{pmatrix}$$

(2.2.17)

In symbolic form, Equation 2.2.17 is

$$\tilde{q} = \tilde{\tilde{K}} \cdot \tilde{I}$$

(2.2.18)

Equations 2.2.17 and 2.2.18 are evaluated using the rules of matrix multiplication. For example, for the x-component of the velocity, we have

$$q_x = K_{xx} I_x + K_{xy} I_y + K_{xz} I_z$$

with similar results for the other components.

To obtain a physical interpretation of the components of the hydraulic conductivity tensor, consider the layered system in Figure 2.2.8 with the y-axis chosen to be parallel with the direction of $\tilde{I}$. The only component of the vector $\tilde{I}$ is I_y. According to Equation 2.2.17, the y-component of $\tilde{q}$ is given by

$$q_y = K_{yy} I_y$$

This equation looks just like Darcy's Law. The x-component of $\tilde{q}$, however, is

$$q_x = K_{xy} I_y$$

which shows that K_{xy} relates the y-component of the hydraulic gradient to the x-component of the specific discharge. Other components carry a similar interpretation.

While the coordinate system chosen in Figure 2.2.8 appears to offer some advantages, it is not **natural**. A more appropriate approach is to choose one coordinate to be parallel with the layers and one to be perpendicular to them. The resulting system has one coordinate in the direction of maximum hydraulic conductivity and the other in the direction of its minimum. This unique selection for a coordinate system is referred to as the choice of **principal axes**. Moreover, in selecting the principal axes as the coordinate base, the hydraulic conductivity tensor becomes a diagonal tensor—as shown in Equation 2.2.19.

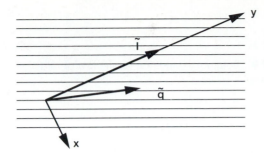

FIGURE 2.2.8 Layered porous medium with y-axis parallel with $\tilde{I}$

$$\begin{pmatrix} q_x \\ q_y \\ q_z \end{pmatrix} = \begin{bmatrix} K_{xx} & 0 & 0 \\ 0 & K_{yy} & 0 \\ 0 & 0 & K_{zz} \end{bmatrix} \cdot \begin{pmatrix} I_x \\ I_y \\ I_z \end{pmatrix} \qquad (2.2.19)$$

This form is always valid for principal axes, and one can write Darcy's Law simply as

$$q_x = K_{xx} I_x$$
$$q_y = K_{yy} I_y \qquad (2.2.20)$$
$$q_z = K_{zz} I_z$$

The important feature of Equation 2.2.20 is that the principal components K_{xx}, K_{yy}, and K_{zz} are not all equal.

A common notation that appears in the literature is tensor notation, according to which Darcy's Law takes the form

$$q_i = -K_{ij} \frac{\partial h}{\partial x_j} \qquad (2.2.21)$$

In this form, the repeating index j on the right-hand side is to be summed over x, y, and z. This operation is the so-called **Einstein summation convention**. All of the equations in this section could be written in this form. For a further discussion of the hydraulic conductivity tensor and its properties, see Section 5.6 of Bear (1972).

Refraction of Streamlines

An important topic in the analysis of layered porous media is the boundary condition for flow between layers. Consider the flow path shown in Figure 2.2.9 that arises from fluid movement from a layer with hydraulic conductivity K_1 to one with conductivity K_2, where for the figure shown $K_2 > K_1$. To find the appropriate boundary condition, we impose two physical conditions. The first is derived from mass conservation. The vertical components of the velocity must be

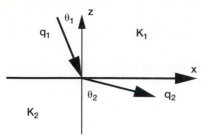

FIGURE 2.2.9   Boundary condition for the law of refraction of streamlines

equal on both sides of the interface; otherwise, the boundary will act as a source or sink. With reference to Figure 2.2.9, we have $q_{z1} = q_{z2}$, or

$$q_1 \cos(\theta_1) = q_2 \cos(\theta_2)$$

The second condition is derived from the pressure balance along the boundary between the layers. If the pressures do not balance, there will be a pressure discontinuity at the interface with a resulting infinite force acting on the fluid. Thus, $h_1 = h_2$ along $z = 0$ (the elevation of the interface), and along this same elevation,

$$\frac{\partial h_1}{\partial x} = \frac{\partial h_2}{\partial x}$$

Recall that

$$q_z = -K\frac{\partial h}{\partial x} = q \sin(\theta)$$

so

$$\frac{q_1 \sin(\theta_1)}{K_1} = \frac{q_2 \sin(\theta_2)}{K_2}$$

Combining these two conditions gives

$$\frac{K_1}{K_2} = \frac{\tan(\theta_1)}{\tan(\theta_2)} \tag{2.2.22}$$

Equation 2.2.22 is the **law for refraction of streamlines** or flow lines.[1] The following example shows the implications of this relationship.

---

[1] The law for refraction of streamlines is similar to Snell's law for the refraction of light, except the latter appears with the sine function rather than the tangent.

EXAMPLE   **2.2.1   Refraction of streamlines**
PROBLEM   Consider a case with $K_1 = 10^{-6}$ cm/s, $K_2 = 10^{-3}$ cm/s, and $\theta_1 = 5°$ ($\tan(\theta_1) = 0.0875$).
Then, Equation 2.2.22 gives

$$\tan(\theta_2) = \frac{K_2}{K_1} \tan(\theta_1) = 87.5$$

so $\theta_2 = 89.3$ degrees. The point of interest from this example is that the flow in
layer one is nearly vertical, while that in layer two is nearly horizontal. Equation 2.2.22 states that the refraction of the flow lines is proportional to the ratio
of the permeabilities of the layers. This example is relevant in that it provides
an explanation of the role of aquifers and aquitards in regional flow systems. Because the permeability of an aquifer is generally a few orders of magnitude larger than that of the bounding aquitards, one will find, in general, horizontal flow
in aquifers and vertical flow through aquitards.

Natural Heterogeneity

In this section, we have focused on uniform layered heterogeneities which are
somewhat artificial compared with most geologic depositional systems. We will
now examine the relationship between natural heterogeneity and anisotropy
under specific field conditions through review of a hydrogeologic setting that
has been studied in detail. The site is located at an inactive sand quarry at the
Canadian Forces Base, Borden, Ontario (Mackay et al., 1986). This site was the
location of an extensive natural gradient tracer test performed to improve our
understanding of how dispersion and sorption processes control the transport of
chemicals released in the subsurface environment. The contaminant transport
characteristics will be discussed in Chapter 8, "Advection-Dispersion Transport
and Models." Here, however, we focus on the site's hydraulic characteristics. The
purpose of this discussion is to show the procedures that are used to describe natural heterogeneity, and they require the use of statistical correlations.

The Borden aquifer is composed of clean, well-sorted, fine- to medium-
grained sand. Although the aquifer is relatively homogeneous (this hydrogeologic system is much more homogeneous than most systems), undisturbed cores
reveal distinct bedding features which were primarily horizontal and parallel,
although some cross and convolute bedding was observed. The median grain
sizes from a set of 846 samples taken from 11 undisturbed cores ranged from
0.070 to 0.69 mm, with the clay size fraction missing or low in most samples.
The magnitude and variability of the porosity, bulk density, and solid density
were determined from four core samples (approximately 1–1.5 m long, 5 cm in
diameter) which were subdivided into short vertical subsections (generally 15
cm long). After drying at 105 degrees C, the volume-weighted arithmetic mean
bulk density of the 36 available samples was measured to be 1.81 g/cm^3, while
the standard deviation of the spatial distribution of the measured values was
0.045 g/cm^3. The solid density of subsamples of the aquifer solids was measured

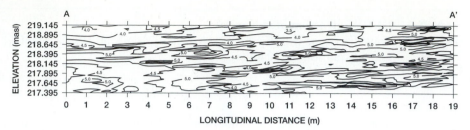

FIGURE 2.2.10 Distribution of -ln(K) with contour interval 0.5 and vertical exaggeration 2 (from Sudicky, 1986)

by water displacement, giving a mean value of 2.71 g/cm³ and a standard error of 0.01 g/cm³. The porosity was calculated from the values of the bulk density and solid density. The volume-weighted arithmetic mean of the 36 samples gave a porosity of 0.33 with a standard deviation of 0.017. Relative to the hydraulic conductivity, the spatial variability of porosity was noted to be small (coefficient of variation = 0.05).

Figure 2.2.10 shows the hydraulic conductivity distribution based on falling-head permeameter tests of 720 repacked subsamples taken from 20 2 m-long cores (Sudicky, 1986). A second set of 12 cores was taken along a perpendicular crossing line and showed similar characteristics. An immediate question concerns how to characterize the hydraulic conductivity distribution. The observed pattern is not random; values at points close together are correlated. Furthermore, it appears that the correlation is greater in the horizontal direction than in the vertical direction.

One approach is to represent the distribution of hydraulic conductivity as a spatially correlated random function that is described by its statistical moments. With $Y = ln(K)$, the most important statistics are the mean μ_Y^2, variance σ_Y^2, and covariance $B_Y(\tilde{x}_1, \tilde{x}_2)$ or correlation function $R_Y(\tilde{x}_1, \tilde{x}_2) = B_Y(\tilde{x}_1, \tilde{x}_2)/\sigma_Y^2$. If the field is **statistically homogeneous**, then the covariance or correlation function depends only on the distance and direction between two points, and not on their location within the field. This situation means, for example, that

$$B_Y(\tilde{x}_1, \tilde{x}_2) = B_Y(\Delta\tilde{x})$$

where $\Delta\tilde{x} = \tilde{x}_2 - \tilde{x}_1$.

The estimated mean $\hat{\mu}_Y$ and variance $\hat{\sigma}_Y^2$ of $ln(K)$ at 22 degrees C determined from all 1,279 measurements of hydraulic conductivity was found to be -4.63 and 0.38, respectively (Sudicky, 1986), with a geometric mean of 9.75×10^{-3} cm/s. For the correlation function, Sudicky used

$$R_Y(\Delta\tilde{x}) = \eta \exp\left(-\sqrt{\frac{x^2 + y^2}{\lambda_h^2} + \frac{z^2}{\lambda_v^2}}\right)$$

When this exponential function was fit to the data for $\Delta x_i \neq 0$, the fitted curve gives $\eta = 0.75$, where η represents the ratio $\sigma_Y^2/\hat{\sigma}_Y^2$ with $\sigma_Y^2 = \hat{\sigma}_Y^2 - \sigma_o^2$, where σ_Y^2

is an estimate of the variance of the structured $\ln(K)$ variability and σ_o^2 is an estimate of the variance due to essentially uncorrelated variations below the scale of measurement. For this data, $\sigma_Y^2 = 0.75 \times 0.38 = 0.285$, and $\sigma_o^2 = 0.095$. The parameter λ_i is a measure of the length scale associated with the correlation in the ith direction. Formally, λ_i is called the **integral scale**. For the horizontal direction, Sudicky found $\lambda_h = 2.8$ m—while for the vertical direction it was determined that $\lambda_v = 0.12$ m. Thus, the covariance function for the $\ln(K)$ field of Figure 2.2.10 is given by

$$B_Y(\tilde{x}) = \sigma_o^2 \left(1 - H(\tilde{x})\right) + \sigma_Y^2 \exp\left(-\sqrt{\frac{x^2 + y^2}{\lambda_h^2} + \frac{z^2}{\lambda_v^2}}\right)$$

$$= 0.0095 \left(1 - H(\tilde{x})\right) + 0.285 \exp\left(-\sqrt{\frac{x^2 + y^2}{(2.8 \text{ m})^2} + \frac{z^2}{(0.12 \text{ m})^2}}\right)$$

where $H(\tilde{x}) = 0$ for $\tilde{x} = 0$ and $H(\tilde{x}) = 1$ for $\tilde{x} > 0$, and $\tilde{x}$ is measured in meters.

This statistical description of the porous medium is important in that it leads to a description of the effective hydraulic parameters. The effective hydraulic conductivity tensor satisfies the relations

$$\langle \tilde{q} \rangle = -\tilde{\tilde{K}}_{ef} \cdot \nabla \langle h \rangle = \tilde{\tilde{K}}_{ef} \cdot \langle \tilde{I} \rangle$$

where $\langle \, \rangle$ represents the ensemble mean value (average). For small $\sigma_Y^2 < 1$, and with $e = \lambda_v / \lambda_h$, Dagan (1989) shows that with small e

$$K_{efh} \cong K_g \left(1 + \sigma_Y^2 \left(\frac{1}{2} - \frac{\pi e}{4}\right)\right)$$

$$K_{efv} \cong K_g \left(1 - \sigma_Y^2 \left(\frac{1}{2} - \frac{\pi e}{4}\right)\right)$$

For this data set, we have $e = 4.29 \times 10^{-2}$ and $\sigma_Y^2 = 0.285$, so

$$K_{efh} \cong 1.10 \times 10^{-2} \text{ cm/s}$$

$$K_{efv} \cong 8.55 \times 10^{-3} \text{ cm/s}$$

The degree of anisotropy is

$$\frac{K_{efh}}{K_{efv}} = 1.29$$

Note that as $e \to 0$ (which is the stratified aquifer case) we again capture Equations 2.2.11 and 2.2.12 since, for the lognormal distribution,

$$K_a = K_g \left(1 + \frac{\sigma_Y^2}{2}\right)$$

$$K_h = K_g \left(1 - \frac{\sigma_Y^2}{2}\right)$$

We contrast these results with those for an isotropic aquifer, where one finds (Dagan, 1989)

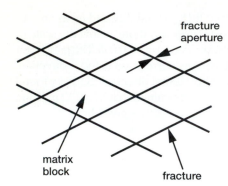

fracture
aperture

matrix
block

fracture

FIGURE 2.2.11   Structure of a typical fractured porous medium

$$K_{eff} = K_g \left(1 + \frac{\sigma_Y^2}{6} \right)$$

Again, it is noted that these results apply for small σ_Y^2, typically $\sigma_Y^2 < 1$. For a further discussion, see Dagan (1989) and references therein.

2.2.4   Fractured Media

We now turn to the question of flow in **fractured media**, such as the examples shown schematically in Figure 2.2.11. The figure identifies the fracture system, the rock or block porous matrix, and the fracture aperture (or width of the fracture openings). There are at least three approaches to modeling groundwater flow in fractured systems. First, one can attempt to model flow in the individual (discrete) fractures. The basic idea is that one would identify the most significant fractures within a location and apply principles of fluid mechanics in modeling the flow in the resulting discrete fracture system. At well-instrumented field sites and after tracer tests have been completed, this approach may be applicable, but in general the identification of the location, size, shape, and orientation of the individual fractures is beyond current measurement capabilities.

The second approach surrounds the use of what are called **dual porosity models**. First introduced by Barenblatt, et al. (1960), the basic idea is to assign two conceptual sets for values of porosity, permeability, etc., to all points within the domain—one representing the block matrix of the rock and the other representing the fracture system. The key element of this approach is founded on a continuum hypothesis for both the fracture set and porous block matrix. Water is allowed to flow within the porous matrix of the rock, within the fracture system, and between the matrix and fractures. There are two fundamental issues to be addressed when employing this approach: (1) identifying the existence of a physical averaging volume (REV) for the fracture set, and (2) quantifying the fluid transfer between the two porosity systems. The problem is especially difficult when dealing with transient flow and with solute transport in that the fracture system contains most of the permeability of the system, while the matrix contains most of the fluid storage.

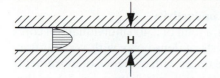

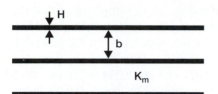

FIGURE 2.2.12 Viscous flow through an individual fracture

FIGURE 2.2.13 Parallel system of fractures

The third approach is to model the fractured porous medium as an equivalent continuum. Using this approach, flow through an individual fracture may be modeled by solving the Navier-Stokes equations for flow between two parallel plates, as shown in Figure 2.2.12. The velocity profile is parabolic, and the average velocity is given by (see Equation 2.1.9)

$$v = \frac{H^2 \rho g}{12\mu} I \tag{2.2.23}$$

where v is the average velocity within the fracture, I is the hydraulic gradient parallel to the fracture, and H is the fracture aperture.

Consider the parallel system of fractures shown in Figure 2.2.13 with mean fracture spacing b and fracture aperture H. A direct calculation of the discharge per unit width gives

$$U = \Sigma b K_m I + \Sigma H v = \Sigma b K_m I + \Sigma \frac{H^3 \rho g}{12\mu} I$$

using Equation 2.2.23, where K_m is the block matrix hydraulic conductivity. The same calculation for the effective continuum with effective hydraulic conductivity K_{eff} gives

$$U = \Sigma(b + H)K_{eff} I \cong \Sigma b K_{eff} I$$

because $H \ll b$. Comparing these two calculations, we find

$$K_{eff} = K_m + \frac{H^3 \rho g}{12 b \mu} \tag{2.2.24}$$

Equation 2.2.24 is called the **cubic law** because of the dependency of the effective hydraulic conductivity on the cube of the fracture aperture.

**EXAMPLE
PROBLEM**

2.2.2   Estimation of Fracture Hydraulic Conductivity

During a site characterization investigation, a total of 2,400 individual soil cores were examined. The porous medium consists of dense clay with vertical fractures, and fractures or slickensides (fracture traces with evidence of movement) were identified in 3.4 percent of the cores. The soil core diameter is 7.4 cm. Estimate the effective vertical hydraulic conductivity for this medium.

It is assumed that the cores represent independent samples and that the human eye can identify fractures of aperture size 100 microns or larger. If D is the core diameter size and b is the average fracture spacing, then the probability of a randomly placed core hitting a fracture is $P = D/b$. Thus, the data suggests that the mean fracture spacing is $b = D/P = 7.4$ cm/0.034 = 220 cm. With $H = 0.01$ cm, Equation 2.2.24 gives the effective fracture hydraulic conductivity

$$K_{\text{fracture}} = \frac{(0.01 \text{ cm})^3 \times (1 \text{ g/cm}^3) \times (981 \text{ cm/s}^2)}{12 \times (220 \text{ cm}) \times (0.01 \text{ g/cm} \cdot \text{s})} = 4 \times 10^{-5} \text{ cm/s}$$

Because the matrix permeability of the dense clay is likely to be 10^{-7} cm/s or less, it is clear that most of the medium permeability is associated with the fracture system.

For a discussion of more general fracture systems and additional references, see Marsily (1986).

2.2.5   Limitations to the Validity of Darcy's Law

Meinzer (1942) notes that there does not appear to be a lower limit to the range of applicability of Darcy's Law for aquifer materials. With hydraulic gradients as small as a few inches per mile, the linear relationship remains valid. For clay materials, on the other hand, the situation is not so clear. With the large charged surface area of clays, there is some suggestion that water does not act as a Newtonian fluid, at least for small gradients, and that there is a threshold gradient, I_o, which must be exceeded before flow occurs. Figure 2.2.14 shows the inferred

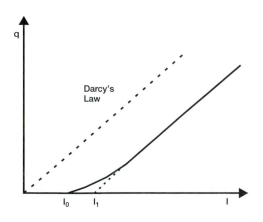

FIGURE 2.2.14   Non-Darcy flow in clay materials

relationship between the volume flux and energy gradient. If we extrapolate the linear portion to I_1, then the following empirical model should apply approximately over the entire range for $I > I_1$:

$$q = K(I - I_1) \qquad (2.2.25)$$

with $q = 0$ for $I < I_1$.

For larger hydraulic gradients and velocities, there is ample evidence that the flow does depart from the linear relationship of Darcy's Law. In hydraulics, the nature of the flow is characterized by the relationship between the friction factor and the Reynolds number. Rose (1945, reported by Todd, 1959) has summarized the experimental results from a number of investigators in terms of the Fanning friction factor. The Fanning friction factor, f_f, is defined by the relation

$$f_f = \frac{d\Delta p}{2\rho L q^2} \qquad (2.2.26)$$

where d is the average grain diameter, Δp is the pressure difference over a length of porous media L measured along the line of flow, and q is the Darcy velocity. Comparison of Equation 2.2.26 with the Darcy-Weisbach equation (see White, 1986) shows that the Fanning friction factor is related to the more familiar Darcy friction factor through $f_f = f_d/4$.

Figure 2.2.15 shows the experimental results for f_f as a function of the Reynolds number, N_R, where

$$N_R = \frac{q\rho d}{\mu} \qquad (2.2.27)$$

Figure 2.2.15 shows that Darcy's Law remains valid for N_R smaller than 10, with the friction factor given by

$$f_f = \frac{1000}{N_R} \qquad (2.2.28)$$

Combining Equations 2.2.26 through 2.2.28 with $\Delta h = \Delta p/\rho g$ gives

$$q = \frac{\rho g d^2}{2000\mu} \frac{\Delta h}{L} \qquad (2.2.29)$$

The leading term on the right side of Equation 2.2.29 corresponds to the hydraulic conductivity, and comparison with Equation 2.1.11 shows that his result gives for the intrinsic permeability

$$k = \frac{d^2}{2000} \qquad (2.2.30)$$

The data shown in Figure 2.2.15 is from experiments with unconsolidated media. Muskat (1946, Figure 8, p. 60) provides results from other experiments that show the inverse relation between friction factor and Reynolds number from Equation 2.2.28 remains valid for consolidated media for small N_R values, except that the factor 1,000 can vary by many orders of magnitude.

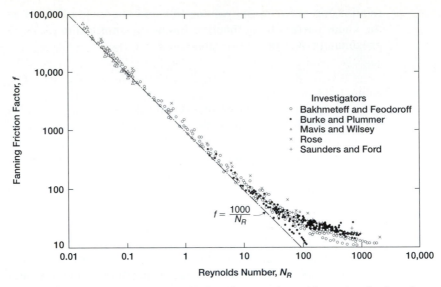

FIGURE 2.2.15   Relation of Fanning fraction factor to Reynolds number for flow through granular porous media (from Todd, 1959)

The departure from Darcy's Law at high N_R is associated with **inertial effects** rather than turbulence. These inertial effects show up in the convective acceleration term $\tilde{u} \cdot \nabla \tilde{u}$ of the total acceleration from the Navier-Stokes equations:

$$\frac{d\tilde{u}}{dt} = \frac{\partial \tilde{u}}{\partial t} + \tilde{u} \cdot \nabla \tilde{u}$$

This term suggests that the gradient should vary as the velocity squared, a relationship that is commonly found for **turbulent flow** in open channels and closed conduits. A model suggested by Forchheimer (1901) gives

$$I = \alpha q + \beta q^2 \qquad \qquad \textbf{(2.2.31)}$$

where α and β are empirical coefficients. From experiments, Ergun (see Bird et al., 1960) suggests the model

$$I = \frac{150(1-n)^2\mu}{n^3 d^2 \rho g} q + \frac{1.75(1-n)}{gn^3 d} q^2 \qquad \qquad \textbf{(2.2.32)}$$

where d is the mean grain diameter.[2] From experiments, Ward (1964) has suggested the following model

$$I = \frac{\mu}{k\rho g} q + \frac{0.55}{g\sqrt{k}} q^2 \qquad \qquad \textbf{(2.2.33)}$$

with the intrinsic permeability $k = d^2/360$.

---

[2]Equation 2.2.32 differs from the Kozeny-Carman equation by a factor of 150, rather than 180 in the first term.

2.2.6   Laboratory Measurement

Laboratory measurement of hydraulic conductivity involves the use of constant head and **falling head permeameters**. A constant head device is similar to that used originally by Darcy and is shown in Figure 2.2.1. A constant head difference is maintained across the soil sample, and the flow rate is measured (alternatively, the flow rate is fixed and the head difference is measured). Such a device is most often used for samples with appreciable permeability so that the flow rate is easily measured. The integrated form of Darcy's Law for a laboratory column gives

$$K = \frac{QL}{A(h_1 - h_2)} = \frac{qL}{h_1 - h_2} \qquad (2.2.34)$$

where h_1 and h_2 are the upstream and downstream hydraulic heads, respectively, and L is the length of the packed soil column.

A falling head permeameter, such as that shown in Figure 2.2.16, is often used for hydraulic conductivity measurement in low permeability soils where the discharge rate is small. A constant effluent head is maintained, and the water level in the small standpipe (or burette) is measured as a function of time. Measurement of permeability follows from application of continuity principles.

Applying continuity and Darcy's Law to the permeameter shown in Figure 2.2.14, one finds

$$Q = KA\frac{H}{L} = -a\frac{dH}{dt}$$

Separating variables and integrating gives

$$K = \frac{La}{A(t - t_o)} \, ln\left(\frac{H_o}{H}\right) \qquad (2.2.35)$$

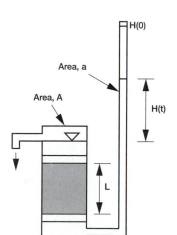

FIGURE 2.2.16 Falling head permeameter used to measure hydraulic conductivity of low permeability materials

The hydraulic conductivity may be estimated either from individual measurements of the water level in the standpipe or from the slope of a plot of the logarithm of H versus time. Field methods for measurement of hydraulic conductivity are discussed in Chapter 3, "Groundwater and Well Hydraulics."

2.3   Continuity Relations for Flow in Porous Media

In Section 2.2, we looked at Darcy's Law and its various generalizations. We saw that Darcy's Law represents an application of the momentum equation from fluid mechanics to porous media flow. In this section, we consider the application of another basic principle, the conservation of mass, or the continuity equation. Together, Darcy's Law and the continuity equation, along with appropriate initial and boundary conditions, provide the mathematical framework to find the head distribution within a region as a function of location and time. In this section, we will primarily be concerned with formulation of the basic equations. Later chapters will focus on various solution scenarios and applications.

2.3.1   Unconfined Aquifers

We first consider flow in an unconfined aquifer, as shown in Figure 2.3.1. The level H above the datum represents the elevation of the **water table** (i.e., the locus of all points at atmospheric pressure). Above the water table the (gauge) pressure in the water phase is negative, while below the water table the pressure is positive. The elevation of the base of the aquifer is represented by ζ. While H is a

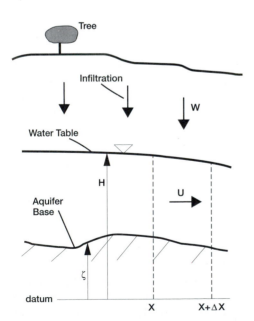

FIGURE 2.3.1   Cross-section of an unconfined aquifer

function of $\tilde{x}$ and t, ζ is considered to be a function of $\tilde{x}$ alone. Additionally, the inflow rate W is from infiltration or diffuse recharge and has the units of volume/area/time (or L/T).

Storage in Unconfined Aquifers: Specific Yield

If a saturated soil is left to drain under the force of gravity, not all of the water is removed. Capillary forces hold a fraction of the interstitial water. The fraction of the volume that drains is called the **specific yield**, or drainage porosity. The former term was introduced by Meinzer (1923), while the latter is used by Marsily (1986). The specific yield is defined by

$$S_y = \frac{\text{gravity drainage volume}}{\text{bulk volume}} \qquad (2.3.1)$$

This definition may be interpreted as follows. If the water table is lowered by an amount Δh over an area A, then the volume of water that is drained from the soil is

$$V_{\text{drained}} = S_y \Delta h A \qquad (2.3.2)$$

The fraction which is held by the capillary forces is called the **specific retention** and is given by the relation

$$S_r = n - S_y \qquad (2.3.3)$$

As used in groundwater hydrology, the specific retention is the same as the term **field capacity** used in the agricultural literature.

Values of the specific yield will vary with the soil texture. For coarse sand and gravel, typical values range from 0.2 to 0.3. For fine sands, the range is 0.1 to 0.15, and for clays the specific yield is 0.05 or less (modified from Davis and De Wiest, 1966). Drainage of water associated with lowering the water table will also depend on time. If the water table is lowered quickly, such as through operation of a pumping well, the downward movement of water may lag the rapid adjustment of the water table (pressure field), and the ultimate specific yield will only be reached after a period of drainage. The time delay of specific yield is important when analyzing the drawdown during aquifer (pumping) tests and is discussed in Chapter 3. For most analyses of long-term water budget, however, the time dependency of the specific yield is not considered.

Hydraulic Approach to Groundwater Flow

The flow of water in the subsurface is three-dimensional, which means that the groundwater velocity, $\tilde{q}$, has components in three directions. The hydraulic head depends on the spatial location and time; that is, $h(x,y,z,t)$. Most aquifers are of a much greater lateral extent (kilometers) than thickness (meters or tens of meters), however. Furthermore, the law of refraction of flow lines, Equation 2.2.22, suggests that flow is nearly horizontal in regions with large permeability and is nearly vertical in regions with small permeability. These ideas suggest that on a large scale, flow in aquifers is essentially horizontal, while flow in confining beds is essentially vertical. The analysis of groundwater flow based on the assumption that flow in aquifers is horizontal is called the **hydraulic approach**. This assumption implies that the aquifer flow velocity has only two components and that the head depends on only the horizontal location and time. Vertical flow and vertical variations in head within an aquifer are neglected. Vertical flow can occur from one aquifer to another through a semi-confining layer, but the flow in each aquifer is considered to be horizontal.

The assumption of essentially horizontal flow fails in regions where the flow has a large vertical component, such as in the vicinity of partially penetrating wells, springs, rivers, etc. Within the hydraulic approach to analyzing groundwater flow, these effects are included through additional head loss terms that account for the vertical flow components. Bear (1979) suggests that even near these features, essentially horizontal flow occurs at distances from the features of 1.5 to 2 times the thickness of the aquifer. Thus, vertical flow tends to be a localized phenomenon in aquifers.

Use of the hydraulic approach reduces the dimensionality of aquifer flow problem analysis, which makes the analysis much easier. Regions where vertical flow is significant are addressed through additional head loss terms. In addition, aquifer parameters are defined through a vertical averaging process, so that at least implicitly, vertical variations in properties within the aquifer are included. This vertical averaging approach, however, introduces difficulties in the analysis of chemical transport through aquifers, because the transport will occur faster through the more permeable strata and slower through the less-permeable layers. In this averaging process, the recognition of these regions with different permeabilities is lost. These issues are discussed in Chapter 8.

The Dupuit-Forchheimer Assumptions

The hydraulic approach to analysis of groundwater flow in an unconfined aquifer is based on two assumptions, which together are known as the **Dupuit-Forchheimer assumptions**.[3] The first of these assumptions is that within an unconfined aquifer, the head is independent of depth. Mathematically, this means that h is a function of three variables rather than four:

$$h(x, y, t) \rightarrow h(x, y, t) \tag{2.3.4}$$

Equation 2.3.4 implies that the flow can only be horizontal and no vertical gradients are present. Such an approach allows one to employ the principle of "hydrostatics" in determining the piezometric head within the aquifer. The second assumption is that the discharge is proportional to the slope of the water table. Along with the first assumption, this statement implies that the function

$$h(x, y, t) \rightarrow H(x, y, t) \tag{2.3.5}$$

where H is the water table elevation. Moreover, it is assumed that ρ and n are constant throughout the unconfined unit.

---

[3]Dupuit is the French scientist who first suggested the ideas, and Forchheimer is the German engineer who showed their ramifications and applications in a variety of field problems.

These assumptions imply that the **aquifer flux**, which is the product of the Darcy velocity and the saturated thickness with units of L^2/T, is equal to

$$U_x = -K(H - \zeta)\frac{\partial H}{\partial x} \tag{2.3.6}$$

and similarly for U_y, where ζ is the elevation of the base of the aquifer.

One-Dimensional Continuity Equation

The principle of conservation of mass states that for an arbitrary control volume, the rate of mass accumulation within the volume plus the net mass flux out of the volume must equal the rate of mass generation within the volume (Bird et al., 1960). Because the density is assumed constant, we have

$$\begin{array}{ccc}
\text{rate volume} & \text{net} & \text{volume} \\
\text{storage} & + \text{ volume} = & \text{source} \\
\text{increase} & \text{flux out} & \text{strength}
\end{array} \tag{2.3.7}$$

Consider a slice of Figure 2.3.1 of thickness Δx between the locations x and $x + \Delta x$. Applying the continuity principle to this slice, we find

$$S_y \frac{\Delta H}{\Delta t}\Delta x + U_{x+\Delta x} - U_x = W\Delta x$$

where S_y is the specific yield and W is the infiltration or diffuse recharge rate with units of L/T. For small Δx, we have from Taylor's theorem,

$$U_{x+\Delta x} \cong U_x + \frac{\partial U}{\partial x}\Delta x$$

Combining this information with the continuity principle of Equation 2.3.7 and canceling Δx, we find

$$S_y \frac{\partial H}{\partial t} + \frac{\partial U}{\partial x} = W \tag{2.3.8}$$

Equation 2.3.8 is the general one-dimensional continuity equation for flow in an unconfined aquifer. If we combine Equation 2.3.8 with Darcy's Law in the form of Equation 2.3.6, we find

$$S_y \frac{\partial H}{\partial t} = \frac{\partial}{\partial x}\left(K(H - \zeta)\frac{\partial H}{\partial x}\right) + W \tag{2.3.9}$$

Equation 2.3.9 is a nonlinear, second order *partial differential equation* (PDE) which may be used with appropriate initial and boundary conditions to find $H(x,t)$. Unfortunately, Equation 2.3.9 cannot be solved in general except through application of numerical methods of approximation. In order to see the value of the continuity equation, we consider simplifications that lead to elementary but useful solutions.

Steady Flow

The first assumption is that the flow is steady; that is, at a location the water table, elevation does not change with time. This idea reduces Equation 2.3.9 to a nonlinear, second order *ordinary differential equation* (ODE). Furthermore, assume that ζ and K are constant with $\zeta = 0$ (by choosing the elevation datum as the uniform base of the aquifer). Then, Equation 2.3.9 reduces to

$$K \frac{d}{dx} \left(H \frac{dH}{dx} \right) + W = 0$$

This equation remains nonlinear in H. If we recognize, however, that

$$H \frac{dH}{dx} = \frac{1}{2} \frac{dH^2}{dx}$$

we find

$$\frac{d}{dx} \left(\frac{dH^2}{dx} \right) + \frac{2W}{K} = 0 \qquad (2.3.10)$$

which is linear in H^2. If W is also constant, we can integrate Equation 2.3.10 to find

$$H^2 + \frac{Wx^2}{K} = Ax + B \qquad (2.3.11)$$

where A and B are integration constants that are determined by appropriate boundary conditions. Equation 2.3.11 is an important result and we now consider a few of its applications.

EXAMPLE PROBLEM

2.3.1   Dupuit Equation (Flow Through an Embankment)

As a first example, we consider flow through an embankment without recharge. For this case, $W = 0$, so $H^2 = Ax + B$. We need two boundary conditions. These are given by $H(0) = H_0$ and $H(L) = H_L$. The first dictates that $B = H_0^2$, while the second gives

$$A = \frac{H_L^2 - H_0^2}{L}$$

Thus,

$$H^2 = \frac{H_L^2 - H_0^2}{L} x + H_0^2. \qquad (2.3.12)$$

The discharge (per unit width) is given by

$$U = Hq = H \left(-K \frac{dH}{dx} \right) = -\frac{K}{2} \frac{dH^2}{dx} = \frac{K}{2} \left(\frac{H_0^2 - H_L^2}{L} \right) \qquad (2.3.13)$$

Equation 2.3.13 is one of the more common and useful forms of the **Dupuit equation** for steady, unconfined flow.

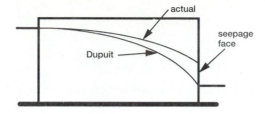

FIGURE 2.3.2 Comparison of the Dupuit and actual flow profile through an embankment

Discussion

Figure 2.3.2 shows a comparison of the actual water table in an embankment and the water table predicted by the Dupuit equations. The actual water table intersects the outflow boundary above the level of freestanding water, and the region in-between is called a **seepage face**. It is apparent that the Dupuit equation may not yield useful results if one is interested in the actual water table elevation and pore water pressures (such as one would need for slope stability calculations). Charni, however (see Bear, 1972), has shown that the Dupuit assumptions lead to the correct formula for the discharge (Equation 2.3.13), taking into account the actual water table elevation and the seepage face. Thus, as long as our interests are focused on the hydraulics of groundwater rather than its geotechnical ramifications, the Dupuit equations lead to useful results.

EXAMPLE PROBLEM

2.3.2 Agricultural Drains

Agricultural drains are used to control the groundwater elevation in a field used for crop production so that the plant roots may have appropriate aeration. Consider the pair of agricultural drains shown in Figure 2.3.3. The height of the drain above an impermeable boundary is H_d, while the height of the water table at the midpoint between drains is H_m. For the problem of predicting the water table elevation, the general solution is given by Equation 2.3.11. We need two boundary conditions to determine the integration constants A and B. Two choices present themselves; we could use either of the following:

$$H^2(0) = H_d^2 \; ; H^2(L) = H_d^2 \tag{a}$$

$$H^2(0) = H_d^2 \; ; \frac{dH^2}{dx}\left(\frac{L}{2}\right) = 0 \tag{b}$$

For case (a), we have $H^2(0) = B = H_d^2$ and $H^2(L) = H_d^2 = -\dfrac{WL^2}{K} + AL + H_d^2$ so $A = \dfrac{WL}{K}$ Thus, the water table elevation is given by

$$H^2 = H_d^2 + \frac{Wx}{K}(L - x) \tag{2.3.14}$$

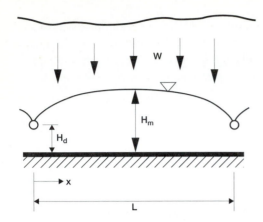

FIGURE 2.3.3 Cross-section between two agricultural drains

Now consider case (b). The same result is found for B. The second condition for case (b) gives $\dfrac{dH^2(\frac{L}{2})}{dx} = -\dfrac{2W(L/2)}{K} + A = 0$ which also gives $A = \dfrac{WL}{K}$ The conclusion from these two approaches is that as long as the problem is correctly posed, the mathematics will lead to the same (and correct) result. The maximum saturated thickness is found from

$$H_m = \sqrt{H_d^2 + \frac{WL^2}{4K}} \qquad (2.3.15)$$

Equation 2.3.15 shows the relationship between the infiltration or irrigation rate, the drain spacing, the soil permeability, and the drain height above a layer of clay or other impermeable boundary.

Two-Dimensional Flows in Unconfined Aquifers

The continuity equation for unconfined aquifers in two dimensions is similar to that for one dimension. The two-dimensional version of Equation 2.3.9 is

$$S_y \frac{\partial H}{\partial t} = \frac{\partial}{\partial x}\left(K(H - \zeta)\frac{\partial H}{\partial x}\right) + \frac{\partial}{\partial y}\left(K(H - \zeta)\frac{\partial H}{\partial y}\right) + W \qquad (2.3.16)$$

Again, this equation is too difficult to solve in general. The assumptions required to obtain the simplest tractable form are ζ = constant = 0, the flow is steady-state, and K is constant. For this case, Equation 2.3.16 reduces to

$$\frac{\partial}{\partial x}\left(\frac{\partial H^2}{\partial x}\right) + \frac{\partial}{\partial y}\left(\frac{\partial H^2}{\partial y}\right) + \frac{2W}{K} = 0$$

or in terms of the Laplacian operator, ∇^2,

$$\nabla^2(H^2) + \frac{2W}{K} = 0 \qquad\qquad (2.3.17)$$

For general W, Equation 2.3.17 is an example of Poisson's equation in H^2. One case of special interest is flow along a radial direction either towards or away from a well, which is really one-dimensional flow in the radial direction.

Radial Flow

For **radial flow**, the Laplacian ∇^2 takes the following form:

$$\nabla^2 = \frac{1}{r}\frac{d}{dr}\left(r\frac{d}{dr}\right) = \frac{d^2}{dr^2} + \frac{1}{r}\frac{d}{dr}$$

Hence, Equation 2.3.17 may be written as:

$$\frac{1}{r}\frac{d}{dr}\left(r\frac{dH^2}{dr}\right) + \frac{2W}{K} = 0$$

If W is uniform, this equation is easily integrated to give

$$H^2 + \frac{Wr^2}{2K} = A\,ln(r) + B \qquad\qquad (2.3.18)$$

Equation 2.3.18 is the general form of the **Dupuit equation** for radial flow.

EXAMPLE PROBLEM **2.3.3   Radial Flow to a Well with Recharge**

Figure 2.3.4 shows a view of a well pumping from an unconfined aquifer that receives recharge from above at a rate W. The saturated thickness beyond the radius R of influence of the well is H_o. We are interested in the drawdown distribution and the radius of influence (distance to which the drawdown is negligible). We need boundary conditions to evaluate the integration constants A and B from Equation 2.3.18. Continuity dictates that all of the recharge received within the radius R is recovered by the well. This gives $Q = \pi R^2 W$ so $H = H_o$ at

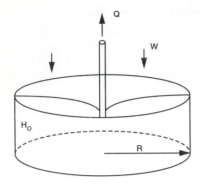

FIGURE 2.3.4 Drawdown distribution around a well receiving vertical recharge

$$R = \sqrt{\frac{Q}{\pi W}}$$

A second boundary condition is given by $U_r = 0$ at $r = R$, so

$$\left.\frac{dH^2}{dr}\right|_R = 0 \text{ and } A = \frac{WR^2}{K} = \frac{Q}{\pi K}$$

With these results, we then have

$$B = H_o^2 + \frac{WR^2}{2K} - \frac{Q}{\pi K} ln(R)$$

so that the water level is given by

$$H_o^2 - H^2 = \frac{Q}{\pi K} ln\left(\frac{R}{r}\right) - \frac{W}{2K}(R^2 - r^2) \tag{2.3.19}$$

R is the radius of capture for the well, or the **radius of influence**. The water level at the well having radius r_w is found from

$$H_w = \sqrt{H_o^2 - \frac{Q}{\pi K} ln\left(\frac{R}{r_w}\right) + \frac{W}{2K}(R^2 - r_w^2)}$$

As long as H_w is positive, then steady-state production from the aquifer at a rate Q is possible.

Discussion

There are a couple of points of interest brought out by this last example. One should note that as Q increases, so do R and $\dfrac{dH}{dr}$. Thus, a maximum possible production rate exists, which is found by setting $H_w = 0$ in the last equation and solving for Q $\left(\text{with } R = \sqrt{\dfrac{Q}{\pi W}}\right)$. Also, as $W \to 0$ with Q constant, $R \to \infty$, and steady flow is not possible. Steady flow conditions are possible only when a source of recharge exists that can supply the well discharge from the aquifer.

2.3.2   ### Storage in Confined Aquifers

The mechanism of water storage and release in an unconfined aquifer is drainage and filling of the pore space above the water table. In confined aquifers, however, the situation is much different. Production from a confined aquifer decreases the pressure in the water within the aquifer, which in turn increases the stress that must be maintained by the soil matrix. The decrease in water pressure allows for a small expansion of water itself. Thus, the mechanism of storage in confined aquifers is associated with the compressibility of the water and the porous matrix. The mechanism of water release within a confined unit is discussed in greater detail below.

Effective stress

The concept of effective stress, as introduced by Terzaghi (1923), is important in understanding the mechanism of storage in confined aquifers. Consider the surface with area A shown in the soil profile in Figure 2.3.5 with the weight W_2 initially equal to zero. The initial stress on the surface A is equal to the total weight supported by the surface (i.e., the "overburden") divided by the area. Terzaghi defined the effective stress as the difference between the total stress and the stress supported by the fluid (the fluid pressure):

$$\sigma = \frac{W_1}{A} = \sigma_e + p \tag{2.3.20}$$

As sketched in Figure 2.3.5, the effective stress is the stress maintained by the soil matrix and transmitted between grains at contact points.

If the weight W_2 is applied at the ground surface in Figure 2.3.5, and the shear stress along the sides of the soil column is negligible, then the total stress on the surface increases to

$$\sigma = \frac{W_1 + W_2}{A}$$

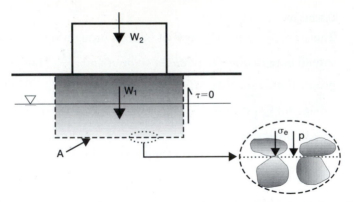

FIGURE 2.3.5 Total stress is equal to the weight supported by the surface divided by the surface area. The effective stress of the soil matrix is transmitted from grain-to-grain at contact points.

How is the additional stress supported? It is carried by the soil matrix through an increase in σ_e, and by the water through an increase in p. The amount carried by each depends on their relative compressibility. Since water is usually less compressible, it initially carries most of the load. However, this creates a region with fluid pressures larger than the ambient ones. This excess pressure causes the water to move away from the load, thereby decreasing the fluid pressure while increasing the effective stress on the soil. This redistribution of pressure is, however, a time-dependent process. The added stress causes the soil to consolidate over time. The speed with which these changes occur depends on the permeability of the soil. This relationship between fluid pressure, effective stress, and flow is essential to understanding the storage characteristics in confined aquifers.

Specific Storage

If water is injected into an aquifer then the pressure increases, although the total stress remains the same. If σ is approximately constant, then

$$dp = -d\sigma_e \qquad \textbf{(2.3.21)}$$

An increase in pressure means that the water is compressed. At the same time, a decrease in the effective stress implies that the soil matrix expands. Both of these make it look like the volume of water per unit bulk volume increases. Accordingly, we say that there has been a *storage increase*.

To quantify these relationships, consider the volume V_t in Figure 2.3.6 that contains a mass M and a fixed set of soil grains. The volume is allowed to deform (1D vertical consolidation) but remains fixed on the set of grains. We are interested in the change in mass contained within the volume.

The total volume of this control volume, volume of solids it contains, and volume of voids are given by

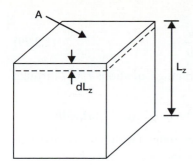

FIGURE 2.3.6   Control volume containing a fixed set of grains from the soil matrix

$$V_t = AL_z \tag{2.3.22}$$

$$V_s = (1 - n)AL_z \tag{2.3.23}$$

$$V_v = nAL_z \tag{2.3.24}$$

The total mass of water contained within the control volume is

$$M = \rho n A L_z \tag{2.3.25}$$

Injection of water into the aquifer will cause a change in the stress and a corresponding change in the mass of water contained within the volume. Under the assumption that A remains constant during these changes (1D consolidation), we have from Equation 2.3.25

$$dM = nAL_z d\rho + \rho AL_z dn + \rho n A dL_z \tag{2.3.26}$$

At the same time, the soil grains are considered to be incompressible ($dV_S = 0$), so from Equation 2.3.23 we see that

$$L_z dn = (1 - n)dL_z$$

We can use this result to eliminate $L_z \, dn$ in Equation 2.3.26 and find

$$dM = nAL_z d\rho + \rho A dL_z$$

With Equations 2.3.22 and 2.3.24, this gives

$$dM = V_v d\rho + \rho dV_t \tag{2.3.27}$$

We now introduce the water and soil matrix compressibilities. The *water compressibility*, β, is defined by

$$\beta = -\frac{\dfrac{dV_w}{V_w}}{dp} = \frac{1}{\rho}\frac{d\rho}{dp} \tag{2.3.28}$$

Similarly, the *soil matrix compressibility*, α, is defined by

$$\alpha = -\frac{\dfrac{dV_t}{V_t}}{d\sigma_e} = \frac{1}{V_t}\frac{dV_t}{dp} \tag{2.3.29}$$

where Equation 2.3.21 has also been used. Combining Equations 2.3.27 through 2.3.29 we find

$$dM = V_v\beta\rho dp + \rho\alpha V_t dp = \rho V_t(\alpha + \beta n)dp = \rho V_t(\alpha + \beta n)\rho g dh$$

Instead of working with the mass of water contained within the control volume, it is more convenient to work with the volume of water. We have

$$dV_w = \frac{dM}{\rho} = V_t(\alpha + \beta n)\rho g dh$$

The *specific storage* is the volume of water removed or added to *storage*, per unit volume, per unit change in head. Thus, the specific storage (units, L^{-1}) is

$$S_s = \frac{1}{V_t}\frac{dV_w}{dh} = (\alpha + \beta n)\rho g \tag{2.3.30}$$

Jacob (1940,1950) presented this relationship.

Table 2.3.1 shows the range of compressibility values for a number of media. Comparison of the values in this table suggests that most of the compressibility comes from that of the porous matrix, rather than from water.

Coefficient of Storage

For many problems we are interested in the storage over the entire thickness of an aquifer. The *coefficient of storage* or storage coefficient, S, provides this information. The relationship between the specific storage and storage coefficient is

$$S = S_s b = \frac{1}{A}\frac{dV_w}{dh} \tag{2.3.31}$$

The storage coefficient is a dimensionless quantity that gives the volume of water added to or removed from storage per unit surface area per unit change in the head.

TABLE 2.3.1 Range of Values of Compressibility (after Freeze and Cherry, 1979)

Medium	Compressibility, α
	$(\mathbf{m^2/N}$ or $\mathbf{Pa^{-1}})$
Clay	$10^{-6}-10^{-8}$
Sand	$10^{-7}-10^{-9}$
Gravel	$10^{-8}-10^{-10}$
Jointed rock	$10^{-8}-10^{-10}$
Sound rock	$10^{-9}-10^{-11}$
Water (β)	$4.4(10)^{-10}$

EXAMPLE   **2.3.4   Storage in a sandstone aquifer**

PROBLEM   The following data is typical for a cemented sandstone aquifer: $n = 0.1$, $\alpha = 4(10^{-7})$ ft²/lb, $\beta = 2.3(10^{-8})$ ft²/lb. These give

$$\rho g \alpha = 2.5 \times 10^{-5}\,\text{ft}^{-1}$$

$$\rho g \beta n = 1.4 \times 10^{-7}\,\text{ft}^{-1}$$

Thus most of the storage is associated with compressibility of the sandstone matrix rather than compressibility of water. For a formation having a thickness of 100 feet and underlying a surface area of 10 miles² = 279,000,000ft², the storage coefficient is

$$S = (2.5 \times 10^{-5} + 1.4 \times 10^{-7}) \times 100 = 0.00251$$

If the head is lowered an average of 3 feet then the volume of water removed from storage is

$$\Delta V_w = SA\Delta h = 2.1 \times 10^{6}\,\text{ft}^{3}$$

The fraction of this volume associated with expansion of water is 0.56 percent, or 12,000 ft³. Finally, for comparison, if the aquifer had been unconfined with a specific yield of 0.05, then the 3 foot drop in the water table would give a volume of water removed from storage associated with drainage of the pore space of $\Delta V_w = S_y A \Delta h = 42 \times 10^{6}$ ft³, larger by a factor of 20. For unconsolidated media the ratio $(\Delta V_w)_{\text{unconfined}}/(\Delta V_w)_{\text{confined}}$ can easily reach values of 100 to 1000, emphasizing the importance of the presence of a free water table in estimating the magnitude of storage for a hydrogeologic formation.

2.3.3   General Continuity Equation

Earlier in this section we have seen the continuity equation for an unconfined aquifer, Equation 2.3.8. In words this continuity equation states that the rate of storage increase plus the net volume flux (flow) out of the volume is equal to the recharge (source) rate. We now consider various generalizations of this relation.

General Three-Dimensional (3D) Equation

The general three-dimensional form of the continuity equation is

$$S_s \frac{\partial h}{\partial t} + \nabla \cdot \tilde{q} = W' \tag{2.3.32}$$

In Equation 2.3.32, W' is the source strength in units of volume/time/volume. The derivation of Equation 2.3.32 follows that of Equation 2.3.8, except for the increase in number of dimensions. Combining Equation 2.3.32 with Darcy's Law Equation 2.2.20 and choosing principal axes gives

$$S_s \frac{\partial h}{\partial t} = \frac{\partial}{\partial x}\left(K_{xx}\frac{\partial h}{\partial x}\right) + \frac{\partial}{\partial y}\left(K_{yy}\frac{\partial h}{\partial y}\right) + \frac{\partial}{\partial zz}\left(K_{zz}\frac{\partial h}{\partial z}\right) + W' \quad \textbf{(2.3.33)}$$

Equation 2.3.33 is the general **3D continuity equation** for a porous medium.

Discussion

The correct Equations 2.3.32 and 2.3.33 with specific storage given by Equation 2.3.30 are presented with some slight-of-hand. The divergence of the Darcy velocity field is calculated using a fixed control volume while for the analysis leading to Equation 2.3.30, the control volume was allowed vertical deformation. This inconsistency was first noted by DeWiest (1966), who presented an alternative derivation leading to a different form for the specific storage. Cooper (1966) showed that Jabob's result is essentially correct when the vertical coordinate is taken as a deforming coordinate. A careful derivation (see Remson et al., 1971; Bear, 1972) using conservation statements for the fluid and solid medium, with Darcy's velocity expressed relative to the moving grains, shows that Equation 2.3.33 with 2.3.30 is mathematically correct.

Confined and Semiconfined Aquifers

For many applications, we are interested in the flow within aquifers without consideration of the vertical motion of the fluid (the hydraulic approach for analyzing groundwater flow). The resulting aquifer equations are two-dimensional and are found by integrating the three-dimensional equation over the thickness of aquifer, under the assumption that the head, h, does not depend on z. A cross-section is shown in Figure 2.3.7.

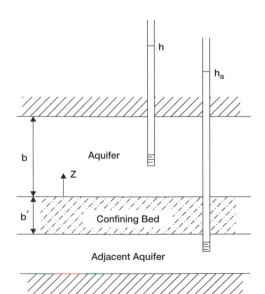

FIGURE 2.3.7 Cross-section of an aquifer of thickness b with confining bed and adjacent aquifer

If Equation (2.3.33) is integrated over the thickness of the section shown in Figure 2.3.7, and if h is independent of z, then one finds

$$\int_0^b S_s dz \frac{\partial h}{\partial t} = \frac{\partial}{\partial x}\left(\int_0^b K_{xx} dz \frac{\partial h}{\partial x}\right) + \frac{\partial}{\partial y}\left(\int_0^b K_{yy} dz \frac{\partial h}{\partial y}\right) + \int_0^b \frac{\partial}{\partial z}\left(K_{zz}\frac{\partial h}{\partial z}\right)dz + \int_0^b W' dz \quad \textbf{(2.3.34)}$$

This equation suggests that we may introduce the following aquifer parameters:

$$S = \int_0^b S_s\, dz \;\; ; \;\; Q\delta(\tilde{x}) = \int_0^b W' dz \;\; ; \;\; T_{xx} = \int_0^b K_{xx} dz \;\; ; \;\; T_{yy} = \int_0^b K_{yy} dz \quad \textbf{(2.3.35)}$$

These are the storage coefficient, injection well strength (volume per time per area[4]), and *aquifer transmissivities*, respectively. The remaining term gives

$$\int_0^b \frac{\partial}{\partial z}\left(K_{zz}\frac{\partial h}{\partial z}\right)dz = K_{zz}\frac{\partial h}{\partial z}\bigg|_b - K_{zz}\frac{\partial h}{\partial z}\bigg|_0 = -q_z|_b + q_z|_0$$

This relation quantifies the upward flow into the aquifer from below and the downward flow into the aquifer from above, both evaluated at the confining bed interface. More formally, we recognize these relations as the leakage terms. If we assume steady flow across the confining layer, then

$$q_z|_0 = -\frac{K'}{b'}(h - h_a)$$

$$q_z|_b = -\frac{K'}{b'}(h_a - h)$$

where K' and b' are the hydraulic conductivity and thickness of the confining bed and h_a is the head in the adjacent aquifer. With these changes, the continuity equation for the aquifer becomes

$$S\frac{\partial h}{\partial t} = \frac{\partial}{\partial x}\left(T_{xx}\frac{\partial h}{\partial x}\right) + \frac{\partial}{\partial y}\left(T_{yy}\frac{\partial h}{\partial y}\right) - \frac{K'}{b'}(h - h_a) + Q\delta(\tilde{x}) \quad \textbf{(2.3.36)}$$

Here, the parameter K'/b' is often called the '*leakance*'. The leakage term vanishes for confined aquifers. In Chapter 3 we consider analytical solutions to the various continuity equations and their application to practical problems.

2.3.4 Continuity with a Change in Total Stress on the Aquifer

For a number of applications it is of interest to investigate the effects that a change in the total stress (vertical load) acting on an aquifer will have on the water levels in observation wells. Examples of relevant applications include the

---

[4]The delta function $\delta(\tilde{x})$ fixes the area to the point of the well within the aquifer. The only property of this function which is required here is that its integral over the area of the aquifer is unity, implying that the delta function has units of L^{-2}. This makes equation (2.3.36) dimensionally consistent.

influence of changes in the level of the tide on coastal aquifers and the change in barometric pressure on confined aquifers in general. In this section, we proceed with a simplified examination of these questions using the developments of Bredehoeft and Cooley (1983). For a more detailed discussion, including flow under drained and undrained conditions, one might consult Narasimhan and Kanehiro (1980) and Narasimhan (1983).

To examine these effects, let us first write Equation 2.3.32 in the form

$$-\nabla \cdot (\rho \tilde{q}) = \rho(\alpha + \beta n) \frac{\partial p}{\partial t} \qquad (2.3.37)$$

where the source term W' is assumed to vanish. The new feature that appears is that changes in p are not directly balanced by changes in effective stress, σ_e, as suggested by Equation 2.3.21. However, the relationships between changes in p and σ_e, as developed in Section 2.3.2, are assumed to remain correct. The only requirement is to add on the additional effect of the change in total stress σ on the left side of Equation 2.3.37. This is accomplished by replacing Equation 2.3.29 by

$$dV_t = \alpha V_t dp - \alpha V_t d\sigma \qquad (2.3.38)$$

With this change Equation 2.3.32 becomes

$$-\nabla \cdot (\rho \tilde{q}) = \rho(\alpha + \beta n) \frac{\partial p}{\partial t} - \rho \alpha \frac{\partial \sigma}{\partial t} \qquad (2.3.39)$$

Equation 2.3.39 is the general form of the continuity equation. It is interesting to note that in comparison with Equation 2.3.32, we see that the change in vertical load appears as a source term. Thus, if a local vertical load is applied, then both $\frac{\partial p}{\partial t} > 0$ and $\nabla \cdot (\rho \tilde{q}) > 0$ that is, both the pressure increases and flow diverges from the location. However, in a compressible medium, a true source term is associated with an increase in porosity (the local pressure increases so that the effective stress decreases and the porous matrix expands). Bredehoeft and Cooley (1983) show that this is not the case for an increase in vertical load. Indeed, since both $\frac{\partial p}{\partial t} > 0$ and $\nabla \cdot (\rho \tilde{q}) > 0$ then Equation 2.3.39 shows that

$$\rho(\alpha + \beta n) \frac{\partial p}{\partial t} < \rho \alpha \frac{\partial \sigma}{\partial t}$$

which implies that

$$\frac{\partial \sigma}{\partial t} > \frac{\partial p}{\partial t}$$

With Equation 2.3.20, we then see that $\dfrac{\partial \sigma_e}{\partial t} > 0$ and that the porosity decreases with an increase in effective stress. Thus the vertical load term is physically distinct from a source term, even though they occupy the same position mathematically.

Jacob (1950) describes a situation in which a tidal load is uniformly placed over a wide area of the aquifer so that no gradients are set up to induce flow. For this situation Equation 2.3.39 gives

$$\rho(\alpha + \beta n)\frac{dp}{dt} = \rho\alpha\frac{d\sigma}{dt}$$

or

$$\frac{dp}{d\sigma} = \frac{\alpha}{\alpha + \beta n} \tag{2.3.40}$$

This is Jacob's expression for the *tidal efficiency*, which is the ratio of the increase in water level in an observation well to the to the size (depth) of the tidal loading. The assumption is that the tidal water does not enter the observation well, which remains open to atmospheric pressure.

A related concept to tidal efficiency is the *barometric efficiency*, which is the decrease in the water level in an observation well caused by an increase in the barometric pressure. Here, the increase in vertical load is applied both to the ground surface and to the water within the observation well. However, part of the load applied to the ground surface is carried by the effective stress in the aquifer. Hence, the net load applied to the observation well water column exceeds that to the aquifer water, and the water level actually declines. Equation 2.3.39 still gives

$$\rho(\alpha + \beta n)\frac{dp}{dt} = \rho\alpha\frac{d\sigma}{dt}$$

Because of the loading we have $dp = d\sigma - d\sigma_e$ which gives

$$\frac{d\sigma_e}{d\sigma} = \frac{\beta n}{\alpha + \beta n} \tag{2.3.41}$$

Equation 2.3.41 is Jacob's expression for the barometric efficiency. With Equations 2.3.40 and 2.3.41 one finds that the sum of the barometric efficiency plus the tidal efficiency is equal to unity.

EXAMPLE   **2.3.5.   Tidal efficiency (from Jacob, 1950)**
PROBLEM   An extensive confined sand (the Lloyd sand on Long Island, New York) having a thickness of about 200 ft and an assumed porosity of 0.35 is overlain by an
extensive body of tidewater in which there is an almost uniform tide. The water
level in a tightly cased well tapping the sand fluctuates simultaneously with amplitude 42 percent of that of the tide. Determine the compressibility of the sand
under this stress variation and frequency, neglecting the small volume of water
that flows into and out of the well. With Equation 2.3.40 and with $\beta = 3.3 \times 10^{-6}$ in^2/lb, we have

$$0.42 = \frac{\alpha}{\alpha + 3.3 \times 10^{-6} \times 0.35}$$

Solving for α one finds $\alpha = 0.84 \times 10^{-6}$ in/lb $= 1.22 \times 10^{-10}$Pa^{-1}. For this
consolidated sand the compressibility of the porous matrix is nearly equal to
that of water.

Seepage Forces and "Quick" Sand

Water inflow to excavations is detrimental. The water must be pumped out or
else the excavation will become flooded. In addition, the upward hydraulic gradient decreases the effective stress in the soil matrix, and the soil as a whole
looses strength. If the upward gradient exceeds a critical value, then the soil
cannot hold any surface load (the effective stress vanishes). Under this critical
condition the net weight of the soil just balances the upward seepage force.

Consider the set-up shown in Figure 2.3.8. Water moves from the upper
reservoir into the soil sample and discharges in the upward direction. The seepage force (per unit weight of water) is $-\nabla h$. The seepage force per unit volume
is given by

$$f_{\text{seepage}} = -\rho g \nabla h$$

The net weight (vertical stress) is equal to the weight of soil plus water
above a given location minus the buoyant force on the soil plus water. This gives

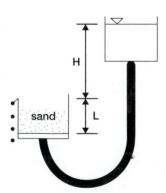

FIGURE 2.3.8   Experiment for seepage forces due to upward flow
through soil

$$f_{\text{gravity}} = (1 - n)\rho_s g + n\rho g - (1 - n)\rho g - n\rho g = (1 - n)(\rho_s - \rho)g$$

Thus the net downward force is

$$f_{z\text{net}} = (1 - n)(\rho_s - \rho)g - \rho g \nabla h$$

The critical condition occurs when the net force (effective stress) vanishes, and so we have

$$\nabla h\big|_{\text{critical}} = (1 - n)(SG_{\text{soil}} - 1) \tag{2.3.42}$$

where SG_{soil} is the specific gravity of the soil. Using typical values we have

$$\nabla h\big|_{\text{critical}} = (1 - 0.35)(2.65 - 1) = 1.07 \cong 1$$

Thus with an upward unit gradient the soil has no bearing capacity for loads applied at the ground surface.

2.4   Flow of Air through Porous Media

The flow of air through soil is of interest for a number of applications. Air may be used as the fluid in laboratory columns for measurement of permeability of a porous media. Also, the forced flow of air is very important in design of soil vapor extraction systems for remediation of unsaturated soils that have become contaminated by volatile organic substances. There are many other examples that arise in practice.

For airflow, it is still generally assumed that Darcy's Law applies. Again, with reference to Section 2.1, this implies that the viscous resistance force associated with the flow balances the driving force from pressure gradients and gravity. Furthermore, this viscous resistance force is proportional to the velocity to the first power. Nevertheless, since the density of air is not constant, we cannot define a hydraulic head, as we did for flow of water. Hubbert (1940) defines a fluid potential (energy per unit mass) by

$$\Phi^* = gz + \int_{p_o}^{p} \frac{dp}{\rho} + \frac{v^2}{2} \tag{2.4.1}$$

where the density, ρ, is considered to be a function only of the fluid pressure, $\rho(p)$, and p_o is a reference pressure. With the velocities relevant for groundwater problems, the kinetic energy term is negligible and the potential simplifies to

$$\Phi^* = gz + \int_{p_o}^{p} \frac{dp}{\rho} \tag{2.4.2}$$

The gradient of the potential is

$$\nabla \Phi^* = g\hat{k} + \frac{\nabla p}{\rho}$$

Comparison with Equation 2.1.5 shows that $-\nabla\Phi^*$ is the driving force (pressure gradient plus gravity) per unit mass. From a physical and mathematical point of view, the advantage of the potential defined by Equation 2.4.2 over the usual hydraulic head is that it defines the energy content of the fluid in a way which is independent of any assumptions of fluid compressibility.

Repeating the arguments of Section 2.1, we can again show that the general form of Darcy's Law is

$$\tilde{q} = -\frac{k}{\mu}\left(\nabla p + \rho g\hat{k}\right) \qquad (2.4.3)$$

Thus, the form of Equation 2.2.8 is valid for both incompressible and compressible fluids. Under most conditions the force associated with pressure gradients is much larger than that due to gravity (the density of air is so small). Hence, Darcy's Law may be written in the approximate form

$$\tilde{q} = -\frac{k}{\mu}\nabla p \qquad (2.4.4)$$

Equation 2.4.4 is exact for horizontal flow and will be accepted, in general, for flow of air in porous media.

Note that while the use of the potential Φ^* leads to the same form of Darcy's Law, the potential Φ^* will lead to different relationships between the pressure drop and permeability of the porous media. The explanation here is derived from the fact that as the fluid moves along the flow path, the gas density will not remain constant. These ideas are briefly discussed in the next subsection.

Laboratory Measurement with Air Flow

In a constant head permeameter, if the fluid is air or another compressible gas, then the approach leading to Equation 2.2.32 is no longer valid because neither q nor ρ are constant along the length of the column. However, the mass flux must remain constant and we have for a column of cross-section area A

$$\dot{m} = \rho q A = \text{constant}$$

With Equation 2.4.4 this gives

$$\dot{m} = -\rho A \frac{k}{\mu}\frac{dp}{dL} = \text{constant} \qquad (2.4.5)$$

where L is the distance along the length of the column. In order to integrate this equation, we need the equation of state, $\rho(p)$. Two simple cases arise. First, if the air expansion is *isothermal*, then Boyle's Law holds and

$$p = \rho\frac{p_o}{\rho_o} \qquad (2.4.6)$$

where p_o and ρ_o are the pressure and density at some reference or standard state (such as standard atmospheric conditions). With the isothermal equation of state given by 2.4.6, Equation 2.4.5 may be integrated to find

$$\dot{m}_{iso} = \frac{A\rho_o k}{2p_o\mu}\frac{p_1^2 - p_2^2}{L} \tag{2.4.7}$$

A second result arises if the gas expansion is *adiabatic* (constant entropy) where the equation of state is

$$p = p_o\left(\frac{\rho}{\rho_o}\right)^\gamma \tag{2.4.8}$$

where $\gamma = c_p/c_v$, the ratio of the specific heats at constant pressure and volume, and p_o and ρ_o are reference pressure and density (see White, 1986).[5] Combining Equation 2.4.8 with 2.4.5 and integrating, we find

$$\dot{m}_{adia} = \frac{\gamma A\rho_o k}{(1+\gamma)p_o^{1/\gamma}\mu}\frac{p_1^{1+1/\gamma} - p_2^{1+1/\gamma}}{L} \tag{2.4.9}$$

Muskat (1946) shows that Equations 2.4.7 and 2.4.9 may combined with $p_o = p_1$ to give

$$\frac{\dot{m}_{adia}}{\dot{m}_{iso}} = \frac{2\gamma}{1+\gamma}\frac{1 - \left(\frac{p_2}{p_1}\right)^{1+1/\gamma}}{1 - \left(\frac{p_2}{p_1}\right)^2} \tag{2.4.10}$$

It follows that the mass flux rate for nonisothermal flow ($\gamma > 1$) exceeds that for isothermal flow with $\dot{m}_{adia}/\dot{m}_{iso}$ increasing as γ increases. Furthermore, for a specific gas (fixed γ), $\dot{m}_{adia}/\dot{m}_{iso}$ increases as p_2/p_1 decreases, with a maximum value for expansion of the gas into a vacuum:

$$\frac{\dot{m}_{adia}}{\dot{m}_{iso}} = \frac{2\gamma}{1+\gamma} \tag{2.4.11}$$

For air this ratio is 1.167.

EXAMPLE PROBLEM

2.4.1   Air flow permeability measurements

Air expands from an absolute pressure of 4 atm to 1 atm across a limestone core sample of length 20 cm and cross-section area 12 cm^2 at a temperature of 25°C. The airflow rate is 3×10^{-6} kg/s. Estimate the intrinsic permeability of the sample as well as its hydraulic conductivity to water. Also determine the exit velocity of air as it enters the outflow chamber.

One atmosphere corresponds to a pressure of 101,300 Pa. The density corresponding to a given pressure is found from the ideal gas law ($p = \rho RT$). The gas constant R varies, however, for each gas as follows

$$R_{gas} = \frac{\Lambda}{\omega_{gas}}$$

[5]For air, $\gamma = 1.40$.

where L is the universal gas constant $\left(8{,}310\,\dfrac{m^2}{s^2K}\right)$ and ω_{gas} is the molecular weight of the gas. For air with a molecular weight of $\omega_{air} = 28.97$, the gas constant is $R = \dfrac{m^2}{s^2K}$. Thus, the density corresponding to a pressure of 4 atmospheres is

$$\rho = \frac{p}{RT} = \frac{4 \times 101300}{287 \times (273 + 25)} = 4.74\,\frac{kg}{m^3}$$

Equation 2.4.7 may be written

$$k = \frac{2p_1 \mu \dot{m} L}{A\rho_1(p_1{}^2 - p_2{}^2)}$$

which gives $k = 1\,3\,10214$ m^2 when the viscosity of air is about $1.8 \times 10^{-5}\,\dfrac{kg}{m \cdot s}$. The corresponding hydraulic conductivity to water is found from

$$K = \frac{k\rho g}{\mu} = \frac{(1 \times 10^{-14}) \times 1000 \times 9.81}{10^{-3}} = 9.8 \times 10^{-8}\,m/s = 0.0085\,m/d$$

The outflow velocity is found from continuity. We have $\dot{m} = Aq\rho$, and at the outflow face, because $p = 1$ atm, the density is $\rho = 1.23$ kg/m^3. The velocity is given by

$$q = \frac{\dot{m}}{A\rho} = \frac{3 \times 10^{-6}}{0.0012 \times 1.23} = 0.0020\,m/s = 0.2\,cm$$

Note that at the entrance to the outflow chamber the porosity is $n = 1$, so $v = 0.2$ cm/s also.

Continuity for Air Flow

As noted above, the flow of air is important for laboratory testing of soil permeability, for contaminated soil remediation through soil vapor extraction technologies, and for other applications. When considering continuity for soil air, one may generally neglect compressibility of the porous matrix and write the continuity equation in the form

$$\theta_a \frac{\partial p}{\partial t} + \nabla \cdot (\rho \tilde{q}) = 0 \tag{2.4.12}$$

where θ_a is the volumetric air content and sources and sink have been neglected. For the flow of air, the force of gravity is usually small compared to those forces due to pressure gradients, and Equation 2.4.4 may be used for the volume flux. In addition, if the transport process is isothermal then Equation 2.4.6 applies as an equation of state, and Equation 2.4.12 reduces to

$$\theta_a \frac{\partial p}{\partial t} + \nabla \cdot \left(p \frac{k}{\mu} \nabla p \right) \tag{2.4.13}$$

Equation 2.4.13 is obviously nonlinear, and indeed its form is similar to that for continuity in an unconfined aquifer, Equation 2.3.16. As noted for flow in an unconfined aquifer, the analytic solution to this equation is a difficult and unattractive problem.

For some applications, especially those associated with soil vapor extraction systems, Equation 2.4.13 may be linearized by assuming that the pressure is equal to atmospheric pressure plus a small deviation:

$$p = p_{atm} + p^* \tag{2.4.14}$$

where $p^* \ll p_{atm}$. Under these conditions, the viscosity of a vapor is also nearly constant and Equation 2.4.13 reduces to

$$\frac{\theta_a \mu}{p_{atm}} \frac{\partial p^*}{\partial t} = \nabla \cdot (k \nabla p^*)$$

If we may further assume that the intrinsic permeability is constant, then we obtain the simplified form

$$\frac{\theta_a \mu}{k p_{atm}} \frac{\partial p^*}{\partial t} = \nabla^2 (p^*) \tag{2.4.15}$$

In Equation 2.4.15 p^* represents the deviation of the air pressure from atmospheric.

Under steady flow conditions, Equation 2.4.13 simplifies to

$$\nabla \cdot \left(\frac{k}{\mu} \nabla p^2 \right) = 0$$

If we may further neglect variations in the intrinsic permeability as well as those in the viscosity, then we have

$$\nabla^2(p^2) = 0 \qquad\qquad (2.4.16)$$

In Equation 2.4.16, p is the absolute pressure. This equation shows that under steady-isothermal conditions, the square of the absolute pressure is a potential function.

2.5   Groundwater Management Models

The continuity equations developed in the previous sections serve as the starting point for analysis of groundwater flow systems. If a field problem of can be simplified so that only a single aquifer unit is involved and one can assign a constant effective aquifer transmissivity, and if the domain geometry and boundary conditions are also simple, then one may be able to develop analytical solutions to the flow equations. These are the subjects of the next chapter that focuses on well hydraulics. In general, however, the parameter fields are heterogeneous and sometimes anisotropic, the domains are of irregular geometry, and the boundary conditions are often not simple. For such cases groundwater management models are formulated and solved numerically. This section presents one approach for developing groundwater hydraulics models using the finite difference method. There are other approaches and numerical methods (in particular, the finite element method), and many important issues for applications of numerical models that are left to other references for discussion.

Even when numerical methods are to be applied to solve a flow problem, the continuity equation and Darcy's Law still provide the starting point for the mathematical modeling of the aquifer. One approach is to take the continuity equations in their form of partial differential equations, such as Equation 2.3.33, and apply Taylor's theorem to arrive at an approximating set of algebraic equations that are solved by a computer. This approach also provides estimates of the accuracy of the resulting equations and serves as a starting point for analysis of consistency, convergence and stability of the numerical method. A second approach, and the one which is followed here, is to apply the continuity principle directly to a control volume of finite size and develop the set of approximating algebraic equations in that fashion.

Consider the aquifer layout shown in Figure 2.5.1. The aquifer is represented within a grid of cells of finite size. The columns of the grid of cells are labeled by an index j that runs from 1 through N_C, while the rows are designated by an index i that runs from 1 at the top of the grid to N_R at the bottom. The N_L layers are labeled by index k. The head in row i, column j, and layer k at time $t = m\Delta t$ is specified as $h_{i,j,k}{}^m$. The groundwater models are used to calculate the heads at time level $m + 1$, given the set of heads at time level m. Repeating this procedure for each successive time step constitutes a time-dependent solution of the numerical groundwater management model.

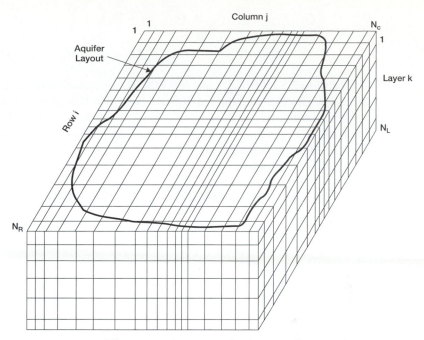

FIGURE 2.5.1   Finite-difference grid covering the domain of an aquifer

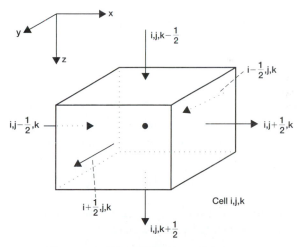

FIGURE 2.5.2   Finite difference cell for application of the continuity principle

We want to apply the continuity principle to each cell of the overlying grid. The continuity principle states that the flow into a cell is equal to the storage increase within the cell plus the flow out of the cell. With reference to Equation 2.3.33 and Figure 2.5.2, this may be written as:

$$Q_{i-1/2,j,k} + Q_{i,j-1/2,k} + Q_{i,j,k-1/2} + W_{i,j,k}$$

$$= SS_{i,j,k} \frac{h_{i,j,k}{}^{m+1} - h_{i,j,k}{}^m}{\Delta t} \Delta V_{i,j,k} + Q_{i+1/2,j,k} + Q_{i,j+1/2,k} + Q_{i,j,k+1/2} \quad \textbf{(2.5.1)}$$

In Equation 2.5.1 $Q_{i-1/2,j,k}$ represents the volume discharge from cell $(i-1,j,k)$ into cell (i,j,k), etc., and the notation suggests that this discharge is evaluated at the boundary between the two cells. $W_{i,j,k}$ is the source strength (volume per unit time), $SS_{i,j,k}$ is the specific storage for the cell, $\Delta V_{i,j,k}$ is the cell volume, and Δt is the computation time step. Equation 2.5.1 provides the required form of the continuity equation.

Darcy's Law is used to evaluate the discharge between cells. For the discharge between cell $(i-1,j,k)$ and cell (i,j,k) Darcy's Law gives

$$Q_{i-1/2,j,k} = -KY_{i-1/2,j,k} \Delta x_j \Delta z_k \frac{h_{i,j,k} - h_{i-1,j,k}}{\Delta y_{i-1/2}} \quad \textbf{(2.5.2)}$$

In Equation 2.5.2 $KY_{i-1/2,j,k}$ is the effective hydraulic conductivity for flow in the y-direction (column) between the cells, Δx_j and Δz_k are the cell row width and layer height, and $\Delta y_{i1/2}$ is the distance between the centers of cell $(i-1,j,k)$ and cell (i,j,k). The magnitude of $\Delta y_{i-1/2}$ is $\Delta y_{i-1/2} = (\Delta y_{i-1} + \Delta y_i)/2$. A value of the hydraulic conductivity (for each direction) is assigned to each cell. As the water moves from the middle of one cell to the middle of the next, it moves part of the distance through a medium with one K value and the rest through a medium with a second K value. Since the flow is across the media (rather than parallel within them), the appropriate effective K between cells is found from the harmonic average (see section 2.2.3). Thus, for the flow between cell $(i-1,j,k)$ and cell (i,j,k)

$$KY_{i-1/2,j,k} = \frac{(\Delta y_{i-1} + \Delta y_i)/2}{\left(\dfrac{\Delta y_{i-1}/2}{KY_{i-1,j,k}} + \dfrac{\Delta y_i/2}{KY_{i,j,k}}\right)} = \frac{KY_{i-1,j,k}KY_{i,j,k}(\Delta y_{i-1} + \Delta y_i)}{KY_{i,j,k}\Delta y_{i-1} + KY_{i-1,j,k}\Delta y_i} \quad \textbf{(2.5.3)}$$

If Δy is constant then Equation 2.5.3 may be written in the simpler form

$$KY_{i-1/2,j,k} = \frac{2KY_{i,j,k}KY_{i-1,j,k}}{KY_{i,j,k} + KY_{i-1,j,k}} \quad \textbf{(2.5.4)}$$

Combining Equations 2.5.2 and 2.5.3, one finds that the discharge between cell $(i-1,j,k)$ and cell (i,j,k) is calculated from

$$Q_{i-1/2,j,k} = \frac{2KY_{i-1,j,k}KY_{i,j,k}\Delta x_j \Delta z_k}{KY_{i,j,k}\Delta y_{i-1} + KY_{i-1,j,k}\Delta y_i} (h_{i-1,j,k} - h_{i,j,k}) \quad \textbf{(2.5.5)}$$

To simply notation we write this equation as

$$Q_{i-1/2,j,k} = CY_{i-1/2,j,k}(h_{i-1,j,k} - h_{i,j,k})$$

$$CY_{i-1/2,j,k} = \frac{2KY_{i-1,j,k}KY_{i,j,k}\Delta x_j \Delta z_k}{KY_{i,j,k}\Delta y_{i-1} + KY_{i-1,j,k}\Delta y_i} \quad \textbf{(2.5.6)}$$

In Equation 2.5.6 $CY_{i1/2,j,k}$ is the y-direction (column) hydraulic conductance (L^2/T), or simply conductance, between cell $(i-1,j,k)$ and cell (i,j,k). Equations similar to 2.5.6 are developed for the inter-cell discharge for the other faces of cell (i,j,k), with row and layer conductance values designated CX and CZ, respectively.

The source strength term, $W_{i,j,k}$, includes injection wells, infiltration, leakage from rivers or surface water bodies, and other sources. Production wells, base flow from aquifers to rivers, and other sink terms are represented as negative sources. For injection wells the source strength is represented as

$$W_{i,j,k} = QW_{i,j,k}(t^m) \tag{2.5.7}$$

where $QW_{i,j,k}(t^m)$ is the well injection discharge (L^3/T) during the stress period at time step t^m. For diffuse recharge from infiltration the source term may be written

$$W_{i,j,1^*}=R_{i,j}(t^m)\Delta x_j \Delta y_i \tag{2.5.8}$$

In Equation 2.5.8 the notation $W_{i,j,1^*}$ designates the uppermost active cell in (row, column) = (i,j), and $R_{i,j}$ is the specified infiltration recharge rate (L/T) during the stress period.

For diffuse recharge from river infiltration, the river discharge depends on the effective length and width of the riverbed within the cell, the hydraulic conductivity of the riverbed material, the average river stage (elevation), and the aquifer head within the cell. With reference to Figure 2.5.3, as long as $h_{i,j,k} > H_{BOT}$, then the river discharge depends on the linear distance between the cell head and the river stage, H_{RIV}. However, if $h_{i,j,k} < H_{BOT}$, the river discharge is constant. The model for river recharge may be written

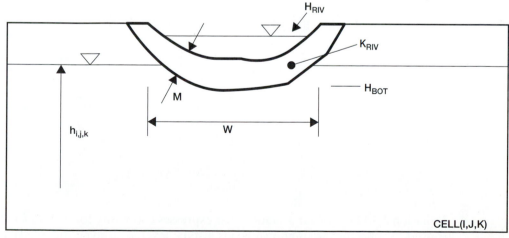

FIGURE 2.5.3 Configuration for the MODFLOW River Package—Channel Cross Section

$$W_{i,j,k} = \frac{WLK_{RIV}}{M}(H_{RIV} - h_{i,j,k}) = CR_{i,j,k}(H_{RIV} - h_{i,j,k}) \; ; \; h_{i,j,k} > H_{BOT}$$

$$W_{i,j,k} = CR_{i,j,k}(H_{RIV} - H_{BOT}) \; ; \; h_{i,j,k} < H_{BOT} \qquad \textbf{(2.5.9)}$$

where L is the length of the river reach contained within the cell. The river conductance CR is analogous to the aquifer conductance introduced through Equation 2.5.6.

The evapotranspiration model has a similar structure to the river recharge model. The evapotranspiration rate has a maximum value when the water table is near the ground surface. If the water table is sufficiently deep then the evapotranspiration rate vanishes. If $ETM_{i,j}$ is the maximum evapotranspiration rate, $HS_{i,j}$ is the water table elevation at which is maximum value occurs, and $d_{i,j}$ is the cutoff or extinction depth, then the source term representation is

$$W_{i,j,k} = -ETM_{i,j}\Delta x_j \Delta y_i \; ; \; h_{i,j,k} \geq HS_{i,j}$$

$$W_{i,j,k} = -ETM_{i,j}\left(\frac{h_{i,j,k} - (HS_{i,j} - d_{i,j})}{d_{i,j}}\right)\Delta x_j \Delta y_i \; ; \; (HS_{i,j} - d_{i,j}) \leq h_{i,j,k} < HS_{i,j} \quad \textbf{(2.5.10)}$$

$$W_{i,j,k} = 0 \; ; \; h_{i,j,k} \leq (HS_{i,j} - d_{i,j})$$

In general, the combination of source term models may be written for an individual cell as

$$W_{i,j,k} = \sum_{n=1}^{N} p_{i,j,k,n} h_{i,j,k} + \sum_{n=1}^{N} q_{i,j,k,n} = P_{i,j,k} h_{i,j,k} + Q_{i,j,k} \qquad \textbf{(2.5.11)}$$

where N is the number of external sources or stresses affecting a single cell, and $p_{i,j,k,n}$ and $q_{i,j,k,n}$ are constants that describe the individual external sources or stresses.

With these results Equation 2.5.1 may be written

$$CY_{i-1/2,j,k}\left(h_{i-1,j,k}{}^{m+1} - h_{i,j,k}{}^{m+1}\right) + CY_{i+1/2,j,k}\left(h_{i+1,j,k}{}^{m+1} - h_{i,j,k}{}^{m+1}\right)$$

$$+ \; CX_{i,j-1/2,k}\left(h_{i,j-1,k}{}^{m+1} - h_{i,j,k}{}^{m+1}\right) + CX_{i,j+1/2,k}\left(h_{i,j+1,k}{}^{m+1} - h_{i,j,k}{}^{m+1}\right) \qquad \textbf{(2.5.12)}$$

$$+ \; CZ_{i,j,k-1/2}\left(h_{i,j,k-1}{}^{m+1} - h_{i,j,k}{}^{m+1}\right) + CZ_{i,j,k+1/2}\left(h_{i,j,k+1}{}^{m+1} - h_{i,j,k}{}^{m+1}\right)$$

$$+ \; P_{i,j,k} h_{i,j,k}{}^{m+1} + Q_{i,j,k} = SS_{i,j,k}\frac{\left(h_{i,j,k}{}^{m+1} - h_{i,j,k}{}^{m}\right)}{\Delta t_{m+1}}\Delta x_j \Delta y_1 \Delta z_k$$

Equation 2.5.12 is a linear equation that expresses continuity for cell (i,j,k). It is used to solve for the unknown heads at time level $m+1$ given the heads at time level m and the corresponding external sources and stress terms. The problem formulation starts with *initial conditions* that provide the starting values ($m = 0$) of the head in each cell, $h_{i,j,k}{}^0$. These values are used in Equation 2.5.12 to find the head values $h_{i,j,k}{}^1$, which in turn ($m = 1$) are used to find $h_{i,j,k}{}^2$, etc.

MODFLOW

The model formulation outlined above is implemented in the MODFLOW model that was developed by the U.S. Geological Survey (McDonald and Harbaugh, 1988, 1996). Table 2.5.1 lists the World Wide Web address from which one may download copies of the model and its documentation. The user should also consult McDonald and Harbaugh (1988) for model details.

MODFLOW is finite difference model that may be used to solve groundwater flow problems in one-, two-, or three-dimensions. The MODFLOW program is divided into a main program and a series of independent subroutines called modules. The modules are grouped into "packages", each of which is a group of modules that deals with a single aspect of the simulation. For example, the Well Package simulates the effects of injection and production wells, the River Package simulates the effect of rivers, etc. Individual packages may or may not be required, depending on the problem being simulated. Table 2.5.2 lists the MODFLOW packages and their main attributes. The Basic, Block-Centered Flow, and Solution Procedure packages are required for all simulations.

TABLE 2.5.1 WWW Address for MODFLOW Model and Documentation: USGS Ground-Water Software may be downloaded from http://water.usgs.gov/software/ground_water.html

Application	Description	Comments
MODFLOW–96	Modular three-dimensional finite-difference groundwater flow model	Includes executable, source code, and PDF version of User's and Programmer's guides with installation instructions. Additional documentation in McDonald and Harbaugh (1988).
MFI	Data input program for MODFLOW, MODPATH, and MOC3D	Includes executable, source code, and PDF version of User's guide with installation instructions. Simple to use application to develop input files.
MOC3D	Three-dimensional method-of-characteristics groundwater flow and transport model (discussed in Chapter Eight, "**Advection-Dispersion Transport and Models**")	

TABLE 2.5.2   MODFLOW Packages and Their Attributes

Package	Purpose*	Major Arrays
Basic Package	Handles tasks that are required for each simulation, including specification of boundaries, determination of time-step length, establishment of initial conditions, and printing of results.	Number of layers, rows and column: NLAY, NROW, NCOL; IBOUND(I,J,K) to specify status of cell (see text); Initial heads, Shead (I,J,K); Stress period length, number of steps, and time step multiplier: PERLEN, NSTP, TSMULT; Time unit indicator, ITMUNI (0—undefined, 1–seconds, 2–minutes, 3–hours, 4–days, 5–years). IUNIT (24) array is not required for MODFLOW (1996).
Block-Entered Flow Package	Calculates hydraulic conductance and external source terms of finite-difference equations that represent flow from cell to cell and storage.	Flag for steady state, ISS (0—transient); Layer type indicator, LAYCON(K) (0—no provision for transmissivity modification or storage term conversion, 1—unconfined aquifer, 2—storage term conversion, 3—both transmissivity modification and storage term conversion); Arrays for storage coefficients, transmissivities, hydraulic conductivities, layer top and bottom elevations, vertical leakance, depending on LAYCON values.
River Package	Stress package. Adds terms representing flow to rivers to the finite-difference equations.	Number of active riber reaches during stress period; For each reach its cell (I,J,K), river stage, conductance and bottom elevation.
Recharge Package	Stress package. Adds terms representing diffuse recharge to the finite-difference equations.	Recharge option code, NRCHOP (1—recharge only to top grid layer, 2—user specified vertical distribution of recharge, 3—recharge to highest active cell); Recharge rate array (L/T); Location array.
Well Package	Stress package. Adds terms representing flow to wells to the finite-difference equations.	Number of active wells; For each well its cell location (I,J,K) and discharge (L^3/T) (positive for recharge wells).
Drain Package	Stress package. Adds terms representing flow to drains to the finite-difference equations.	Number of active drain cells; For each drain cell its cell location (I,J,K) and the drain elevation and hydraulic conductance values.
Evapotranspiration Package	Stress package. Adds terms representing ET to the finite-difference equations.	Evapotranspiration option code, NEVTOP (1—ET calculated only for top grid layer, 2—user specified distribution of ET cells); ET surface elevation, potential rate (L/T), extinction depth arrays; Location array.

TABLE 2.5.2 Continued

Package	Purpose*	Major Arrays
General-Head Boundary Package	Stress package. Adds terms representing general-head boundaries to the finite-difference equations.	Number of active general-head boundaries; For each boundary its cell location (I,J,K), boundary head and hydraulic conductance.
Solution Procedure Package	MODFLOW (1996) supports preconditioned-conjugate gradient, strongly-implicit, slice-successive over relaxation, and direct solver using diagonal ordering procedures.	Data specific to solver procedure including maximum number of iterations, closure criteria, and convergence acceleration parameters.

* after McDonald and Harbaugh, 1988

For each simulation the user specifies the characteristics of the domain and information about the simulation time interval. The continuity principle is applied to each active cell during each time step of the simulation. The general form of the continuity equation for each cell is given by Equation 2.5.12. The simulation time interval is specified in terms of stress periods, during which the pumping rate, recharge rate, boundary conditions, etc., are constant, and time steps during each stress period. For each time step the continuity equations are solved numerically by iteration techniques. Thus the general computation algorithm consists of three loops: an outer loop for each stress period, a second loop for each time step, and an inner loop for iterations in solving the continuity equations.

For a given simulation, it is not necessary to formulate an equation of the form of Equation 2.5.12 for every cell in the model mesh. The status of certain cells is specified in advance in order to simulate the **boundary conditions** of the problem. The status of each cell is specified by values in the IBOUND(I,J,K) array from the Basic Package. Cells are designated as either constant-head (IBOUND < 0), no-flow, or inactive (IBOUND = 0), or variable-head (IBOUND > 0). Equation 2.5.12 is used for variable-head cells. More general boundary conditions may be specified using external source terms or the General-Head Boundary Package.

The main purpose of the Block-Centered Flow Package is to formulate the cell continuity Equation 2.5.12 for solution. This involves calculation of conductance values and storage and external source terms. Horizontal conductance values for saturated layers (confined aquifers) are entered through the layer transmissivity values so that the layer thickness is not explicitly required. The conductance in Equation 2.5.6 is calculated as

$$CY_{i-1/2,j,k} = \frac{2\Delta x_j TY_{i-1,j,k} TY_{i,j,k}}{TY_{i,j,k}\Delta y_{i-1} + TY_{i-1,j,k}\Delta y_i}$$

(2.5.13)

where TY is the column-direction transmissivity. A similar formulation is used for the row-direction conductance values using the row-direction transmissivity TX. For unconfined aquifers (layers that become partially dewatered) the layer transmissivity value is updated during each iteration based on the product of the cell hydraulic conductivity and the layer saturated thickness, ($h_{i,j,k}$ − $BOT_{i,j,k}$), where $BOT_{i,j,k}$ is the elevation of the base of the cell.

The vertical conductance values are also entered in a fashion that does not require explicit specification of the layer thickness. Data is entered through values of the "vertical leakance," which is defined through

$$\text{VCONT}_{i,j,k+1/2} = \frac{1}{\displaystyle\sum_{g=1}^{n} \frac{\Delta z_g}{KZ_g}} = \frac{CZ_{i,j,k}}{\Delta x_j \Delta y_i} \qquad (2.5.14)$$

where Δz_g is the thickness of the layer that contributes to the vertical leakance, KZ_g is the layer vertical hydraulic conductivity, and CZ is the vertical conductance. For example, with reference to Figure 2.5.4, the vertical leakance is calculated from

$$\text{VCONT}_{i,j,k+1/2} = \frac{1}{\dfrac{\Delta z_k/2}{KV_k} + \dfrac{\Delta z_{CB}}{KV_{CB}} + \dfrac{\Delta z_{k+1}/2}{KV_{k+1}}} \qquad (2.5.15)$$

Values of the VCONT array are entered to the simulation and the program multiplies these by the horizontal cell area to calculate the vertical conductance values. The calculations shown in Equation 2.5.15 must be performed outside of the MODFLOW program for data entry. With this added step, the resulting program is much more versatile.

The MODFLOW model also has capabilities for handling situations where the layer storage characteristics change from those of a confined aquifer to an unconfined aquifer, or vice versa. In addition, if this happens within a layer below the uppermost saturated layer then the vertical flow model must be modified. McDonald and Harbaugh (1988) describe these features of MODFLOW.

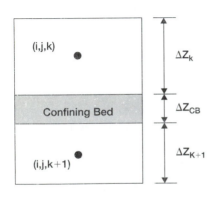

FIGURE 2.5.4   Calculation of MODFLOW Vertical Leakance

The General-Head Boundary Package is similar to the river, drain and ET packages in that flow into or out of a cell from an external source is proportional to the head difference between the cell and the external source. It may be written

$$QB_{i,j,k} = CB_{i,j,k}(HB_{i,j,k} - h_{i,j,k}) \qquad (2.5.16)$$

where QB is the discharge into the cell from the external source, CB is the conductance between the cell and the external source, and HB is the head assigned to the external source.

Application of Groundwater Management Models

Groundwater management models have many different types of applications. They are used in parameter identification studies, in simulation investigations of various management schemes, as constraint sets for optimization models, and in other types of studies. A few examples are briefly noted below.

The most obvious application of groundwater models is for simulation of local and regional flow systems. If the parameter fields are known or have been estimated through other studies, then numerical models may be used to evaluate the response of the groundwater system to changes in operating conditions (changes in pumping rates or addition of wells). Simulation of local flow systems is very important when evaluating subsurface flow near waste disposal sites and sites of groundwater contamination. Often the primary goal of a groundwater remediation project is prevention of off-site migration of contaminants. Here, numerical models may be used to estimate the best locations for pumping wells and the required production rates in order to meet these objectives. It is important to note, however, that such applications require the necessary parameter fields be quantified to an appropriate level of certainty prior to executing the model. Accordingly, one of the major difficulties in carrying out such analyses is lack of sufficient field data to characterize the site hydrogeology with an acceptable degree of certainty. Such studies are important on both the local and regional scales.

Numerical models are also used for parameter identification for both local and regional flow systems. Simulation models may be used for local flow systems when evaluating the results of aquifer tests where known heterogeneities exist or boundary conditions are not simple, and the analytical results from Chapter 3 are not applicable. The simulated head levels are compared with measured values from observation wells, and the parameter fields are adjusted so that a best fit is obtained. The resulting fields of permeability, storage, etc. may then be used in further investigations.

In analysis of regional and multi-aquifer systems, numerical models play an important role in identifying the hydrogeologic controls on a system. The results from simulations may be compared with regional hydrogeologic field data to estimate the relative permeabilities of various formations, controls on recharge and discharge areas, and the possible flow paths and travel times through the regional systems. Such investigations are very important in evaluation of potential facility locations for long-term waste disposal projects.

Groundwater flow modeling is also a necessary first step in evaluation of subsurface transport and fate of chemicals and radionuclides. One cannot know where the chemicals are being transported unless one understands the movement of water. In this sense, groundwater flow modeling is a necessary prerequisite to analysis of transport.

The groundwater flow equations, in their numerical form, may also be used as a constraint set in groundwater optimization models. An objective might be to produce the most water from an aquifer system without dropping the heads below given levels at specified locations. The problem would be stated as maximize the discharge subject to given head constraints. In setting up such a problem for simulation, the flow equations act as an implicit constraint set governing the physical behavior of the system. However, one of the major limitations to the widespread use of optimization models in the groundwater field is that they generally assume that the parameter fields are known precisely. Moreover, through poor interpretation by the user, the results of such models have been reported as absolute truths without recognition of the means used in generating these solutions. Such misuse by users has produced an unwarranted distrust of this type of analysis. Most models, if used properly with a clear recognition of their capabilities and limitations, provide a platform from which to formulate an ordered decision response to a variety of complex problems and issues.

EXAMPLE PROBLEM

2.5.1. Simulation of groundwater recovery in a valley aquifer

An alluvial valley contains a river that is in direct communication with an unconfined aquifer ($K_h = 5$ m/d, $K_v = 1$ m/d), which in turn overlies a semi-confined aquifer ($T = 100$ m²/d, $K_v = 1$ m/d). A series of water supply wells with a combined discharge of 1200 m³/d are to be placed in an undeveloped region of the semi-confined aquifer. Groundwater simulation is used to assess the potential impacts of the wells on groundwater resources. The simulation uses the MODFLOW model, and for the model representation, the wells are treated as a single well.

The MODFLOW representation of the physical system is shown in Figure 2.5.5. The length of the aquifer valley is represented through 11 rows of cells, each with a cell size $\Delta y = 300$ m. The width of the valley is represented by 8 columns with cell size $\Delta x = 200$ m. The model uses two layers with the confining bed represented in the vertical conductance between the unconfined and semi-confined aquifers. The river extends along column 3 of layer 1 of the grid. The well

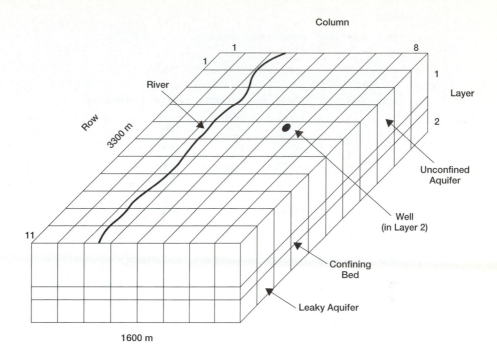

FIGURE 2.5.5 MODFLOW Representation of Alluvial Valley Aquifer System

field is located in cell $(i,j,k) = (5,6,2)$. Diffuse recharge occurs at a uniform rate of 20 cm/yr. The hydraulic heads along row 1 are approximately 22 m above the confining bed, while the heads along row 11 are 18 m above the confining bed. The slope of the riverbed is uniform following the regional aquifer gradient.

An unconfined aquifer hydraulic conductivity of 5 m/d is used, along with a semi-confined aquifer transmissivity of 100 m²/d (based on $K_h = 10$ m/d and $b = 10$m). The vertical leakance is calculated using Equation 2.5.15 to be VCONT $= 0.0087 d^{-1}$, assuming a value of $K'/b' = 0.001 \ d^{-1}$ for the confining bed.

The selected model representation uses the River Package, Well Package, Recharge Package and General-Head Boundary Package. The river conductance is calculated using Equation 2.5.9 with $W = 10$ m, $L = 300$ m, $K_{RIV} = 0.1$ m/d, and $M = 0.2$ m, which gives $CR = 1,500$ m²/d. The boundary conductance is more difficult to estimate. Treating the external source as a horizontally adjacent cell and using Equation (2.5.6) gives $CB = 67$ m²/d. This value is appropriate for connecting cells within the flow domain, but it is not large enough to maintain the desired boundary heads. A value of $CB = 600$ m²/d is assumed,

1 WELLS

LAYER	ROW	COL.	STRESS RATE	WELL NO.
2	5	6	1200.0	1

11 RIVER REACHES

LAYER	ROW	COL.	STAGE	CONDUCTANCE	BOT. ELEV.	REACH NO.
1	1	3	22.00	1500.	20.00	1
1	2	3	21.60	1500.	19.60	2
1	3	3	21.20	1500.	19.20	3
1	4	3	20.80	1500.	18.80	4
1	5	3	20.40	1500.	18.40	5
1	6	3	20.00	1500.	18.00	6
1	7	3	19.60	1500.	17.60	7
1	8	3	19.20	1500.	17.20	8
1	9	3	18.80	1500.	16.80	9
1	10	3	18.40	1500.	16.40	10
1	11	3	18.00	1500.	16.00	11

16 HEAD-DEPENDENT BOUNDARY NODES

LAYER	ROW	COL.	ELEVATION	CONDUCTANCE	BOUND NO .
1	1	1	22.00	600.0	1
1	1	2	22.00	600.0	2
1	1	3	22.00	600.0	3
1	1	4	22.00	600.0	4
1	1	5	22.00	600.0	5
1	1	6	22.00	600.0	6
1	1	7	22.00	600.0	7
1	1	8	22.00	600.0	8
1	11	1	18.00	600.0	9
1	11	2	18.00	600.0	10
1	11	3	18.00	600.0	11
1	11	4	18.00	600.0	12
1	11	5	18.00	600.0	13
1	11	6	18.00	600.0	14
1	11	7	18.00	600.0	15
1	11	8	18.00	600.0	16

RECHARGE = 0.5600000E 03

FIGURE 2.5.6   MODFLOW Data for Stress Packages

which is nearly an order-of-magnitude larger. The data for the stress packages is shown in Figure 2.5.6.

A summary printout from the MODFLOW simulation is shown in Figure 2.5.7. The upper part of this figure provides the head values in two tables for the 8 columns, 11 rows and 2 layers. A review of these tables shows that the average head within the pumping cell is 17.93 m, while the overlying head in the unconfined aquifer is 18.89 m. These values along with the vertical conductance (Equation 2.5.15) of $CZ =500$ m^2/d show that the vertical flow from the upper to lower aquifers in this region equals 500 m^3/d, which is a significant fraction of the total well system discharge. Comparison of the head values for column 3, layer 1, with those for the river stage in Figure 2.5.5 shows that the river looses water to the aquifer only in the upper cell, and over most of the reach the river is recharged by the unconfined aquifer. The summary table at the bottom of Figure 2.5.6 shows that the net recharge to the river from the aquifer over the simulated river reach is approximately 870 m^3/d (916 m^3/d–50 m^3/d).

1

HEAD IN LAYER 1 AT END OF TIME STEP 1 IN STRESS PERIOD 1

	1	2	3	4	5	6	7	8
1	21.98	21.96	21.97	21.91	21.88	21.86	21.88	21.91
2	21.84	21.77	21.62	21.61	21.53	21.44	21.53	21.61
3	21.63	21.51	21.24	21.25	21.08	20.86	21.09	21.25
4	21.36	21.21	20.85	20.85	20.54	20.00	20.56	20.86
5	21.04	20.87	20.45	20.45	20.02	18.79	20.05	20.50
6	20.69	20.50	20.06	20.13	19.86	19.33	19.91	20.23
7	20.31	20.12	19.67	19.83	19.73	19.55	19.80	19.99
8	19.90	19.72	19.27	19.51	19.53	19.49	19.62	19.72
9	19.46	19.29	18.88	19.16	19.24	19.25	19.32	19.38
10	18.95	18.83	18.48	18.76	18.84	18.87	18.91	18.94
11	18.32	18.27	18.08	18.25	18.28	18.30	18.31	18.32

1

HEAD IN LAYER 2 AT END OF TIME STEP 1 IN STRESS PERIOD 1

	1	2	3	4	5	6	7	8
1	21.94	21.91	21.89	21.84	21.80	21.77	21.80	21.83
2	21.81	21.74	21.61	21.58	21.49	21.40	21.49	21.57
3	21.60	21.48	21.26	21.22	21.05	20.82	21.05	21.22
4	21.33	21.18	20.88	20.82	20.50	19.88	20.52	20.83
5	21.01	20.84	20.49	20.42	19.95	17.93	19.98	20.47
6	20.66	20.47	20.11	20.10	19.82	19.21	19.87	20.20
7	20.28	20.09	19.72	19.80	19.70	19.51	19.77	19.96
8	19.87	19.69	19.34	19.49	19.50	19.46	19.58	19.69
9	19.42	19.26	18.94	19.13	19.20	19.22	19.29	19.35
10	18.94	18.82	18.55	18.74	18.82	18.85	18.89	18.92
11	18.45	18.38	18.22	18.35	18.40	18.42	18.44	18.45

1

VOLUMETRIC BUDGET FOR ENTIRE MODEL AT END OF TIME STEP 1 IN STRESS PERIOD 1

CUMULATIVE VOLUMES	L**3	RATES FOR THIS TIME STEP	L**3/T
IN:		IN:	
CONSTANT HEAD =	0.0000	CONSTANT HEAD =	0.0000
WELLS =	0.0000	WELLS =	0.0000
RIVER LEAKAGE =	49.5280	RIVER LEAKAGE =	49.5280
HEAD DEP BOUNDS =	383.8823	HEAD DEP BOUNDS =	383.8823
RECHARGE =	2956.8003	RECHARGE =	2956.8003
TOTAL IN =	3390.2104	TOTAL IN =	3390.2104
OUT:		OUT:	
CONSTANT HEAD =	0.0000	CONSTANT HEAD =	0.0000
WELLS =	1200.0000	WELLS =	1200.0000
RIVER LEAKAGE =	915.5662	RIVER LEAKAGE =	915.5662
HEAD DEP BOUNDS =	1274.6294	HEAD DEP BOUNDS =	1274.6294
RECHARGE =	0.0000	RECHARGE =	0.0000
TOTAL OUT =	3390.1956	TOTAL OUT =	3390.1956
IN - OUT =	1.4893E—02	IN - OUT =	1.4893E—02
PERCENT DISCREPANCY =	0.00	PERCENT DISCREPANCY =	0.00

FIGURE 2.5.7 Summary Printout from MODFLOW Simulation

Problems

2.1.1. A capillary tube containing water has a radius 0.1 mm. If the hydraulic gradient is 0.01, what is the average velocity and how far would water move during a one-year period?

2.1.2. For flow between parallel plates, Equation 2.1.6 reduces to $-\dfrac{d^2u}{dy^2} = \dfrac{\rho g I}{\mu}$

for the function $u(y)$. The boundary conditions are $u(0) = 0$ and $u(H) = 0$. Verify that Equation 2.1.9 gives the average velocity within the flow space.

2.1.3. The sand in a filter that is 1.2 m thick with a cross-section area of 40 m² has a mean grain size of 0.4 mm and a porosity of 0.35. When water is ponded to a depth of 0.5 m on the filter and the base is free draining (the pressure is atmospheric), what is the discharge across the filter? Use the Kozeny-Carman equation to estimate the hydraulic conductivity.

2.2.1. The manometers shown in Figure 2.2.1 measure a water head difference of 3 ft across the sand column. The cross-section area is 0.8 ft² and the length of the sand column is 3 ft. If the measured discharge is 8 ft³/d, what is the hydraulic conductivity in ft/d?

2.2.2. A semiconfined aquifer is recharged from an overlying unconfined aquifer through an aquitard. From water balance studies it is estimated that the recharge rate is 2.9 cm/year. If the average piezometric surface of the confined aquifer is 10 m below the water table in the unconfined aquifer and the aquitard is 1 m thick, what is K_v of the aquitard in m/d?

2.2.3. The experiment of problem 2.2.1 is repeated with the same set-up and soil but with oil with a dynamic viscosity of 2 cp (centipoise) and specific gravity of 0.8. What is the daily discharge through the column for an oil head drop of 3 ft?

2.2.4. Three piezometers monitor water levels (piezometric heads) in a confined aquifer. Piezometer A is located 3,000 ft due south of piezometer B. Piezometer C is located 2,000 ft due west of piezometer B. The surface elevations of A, B, and C are 480, 610 and 545 ft, respectively. The depth to water in A is 40 ft, in B is 140 ft, and in C is 85 ft. Determine the direction of groundwater flow through the triangle ABC and calculate the hydraulic gradient.

2.2.5. Two piezometers are situated side-by-side and penetrate an unconfined aquifer. The first piezometer is screened at a depth of 35 ft while the second is screened at a depth of 60 ft. Downhole pressure transducers which are located adjacent to the screened intervals record pressures of 12.88 and 23.98 psi in piezometers one and two, respectively. Determine whether this is a recharge or discharge zone and estimate the depth to the water table in feet below the land surface. If the vertical conductivity is estimated to be 0.1 ft/d, calculate the yearly infiltration or evaporation in inches per year.

2.2.6. A piezometer is screened at a depth of 20 m below land surface and records a pressure of 120 kPa on a pressure transducer. An immediately adjacent piezometer is screened at a depth of 10 m below land surface, and the depth to water in this piezometer is 7 m below land surface. Is the vertical component of flow upward or downward at this location, and what is the depth of the water table below land surface? Assume the specific weight of water is 9810 N/m³.

2.2.7. At a location $(x_1, y_1, z_1) = (1,000\ m, 800\ m, 120m)$ the pressure head is 130 m and the hydraulic gradient is $\tilde{I} = 0.001\hat{\imath} - 0.008\hat{\jmath} + 0.05\hat{k}$. What are the hydraulic head and pressure (in Pa) at location $(x_2, y_2, z_2) = (2000\ m, -200\ m, 220\ m)$ if the head gradient is assumed to be uniform? The specific weight of water is $9,810\ N/m^3$.

2.2.8. A uniform granular media has a mean grain diameter of 0.3 mm and is classified as medium sand. This sand will be packed to a porosity of about 0.4 to make a 2 m thick sand filter. If the maximum allowable depth of water ponding on top of the filter is 0.75 m, what radius of the circular filter is required to pass a discharge of 200 L/s if the water temperature is 25 degrees C and the filter is freely draining at its base? How would the required radius change if the water temperature was 5 degrees C?

2.2.9. Four horizontal, homogeneous, isotropic geologic strata overlie one another. In descending order, they have thickness and hydraulic conductivity values of 15 ft and 12 ft/d; 6 ft and 0.1 ft/d; 5 ft and 6 ft/d; and 15 ft and 0.3 ft/d. Calculate the horizontal and vertical components of hydraulic conductivity for the equivalent homogeneous-but-anisotropic formation. If the regional hydraulic gradient makes an angle of 60 degrees with the horizontal and has a magnitude of 0.004, estimate the magnitude and inclination of the mean Darcy flow.

2.2.10. In an area of groundwater recharge, vertical flow occurs through an unconfined perched aquifer with 5-m saturated thickness and $K = 5$ m/d, through a silty-clay confining layer of thickness 2 m and $K = 0.2$ m/d, and into a lower leaky aquifer. The pressure head measured in a piezometer screened just below the confining layer is 6 m. What is the vertical recharge rate through the confining layer and into the leaky aquifer?

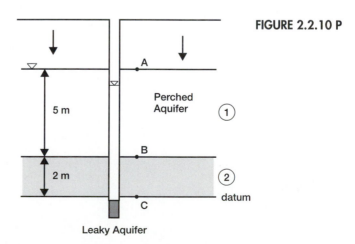

FIGURE 2.2.10 P

A

Perched Aquifer ①

5 m

B

2 m ②

datum

C

Leaky Aquifer

2.2.11. Heavy infiltration (due to excess irrigation) of 3 cm/d causes a perched water table to form above a low permeable, flow-restricting layer. The top of the restricting layer is at a depth of 2 m, it is 0.4 m thick, and it has a K_v of 0.01 m/d. The material above the restricting layer is a silt loam with a K_v of 0.12 m/d. Coarse sand and gravel occurs below the restricting layer. After flowing through the restricting layer, the water moves as unsaturated flow through the sand and gravel to an unconfined aquifer. What is the height of the perched water table above the top of the restricting layer? The setting is shown below.

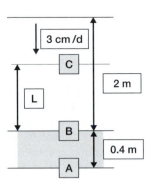

FIGURE 2.2.11 P

2.2.12. A set of parallel fractures has a mean aperture of 0.2 mm and an average fracture spacing of 1 m. If the hydraulic conductivity of the rock matrix is 10^{-8} cm/s, estimate the effective conductivity of the medium along the direction parallel to the fractures. What percent of the conductivity is associated with the fractures?

2.2.13. What is the effective porosity of the porous medium of Figure 2.2.15 and Equation 2.2.30 if the Kozeny-Carman theory applies?

2.2.14. Water is produced at a rate of 100 gpm from a well with a screen length of 15 ft and radius 4 in. The aquifer is fine sand with a mean grain diameter of 0.2 mm. Is Darcy's law valid for the entire region of flow? If not, to what radius from the center of the well is Darcy's law valid?

2.2.15. Water moves through a packed bed under hydraulic gradients of 1, 10, and 1000. If the mean grain diameter is 0.25 mm and $n = 0.4$, estimate the corresponding flow rates (Darcy velocities) in cm/s using the Ergun Equation 2.2.32.

Hint: to solve this equation iteratively, it is useful to write it as $q = \dfrac{I}{\alpha + bq}$

2.2.16. The falling head piezometer of Figure 2.2.16 has the soil sample of length 20 cm held in the 8-cm diameter cylinder, and the diameter of the burette is 8 mm. After 1 hr of operation, the water level in the standing tube is 40 cm above the elevation of the lower reservoir, while after 24 hours the elevation is 30 cm above the reservoir. Find the hydraulic conductivity of the soil sample.

2.3.1. The Salt River Valley in central Arizona is an alluvial valley with an unconfined aquifer system that is heavily pumped but receives little replenishment. The area is about 100,000 ha ($1\,ha$ = 10,000 m^2), the groundwater pumping about 500 million m^3 per year, and the water table drop about 3 m per year. Assuming no replenishment, what is the specific yield of the aquifer material? (After Bouwer, 1979.)

2.3.2. Within a field that is used for agricultural production, a clay unit is found at a depth 8 ft below the land surface. The surface soil has a hydraulic conductivity of 0.6 ft/d. If parallel drains are placed 2 ft above the clay unit and the maximum prolonged infiltration rate is expected to be 2 inches per day, and if the water level is to be maintained below 18 inches from the ground surface, what is the maximum allowable lateral drain spacing? Is there any appreciable advantage to placing the drains 1 ft above the clay pan?

2.3.3. A well with a radius of 0.25 m is to be placed in an unconfined aquifer which has an effective hydraulic conductivity of 3 m/d and is recharged at a rate of 15 cm/yr. If the initial saturated thickness of the aquifer is 25 m, what is the maximum long-term production rate in m^3/d? If a 2 m saturated thickness must be maintained within the well, what is the maximum production rate?

2.3.4. A well with radius 0.2 m is to be placed in an unconfined aquifer with saturated thickness 25 m and hydraulic conductivity 5 m/d. Provide a graph showing the maximum long-term production rate as a function of the rate of recharge for recharge values ranging from 5 cm/yr to 50 cm/yr. Discuss your results.

2.3.5. An 8-m thick confined aquifer consists of sand with a compressibility of 10^{-8} m^2/N and a porosity of 0.35. Estimate the specific storage and storage coefficient for the formation. If the piezometric surface drops by 10 m, how much water is released from storage per km^2? What is the corresponding change in thickness of the aquifer and percentage change in porosity?

2.3.6. An excavation for a construction project requires dewatering of an underlying aquifer as shown below. If $L = 15$ m, what is the maximum allowable gage pressure (in kPa) at point B which will still ensure that potential seepage forces will not result in a condition of "quick" sand and fluid entering the excavation at a dangerous rate?

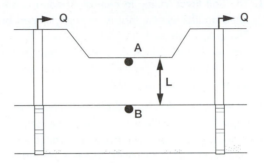

FIGURE 2.3.6 P

Chapter Three
Groundwater and Well Hydraulics

In this chapter, we apply the principles developed in Chapter Two, "Darcy's Law and Continuity Relations," to the hydraulics of groundwater, with emphasis on groundwater flow to wells. The focus is on well hydraulics because of the role groundwater wells play in developing site characterization data through aquifer tests, their use in groundwater development and dewatering operations, and their application in pump-and-treat systems for contaminated aquifer remediation. After a review of the configuration of a groundwater well and its construction, the steady-state hydraulics of flow to wells is discussed. This description is followed by a discussion of aquifer tests under unsteady flow conditions. Aquifer tests are important field methods for quantifying aquifer parameters, such as the transmissivity and field scale (average) hydraulic conductivity. We will first consider an "ideal" well in a confined aquifer. As the term implies, the characteristics of this ideal well are greatly simplified, and many of the features of a "real" well are put aside for the time being. Nevertheless, this look at an ideal well proves to be powerful because the resulting model is robust; the model applies to a wide range of conditions that at first glance do not appear to fit the assumptions. In addition, through comparison of more general solutions with this idealized configuration, we are provided greater quantitative insight into the various processes that control the response of a groundwater well to pumping.

In the remaining sections of this chapter, we will relax some of the initial assumptions and learn how to deal with various complications that arise in practice. We will see how to deal with boundaries, leaky aquifers, finite-sized well casings, unconfined aquifers, partially penetrating wells, and multiple well systems. In doing so, the emphasis is placed on the assumptions implicit within the various solutions and their application to problems of engineering significance. Slug tests, which are useful in site characterization at hazardous waste sites, are reviewed. In addition, hydraulics of flow in stratified aquifers, salt-water intrusion problems, and other transient flow problems are discussed.

3.1   Wells and Their Placement

As shown in Figure 3.1.1, there are essentially two main elements to the configuration of a groundwater well: the casing and the screen (i.e., intake portion). The **casing** houses and protects the pumping equipment, and for deep formations it serves as a vertical conduit for upward flow from the aquifer to the pump intake. When the well is placed in consolidated rock, a casing might not be required over part or all of the length of the well. The screen acts to prevent formation material from entering the well along with the water (except, again, in consolidated formations a screen might not be necessary). In general, the screen must prevent sediments from moving into the borehole, and yet not obstruct the flow of water into the well to such an extent that well efficiency is impeded excessively. For unconfined aquifers where the saturated zone is screened, the pump is installed in the screened section. The casing and screen may be attached with a **packer** made typically of either lead or neoprene rubber, or they may be installed as a continuous string of pipe and screen.

The steps involved in construction of a well include (1) drilling of the hole; (2) completion of the well, which involves placement of the screen, casing, packer, and filter pack (if one is used); (3) development of the well; (4) well testing; (5) pump selection; and (6) disinfection. Drilling methods include the use of cable tools or air drilling for consolidated or dry formations, direct and reverse rotary methods that require the circulation of drilling fluids for wet and unconsolidated formations, and jetting. Well development includes removal of fine-grained materials from the vicinity of the screen and gravel pack (if one is used) through hydraulic action or with chemicals. The gravel, or filter pack, consists of smooth, uniform, clean sand or gravel that is placed in the annulus of the well between the borehole wall and the well screen. The purpose of the filter pack is to prevent formation material from entering the screen. Testing involves evaluation of the well's completion and development in terms of its hydraulic behavior. Step-drawdown tests are often used for this evaluation, and these

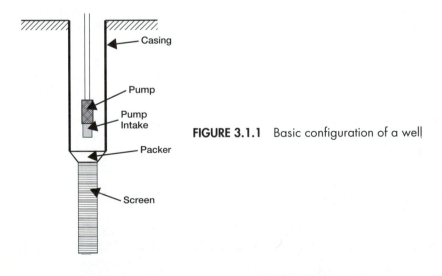

FIGURE 3.1.1   Basic configuration of a well

Casing

Pump

Pump Intake

Packer

Screen

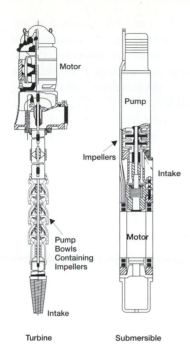

Turbine   Submersible

FIGURE 3.1.2   Deep well turbine and submersible pumps (after Tood, 1980)

tests also aid in pump selection through evaluation of well losses. Driscoll (1986) provides more information on groundwater well characteristics, placement, and evaluation.

Pumping Equipment

For shallow wells where only a small discharge is needed, one may use hand-operated, gear, reciprocating, or other types of low-capacity, low-head pumps. For installations serving the needs of irrigation, municipal, and industrial water use, however, deep-well turbine and submersible pumps are the most appropriate pumping equipment. Deep-well turbine pumps are widely used for large, high-capacity wells. As shown in Figure 3.1.2, the pump impellers are contained in bowls and are suspended vertically on a long drive shaft driven by a surface-mounted motor. Several bowls may be connected in series, forming a multi-stage pump to deliver higher heads. The submersible pump has the pump bowl and impellers close-coupled to a small-diameter submersible electric motor, as shown in Figure 3.1.2. Submersible pumps range in size from small units that can fit inside an 8 cm casing to large-capacity units involving numerous stages (Todd, 1980).

3.2   Steady Flow to a Well

In order to develop analytical solutions to groundwater flow problems, one must generally assume that an effective, uniform hydraulic conductivity or transmissivity value can be identified. For steady flow problems, the water level in observation wells does not change with respect to time, and the storage change (**capacitance**) term from the continuity equation vanishes. The continuity equation reduces to Laplace's equation or Poisson's equation if infiltration recharge is considered. The problem of steady flow to a well is essential in applications, and there are a number of useful results that are presented in this section. For all results, it is assumed that the aquifer is homogeneous and isotropic, that wells are pumped at constant rates, and that wells fully penetrate the aquifer—resulting in horizontal flow.

3.2.1   Confined Aquifers

Consider steady flow to a production well in a confined aquifer. For steady radial flow in a confined aquifer with a constant effective transmissivity, the continuity equation is the following:

$$\nabla^2(h) = \frac{1}{r}\frac{d}{dr}\left(r\frac{dh}{dr}\right) = 0 \tag{3.2.1}$$

Straightforward integration gives

$$h(r) = A\,ln(r) + B \tag{3.2.2}$$

where A and B are two integration constants. The radial aquifer flux is determined by applying Darcy's Law to the total aquifer thickness and is given by

$$U_r(r) = -T\frac{dh}{dr} = -T\frac{A}{r}$$

where U_r is the radial aquifer flux and T is the aquifer transmissivity. Because the flow is steady, the total discharge crossing each concentric cylinder surrounding the well must be the same. That is, the product of the aquifer flux and the circumference of the cylinder are equal to the discharge. This condition gives the following equation:

$$Q = -2\pi r U_r(r) = 2\pi T A$$

Solving for A gives the following for the head:

$$h(r) = \frac{Q}{2\pi T}\,ln(r) + B$$

With the added condition that $h(R) = h_R$, where R is the **radius of influence**, one finds

$$s(r) = h_R - h(r) = \frac{Q}{2\pi T}\,ln\left(\frac{R}{r}\right) \tag{3.2.3}$$

The configuration described by Equation 3.2.3 is shown in Figure 3.2.1. In Equation 3.2.3, $s = h_R - h$ is called the *drawdown*, and it represents the decline in the head associated with groundwater pumping. Equation 3.2.3 is called the

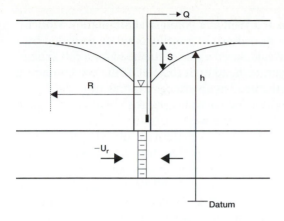

FIGURE 3.2.1 Steady flow to a well in a confined aquifer

Thiem equation[1]. It is one of the two or three most important equations in groundwater hydraulics, appearing in many different places when dealing with radial flow problems. Somewhat more general forms of Equation 3.2.3 give the difference in drawdown at two radii—or equivalently, the head difference at two locations, r_1 and r_2.

$$s_1 - s_2 = \frac{Q}{2\pi T} \ln\left(\frac{r_2}{r_1}\right) \tag{3.2.4}$$

$$h_2 - h_1 = \frac{Q}{2\pi T} \ln\left(\frac{r_2}{r_1}\right) \tag{3.2.5}$$

EXAMPLE PROBLEM

3.2.1 Estimation of Aquifer Transmissivity

A well completed in a confined aquifer is pumped at a rate of 200 m³/d, and the drawdown is measured in two monitoring wells located at distances $r_1 = 10$ m and $r_2 = 25$ m from the pumping well. After a period of pumping, the observed heads in the monitoring wells stabilize (only decrease slowly with time) with measured values $h_1 = 86$ m and $h_2 = 88$ m. What is the aquifer transmissivity?

For this problem, we can use Equation 3.2.5 to solve directly for T:

$$T = \frac{Q}{2\pi(h_2 - h_1)} \ln\left(\frac{r_2}{r_1}\right) = \frac{200\dfrac{\text{m}^3}{\text{d}}}{2\pi(88\text{ m} - 86\text{ m})} \ln\left(\frac{25\text{ m}}{10\text{ m}}\right) = 15\text{ m}^2/d$$

As we will see in Section 3.3, "Transient Flow to a Well in an Ideal, Confined Aquifer," this same approach may be used under unsteady conditions, because after a period of pumping, the water levels in monitoring wells will decline at the same rate.

---

[1]Although Dupuit first developed it in 1863, Thiem (1906) used Equation 3.2.3 for the first aquifer test to evaluate T from known discharge and drawdown.

EXAMPLE **3.2.2   Estimation of Drawdown at a Well in a Numerical Simulation Model**
PROBLEM In this example, we estimate the drawdown at a well from data provided by a numerical simulation model. The calculated value of the head for a cell in a numerical model represents the average head over the cell. If the cell contains a well, then the calculated head is the head at an effective radius from the center of the cell. We want to calculate the effective radius, r_e, as shown schematically in Figure 3.2.2, and estimate the head at the well.

The left side of Figure 3.2.2 shows the flow pattern calculated by the numerical model, while the right side shows the flow pattern for radial flow. The cell is square with $\Delta x = \Delta y$ the cell size, r_2 is the radius between cell centers ($r_2 = \Delta x$), and $r_1 = r_e$, the effective radius of the cell containing the well. We find the effective radius by requiring that the hydraulics of these two geometries be the same. With reference to the numerical model, the flow into the cell containing the well is given by

$$Q = T\frac{h_N - h_C}{\Delta y}\Delta x + T\frac{h_W - h_C}{\Delta x}\Delta y + T\frac{h_S - h_C}{\Delta y}\Delta x + T\frac{h_E - h_C}{\Delta x}\Delta y$$

$$= T(h_N + h_W + h_S + h_E - 4h_C) = 4T(h_{ave} - h_C)$$

because $\Delta x = \Delta y$, and h_{ave} is the average head in the surrounding cells. To find the equivalent radius of the cell, we use the Thiem equation to relate the head, h_2, at radius $r_2 = \Delta x$ to the head, h_1, at the effective cell radius $r_1 = r_e$:

$$Q = \frac{2\pi T(h_R - h_{re})}{ln(r_2/r_1)}$$

For these two representations to be equivalent, we require the same Q and T, and that $h_{ave} - h_C = h_r - h_{re}$. Thus,

$$4 = \frac{2\pi}{ln(r_2/r_1)} = \frac{2\pi}{ln(\Delta x/r_e)}$$

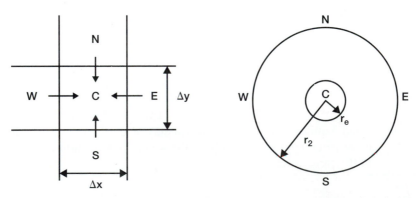

FIGURE 3.2.2   Comparison of flow patterns in numerical model (left) and radial flow

This result gives $r_e = 0.208 \, \Delta x$. The head at the well is then found from

$$h_{\text{well}} = h_c - \frac{Q}{2\pi T} \ln\left(\frac{0.208\Delta x}{r_{\text{well}}}\right) \tag{3.2.6}$$

Prickett and Lonnquist (1971) found this result. Also see Peaceman (1978) for a discussion of well drawdown in numerical models.

3.2.2 Unconfined Aquifers

The continuity equation for radial flow to a well in an unconfined aquifer with infiltration at a constant rate W has already been solved, and the general solution is given as Equation 2.3.18. With this solution, the radial aquifer flux is given by

$$U_r(r) = -KH\frac{dH}{dr} = -\frac{K}{2}\frac{dH^2}{dr} = -\frac{AK}{2r} + \frac{Wr}{2}$$

where H is the saturated thickness. At the production well, continuity gives ($W = 0$ within the well)

$$U_r(r_w) = \frac{Q}{2\pi r_w} = -\frac{AK}{2r_w}$$

Thus, $A = Q/(\pi K)$. For a production well in an unconfined aquifer, the solution becomes

$$H^2(r) = \frac{Q}{\pi K}\ln(r) - \frac{Wr^2}{2K} + B \tag{3.2.7}$$

The most important case that occurs has negligible diffuse recharge ($W \cong 0$) compared with the flow induced by the well. With the further condition that $H = H_R$ at $r = R$, the radius of influence, then Equation 3.2.7 reduces to

$$H^2(r) = H_R^2 - \frac{Q}{\pi K}\ln\left(\frac{R}{r}\right) \tag{3.2.8}$$

Equation 3.2.8 is the **Dupuit equation** for radial flow. It plays the same role for unconfined aquifers that the Thiem equation plays for confined aquifers.

We can write the Dupuit equation in a form that looks like the Thiem equation and presents a useful approximation for analysis of pumping test data. To achieve this result, we introduce the **corrected drawdown** by substituting $H = H_R - s$ in the Dupuit equation (s is the drawdown).

$$H_R^2 - (H_R - s)^2 = \frac{Q}{\pi K}\ln\left(\frac{R}{r}\right) = H_R^2 - H_R^2 + 2H_R s - s^2 = 2H_R\left(s - \frac{s^2}{2H_R}\right)$$

Divide by $2H_R$ to find

$$s'(r) \equiv \left(s - \frac{s^2}{2H_R}\right) = \frac{Q}{2\pi KH_R}\ln\left(\frac{R}{r}\right) \tag{3.2.9}$$

In Equation 3.2.9, s' is the corrected drawdown. Calculation of s' from s accounts for the changes in saturated thickness, allowing use of equations that treat the aquifer as if it has a uniform saturated thickness. In this sense, the drawdown has been **corrected**. Given s', one can find s from

$$s = H_R \left(1 - \sqrt{1 - \frac{2s'}{H_R}} \right) \qquad (3.2.10)$$

which comes from solving the quadratic Equation 3.2.9 for s. The form of Equation 3.2.9 will be useful later in this chapter. Looking ahead, the product $K\,H_R$ in Equation 3.2.9 is the transmissivity based on the aquifer's saturated thickness at the radius of influence, which is also the initial saturated thickness for a transient flow problem that starts with a uniform water table elevation.

There is still another way to write Equation 3.2.8 that is useful in the design of recovery systems for groundwater impacted by free-product petroleum hydrocarbons (see Section 9.8). Charbeneau and Chiang (1995) show that the Dupuit equation may be written

$$s_w = \frac{Q\ln\left(\dfrac{R}{r_w}\right)}{\pi K H_R \left(1 + \sqrt{1 - \dfrac{Q}{\pi K H_R^{\,2}} \ln\left(\dfrac{R}{r_w}\right)} \right)}$$

which expresses the drawdown at the well, $s_w = H_R - H(r_w)$, as a function of the water production rate and water-layer saturated thickness far from the well. Dividing both sides by H_R, this equation may be written as follows:

$$\frac{s_w}{H_R} = \frac{\Gamma}{1 + \sqrt{1 - \Gamma}} \;;\; \Gamma \equiv \frac{Q}{\pi K H_R^{\,2}} \ln\left(\frac{R}{r_w}\right) \qquad (3.2.11)$$

The **fraction discharge**, Γ, is the ratio of the well discharge to the maximum possible well discharge when the aquifer is pumped dry ($H(r_w) = 0$). Equation 3.2.11 expresses the **fraction drawdown**, s_w/H_R, as a function of the fraction discharge. When $s_w/H_R = 1$, $\Gamma = 1$. Equation 3.2.11 can be inverted to express the fraction discharge as a function of the fraction drawdown.

$$\Gamma = 2\left(\frac{s_w}{H_R}\right) - \left(\frac{s_w}{H_R}\right)^2 \qquad (3.2.12)$$

Equation 3.2.12 follows from rearrangement of Equation 3.2.9.

3.2.3   Leaky Aquifers

For steady flow in a semiconfined aquifer, Equation 2.3.35 becomes

$$\frac{\partial}{\partial x}\left(T \frac{\partial h}{\partial x} \right) + \frac{\partial}{\partial y}\left(T \frac{\partial h}{\partial y} \right) - \frac{K'}{b'}(h - h_a) + Q\delta(\tilde{x}) = 0 \qquad (3.2.13)$$

where K' is the vertical hydraulic conductivity of the confining bed, b' is the aquitard thickness, and h_a is the head in the adjacent aquifer. If a constant effective value of the aquifer transmissivity, aquitard thickness, and aquitard hydraulic conductivity can be identified, then Equation 3.2.13 may be written in radial coordinates as

$$\frac{d^2h}{dr^2} + \frac{1}{r}\frac{dh}{dr} - \frac{h}{B^2} = -\frac{h_a}{B^2} \qquad (3.2.14)$$

with the parameter

$$B = \sqrt{\frac{Tb'}{K'}}$$

Equation 3.2.14 is a nonhomogeneous, ordinary differential equation that is known as **Bessel's modified differential equation of order zero**. A particular solution is $h = h_a$. The complementary solution contains modified Bessel functions of the first and second kind of order zero. These are long names, but they may be thought of as radial versions of the exponential function. Using the boundary conditions that $h(r)$ remain finite as r becomes large (infinite) and that

$$U_r(r) = -\frac{Q}{2\pi r}$$

as r approaches the well (goes to zero), the solution takes the following form:

$$s(r) = h_a - h(r) = \frac{Q}{2\pi T} K_o\left(\frac{r}{B}\right) \qquad (3.2.15)$$

Equation 3.2.15 is attributed to De Glee (1930). In Equation 3.2.15, $K_o(r/B)$ is called the **modified Bessel function of the second kind of order zero**. This function behalves like the logarithmic function for small arguments and decreases exponentially for large r/B. Specifically, as $r/B \to 0$, we have

$$K_o\left(\frac{r}{B}\right) \to 1n\left(\frac{1.123}{r/B}\right) \qquad (3.2.16)$$

Equation 3.2.16 is good to within 5 percent for $r/B < 0.33$, and to within 1 percent for $r/B < 0.16$. For $r/B > 1$, we have

$$K_o\left(\frac{r}{B}\right) \to \sqrt{\frac{\pi}{2r/B}}\, e^{-r/B} \qquad (3.2.17)$$

These results are sufficient for our purposes. For a comprehensive presentation of the properties of Bessel functions, see Abramowitz and Stegun (1964). A module for calculating $K_o(\ )$ using a spreadsheet is presented in Appendix D.

Equation 3.2.15 is a general result, and it is valid throughout an extensive aquifer. Equation 3.2.16 suggests a simple approximation that is valid within the vicinity of the well. According to Equation 3.2.16, for small radii we can approximate Equation 3.2.15 by

$$s \cong \frac{Q}{2\pi T} \, ln \left(\frac{1.123B}{r} \right) \tag{3.2.18}$$

Equation 3.2.18 looks like the Thiem equation with an apparent radius of influence $R = 1.123B$. This equation is the first example for which we can give the radius of influence its usual interpretation. If we look at the drawdown distribution in the vicinity of the well and extrapolate it outward, the radius of influence is the distance at which the logarithmically extrapolated drawdown would be zero.

EXAMPLE PROBLEM   **3.2.3   Drawdown Distribution Around a Well in a Semiconfined Aquifer**

Consider a leaky aquifer with the following characteristics: $T = 80$ m²/d, $K' = 0.01$ m/d, $r_w = 0.2$ m, $b' = 5$ m, $Q = 700$ m³/d. With these parameters, B is calculated to be

$$B = \sqrt{\frac{Tb'}{K'}} = 200 \text{ m}$$

At the well, we have from Equation 3.2.18

$$s = \frac{700 \text{ m}^3/\text{d}}{2 \times \pi \times 80 \text{ m}^2/\text{d}} \, ln \left(\frac{1.123 \times 200 \text{ m}}{0.2 \text{ m}} \right) = 9.7 \text{ m}$$

Extrapolating the potentiometric surface outward from the well using the logarithmic approximation, it looks like the radius of influence is $R = 1.23 \times 200 = 225$ m. This approximation suggests that the drawdown should vanish at this distance. There is a potentiometric surface drawdown at radius R, however. At this distance, we can use the approximation given by Equation 3.2.17 $r/B = R/B = 1.123$:

$$s = \frac{700 \text{ m}^3\text{d}}{2 \times \pi \times 80 \text{ m}^2/\text{d}} \sqrt{\frac{\pi}{2 \times 1.123}} \, e^{-1.123} = 0.53 \text{ m}$$

With mathematical tables or the module presented in Appendix D, one finds $K_o(1.123) = 0.36$, so the calculated drawdown is 0.50 m at a radius of 225 m from the pumping well.

3.2.4   Image Well Theory

Application of **image well theory** allows the introduction of simple boundary conditions, such as impermeable faults (no-flow boundaries) and rivers or large surface water bodies that act as constant head boundaries. The basic idea is to conceptually place a well on the opposite side of the boundary so that this imaginary well will hydraulically replace the boundary's influence. This idea is best understood through an example. Consider the case of an injection well and a production well in an unconfined aquifer. The injection well is located at point $(a, 0)$, while the production well is at point $(-a, 0)$, as shown in the upper part of Figure 3.2.3. Using the principle of superposition, the solution for the water level (Equation 3.2.8) gives

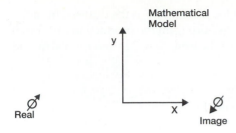

FIGURE 3.2.3 Use of imaginary well to represent a recharge boundary

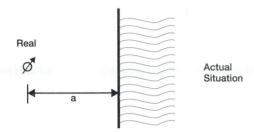

$$H_R^2 - H^2 = \frac{Q}{\pi K} \ln\left(\frac{R}{r_1}\right) - \frac{Q}{\pi K} \ln\left(\frac{R}{r_2}\right) = \frac{Q}{\pi K} \ln\left(\frac{r_2}{r_1}\right) \qquad (3.2.19)$$

where r_1 is the distance from the production well to the point (x, y), and r_2 is the distance from the injection well to the point (x, y):

$$r_1 = \sqrt{(x + a)^2 + y^2} \; ; \; r_2 = \sqrt{(x - a)^2 + y^2} \qquad (3.2.20)$$

The radius of influence has dropped out of the analysis, although the reference head H_R still remains.

Along the y-axis, $r_1 = r_2$, so $H^2 = H_R^2$. Thus, the perpendicular bisector of the line segment between the wells is a line with no drawdown, and the saturated thickness is equal to H_R. This solution is the same solution one would find with a production well next to a large water body. The injection well serves as a means to introduce a **constant head boundary**, as shown in the lower part of Figure 3.2.3. For the case where the constant head boundary is to be represented, the injection well is called an imaginary well or an **image well**.

EXAMPLE   **3.2.4   Application of Image Well Theory**
PROBLEM   A well of radius 0.2 m produces 550 m³/d from an unconfined aquifer with $K = 10$ m/d. The well is located 250 m from a river, and the saturated thickness at the river is 30 m. What is the drawdown at the well?

According to Equation 3.2.19, the saturated thickness is determined from

$$H_R^2 - H^2(x,y) = \frac{Q}{\pi K} \ln\left(\frac{r_i}{r_r}\right)$$

where $H(x,y)$ is the saturated thickness at location (x,y), r_i is the distance from the image well to the point (x,y) and r_r is the distance from the production (real) well to this point, as given by Equation 3.2.20. At the well, $r_r = 0.2$ m and $r_i = 500$ m, and the saturated thickness is

$$H_{well} = \sqrt{\left(30 \text{ m}^2 - \frac{550 \text{ m}^3\text{d}}{\pi \times 10 \text{ m/d}} \, ln \left(\frac{500 \text{ m}}{0.2 \text{ m}}\right)\right)} = 27.6 \text{ m}$$

Near the well, $r_i \cong 2a$, where a is the distance from the well to the river, so that

$$H^2(x,y) = H_R^2 - \frac{Q}{\pi K} \, ln \left(\frac{2a}{r_r}\right) \tag{3.2.21}$$

This result suggests that for a well adjacent to a recharge boundary, the apparent radius of influence is

$$R = 2a \tag{3.2.22}$$

Equations 3.2.21 and 3.2.22 show that the drawdown distribution near a well next to a recharge boundary is the same as that for a well in the middle of a circular island with a radius equal to twice the distance to the recharge boundary. If you were only observing the drawdown distribution in the vicinity of the well, then you could not determine which case you were dealing with. Further from the well, however, the drawdown cone would become asymmetrical. This drawdown distribution, $s(x,y) = H_R - H(x,y)$, is shown in Figure 3.2.4. Figure 3.2.4(a) is a three-dimensional image of this distribution and shows how the drawdown is focused near the well. Figure 3.2.4(b) is a drawdown contour plot with equal drawdown increments between contours. A careful look at this figure shows that the contours are circular, but contours of smaller drawdown have centers located farther from the constant-head boundary (the y-axis).

An image well can also be used to represent a **no-flow boundary**. For a production well next to a boundary where the unconfined aquifer has been cut-off due to a fault, facies change, or other cause, the influence of the boundary is represented through use of an image production well placed at an equal distance beyond the boundary. This process is shown in Figure 3.2.5. The saturated thickness is determined from Equation 3.2.7 with $W = 0$, where B is a constant:

$$H^2(x,y) = \frac{Q}{\pi K} \, ln(r_r) + \frac{Q}{\pi K} \, ln(r_i) + B = \frac{Q}{\pi K} \, ln(r_r r_i) + B \tag{3.2.23}$$

The x-component of the flow is found from

$$U_x = -KH\frac{\partial H}{\partial x} = -\frac{K}{2}\frac{\partial H^2}{\partial x} = -\frac{Q}{2\pi}\left(\frac{d \, ln(r_r)}{dr_r}\frac{\partial r_r}{\partial x} + \frac{d \, ln(r_i)}{dr_i}\frac{\partial r_i}{\partial x}\right)$$

$$= -\frac{Q}{2\pi}\left(\frac{1}{r_r}\frac{x + a}{r_r} + \frac{1}{r_i}\frac{x - a}{r_i}\right) = -\frac{Q}{2\pi}\left(\frac{x + a}{(x + a)^2 + y^2} + \frac{x - a}{(x - a)^2 + y^2}\right)$$

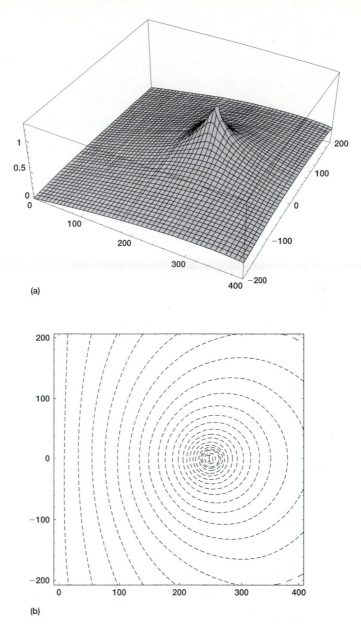

(a)

(b)

FIGURE 3.2.4 Drawdown distribution near a production well adjacent to a constant head boundary

Along $x = 0$, the aquifer flux is

$$U_x(0,y) = \frac{Q}{2\pi}\left(\frac{a}{a^2 + y^2} + \frac{-a}{a^2 + y^2}\right) = 0$$

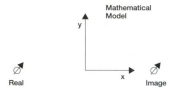

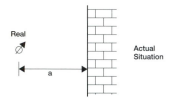

FIGURE 3.2.5 Use of imaginary well to represent a barrier boundary

Thus, the *y*-axis does correspond to a no-flow boundary for discharge in the *x*-direction.

The same principles apply for confined and semiconfined aquifers. A recharge boundary is represented by placing an "image" well with the opposite sign (injection or production) across the boundary at the same distance as the "real" well from the boundary. For a no-flow or barrier boundary, the image well *Q* has the same sign as the real well.

EXAMPLE PROBLEM

3.2.5   Stream Leaving a Mountain Range and Entering an Alluvial Aquifer

Figure 3.2.6 depicts a stream leaving a mountain range and entering an alluvial aquifer. The *y*-axis represents the extent of the mountain range, which appears as a no-flow boundary, while the *x*-axis represents the stream that acts as a recharge boundary. A well that produces from the aquifer is located at a distance *a* from the mountains and a distance *b* from the stream, at the location (*a*,*b*). We want to determine the head distribution within the aquifer and the apparent radius of influence of this boundary combination.

The barrier formed by the mountains alone is represented by an image well located at a distance *a* behind the boundary, as shown by image well i_1 in Figure 3.2.6. The recharge boundary formed by the stream is represented by the image well i_3 located at a distance *b* beyond the stream in Figure 3.2.6. These images alone, however, are not sufficient. Image well i_1 will cause drawdown along the stream, while image well i_3 will cause flow across the mountain front. The image injection well i_2 is required to complete the symmetry of the problem and satisfy all of the boundary conditions. Thus, we have one real well and three image wells for this problem. The head distribution satisfies the equation

$$H^2 = H_R^2 + \frac{Q}{\pi K} \ln\left(\frac{r_r}{R}\right) + \frac{Q}{\pi K} \ln\left(\frac{r_{i1}}{R}\right) - \frac{Q}{\pi K} \ln\left(\frac{r_{i2}}{R}\right) - \frac{Q}{\pi K} \ln\left(\frac{r_{i3}}{R}\right)$$

$$H^2 = H_R^2 + \frac{Q}{\pi K} \ln\left(\frac{r_r r_{i1}}{r_{i2} r_{i3}}\right) \tag{3.2.24}$$

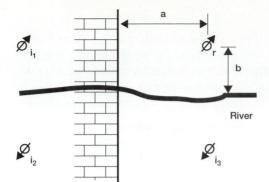

FIGURE 3.2.6 Stream leaving a mountain range and entering an alluvial aquifer

Equation 3.2.24 gives the water table elevation distribution throughout the quadrant of the alluvial aquifer containing the well. This equation may be simplified for the head distribution within the vicinity of the well. For an observation well near the pumping well, we have the approximate relations

$$r_{i1} \cong 2a \; ; \; r_{i2} \cong 2\sqrt{a^2 + b^2} \; ; \; r_{i3} \cong 2b$$

For calculating the drawdown near the well, we can use

$$H^2 = H_R{}^2 - \frac{Q}{\pi K} \ln\left(\frac{R}{r_r}\right)$$

where the apparent radius of influence is given by

$$R = 2\frac{b}{a}\sqrt{a^2 + b^2} \qquad\qquad \textbf{(3.2.25)}$$

3.3 Transient Flow to a Well in an Ideal Confined Aquifer

In this section, we consider the unsteady flow to a well in an ideal confined aquifer. This situation is both the simplest problem to be considered and the most important. Theis (1935) obtained the solution for unsteady groundwater flow to a well through analogy with a problem in heat conduction. If an electric current is passed along a long straight wire embedded within a conducting medium, then because of the resistance to current flow, heat will be given off by the wire—and the heat will be conducted away from the wire by the medium. In this problem, the wire corresponds to a well, the heat flux to the flow of water, and the temperature to the hydraulic potential or head. The solution to the heat conduction problem was well known. Making the appropriate analogy, Theis presented a solution to the problem of transient groundwater flow to a well. To appreciate the significance of this achievement, one need only note that from Chapter Two, formulation of the continuity equation for groundwater flow did not come until a later date.

3.3.1   Assumptions for Ideal Flow to a Well

The configuration for flow to a well in a confined aquifer is shown in Figure 3.3.1. The potentiometric surface is initially horizontal. With pumping at a constant rate Q, the elevation of this surface is lowered and a drawdown cone develops. Groundwater moves horizontally towards the well that is screened over the entire thickness of the aquifer. The major assumptions are listed as follows:

1. The aquifer is homogeneous, isotropic and large (infinite).
2. The original piezometric surface is horizontal.
3. The well is pumped at a constant discharge.
4. The well is fully penetrating, and the flow is horizontal.
5. The well diameter is infinitesimal so that storage within the well can be neglected.
6. The aquifer storage release is instantaneous.

   Contrary to the configuration shown in Figure 3.3.1, the ideal well has a casing and screen of negligible size—so there is no water storage within the well. That is, all of the water produced by the well comes from the formation. While there are many assumptions that are made, each of these can be relaxed, and it is through comparison with the ideal case that one can quantitatively evaluate the importance of other processes and well configurations. These assumptions are gradually relaxed through the later sections of this chapter. The main point is that we have arrived at a problem that is tractable through mathematical methods, and the solution may be expressed in terms of known functions.

3.3.2   Theis Equation

The Theis (1935) solution for transient groundwater flow to an ideal well satisfies the continuity equation

$$\frac{S}{T}\frac{\partial h}{\partial t} = \frac{\partial^2 h}{\partial r^2} + \frac{1}{r}\frac{\partial h}{\partial r} \qquad (3.3.1)$$

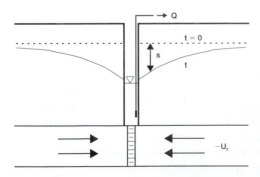

FIGURE 3.3.1   Configuration for flow to an ideal well in a confined aquifer

The initial conditions are that the head is constant everywhere, $h(r,0) = h_0$. This statement provides a boundary condition that says that the head remains constant at infinity, or that the impact of pumping (or recharge) is negligible far away from the well. This idea is stated by the boundary condition $h(r \rightarrow \infty, t) = h_0$. The second boundary condition is provided by continuity at the well. The well discharge is given by

$$Q = \lim_{r \to 0} [2\pi r(-U_r)] = \lim_{r \to 0} \left[2\pi r \left(T \frac{\partial h}{\partial r} \right) \right]$$

The resulting boundary condition at the well is

$$\lim_{r \to 0} \left[r \frac{\partial h}{\partial r} \right] = \frac{Q}{2\pi T} \tag{3.3.2}$$

The mathematical solution to this problem statement is

$$h_0 - h(r,t) = \frac{Q}{4\pi T} \int_u^\infty \frac{1}{\omega} e^{-\omega} d\omega$$

where ω is the variable of integration, and the lower limit of integration is

$$u = \frac{r^2 S}{4Tt} \tag{3.3.4}$$

In mathematical physics, the integral in Equation 3.3.3 is called the **exponential integral**, and it is a tabulated function. In the field of groundwater hydraulics, this function is called the **Theis well function**, and it is designated as $W(u)$ such that

$$W(u) = \int_u^\infty \frac{1}{\omega} e^{-\omega} d\omega \tag{3.3.5}$$

Values of the Theis well function are tabulated in Appendix C, and a module for calculating $W(u)$ for use with a spreadsheet is presented in Appendix D.

The difference $h_o - h(r,t) = s(r,t)$ is called the **drawdown**, and Equation 3.3.3 may be written as

$$s(r,t) = \frac{Q}{4\pi T} W(u) \tag{3.3.6}$$

Equation 3.3.6 is known as the **Theis equation** and is used to describe the development of a **drawdown cone** within a confined aquifer as a function of radial distance and time.

There are properties of Equation 3.3.3 that are important for understanding features of well hydraulics. As $t \rightarrow \infty$, $u \rightarrow 0$ and $W(0) \rightarrow \infty$. This situation implies that at any location, the drawdown will ultimately become infinitely large. This idea simply means that the drawdown cone will continue to spread until a source of recharge is reached.

The previous paragraph shows that if a source of recharge is not reached, the drawdown will continue to increase indefinitely. We are also interested in the rate of increase in drawdown. At any location, we have

$$\frac{\partial s}{\partial t} = \frac{Q}{4\pi T}\left(-\frac{1}{u}e^{-u}\right)\frac{\partial u}{\partial t} = \frac{Q}{4\pi T}\left(-\frac{1}{u}e^{-u}\right)\left(-\frac{r^2 S}{4Tt^2}\right) = \frac{Q}{4\pi T}\left(\frac{1}{t}e^{-u}\right)$$

For large t or small r, $e^{-u} \to 1$, such that

$$\frac{\partial s}{\partial t} \cong \frac{Q}{4\pi T}\frac{1}{t}$$

Thus, the rate of change of drawdown decreases through time, and the water levels appear to stabilize after the initial period of pumping. Furthermore, after a period of pumping, the rate of drawdown increase is independent of the radius from the well. This reason is why the transmissivity could be calculated using Equation 3.2.5 in Example 3.2.1, even if steady-state conditions had not been achieved.

EXAMPLE PROBLEM

3.3.1   Application of the Theis Equation

If a well of radius $r_w = 0.2$ m produces water at a rate $Q = 1500$ m³/d from a confined aquifer with $T = 600$ m²/d and $S = 0.0004$, determine the drawdown after 1 year of pumping at the well and at a radius of $r = 1$ km.

At the well after a 1-year period of time, the parameter u is given by

$$u = \frac{(0.2 \text{ m})^2 \times 0.0004}{4 \times 600 \text{ m}^2/\text{d} \times 365 \text{ d}} = 1.8 \times 10^{-11}$$

With this value of u, Appendix C gives $W(u) = 24.18$. From Equation 3.3.6, we then have

$$s(0.2,365) = \frac{1500 \text{ m}^3/\text{d}}{4 \pi 600 \text{ m}^2/\text{d}} \times 24.18 = 4.81 \text{ m}$$

Similarly, at a radius of 1 km after 1 year, we have

$$u = \frac{(1000 \text{ m})^2 \times 0.0004}{4 \times 600 \text{ m}^2/\text{d} \times 365 \text{ d}} = 4.6 \times 10^{-4}$$

For this value of u, Appendix C gives $W(u) = 7.12$, so $s(1000,365) = 1.42$ m.

3.3.3   Jacob Approximation

The Theis well function can be expressed in terms of the following infinite series:

$$W(u) = -\gamma - ln(u) + u - \frac{u^2}{2 \times 2!} + \frac{u^3}{3 \times 3!} - \dots \tag{3.3.7}$$

with

$$u = \frac{r^2 S}{4Tt} \text{ and } \gamma = \text{Euler constant} = 0.5772\ldots.$$

For small r or large t, u is small, and Cooper and Jacob (1946) noted that an approximate solution could be written as

$$W(u) \cong -\gamma - ln(u) = ln\left(\frac{e^{-\gamma}}{u}\right) = ln\left(\frac{0.561}{u}\right) = ln\left(\frac{2.25Tt}{r^2 S}\right) \qquad \textbf{(3.3.8)}$$

Thus, the Theis Equation 3.3.6 may be written

$$s(r,t) = \frac{Q}{4\pi T} ln\left(\frac{2.25Tt}{r^2 S}\right) = \frac{2.30Q}{4\pi T} \log\left(\frac{2.25Tt}{r^2 S}\right) \qquad \textbf{(3.3.9)}$$

This equation is called the **Jacob equation**.

In validating the Jacob equation, the question becomes one of identifying the limit on the argument u for which the terms beyond the first two of Equation 3.3.7 are negligible. By arbitrarily selecting small values of u, we may compare the well function value to the logarithmic approximation given by Cooper and Jacob. When $u = 0.01$, the approximation gives

$$W(0.01) \cong -0.5772 - ln(0.01) = 4.03$$

Appendix C shows that $W(0.01) = 4.04$. Thus, the **Jacob approximation** is valid at least for u less than 0.01. Accordingly, we will establish this value as our **criteria for validity** of the Jacob approximation: "$u < 0.01$."

EXAMPLE PROBLEM   **3.3.2   Application of Jacob's Equation**
Repeat Example 3.3.1 using Jacob's equation. For both the well and the 1 km radius, our criteria of $u < 0.01$ is valid, so the approximation should give good results.

$$s(0.2 \text{ m}, 365\text{d}) = \frac{1500 \text{ m}^3/\text{d}}{4\pi \times 600 \text{ m}^2/\text{d}} ln\left(\frac{(0.2 \text{ m}) \times 0.0004}{4 \times 600 \text{ m}^2/\text{d} \times 365\text{d}}\right) = 4.81 \text{ m}$$

$$s(1000 \text{ m}, 365\text{d}) = \frac{1500 \text{ m}^3/\text{d}}{4\pi \times 600 \text{ m}^2/\text{d}} ln\left(\frac{(1000 \text{ m})^2 \times 0.0004}{4 \times 600 \text{ m}^2/\text{d} \times 365\text{d}}\right) = 1.41 \text{ m}$$

Comparison shows that the Jacob equation does work well in this range.

3.3.4   Radius of Influence

The **radius of influence** might be defined as the distance from a pumping well to where the drawdown is negligible. Here, the term "negligible" is problem-dependent, and consequently the radius of influence in not a readily fixed quantity. A useful representation is found through Jacob's equation. For $s = 0$ in Equation 3.3.9, the argument of the logarithmic function must equal unity.

$$\frac{2.25Tt}{r^2 S} = 1$$

Using the radius that satisfies this equation as the radius of influence, R, we have

$$R = 1.5 \sqrt{\frac{Tt}{S}} \qquad (3.3.10)$$

The radius of influence of a well increases as the square root of time and is larger in formations with greater transmissivities and smaller storage coefficients. It may be observed from Equation 3.3.10 that the radius of influence is independent of the discharge. At first glance, this idea may not be intuitively apparent. This radius, however, is a measure of the distance of propagation of small energy (head) disturbances. In the limit, the movement of these small disturbances is independent of the discharge. The Theis equation shows that within the radius of influence, the drawdown is directly proportional to Q.

The definition of the radius of influence from Equation 3.3.10 is derived by extrapolating the distribution of drawdown predicted by the Jacob equation. In so doing, we have exceeded the criteria for validity of the approximation. Namely, Jacob's approximation is not valid for predicting small drawdown. We have seen that the logarithmic approximation is valid only when $u < 0.01$, or

$$r_{\text{Jacob}} = 0.2 \sqrt{\frac{Tt}{S}} \qquad (3.3.11)$$

For $r < r_{\text{Jacob}}$, the Jacob's logarithmic equation provides a good approximation to the exponential integral (Theis well function). In Equation 3.3.10, we have extrapolated the Jacob equation beyond the valid region. Beyond the radius $r = R$, the Jacob equation predicts a negative drawdown (a build-up in head) which is clearly impossible. By contrast, the Theis equation predicts zero drawdown at large distances. Nevertheless, the approximation of R based on Jacob's relation provides a useful estimate of this important parameter. This situation is shown schematically in Figure 3.3.2.

With the logarithmic approximation and the definition of the radius of influence given by Equation 3.3.10, the Theis equation becomes

Jacob
Approximation
Region

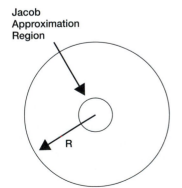

R

FIGURE 3.3.2 Jacob's logarithmic approximation is extrapolated out to where it predicts zero drawdown to define the raius of influence of a well

$$s(r,t) = \frac{Q}{4\pi T} \ln\left(\frac{R^2}{r^2}\right) = \frac{Q}{2\pi T} \ln\left(\frac{R}{r}\right)$$

which is the Thiem equation again.

Equation 3.3.10 provides one method for estimating the radius of influence. There are other empirical methods based on the hydraulic conductivity, recharge rates, and time of pumping (see Bear, 1979). Obtaining a reliable estimate of R remains a difficult problem. Because it appears within the logarithm, however, it is not a sensitive parameter.

EXAMPLE
PROBLEM

3.3.3. Radius of Influence

What is the radius of influence for Example 3.3.1? According to Equation 3.3.10, we have

$$R = 1.5 \sqrt{\frac{600 \text{ m}^2/\text{d} \times 365\text{d}}{0.0004}} = 35{,}100 \text{ m}$$

The actual drawdown at this distance is found from using the Theis equation. At a radius of 35,100 m, we have $u = 0.563$, and from Appendix C the Theis well function is $W(u) = 0.49$. Thus, the drawdown at 35 km from the well is $s = 0.097$ m.

Example 3.3.3 predicts that the radius of influence extends 35 km from a well after a period of one year. This distance is the considerable range from which water must be drawn in order to satisfy the required discharge. In reality, however, a drawdown cone will spread out continuously until it intercepts a source of recharge. After that time, the aquifer acts as a conduit carrying the flow from the recharge source to the well, and the flow field reaches a steady-state condition.

3.3.5 **Superposition with Transient Flow**

The continuity equations for steady state and transient flow in confined aquifers with constant effective parameters are, respectively:

$$\nabla^2(h) = 0 \tag{3.3.12}$$

$$\nabla^2(h) = \frac{S}{T}\frac{\partial h}{\partial t} \tag{3.3.13}$$

Both of these equations are linear in h. Thus, if h_1 and h_2 are solutions, then so is $h = h_1 + h_2$. The principle of **superposition** is not affected by transient conditions for confined and/or leaky aquifers.

For example, with steady uniform flow in the x-direction plus a well, we have for the steady flow

$$h_1 = Ax + B$$

and for the well

$$h_2 = h_o - \frac{Q}{4\pi T} W(u)$$

Due to superposition, we then have

$$h = Ax + B' - \frac{Q}{4\pi T} W(u)$$

The constant A fixes the regional gradient, while the coefficient B' fixes the initial elevation or level of the hydraulic head surface at $x = 0$. We next look specifically at problems with multiple wells and multiple pumping periods.

Multiple Wells

For a problem with multiple wells, the principle of superposition says the following: The drawdown due to multiple wells is equal to the sum of the individual well drawdown values.

EXAMPLE PROBLEM

3.3.4   Superposition with Multiple Wells

In this example, we calculate the drawdown at an observation well due to two pumping wells, as shown in Figure 3.3.3. The aquifer has $T = 300$ m^2/d and $S = 0.0006$. Well number 1, which is located at a distance of $r_1 = 30$ m from the observation well, produces water at a rate of $Q_1 = 550$ m^3/d, while well number 2, at a distance of $r_2 = 40$ m, produces water at a rate of $Q_2 = 1100$ m^3/d. The observation well drawdown at a time of $t = 200$ d is required.

According to the principle of superposition, we have

$$s_{obs} = \frac{Q_1}{4\pi T} W(u_1) + \frac{Q_2}{4\pi T} W(u_2)$$

For the given data,

$$u_1 = \frac{(30 \text{ m})^2 \times 0.0006}{4 \times 300 \text{ m}^2/\text{d} \times 200 \text{ d}} = 2.2 \times 10^{-6} \rightarrow W(u_1) = \ln\left(\frac{0.561}{u_1}\right) = 12.43$$

$$u_2 = \frac{(40 \text{ m})^2 \times 0.0006}{4 \times 300 \text{ m}^2/\text{d} \times 200 \text{ d}} = 4 \times 10^{-6} \rightarrow W(u_2) = \ln\left(\frac{0.561}{u_2}\right) = 11.85$$

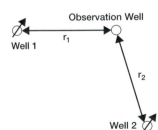

FIGURE 3.3.3   Drawdown at an observation well due to two pumping wells

Thus,

$$s_{obs} = \frac{550 \text{ m}^3/\text{d}}{4\pi(300 \text{ m}^2/\text{d})} (12.43) + \frac{1100 \text{ m}^3/\text{d}}{4\pi(300 \text{ m}^2/\text{d})} (11.85) = 1.81 \text{ m} + 3.46 \text{ m} = 5.27 \text{ m}$$

Well 1 causes 1.81 m of drawdown at the observation well, while well 2 causes 3.46 m, for a total drawdown of 5.27 m. Under these circumstances, we say that the drawdown cones of the wells overlap, or that the wells **interfere** with each other.

Multiple Pumping Rates

For a problem with multiple pumping rates, the principle of superposition says: The drawdown due to a changing pumping rate is equal to the sum of the drawdown values due to the "incremental changes" in rates.

The following example will help explain what incremental changes and incremental duration mean.

EXAMPLE
PROBLEM

3.3.5 Superposition with Multiple Pumping Rates

A well with an effective radius of 0.3 m produces from an aquifer with $T = 300 \text{ m}^2/\text{d}$ and $S = 0.0006$. The well pumps at a rate of 550 m³/d for 30 days, and then at a rate of 800 m³/d for an additional 30 days. The well is then shut off. Calculate the residual drawdown after 80 days from the start of pumping (20 days after the well was shut off).

The table below shows what is meant by incremental changes. Each change in rate is implemented by a new well function with the incremental discharge added to the existing set of well functions. The solution for this problem is represented as the result of three wells pumping at different rates and for different durations at the same location. The **Basic Rule** is that *Once a well function has been "turned on," it cannot be turned off.*

Pumping Schedule for Prediction of Drawdown at 80 Days

	Period 1	Period 2	Period 3
Time (days)	0–30	30–60	60–80
Q (m³/d)	550	800	0
ΔQ (m³/d)	550	250	−800
Duration Δt (days)	80	50	20

For the incremental rates and durations we have

$$u_1 = \frac{(0.3\ m)^2 \times 0.0006}{4 \times 300\ m^2/d \times 80\ d} = 5.6 \times 10^{-10}\ ;$$

$$u_2 = \frac{(0.3\ m)^2 \times 0.0006}{4 \times 300\ m^2/d \times 50\ d} = 9.0 \times 10^{-10}\ ;$$

$$u_3 = \frac{(0.3\ m)^2 \times 0.0006}{4 \times 300\ m^2/d \times 20\ d} = 2.2 \times 10^{-9}\ ;$$

and the residual drawdown is

$$s_{well} = \frac{550\ m^3/d}{4\pi \times 300\ m^2/d}\, ln\left(\frac{0.561}{5.6 \times 10^{-10}}\right) + \frac{250\ m^3/d}{4\pi \times 300\ m^2/d}\, ln\left(\frac{0.561}{9.0 \times 10^{-10}}\right)$$

$$- \frac{800\ m^3/d}{4\pi \times 300\ m^2/d}\, ln\left(\frac{0.561}{2.2 \times 10^{-9}}\right)$$

$$= 3.02\ m + 1.34\ m - 4.10\ m = 0.26\ m$$

Thus, a drawdown of 26 cm remains 20 days after the well has stopped pumping.

3.3.6 Image Well Theory

Image well theory is an application of the principle of superposition that allows us to represent simple types of boundaries for an aquifer. The theory is also applicable for transient problems. For example, if a production well is adjacent to a constant head boundary, the drawdown is given by

$$s_{well} = \frac{Q}{4\pi T}\left(W(u_r) - W(u_i)\right) \tag{3.3.14}$$

where u_r and u_i represent the real well and its image well, which is located opposite to the recharge boundary. As time increases, the Jacob approximation becomes valid for the real well ($u_r < 0.01$), and

$$s_{well} = \frac{Q}{4\pi T}\left(ln\left(\frac{2.25Tt}{r_r^2 S}\right) - W(u_i)\right)$$

Ultimately, the Jacob approximation becomes valid for the image well, and

$$s_{well} = \frac{Q}{4\pi T}\left(ln\left(\frac{2.25Tt}{r_r^2 S}\right) - ln\left(\frac{2.25Tt}{r_i^2 S}\right)\right) = \frac{Q}{2\pi T}\, ln\left(\frac{r_i}{r_r}\right)$$

This result is the same result found in Section 3.2.4 for steady flow conditions near a recharge boundary. Thus, by the time the Jacob approximation becomes valid for both the real and image wells, the drawdown has reached its steady value. Near the well, this occurs when $u_i < 0.01$, or

$$t_{\text{steady}} > 100 \frac{r_i^2 S}{4T} = 100 \frac{a^2 S}{T} \qquad \text{(3.3.15)}$$

where a is the distance to the body of water ($r_i = 2a$).

3.4   Pumping Tests

This section and those that follow provide a more detailed examination of **in situ** methods for estimating aquifer and well characteristics. One usually distinguishes two different types of pumping tests. **Aquifer tests** are used to determine field scale hydraulic characteristics of the aquifer, particularly its transmissivity and storage coefficient. *Well tests* are used to determine yield and well loss information for a particular well. Aquifer tests generally require an observation well or piezometer in which the water level response to an induced aquifer stress (pumping of a well) is recorded, although the aquifer transmissivity can be determined accurately with only pumped well data. Well tests require only pumped well data and are discussed later in this chapter.

3.4.1   Aquifer Tests

Estimation of aquifer characteristics through analysis of aquifer test data is a standard practice in the evaluation of groundwater resources. There is much art to successful aquifer testing and data interpretation, and an excellent reference is Kruseman and de Ridder (1991). In addition, Walton (1970, 1987) presents various case histories. The basic setup for an aquifer test is to pump one well with a known discharge and measure the drawdown in nearby observation wells that are generally located at distances from 10 m to 100 m from the pumped well. Figure 3.4.1 shows a possible well configuration, in addition to the type of data one typically obtains.

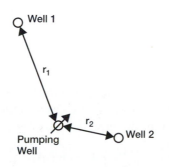

Time	Drawdown Well 1	Well 2
--	--	--
--	--	--
--	--	--
--	--	--
--	--	--
--	--	--
--	--	--

Data Sheet

FIGURE 3.4.1   Aquifer test data

Before the aquifer test is initiated, the water levels in monitoring wells are measured to establish baseline trends for natural changes that are occurring independent of the test. The measured data from the test are adjusted according to the observed baseline trends. For long-term aquifer tests, the barometric pressure is also measured. The aquifer piezometric levels naturally respond to atmospheric pressure changes, and one may use the barometric efficiency of the aquifer to correct for significant atmospheric pressure changes during the test. Water levels in monitoring wells are measured frequently at the beginning of the test, usually every 30 seconds or so, and the monitoring frequency decreases over the test duration with perhaps one measurement every 30 minutes after a few hours. Water levels may be measured using electric sounders, air lines and pressure gauges, calibrated steel tapes, and pressure transducers. The pumping well discharge must also be monitored, although not with the same frequency as water level measurements. Reliable results from analysis of the aquifer test measurements generally requires that the discharge should not vary more than plus or minus 5 percent.

The hydraulic characteristics that can be estimated from an aquifer test include the transmissivity, coefficient of storage, specific yield, horizontal and vertical hydraulic conductivity, and confining layer leakage. Estimation of the aquifer's transmissivity, storage coefficient, and other characteristics from an aquifer test involves direct application of the equations developed in this chapter. The response of an aquifer to pumping will depend on its characteristic parameters. The method of analysis of aquifer test data involves either graphical or computer estimation of parameter values using observed drawdown and well production rate records. With computer methods, the curve fitting is done automatically, although the analyst must choose which portions of the curve to use to avoid the inevitable non-idealities. Graphical methods are still important for practical and educational purposes. In particular, the Jacob or semi-logarithmic method presented in the following paragraph should be used as a standard tool while in the field to assess the performance of the aquifer test and choose its time of completion and recovery time. For confined aquifers, the two graphical methods include the log-log plot of drawdown versus time for the Theis method and the semilog plot for the Jacob method. Both methods are discussed here.

3.4.2   Theis Method

The **Theis method** is used for analysis of data from an aquifer test of a confined aquifer using a constant well discharge rate. The Theis method is based on Equation 3.3.6. Taking the logarithm of this equation gives

$$ln(s) = ln\left(\frac{Q}{4\pi T}\right) + ln(W(u)) \qquad \textbf{(3.4.1)}$$

The middle term in Equation 3.4.1 is a grouping of constants, so it is constant. Also, the logarithm of Equation 3.3.4 may be written

$$ln(t) = ln\left(\frac{r^2 S}{4T}\right) + ln\left(\frac{1}{u}\right)$$ (3.4.2)

The middle term in Equation 3.4.2 is also constant. Comparison of Equations 3.4.1 and 3.4.2 shows that the relation between $ln(s)$ and $ln(t)$ has the same form as the relation between $ln(W(u))$ and $ln(1/u)$. The two relations differ from each other by a constant factor. It follows that a plot of $ln(s)$ versus $ln(t)$, which may be obtained from field data, should look the same as a plot of $ln(W(u))$ vs. $ln(1/u)$. This correspondence is used in the Theis method to estimate the transmissivity and storage coefficient of an aquifer. The procedure for this graphical method is as follows:

i) Plot the function $W(u)$ versus $1/u$ on log-log paper. 3×5 cycle log-log paper is generally used. For typical aquifer test duration and parameters, the range of $1/u$ should extend from 0.1 to 10^4. The resulting curve is called a **type curve**. The Theis type curve is shown in Figure 3.4.2.

ii) Plot the measured drawdown versus time on log-log paper of the same size and scale as the type curve. This plot is called the **data curve**.

iii) Superimpose the field data curve on the type curve keeping the coordinate axes parallel. Adjust the curves up-and-down and right-and-left until most of the field data points fall on the type curve.

iv) Select an arbitrary **match point** and read off a paired set of values of $W(u)$, $1/u$, s, and t at the match point. For example, one could take the point

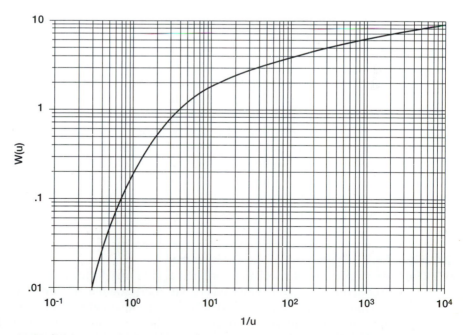

FIGURE 3.4.2 Type curve for aquifer testing using the Theis method for a confined aquifer

$W_{mp}(u) = 1$ and $1/u_{mp} = 1$ on the type curve as the match point and find the corresponding $s_{mp} - t_{mp}$ pair on the data curve.

(v) Using these values together with the pumping rate Q and the radial distance r from the pumping well to the observation well, calculate T from the Theis equation written as

$$T = \frac{QW_{mp}}{4\pi s_{mp}} \tag{3.4.3}$$

(vi) Calculate S from Equation 3.2.4 written as

$$S = \frac{4Tt_{mp}u_{mp}}{r^2} \tag{3.4.3}$$

The procedure that has just been described is of historical interest. A great deal of literature has been devoted to type curve methods. Today, however, these procedures are seldom used except for teaching purposes. In practice, software for *personal computers* (PCs) is usually used in the analysis of aquifer and well test data. Nevertheless, the important concepts remain the same. The following discussion shows how one may develop an analysis program using PC spreadsheet software.

A straightforward approach for analysis of aquifer test data using a PC spreadsheet application is to select different values of T and S and use the Theis model Equation 3.3.6 to predict the drawdown. Different T and S values are selected until a good fit between the predicted and observed drawdown is achieved. Adequacy of fit may be evaluated visually by plotting the observed data and the predicted curve, or numerical measures may be utilized, such as the sum of the squared differences between observed and predicted drawdown values. Once a plot of observed and predicted drawdown values is set up, it is automatically updated each time a new value of T or S is selected—allowing the user to quickly approach best-fit values.

In order to apply this approach, the Theis well function $W(u)$ must be evaluated for arbitrary values of u. The polynomial and rational approximations from Abramowitz and Stegun (1964) are presented in Table 3.4.1 in terms of a Basic language module. The magnitude of the approximation error for $u < 1$ and $u > 1$ is 2×10^{-7} and 2×10^{-8}, respectively. The module is written to define the function $W(u)$, and returns the value of this function when it is entered with a value of u from a spreadsheet cell.

EXAMPLE PROBLEM

3.4.1   Application of the Theis method

An aquifer test of a well in a confined aquifer was performed. The average well production rate was 18,000 ft³/d (12.5 ft³/min), and the drawdown was observed in a monitoring well located 60 ft from the production well. Pumping continued for four hours, and then recovery was observed for an additional six-hour period. The drawdown and recovery data is listed in Table 3.4.2. Estimate the transmissivity and storage coefficient using the Theis method.

TABLE 3.4.1 Numerical Evaluation of the Theis Well Function (Exponential Integral)

Function W(u)
  If $U < 1$ Then
    $t1 = -Log(u) - 0.57721566 + 0.99999193 \times u - 0.24991055 \times u \wedge 2$
    $t2 = 0.05519968 \times u \wedge 3 - 0.00976004 \times u \wedge 4 + 0.00107857 \times u \wedge 5$
    $W = t1 + t2$
  Else
    $t1 = u \wedge 4 + 8.5733287401 \times u \wedge 3 + 18.059016973$
        $\times u \wedge 2 + 8.6347608925 \times u + 0.2677737343$
    $t2 = u \wedge 4 + 9.5733223454 \times u \wedge 3 + 25.6329561486$
        $\times u \wedge 2 + 21.0996530827 \times u + 3.9584969228$
    $W = Exp(-u) \times t1 / (u \times t2)$
  End If
End Function

TABLE 3.4.2 Observed Drawdown versus Time for Aquifer Production/Recovery Test

t(min)	s(ft)	t(min)	s(ft)	t(min)	s(ft)	t(min)	s(ft)
0.6	0.12	45	4.05	240	6.03	300	1.91
1	0.35	60	4.41	241	5.71	315	1.68
2	0.78	75	4.64	242	5.25	330	1.52
4	1.36	90	4.85	245	4.48	345	1.41
8	2.18	120	5.12	250	3.78	360	1.32
12	2.45	150	5.35	255	3.29	400	1.11
24	3.25	180	5.71	270	2.58	455	0.91
36	3.85	210	5.91	285	2.18	545	0.69

This problem is analyzed using a PC spreadsheet, part of which is shown in Figure 3.4.3. This spreadsheet contains a graph that shows two data series: one that contains the observed data, and one that presents the Theis model for selected values of T and S. The left side of the sheet contains the observed data for the test. The box in the upper-left corner shows the discharge, observation well radius, and minimum and maximum times in the observed data record. The two lower columns on the left contain the observed time and drawdown records.

The minimum and maximum observation record times, t_{min} and t_{max}, are used to scale the range of the Theis model for a log-log plot. In order to expand the time range of the observed data, the lower time limit of the Theis model plot is $t_o = 0.95t_{min}$, while the upper limit of the plot is $t_{100} = 1.05t_{max}$. The section of the spreadsheet labeled 'MODEL RESULTS' has in the left column an integer series labeled 'Scale' that runs from zero to 100. The middle column contains for each member of this scale-series a corresponding time value that is scaled logarithmically and calculated from

$$t_i = e^{ln(0.95 \times t_{min}) + (ln(1.05 \times t_{max}) - ln(0.95 \times t_{min})) \times i/100}$$

Theis Analysis of Pumping Test Data			
Use Consistent Units for Length and Time			
Enter Data:		Estimate Parameters:	
$Q (L^3/T) =$	12.5	$T (L^2/T) =$	0.83
$r (L) =$	60	$S =$	0.0008
$t_{min} (T) =$	0.6		
$t_{max} (T) =$	545		

Enter Data:		MODEL RESULTS		
time	drawdown	Scale	t	s
0.6	0.12	0	0.5700	0.1160
1	0.35	1	0.6108	0.1351
2	0.78	2	0.6545	0.1561
4	1.36	3	0.7013	0.1791
8	2.18	4	0.7515	0.2042

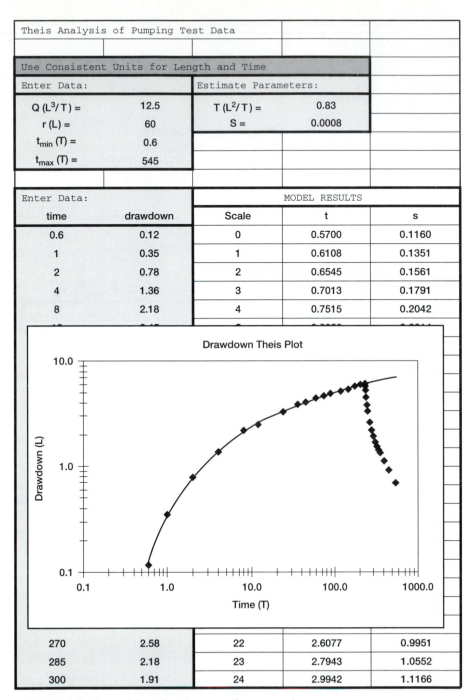

Drawdown Theis Plot

270	2.58	22	2.6077	0.9951
285	2.18	23	2.7943	1.0552
300	1.91	24	2.9942	1.1166

FIGURE 3.4.3 Spreadsheet for Theis aquifer test analysis

The column on the right contains the predicted drawdown values using the Theis model Equation 3.3.6 with $t = t_i$ and the well function W(u) calculated from the function listed in Table 3.4.1. The values to T and S for this calculation are obtained from the upper-right box labeled 'Estimate Parameters.' For each selection of T and S, a different model curve is drawn. Values are adjusted until a visually good fit is achieved. Figure 3.4.3 shows that the data is fit by values $T = 0.83$ ft^2/min $= 1000$ ft^2/d and $S = 0.0008$. It is not expected that the Theis model should fit the recovery data.

3.4.3 Jacob Method

As we have seen in Section 3.3.3, for $u < 0.01$, the Theis equation is approximated by Jacob's Equation 3.3.9. By invoking a simple rearrangement, we obtain

$$s = \frac{Q}{4\pi T} \ln(t) + \frac{Q}{4\pi T} \ln\left(\frac{2.25T}{r^2 S}\right) \tag{3.4.5}$$

Equation 3.4.5 has the form of a straight line ($y = mx + b$) on a semi-log graph with

$$\text{slope} = \text{m} = \frac{Q}{4\pi T} = \frac{s_2 - s_1}{\ln t_2 - \ln t_1}$$

Thus, a plot of drawdown versus time on semilog paper should appear as a straight line. The slope of the line provides the transmissivity:

$$T = \frac{Q}{4\pi} \frac{\ln\frac{t_2}{t_1}}{(s_2 - s_1)} \tag{3.4.6}$$

With the transmissivity provided by Equation 3.4.6, the $s = 0$ intercept may be used to estimate the storage coefficient. For the intercept at t_o, we have

$$\ln\left(\frac{2.25Tt_o}{r^2 S}\right) = 0$$

For the logarithmic function to equal zero, its argument must equal unity. Thus,

$$S = \frac{2.25Tt_o}{r^2} \tag{3.4.7}$$

Determination of T and S from a semilog plot with Equations 3.4.6 and 3.4.7 is called **Jacob's method**. One should note that the early-time data is not expected to fall on the straight line because it may not satisfy the assumptions on which the method is based, namely $u < 0.01$.

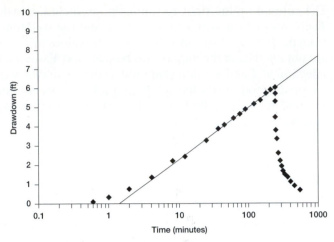

FIGURE 3.4.4   Jacob semilog plot of drawdown

EXAMPLE   **3.4.2   Application of Jacob's Method**
PROBLEM   The data from Example 3.4.1 are plotted on a semilog graph as shown in Figure 3.4.4. A straight line is drawn through the data points, and two points are read off this line. These two data points are $(s_o = 0, t_o = 1.4 \text{ min})$ and $(s_1 = 7.65 \text{ ft}, t_1 = 1000 \text{ min})$. With these two points, Equations 3.4.6 and 3.4.7 give

$$T = \frac{18000 \text{ ft}^3/\text{d}}{4\pi} \frac{ln\left(\dfrac{1000 \text{ min}}{1.4 \text{ min}}\right)}{7.65 \text{ ft} - 0 \text{ ft}} = 1230 \text{ ft}^2/\text{d}$$

$$S = \frac{2.25 \times 1230 \text{ ft}^2/\text{d} \times (1.4 \text{ min}/1440 \text{ min}/\text{d})}{(60 \text{ ft})^2} = 0.00075$$

3.4.4   **Recovery Tests**

Once a pumping well is shut off, the piezometric levels in an aquifer start to recover. Water level measurements during an aquifer test should continue through this period of **recovery,** because heads are not influenced by the possibly variable pumping rate—and because water levels in the pumping well may be used to estimate the transmissivity. The drawdown during the recovery period is calculated using the principle of superposition through the addition of an injection well starting at the shut-off time and having a discharge equal to the average production rate during the test. If the time of pumping is Δt, then

$$s = \frac{Q}{4\pi T}(W(u_1) - W(u_2))$$

$$u_1 = \frac{r^2 S}{4Tt} \; ; \; u_2 = \frac{r^2 S}{4T(t - \Delta t)}$$

A short period after the start of recovery, the Jacob equation becomes valid, and

$$s = \frac{Q}{4\pi T} ln\left(\frac{u_2}{u_1}\right) = \frac{Q}{4\pi T} ln\left(\frac{t}{t - \Delta t}\right) \qquad (3.4.8)$$

A plot of s versus $t/t\text{-}\Delta\tau$ is called a **Horner plot** (Horner, 1951), and this plot can be used to calculate T.

EXAMPLE
PROBLEM

3.4.3   Analysis of water-level recovery data

The recovery data from Example 3.4.1 is presented as a Horner plot in Figure 3.4.5. From the line drawn through the data points, we read off the point ($s = 2.6$ ft, $t/(t - \Delta t) = 9$). Using this value in Equation 3.4.8, we find

$$T = \frac{Q}{4\pi} \frac{ln\left(\dfrac{t}{t - \Delta t}\right)}{s} = \frac{18000 \text{ ft}^3/\text{d}}{4\pi} \frac{ln(9)}{2.6 \text{ ft}} = 1210 \text{ ft}^2/\text{d}$$

We cannot calculate S from the recovery data using this method.

3.4.5   Multiple Observation Wells

Often, one will have drawdown data from more than one observation well. Then, it is possible to carry through the aquifer test analysis for each well separately. If the drawdown data is plotted with s versus t, then the curves for the different wells are parallel. A look at both the Theis equation and the Jacob equation, however, shows that t and r appear only in the combination t/r^2. Thus, if the drawdown

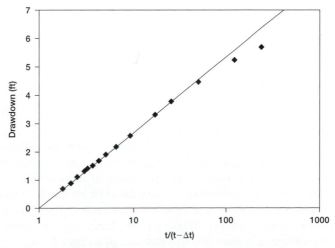

FIGURE 3.4.5   Horner plot of the water level recovery data

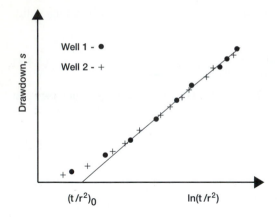

FIGURE 3.4.6 Semilog plot of s versus t/r^2 showing data sets as a single curve

data are plotted as s versus t/r^2, then all of the data curves should fall together. This idea is shown schematically in Figure 3.4.6. If this result does not happen, then one obtains further information as to heterogeneities in the aquifer.

3.4.6   Identification of Boundaries

The observed drawdown versus time data from an aquifer test can also be used to identify the location of any linear boundaries within the formation. Consider the case of a no-flow or barrier boundary. According to image well theory, the drawdown is given by

$$s = \frac{Q}{4\pi T}\left(W(u_r) + W(u_i)\right)$$

Initially, the drawdown at an observation well is caused by the pumping well alone, whereby the drawdown-time curve will correspond to that of a single well in an infinite aquifer. Later, however, the effects of the boundary are observed, and the drawdown exceeds that which would be caused by the well alone. This process is most easily seen on the semilog plot of Figure 3.4.7, where s_r represents the drawdown caused by the pumping well in a large aquifer without boundary effects and s_i represents the drawdown effects of the boundary (image well). Likewise, in the presence of a constant head boundary, the anomalous drawdown is less than that expected for a single well in an infinite aquifer, eventually reaching a steady state (see Equation 3.3.15).

   In order to locate the boundary, we want to analyze the drawdown curve and determine the part of the drawdown that is caused by the pumping well—and that which is due to the presence of the boundary. The line through the data points for times earlier than the time when the influence of the boundary was observed shows the drawdown curve as it would have occurred without the boundaries' presence. The departure from this line is due to the boundary, and its influence is modeled through use of an image well.

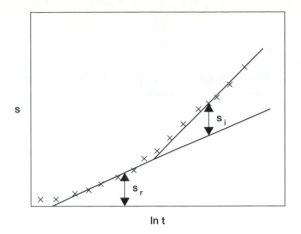

s

s_i

s_r

In t

FIGURE 3.4.7 Drawdown-time curve as influenced by the presence of a barrier boundary

We want to identify times t_r and t_i at which $s_r = s_i$, where s_r is the drawdown caused by the pumping well alone and s_i is the drawdown caused by the boundary or image well. The fact that the drawdown values are equal implies that $u_r = u_i$, or because T and S are constants for the aquifer,

$$\left(\frac{r^2}{t}\right)_r = \left(\frac{r^2}{t}\right)_i$$

This equation shows that the radius from the observation well to the image well is

$$r_i = r_r \sqrt{\frac{t_i}{t_r}} \tag{3.4.9}$$

where r_r is the distance from the observation well to the real well. Thus, the image well lies on a circle of radius r_i centered on the observation well. If two observation wells are available, then points of overlap for their corresponding circles (radii) gives the location of the image well. This analysis is shown in Figure 3.4.8. The boundary is located along the perpendicular bisector between the real well and the image well.

3.5   Other Well Functions

In this section, we will look at a couple of additional problems that will relax some of the assumptions necessary for the Theis solution. Through comparison with the Theis equation, we can identify the role and quantitative importance of various processes in well hydraulics.

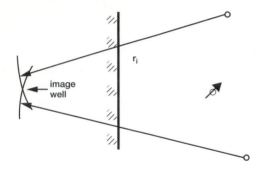

FIGURE 3.4.8 Location of a barrier boundary using aquifer test results

3.5.1   Leaky Aquifer with No Storage in Aquitard

The first problem we consider is **flow to a well in a leaky aquifer** with negligible storage within the confining bed. Further, it is assumed that the water potential in the adjacent aquifer remains constant and that the two aquifers are in equilibrium (same potentials) at the beginning of the pumping period. Neglecting storage in the confining bed means that drawdown in the production aquifer immediately induces flow across the aquitard at a rate proportional to the potential difference. This view is a simplified representation of the leaky aquifer system, in that storage in the confining bed may be significant, and the actual drawdown at any given time will be less than that predicted with this model. Nevertheless, the long-term drawdown is correct (given that the assumptions with regard to the adjacent aquifer are valid).

The continuity equation and initial and boundary conditions for radial flow to a well in a leaky aquifer with uniform properties are

$$\frac{S}{T}\frac{\partial h}{\partial t} = \frac{1}{r}\frac{\partial}{\partial r}\left(r\frac{\partial h}{\partial r}\right) - \frac{h - h_a}{B^2} \tag{3.5.1}$$

$$h(r,0) = h(\infty,t) = h_a$$

$$\lim_{r \to 0}\left[r\frac{\partial h}{\partial r}\right] = \frac{Q}{2\pi T}$$

where

$$B = \sqrt{\frac{Tb'}{K'}}.$$

This problem formulation is identical with the Theis formulation, except that leakage across the aquitard is included. Hantush (1956) has shown that the solution to this problem is given by

$$h_a - h(r,t) = s(r,t) = \frac{Q}{4\pi T}\int_u^\infty \frac{1}{\omega}e^{-\left(\omega + \frac{\beta^2}{4\omega}\right)}d\omega \equiv \frac{Q}{4\pi T}W(u,\beta) \tag{3.5.2}$$

where ω is the dummy variable of integration, u is the same argument as in the Theis well function, $\beta = r/B$, and $W(u,r/B)$ is the Hantush or **"leaky aquifer"** well function that is shown in Figure 3.5.1. For $\beta = r/B = 0$, which corresponds to the case without leakage, the Hantush well function reduces to the Theis well function. This function is the upper curve shown in Figure 3.5.1.

A consideration of our results in Section 3.2.3 for steady flow to a well in a leaky aquifer shows that the Hantush solution of Equation 3.5.2 must approach that of Equation 3.2.14 for long pumping times. Mathematically, this means that for small u (at large times, $u \to 0$)

$$\lim_{u \to 0} W(u,\beta) \to 2K_o(\beta)$$

where $K_o(\ )$ is the modified Bessel function of the second kind of order zero. For the purpose of computations, it is often easiest to calculate the Hantush well function from

$$W(u,\beta) = 2K_o(\beta) - \int_0^u \frac{1}{\omega} e^{-\left(\omega + \frac{\beta^2}{4\omega}\right)} d\omega \qquad (3.5.3)$$

A change of variables provides the following relationship that is also helpful for computations:

$$W(u,\beta) + W\left(\frac{\beta^2}{4u},\beta\right) = 2K_0(\beta) \qquad (3.5.4)$$

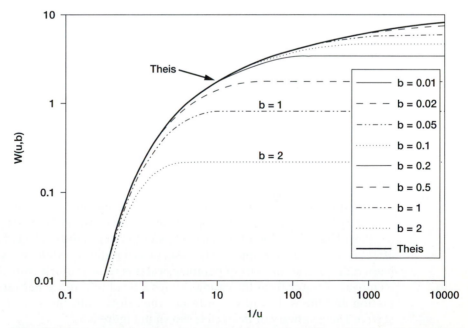

FIGURE 3.5.1   Hantush well function for a leaky aquifer without storage in the aquitard

EXAMPLE   **3.5.1   Application of the Hantush Well Function**
PROBLEM   The following information is available for a leaky aquifer: $T = 100$ m²/d, $S = 0.001$, $K' = 0.020$ m/d, and $b' = 3$ m. If a well is pumped at a rate of $Q = 550$ m³/d, what is the drawdown in an observation well located a distance of $r = 50$ m from the pumping well after two hours of pumping? Also, estimate the time required before the drawdown reaches its steady-state value at the observation well.

Example modules for calculating $K_o(\beta)$, $W(u,\beta)$ and $I_o(\beta)$ using a spreadsheet are presented in Appendix D, where $I_o(\beta)$ is the modified Bessel function of the first kind of order zero, and it is used in the calculation of $K_o(\beta)$. Once these modules have been set up, then analysis of leaky aquifer pumping test data can follow the same procedure outlined in Figure 3.4.3 for the Theis method of analysis. The spreadsheet can also be used for other calculations. For this example, the data give

$$u = \frac{(50 \text{ m})^2 \times 0.001}{4 \times 100 \text{ m}^2/\text{d} \times (2 \text{ hr}/24 \text{ hr/d})} = 0.075,$$

$$B = \sqrt{\frac{100 \text{ m}^2/\text{d} \times 3 \text{ } m}{0.02 \text{ m/d}}} = 122.5 \text{ m},$$

and $\beta = r/B = 0.408$. From the spreadsheet, we find that $W(0.075,0.408) \cong 1.72$. Thus, the drawdown at 2 hours is

$$s = \frac{550 \text{ m}^3/\text{d}}{4\pi \times 100 \text{ m}^2/\text{d}} \times 1.72 = 0.75 \text{ m}$$

After a long period of pumping, the well function is $W(0,0.408) = 2K_o(0.408) = 2.194$, and $s_{\text{steady}} = 0.960$ m. To estimate the time-to-steady state, which is approached asymptotically, we use the time to reach 95 percent of the steady-state value, $W \cong 2.194 \times 0.95 = 2.084$. By trial and error, the value of u for which $W = 2.084$ is found to be $u_{0.95} = 0.0293$. The time to reach 95 percent of steady state is

$$t = \frac{(50 \text{ m})^2 \times 0.001}{4 \times 100 \text{ m}^2/\text{d} \times 0.0293} = 0.213 \text{ days} = 5.12 \text{ hours}$$

3.5.2   Effect of Well Storage

A second relaxation of the assumptions required for the Theis equation is presented by Papadopoulos and Cooper (1967), who considered the drawdown in a well with large diameter for which well storage cannot be neglected (the same effect is seen when T is small). The casing has a radius r_c, while the well screen has a radius r_w. At the start of pumping, water is removed first from the casing. Removal of water from the casing creates a head difference between the formation and the well, and groundwater then flows into the well through the screen. The geometry of the well is shown in Figure 3.5.2.

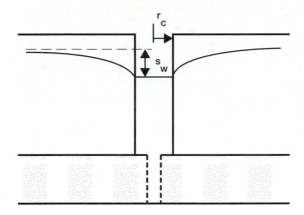

FIGURE 3.5.2 Configuration for drawdown in a large diameter well with well radius r_w and casing radius r_c

The difference between this problem and the Theis formulation concerns the boundary condition at the well. The discharge from the well comes from both decreases in the well casing storage plus inflow from the formation. This concept is expressed through the following boundary condition:

$$Q = \pi r_c^2 \frac{ds_w}{dt} - 2\pi r_w T \frac{\partial s(r_w, t)}{\partial r} \qquad (3.5.5)$$

Equation 3.5.5 is valid for $t > 0$. In this equation, $s_w(t)$ is the drawdown within the well, and $s(r_w, t)$ is the formation drawdown evaluated at the well radius. The first term on the right represents the removal of water from storage within the well casing, while the second term represents the flow into the well from the formation. The continuity equation, initial condition, and boundary condition at infinity remain the same. Papadopoulos and Cooper (1967) give the drawdown within the well as

$$s_w = \frac{Q}{4\pi T} F(u_w, \alpha) \qquad (3.5.6)$$

where

$$u_w = \frac{r_w^2 S}{4Tt} \quad , \quad \alpha = \frac{r_w^2 S}{r_c^2},$$

and $F(u_w, \alpha)$ is the **large diameter well function**. The large-diameter well function is shown in Figure 3.5.3.

For $(u_w/\alpha) < 10^{-3}$, or

$$t > 250 \frac{r_c^2}{T}$$

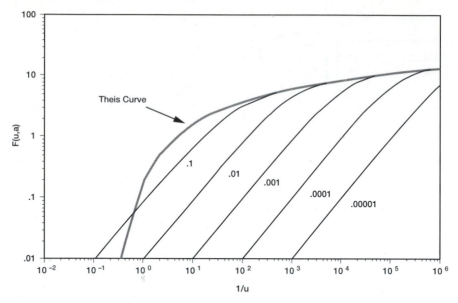

FIGURE 3.5.3 Well function for a large diamter well, $F(u,\alpha)$

the function $F(u_w,\alpha)$ can be closely approximated by the Theis well function: $F(u_w,\alpha) = W(u_w)$. For wells of small diameter and/or aquifers of high transmissivity, this period is small. For wells of large diameter and/or aquifers of low transmissivity, however, this period is considerably larger. For small times (large u_w), the approximations are

$$F(u_w,\alpha) = \frac{\alpha}{u_w}$$

or

$$s_w = \frac{Qt}{\pi r_c^2}$$

which represents conditions under which all of the water pumped is derived from storage within the well casing. The function $F(u_w,\alpha)$ is tabulated by Papadopoulos and Cooper (1967), Marsily (1986), and Kruseman and de Ridder (1991).

EXAMPLE PROBLEM

3.5.2 Influence of Storage in a Large-Diameter Well

The following data is assumed: $Q = 550$ m³/d, $T = 100$ m²/d, $S = 0.001$, and $r_c = r_w = 0.20$ m. How long does well storage influence the observed well drawdown?

Equation 3.5.6 may be replaced by the Theis equation when $t > 250 (0.2 \text{ m})^2/100$ m²/d = 0.1 days = 2.4 hours. Thus, after 2.4 hours, the influence of well storage is negligible. To find the time period within which water is withdrawn primarily from the well casing, note that the linear portion of the curves shown in Figure 3.5.3 corresponds to the condition $F(u_w,\infty) = \infty/u_w$, and this condition is the one for which all of the withdrawn water comes from the well casing. For the specified data, $\alpha = S = 0.001$, and the corresponding curve is linear in Figure 3.5.3 for $1/u$ values up to 10^3. This situation gives $u_w = 0.001$ and

$$t = \frac{(0.2 \text{ m})^2 \times 0.001}{4 \times 100 \text{ m}^2/\text{d} \times 0.001} = 10^{-4} \text{ d} = 0.14 \text{ minutes.}$$

This duration is fairly short.

3.5.3   Unconfined Aquifers

When water is pumped from a confined aquifer, the pumpage creates draw-down in the piezometric surface that induces hydraulic gradients toward the well. The induced flow moves horizontally toward the well; there are no verti-cal components of flow. The water produced from the well comes from storage releases associated with the compressibility of the porous matrix and water. There is no dewatering of the geologic formation.

When water is pumped from an unconfined aquifer, on the other hand, the pumpage causes a drawdown in the water table itself, and the induced hydraulic gradients cause both horizontal and vertical components of flow. The water pro-duced by the well comes from the compressibility of the porous matrix and water —plus the actual dewatering of the unconfined aquifer. The early time storage release comes from the elastic behavior of the aquifer formation and water and is characterized by the coefficient of storage, S. The piezometric surface drops just as it does for a confined aquifer, except that for an unconfined aquifer, the ini-tial piezometric surface coincides with the water table. The drop in the piezo-metric surface sets up a vertical hydraulic gradient in the groundwater near the water table. At moderate times, the vertical gradient near the water table caus-es drainage of the porous matrix; therefore, pumpage is supplied directly from this drainage, and the rate of decline in the piezometric surface slows or stops. Water table decline is a function of both vertical and horizontal hydraulic con-ductivity. Finally, at late times, the rate of water table decline decreases and flow is essentially horizontal. Most of the pumpage is supplied by drainage of the for-mation occurring over a large area and is characterized by the specific yield, S_y.

Boulton (1963) produced a semi-empirical mathematical model that de-scribes the observed drawdown behavior seen during pumping tests in uncon-fined aquifers. His solution has been useful in practice (see Walton, 1970). Boulton's solution, however, required the definition of an empirical **delay index** that was not related clearly to any physical phenomenon. Later, work by Neu-man (1972, 1975) and others showed that the delayed response in unconfined aquifers is related to the vertical components of flow that are induced in the flow system. Rather than being characterized by a constant delay index, the response to pumping is a function of the ratio of horizontal to vertical permeability, ra-dius from the pumping well, and the aquifer saturated thickness. A basic as-sumption in Neuman's model is that the drawdown s is small compared with H.

The model also assumes that $S_y \gg S$. The mathematical problem statement developed by Neuman is given as follows:

$$\frac{S_s}{K_r}\frac{\partial s}{\partial t} = \frac{1}{r}\frac{\partial}{\partial r}\left(r\frac{\partial s}{\partial r}\right) + \frac{K_z}{K_r}\frac{\partial^2 s}{\partial z^2} \tag{3.5.7}$$

$$\frac{\partial s}{\partial z} = 0 \text{ at } z = 0 \tag{3.5.8}$$

$$K_z\frac{\partial s}{\partial z} = -S_y\frac{\partial s}{\partial t} \text{ at } z = H \tag{3.5.9}$$

$$\lim_{r \to 0}\left[r\frac{\partial h}{\partial r}\right] = \frac{Q}{2\pi K_r H} \tag{3.5.10}$$

$$s \to 0 \text{ as } r \to \infty \tag{3.5.11}$$

Equation 3.5.7 is the same as Equation 2.3.33 written in radial coordinates and states the continuity relation for unsteady two-dimensional flow in radial and vertical coordinates. Equation 3.5.8 says that the base of the aquifer is a no-flow boundary. Equation 3.5.9 is the boundary condition at the water table. Because the right of Equation 3.5.9 is the Darcy velocity, it states that

$$q_z = S_y\frac{\partial s}{\partial t} \text{ at } z = H$$

Essentially, this equation provides a diffuse source as a boundary condition and says that the source strength is equal to the rate of drainage at the water table. Equation 3.5.10 is the boundary condition at the well that is assumed to have no well-casing storage. Finally, Equation 3.5.11 says that far from the well, pumping does not influence the water level.

The solution to Equation 3.5.7–3.5.11 may be written as follows:

$$s(r,t) = \frac{Q}{4\pi T}W(u_A,u_B,\eta) \tag{3.5.12}$$

In Equation 3.5.12, $W(u_A,u_B,\eta)$ is the Neuman well function for an **unconfined aquifer**, the parameter u_A corresponds to the early time behavior that is characterized by the storage coefficient, and the parameter u_B corresponds to the late time behavior that is characterized by the specific yield:

$$u_A = \frac{r^2 S}{4Tt} \tag{3.5.13}$$

$$u_B = \frac{r^2 S_y}{4Tt} \tag{3.5.14}$$

The third parameter, η, plays the role of the delay index and is defined by Equation 3.5.15:

$$\eta = \frac{K_z r^2}{K_r H^2} \qquad (3.5.15)$$

The assumption that $S_y \gg S$ means that the solution consists of two separate parts. The first part is associated with the transition from elastic storage release to drainage associated with vertical flow near the water table. This solution may be designated by

$$W(u_A, u_B, \eta) = W(u_A, \eta) \qquad (3.5.16)$$

According to this early time behavior, the drawdown departs from the Theis curve based on the storage coefficient as recharge is obtained from vertical flow near the water table. At a particular location, a steady-state condition is reached in which the water withdrawn by pumpage is supplied by vertical drainage.

The second part of the solution includes the transition from drainage associated with vertical flow to drainage associated with primarily horizontal flow when the water table decline has slowed down sufficiently. This part of the solution may be designated by

$$W(u_A, u_B, \eta) = W(u_B, \eta) \qquad (3.5.16)$$

Both of these solutions are shown in Figure 3.5.4 for values of the delay parameter, η. The early time behavior is shown on the right of the figure with the well function departing from the Theis solution and leveling off, much as it does for

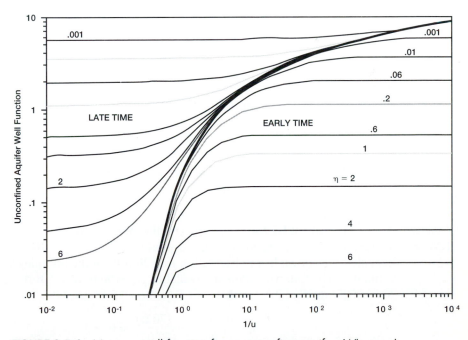

FIGURE 3.5.4 Neuman well function for an unconfine aquifer, $W(u_A, u_B, n)$

the leaky aquifer solution. The late time behavior is shown on the left of the figure with the well function changing from a steady-state condition toward that of the nonequilibrium Theis equation, except that now the storage constant is given by the specific yield of the aquifer.

The assumptions contained within either the solutions of Boulton or Neuman appear to be severe. Questions focus on the role of the unsaturated zone in the storage release at the water table and on the influence of a varying saturated thickness of the aquifer near the pumping well. Analytical results from Kroszynski and Dagan (1975) and numerical mathematical model simulations from Cooley (1971) and Brutsaert et al. (1971) have shown that the position of the water table during pumpage is not substantially affected by the nature of the unsaturated flow above the water table. The influence of the varying saturated thickness near the pumping well is not as easily addressed. Solutions that include the effects of this varying saturated thickness can only be obtained numerically. Nevertheless, as long as the drawdown does not become *too* great, then one may use the corrected drawdown, introduced in Section 3.2, (Equation 3.2.10). With the **corrected drawdown**, the long-term drawdown may be approximated by

$$s' = \frac{Q}{4\pi K_r H_R} W(u_B) \tag{3.5.18}$$

where u_B is given by Equation 3.5.14, H_R is the aquifer saturated thickness at a distance beyond the influence of the pumping well, and the corrected drawdown is given by Equation 3.2.9. In Equation 3.5.18, $W(u)$ is the standard Theis well function. In particular, for $u_B < 0.01$, we have

$$s' = \frac{Q}{4\pi K_r H_R} ln\left(\frac{2.25 K_r H_R t}{r^2 S_y}\right) \tag{3.5.19}$$

Importantly, Equation 3.5.19 also suggests that we may estimate the radius of influence of a well in an unconfined aquifer from

$$R = 1.5 \sqrt{\frac{K_r H_R t}{S_y}} \tag{3.5.20}$$

With Equation 3.5.20, the approximate solution given by Equation 3.5.19 again looks like the Thiem equation.

EXAMPLE PROBLEM **3.5.3   Drawdown History at an Observation Well in an Unconfined Aquifer**

Water is produced from an unconfined aquifer at a rate of $Q = 1000$ ft³/d, and the expected drawdown in an observation well 70 ft from the pumping well is to be estimated. The following data are available: $H_R = 90$ ft, $K_r = 5$ ft/d, $K_z = 0.5$ ft/d, $S = 0.0005$, and $S_y = 0.2$. For this data, Equation 3.5.15 gives

$$\eta = \frac{0.5 \text{ ft/d} \times (70 \text{ ft})^2}{5 \text{ ft/d} \times (90 \text{ ft})^2} = 0.060$$

For early times, Equation 3.5.13 provides the relationship between time and the variable u_A:

$$t = \frac{1}{u_A} \frac{r^2 S}{4T} = 0.00136 \frac{1}{u_A} \text{ days}$$

where the transmissivity is based on the horizontal or radial hydraulic conductivity, $T = K_r H_R$. The drawdown is given by Equations 3.5.12 and 3.5.16 as

$$s = \frac{Q}{4\pi T} W(u_A, \eta) = 1.77 \, W(u_A, \eta) \text{ ft}$$

The relation between $1/u_A$ and $W(u_A, \eta)$ is provided in tables of the Neuman well function. For an example, see Neuman (1975), Kruseman and de Ridder (1991), de Marsily (1986), or Fetter (1994). Similarly, for late times, we have from Equation 3.5.14 that

$$t = \frac{1}{u_B} \frac{r^2 S_y}{4T} = 0.544 \frac{1}{u_B} \text{ days}$$

and with Equations 3.5.12 and 3.5.17, $s = 1.77 W(u_B, \eta)$ ft. The drawdown history in the observation well predicted by these equations is shown in Figure 3.5.5.

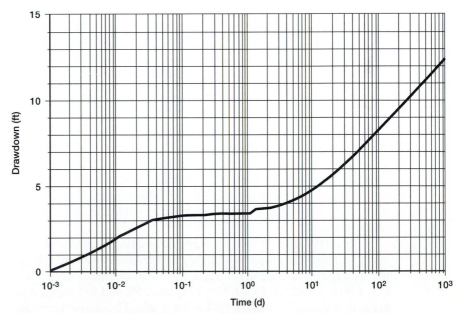

FIGURE 3.5.5 Drawdown history at an observation well in an unconfined aquifer

The early- and late-time behavior is easily distinguishable. The drawdown responds fairly rapidly to pumpage during the early time period reaching a drawdown of more than 3 ft within 1 hour (0.042 days). Local gravity drainage supplies the water produced by the well during the period from about 0.1 to 1.0 days, and the drawdown does not increase during this period of significant vertical flow. This feature is the characteristic delayed yield associated with unconfined aquifers. From about day 1 onward, the drawdown again increases. Here again, the flow is primarily horizontal—with the storage release associated with the specific yield occurring primarily from regions beyond the observation well.

One can distinguish two sections of the drawdown curve shown in Figure 3.5.5 that are nearly linear on this semi-log plot. The early section up to about 0.03 days is associated with elastic storage and is characterized by the coefficient of storage. The later section from about 20 days onward is characterized by the specific yield. For this section, Equation 3.5.19 may be used to predict the drawdown. This equation predicts that the corrected drawdown is 10 ft at about 270 days. The corresponding drawdown is given by

$$s = H_R\left(1 - \sqrt{1 - \frac{2s'}{H_R}}\right) = 10.63 \text{ ft}$$

so that the correction may not be significant at this time for many applications (it is about 6 percent for this case).

Neuman's model required that $S_y >> S$, and for this problem this condition is nearly satisfied. When this condition is not satisfied, the drawdown predicted by the early-time solution does not level off before the drawdown from the late time solution starts to increase again. This situation means that a period dominated by local storage release and vertical flow does not occur, and horizontal flow components are always important.

EXAMPLE PROBLEM   **3.5.4   Evaluation of Approximating Model Equation 3.5.18**

Equation 3.5.18 is an approximating model for estimation of the drawdown in an unconfined aquifer due to well pumpage. The model uses the corrected drawdown to account for changes in aquifer saturated thickness. Evaluate the accuracy of this model through comparison with the MODFLOW numerical simulation model.

The MODFLOW model was set up as a single, unconfined layer with a grid using 15 rows and columns. The row and column sizes are telescoped towards the well that is located in cell 1,1,1, as shown in Figure 3.5.6. The cell sizes are: 1, 1, 2, 4, 8, 16, 32, 32, 64, 64, 128, 128, 256, 256, and 512 m, for a simulation domain that is approximately 1.5 km on a side. The initial saturated thickness is $H_R = 20$ m, and the aquifer hydraulic conductivity and specific yield are 5 m/d and 0.15, respectively. The well produces water at a rate of 1000 m³/d.

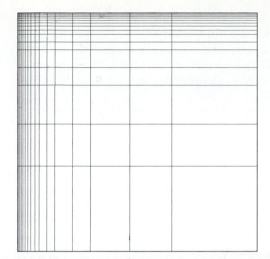

FIGURE 3.5.6 Simulation domain for MODFLOW model

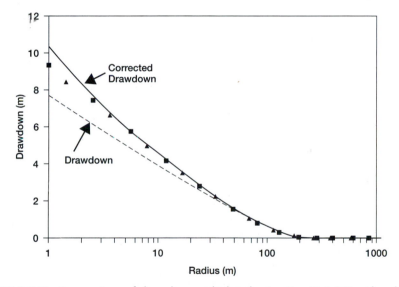

FIGURE 3.5.7 Comparison of drawdown calculated using Eq. (3.5.18) with calculated drawdown using the MODFLOW numercial model at a time of 10 days

The simulation domain shown in Figure 3.5.6 covers one quadrant of the flow domain towards the well. Thus, the actual well discharge specified in the MODFLOW simulation is 250 m³/d. A consideration of the flow geometry near the well, however, shows that the actual cross-section area for flow is greater than that from one quadrant, and the simulated drawdown near the well should be less than one would predict with a radial flow model. The results of the simulation at a time of 5 days are shown in Figure 3.5.7. This figure shows the drawdown as a function of distance from the well from cells located along the domain boundary (squares) and along the 45 degree line that bisects the domain. The drawdown and corrected drawdown distributions calculated using the Theis equation are also shown. At distances beyond a few meters from the well, the comparison between MODFLOW and corrected drawdown distributions is good. As anticipated, the MODFLOW model predicts smaller drawdown values immediately near the well.

3.5.4   Partially Penetrating Wells

A partially penetrating well is a well whose screen does not extend over the entire thickness of an aquifer. The drawdown in such a well is equal to that due to the formation losses associated with horizontal flow to the well, plus the additional losses associated with the converging and vertical flow near the well screen. The total aquifer head losses may be expressed by

$$s_p = s + \Delta s_p \tag{3.5.21}$$

where s_p is the drawdown in a **partially-penetrating well**, s is the drawdown in a fully-penetrating well, and Δs_p is the additional losses associated with the partial penetration. We may define the amount of penetration by

$$p = L/b \tag{3.5.22}$$

where L is the screen length and b is the aquifer thickness, as shown in Figure 3.5.8.

One of the earlier results for partially-penetrating wells is due to Kozeny (1933; see Muskat, 1946, p. 274). Driscoll (1986) presents this result as

$$\frac{Q/s_p}{Q/s} = p\left(1 + 7\sqrt{\frac{r_w}{2L}}\cos\frac{\pi p}{2}\right) \tag{3.5.23}$$

where r_w is the well radius. The ratio Q/s is called the **specific capacity** (see Section 3.6).

Huisman (1972) presents an alternative formulation (attributed to the Netherlands' Hydrological Colloquium, 1964) for the partial penetration losses. These losses are given for Δs_p as

$$\Delta s_p = \frac{Q}{2\pi T}\left(\frac{1-p}{p}\right)\ln\left(\frac{(1-p)L}{r_w}\right) \tag{3.5.24}$$

With the full penetration losses calculated with the Thiem equation, we have

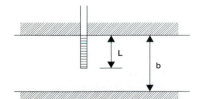

FIGURE 3.5.8 Well partially penetrating the upper part of a confined aquifer

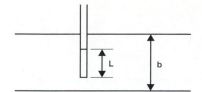

FIGURE 3.5.9 Well with screen centered with a confined aquifer

$$s_p = \frac{Q}{2\pi T}\left(ln\left(\frac{R}{r_w}\right) + \left(\frac{1-p}{p}\right) ln\left(\frac{(1-p)L}{r_w}\right)\right)$$

With this form, we may write

$$\frac{Q/s_p}{Q/s} = \frac{1}{1 + \left(\frac{1-p}{p}\right)\dfrac{ln\left(\dfrac{(1-p)L}{r_w}\right)}{ln\left(\dfrac{R}{r_w}\right)}} \qquad (3.5.25)$$

If the well is screened over the center of the aquifer, as shown in Figure 3.5.9, then a different formula must be used. Huisman (1972) gives

$$\Delta s_p = \frac{Q}{2\pi T}\left(\frac{1-p}{p}\right) ln\left(\frac{(1-p)L}{2r_w}\right) \qquad (3.5.26)$$

Note that this equation follows by symmetry from the formula with penetration at the top of the aquifer. Huisman (1972) also gives the corrections for other placements of the screen within the aquifer.

EXAMPLE PROBLEM **3.5.4 Calculation of Partial Penetration Losses**

Consider the two well completions shown in Figure 3.5.10. For each, the total screen length is 20 m. If we take $r_w = 0.15$ m and $R = 150$ m, then for Well 2, the Kozeny Equation (3.5.23) gives

$$\frac{Q/s_p}{Q/s} = 0.5\left(1 + 7\sqrt{\frac{0.15 \text{ m}}{2 \times 20 \text{ m}}} \cos\frac{\pi}{4}\right) = 0.68$$

For Well 1, by symmetry, the completion corresponds to a 10 m screen centered within a 20 m thick aquifer, and the same equation gives

$$\frac{Q/s_p}{Q/s} = 0.5\left(1 + 7\sqrt{\frac{0.15 \text{ m}}{2 \times 10 \text{ m}}} \cos\frac{\pi}{4}\right) = 0.75$$

A similar calculation shows that for four screen intervals, each of length 5 m, the specific capacity is 86 percent of the maximum value. This example shows that one can achieve a significant part of the maximum potential capacity from a well by spreading the screened intervals across the aquifer thickness.

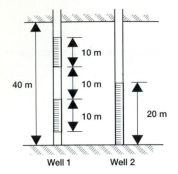

FIGURE 3.5.10   Two well screen configurations for example

One obvious limitation of the Kozeny equation is that it does not take into account the location of the screened interval within the aquifer. Examination of Figures 3.5.8 and 3.5.9 shows that there is less vertically induced flow for Figure 3.5.9, and correspondingly, less partial-penetration losses. These effects are included in Equations 3.5.24 and 3.5.26. For Well 2 in Figure 3.5.10, Equation 3.5.25 gives

$$\frac{Q/s_p}{Q/s} = \frac{1}{1 + \left(\dfrac{1 - .5}{.5}\right)\dfrac{ln\left(\dfrac{(1 - .5)20 \text{ m}}{.15 \text{ m}}\right)}{ln\left(\dfrac{150 \text{ m}}{.15 \text{ m}}\right)}} = 0.62$$

while for Well 1, which corresponds to Figure 3.5.9 rather than Figure 3.5.8 with a screen length of 10 m centered in a 20 m thick aquifer, we find

$$\frac{Q/s_p}{Q/s} = \frac{1}{1 + \left(\dfrac{1 - .5}{.5}\right)\dfrac{ln\left(\dfrac{(1 - .5)10 \text{ m}}{2(.15 \text{ m})}\right)}{ln\left(\dfrac{150 \text{ m}}{.15 \text{ m}}\right)}} = 0.71$$

For comparison, this same equation shows that with a 20 m screen centered within the aquifer, 66 percent of the maximum specific discharge can be obtained.

3.6   Slug Tests

On many occasions, it is not convenient to run an aquifer test requiring continuous pumpage to evaluate aquifer parameters. For example, in low permeability units, the well might not yield sufficient discharge for the pump to continue to operate. When characterizing a hazardous waste site, a conventional aquifer test would require treatment and proper disposal of all of the water produced. In these types of cases, it is much more convenient to develop alternative testing methods.

Slug tests, borehole tests, and rate-of-rise techniques provide alternative testing methods, although with potentially significant limitations. The basic idea is that a volume or slug of water is suddenly removed from or added to a well, and the rate of rise or fall of the water level in the well is measured. From this rate of recovery, one can estimate the permeability character of the formation and possibly its storage characteristics also, although generally with less precision. Essentially, instantaneous lowering of the water level in a well can be achieved by quickly removing the water with a bailer. Another method of lowering the water level, which may be more appropriate at hazardous waste sites where the water may contain various contaminants, is to submerge an object below the water level in the well, let the water level return to equilibrium, and then quickly remove the object. If the aquifer is permeable, the water level in the well may rise rapidly. If this situation is the case, then the level may be measured by using sensitive pressure transducers and fast-response recording equipment. In this section, we will look at three of the most widely used slug tests. For further discussion and references, one might consult Bouwer and Jackson (1974) and Bouwer (1978).

Hvorslev's Method

Hvorslev (1951) provided one of the earlier studies of borehole methods that is still widely used. Hvorslev was primarily concerned with use of water level measurements in boreholes to obtain pore-water pressures for design of foundations and earth structures. His investigations of the time-lag before useful measurements can be made have led to a method for estimation of permeability within the vicinity of the borehole, since the permeability is the primary control on the water flow rate near the observation point and the rate of pressure equilibration. The general configuration for application of Hvorslev's analysis is shown in Figure 3.6.1. For $t < 0$, the water level in the borehole would equal the water table elevation. At time $t = 0$, the water level declines by an amount y_o. The water level will ultimately recover again to the elevation of the water table, while the decline in water level at any time t is $y(t) = y_t$.

An application of the continuity principal shows that the relation between the water level within the borehole and the inflow rate, Q, is

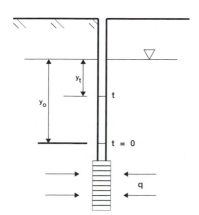

FIGURE 3.6.1 Configuration for Hvorslev method

$$Q = -A\frac{dy}{dt} \qquad (3.6.1)$$

where $A = \pi r_c^2$ is the cross-section area of the borehole, and r_c is the radius of the casing. Application of Darcy's Law gives

$$Q = FKy \qquad (3.6.2)$$

where F is a factor that depends on the shape and dimensions of the intake or well point, K is the hydraulic conductivity in the vicinity of the intake, and y represents the potential difference between fluid outside of the borehole away from the intake and that within the borehole. Combining these equations, we find

$$-\frac{dy}{y} = \frac{FKdt}{A}$$

This equation may be integrated with $y(0) = y_o$ to give

$$ln\left(\frac{y}{y_0}\right) = -\frac{t}{T} \leftrightarrow y = y_o e^{-t/T} \qquad (3.6.3)$$

where Hvorslev defines the "**basic time lag**" as the quantity

$$T = \frac{A}{FK} \qquad (3.6.4)$$

A physical interpretation of the basic time lag is achieved by noting that it may be written

$$T = \frac{Ay_o}{FKy_o} = \frac{V}{Q_o}$$

where V is the initially displaced volume of water and Q_o is the initial inflow rate. Thus, T is the time that would be required to re-establish equilibrium if the inflow rate remained constant at its initial value.

Equation 3.6.3 may be used to solve for K using a semilog graph of $ln(y)$ versus t.

$$K = \frac{A}{Ft} ln\left(\frac{y_0}{y}\right) \qquad (3.6.5)$$

This equation is the basis of the **Hvorslev method** for estimating the permeability. It is similar to the falling-head permeameter considered in Section 2.2.6. In particular, it is noted that when $y/y_o = e^{-1} = 0.37$, then $t = T$, and Equation 3.6.5 takes the simple form $K = A/(FT)$.

Hvorslev (1951) presents the shape factor F for a number of configurations. Configurations for a spherical cavity at the end of a borehole and a borehole extended in a uniform soil are shown in Figure 3.6.2, along with their corresponding shape factors. For $L > 8D$, the case shown in Figure 3.6.2(b) may be approximated by

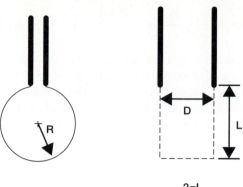

FIGURE 3.6.2   Configuration for shape factors; (a) spherical cavity and (b) hole extended in uniform soil (from Hvorslev, 1951)

$$F = 4\pi R$$

$$F = \frac{2\pi L}{ln\left(\dfrac{L}{D} + \sqrt{1 + \left(\dfrac{L}{D}\right)^2}\right)}$$

(a)   (b)

$$F = \frac{2\pi L}{ln\left(\dfrac{L}{r_c}\right)} \tag{3.6.6}$$

where $r_c = D/2$ is the hole or casing radius. For this case, Equation 3.6.5 may be written

$$K = \frac{r_c^2 ln(L/r_c)}{2Lt} \; ln\left(\frac{y_0}{y}\right) \tag{3.6.7}$$

Cooper's Method

Cooper et al. (1967) present an exact solution for the water level change following an instantaneous addition of water to a fully penetrating well in a confined aquifer. Such a test could be used in determining the hydrologic properties of low transmissivity formations, mostly in relation with deep-well waste disposal studies. The configuration is the same as that shown in Figure 3.5.2. The continuity equation remains unchanged. The problem is defined by the boundary and initial conditions. With the head datum selected as the equilibrium level, the initial and boundary condition far from the well takes the form $h(r,0) = h(\infty,t) = 0$, where $h(r,t)$ is the potential within the aquifer. With $H(t)$ equal to the water level in the well, its initial value is related to the volume of water added to the well, V, and the casing radius through $H(0) = H_o = V/(\pi r_c^2)$. The water level in the well is related to the aquifer potential through

$$h(r_s^+, t) = H(t) \; ; \; 2\pi r_s T \frac{\partial h(r_s^+, t)}{\partial r} = \pi r_c^2 \frac{dH(t)}{dt} \qquad \textbf{(3.6.8)}$$

The first of these equations states that the potential just outside of the screen is equal to the water level within the well. The second equation balances the inflow or outflow from the well with the storage change within the well casing.

Cooper et al. (1967) show that the solution to this problem depends on two dimensionless parameters and may be written

$$\frac{H(t)}{H_o} = \frac{8\alpha}{\pi^2} \int_0^\infty \frac{\exp(-\beta u^2/\alpha)}{u([uJ_o(u) - 2\alpha J_1(u)]^2 + [uY_o(u) - 2\alpha Y_1(u)]^2)} \, du = G(\alpha, \beta) \qquad \textbf{(3.6.9)}$$

where H_o is the initial displacement of the water level in the well, and $H(t)$ is the displacement from equilibrium at a later time t. H_o and $H(t)$ represent displacements from equilibrium and not water level elevations. Also in Equation 3.6.9,

$$\alpha = \frac{r_s^2 S}{r_c^2}$$

is a dimensionless storage parameter and

$$\beta = \frac{Tt}{r_c^2}$$

is a dimensionless time parameter. The functions $J_o(u)$, $J_1(u)$, $Y_o(u)$, and $Y_1(u)$ are the zero and first-order Bessel functions of the first and second kind. The function $G(\alpha, \beta)$ is the **slug test well function**. If the casing and screen are the same diameter, then α is simply equal to the coefficient of storage for the aquifer. The slug test well function $G(\alpha, \beta)$ has been tabulated by Cooper et al (1967) and Papadopulos et al (1973) and is shown in Figure 3.6.3. When water is instantaneously removed from the well and the water level rises during the test, one may replace $H(t)$ and H_o with $y(t)$ and y_o in Equation 3.6.9 to remain consistent with the notation used for Hvorslev's method.

EXAMPLE PROBLEM **3.6.1   Application of Cooper's Slug Test Method**

Cooper et al (1967) report on a slug test that was performed on a well that is cased 24 m with 15.2 cm (6 inch) casing and drilled as a 15.2 cm open hole to a depth of 122 m. A long-weighted float with a displacement volume of 0.01016 m³ was placed in the well, and the water level was allowed to reach equilibrium. The weight was then suddenly withdrawn, and the water level hydrograph was recorded electrically from a pressure transducer that was suspended below the water surface in the well. From the relation $y_o = V/(\pi r_c^2)$, the initial displacement was found to be $y_o = 0.560$ m. The water level data are listed in the two data columns on the left of Figure 3.6.4. Values for the transmissivity and storage coefficient of the formation are desired.

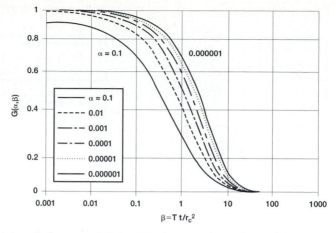

FIGURE 3.6.3 Slug test well function, $G(\alpha, \beta)$, for different values of the storage parameter α

Appendix E presents a module for calculating the slug test well function $G(\alpha,\beta)$ using a spreadsheet. Once the module has been set up, then the procedure for fitting the type curve to the data is the same as that shown in Example 3.4.1 for the Theis method of analysis. A portion of a spreadsheet that can be used for this analysis is shown in Figure 3.6.4. The test variables including the initial head displacement are included in the box on the upper-left. Values of parameters T and S are adjusted until an adequate fit is achieved. The model results box on the right follows the same format as Example 3.4.1. The third column in the box lists the values of

$$G\left(\alpha,\frac{Tt}{r_c^2}\right)$$

that are calculated using the module functions for each time, and these values are shown as the curve on the graph that also contains the data plot. The curve is automatically updated for each selection of T and S. (The module actually calculates the function $G(\alpha,\beta,\epsilon)$, where ϵ specifies the accuracy with which Equation 3.6.9 is calculated.)

Some guidance is provided for selection of initial values of T and S. Storage coefficients often have a magnitude near $S = 0.001$. For this value of S, α can be calculated using the known casing and screen radii, and Figure 3.6.3 shows that $y(t)/y_o = G(0.001,\beta) = 0.5$ at $\beta = 1.3$. The data shows that $y(t)/y_o = 0.5$ at $t = 15$ sec. With the definition of β, this information gives for the initial estimate $T = \beta r_c^2/t = 1.3 \ (7.6 \text{ cm})^2/15 \text{ sec} = 5 \text{ cm}^2/\text{sec}$. The best fit is shown in Figure 3.6.4 and results in $T = 5.26 \text{ cm}^2/\text{s}$ and $S = 0.00085$, with a corresponding (**root mean square**) error of 0.03830. The root mean square error is calculated from

$$\text{RMS Error} = \sqrt{\sum_i \left[\left.\frac{y(t_i)}{y_o}\right|_{\text{data}} - G\left(\alpha,\frac{Tt_i}{r_c^2}\right) \right]^2} \qquad \textbf{(3.6.10)}$$

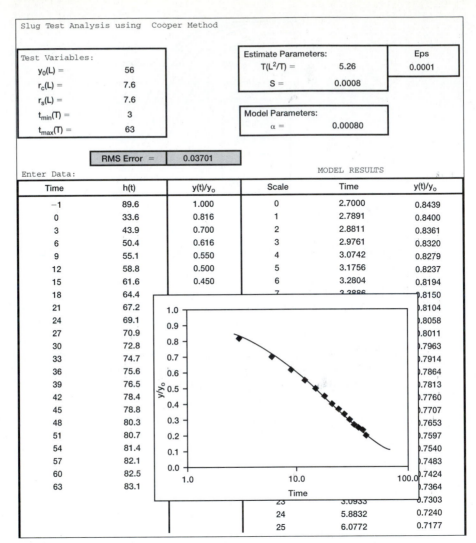

FIGURE 3.6.4   Spreadsheet for slug test analysis using Cooper method

Bouwer and Rice Method

A method for analyzing slug test data from unconfined aquifers was presented by Bouwer and Rice (1976). The geometry for a well in an unconfined aquifer is shown in Figure 3.6.5. For the slug test, the water level in the well is suddenly lowered, and the rate of rise of the water level is measured. The development of the method is similar to that of Hvorslev. The flow rate into the well may be estimated from the Thiem Equation (3.2.3) as

$$Q = \frac{2\pi KL_e y}{ln(R_e/r_w)} \qquad (3.6.11)$$

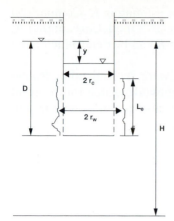

FIGURE 3.6.5 Geometry of a partially penetrating, partially perforated well in an unconfined aquifer with a gravel pack or developed zone around the perforated section

where L_e is the effective height of the portion of the well through which water enters (height of screen or perforated zone), y is the vertical distance between the water level in the well and the equilibrium water table elevation, R_e is the effective radius over which y is dissipated (effective radius of influence), and r_w is the horizontal distance from the well center to the original aquifer (well radius plus thickness of the gravel envelope or developed zone). This equation assumes that the drawdown around the well is negligible, as are the head losses as water enters the well (**well losses**).

The rate of rise of the water level in the well after suddenly removing a slug of water is related to the inflow rate by continuity as

$$\frac{dy}{dt} = -\frac{Q}{\pi r_c^2} \tag{3.6.12}$$

where πr_c^2 is the cross-section area of the well where the water level is rising. The term r_c is the inside radius of the casing if the water level is above the perforated or screened portion of the well. If the water level is rising in the perforated section of the well, and if the hydraulic conductivity of the gravel envelope or developed zone is much larger than that of the aquifer, then allowance must be made for the porosity outside the well casing. Bouwer and Rice (1976) present an example showing that if the radius of the perforated casing is 20 cm and the casing is surrounded by a 10 cm permeable gravel envelope with a porosity of 0.30, then r_c should be calculated as

$$\sqrt{(20 \text{ cm})^2 + 0.30 \times ((30 \text{ cm})^2 - (20 \text{ cm})^2)} = 23.5 \text{ cm}$$

The value of r_w for this well section is 30 cm.

Equations 3.6.7 and 3.6.8 may be combined and integrated with the initial condition that $y = y_o$ at $t = 0$. This equation gives for the permeability

$$K = \frac{r_c^2 ln(R_e/r_w)}{2L} \frac{1}{t} ln\left(\frac{y_0}{y_t}\right) \tag{3.6.13}$$

Equation 3.6.13 is the **Bouwer and Rice model** for estimating K. Comparison with Equation 3.6.7 shows that it is similar to Hvorslev's model, except that R_e appears in place of L. If R_e can be estimated, then K can be determined from the water level rise.

Values of R_e, expressed in terms of $ln(R_e/r_w)$, were determined by Bouwer and Rice (1976) with an electrical resistance network analog for different values of r_w, L_e, D, and H (see Figure 3.6.5). The results from these analog simulations provided the following empirical equation:

$$ln\left(\frac{R_e}{r_w}\right) = \left(\frac{1.1}{ln(D/r_w)} + \frac{A + Bln((H - D)/r_w)}{L_e/r_w}\right)^{-1} \tag{3.6.14}$$

In Equation 3.6.14, A and B are dimensionless coefficients that are functions of L_e/r_w, as shown in Figure 3.6.6. If $H >> D$, Bouwer and Rice find that an increase in D has no measurable effect on $ln(R_e/r_w)$. The analog results indicated that the effective upper limit of $ln((H-D)/r_w)$ is 6. Thus, if H is large so that the calculated value of $ln((H-D)/r_w)$ is greater than 6, a value of 6 should still be used for the term $ln((H-D)/r_w)$ in Equation 3.6.14.

If $D = H$, then the term $ln((H-D)/r_w)$ cannot be used. The analog results of Bouwer and Rice indicated that for this condition, which is the case of a fully penetrating well, Equation 3.6.14 should be modified to

$$ln\left(\frac{R_e}{r_w}\right) = \left(\frac{1.1}{ln(D/r_w)} + \frac{C}{L_e/r_w}\right)^{-1} \tag{3.6.15}$$

where C is a dimensionless parameter that is a function of L_e/r_w, as shown in Figure 3.6.6.

Bouwer and Rice (1976) suggest that Equations 3.6.14 and 3.6.15 yield values of $ln(R_e/r_w)$ that are within 10 percent of the actual analog simulation values if $L_e > 0.4D$ and within 25 percent if $L_e << D$ (for example, $L_e = 0.1D$).

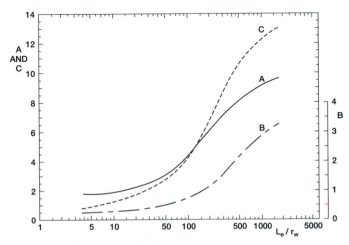

FIGURE 3.6.6 Curves relating coeffecients A, B, and C to L_e/r_w (from Bouwer, 1978)

EXAMPLE   **3.6.2   Application of Bouwer and Rice Method and Hvorslev Method**
PROBLEM   A slug test was performed on a cased well in the alluvial deposits of the Salt River
bed west of Phoenix, Arizona (Bouwer and Rice, 1976). The static water table was
at a depth of 3 m, $H = 80$ m, $D = 5.5$ m, $L_e = 4.46$ m, $r_c = 0.076$ m, and r_w was taken
as 0.12 m to allow for development of the aquifer around the perforated portion
of the casing. A solid cylinder with a volume equivalent to a 0.32 m change in water
level in the well was placed below the water table. After equilibrium levels were
again achieved, the cylinder was quickly removed and the water levels were determined from analysis of the output of a pressure transducer. The results are shown
in Figure 3.6.7. The straight-line portion is the valid part of the readings. The actual y_o value of 0.29 m indicated by the straight line is close to the theoretical value
of 0.32 m calculated from the displacement volume of the cylinder.

For this well, $L_e/r_w = 37$, so from Figure 3.6.6 we estimate that $A = 2.7$ and
$B = 0.45$. Substituting these values into Equation 3.6.10 and using the maximum
value of 6 for $ln\left(\dfrac{H - D}{r_w}\right)$ gives $ln\left(\dfrac{R_e}{r_w}\right) = 2.31$. Using this value in Equation 3.6.13
gives

$$K = \frac{(0.076 \text{ m})^2 \times 2.31}{2 \times 4.56 \text{ m}} \frac{1}{22.5s} ln\left(\frac{0.29 \text{ m}}{0.001 \text{ m}}\right) = 0.00037 \text{ m/s} = 32 \text{ m/d}$$

which agrees well with K values of 10 and 53 m/d obtained by other methods
in this area.

We want to compare this result with the results we would obtain with
Hvorslev's method. We have $y_o = 0.29$ m, so that $t = T$ occurs when $y_t = 0.37 \times$
0.29 m = 0.107 m. According to Figure 3.6.7, this situation occurs at a time of
about 4 seconds. With Equation 3.6.6, we have

$$F = \frac{2\pi(4.46 \text{ m})}{ln\left(\dfrac{4.46 \text{ m}}{0.12 \text{ m}}\right)} = 7.75 \text{ m},$$

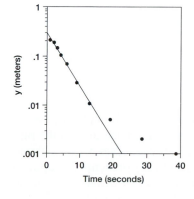

FIGURE 3.6.7   Data from a slug test in a permeable soil (from
Bower and Rice, 1976)

and from Equation 3.6.4 with $A = \pi r_c^2 = 0.18$ m^2, we have

$$K = \frac{0.018 \text{ m}}{7.75 \text{ m} \times 4s} = 0.00058 \text{ m/s} = 50 \text{ m/d}$$

which is on the same order of magnitude.

3.7   Well Tests

Aquifer tests are used to make "in situ" estimates of the characteristic hydraulic parameters of an aquifer such as its transmissivity and storage coefficient. A **well test**, on the other hand, is used to determine the hydraulic characteristics of a well; especially whether the well meets the requirements of a contract (required capacity for maximum allowable drawdown) and for pump selection.

From a well test one can estimate the total energy losses associated with water entry to the well. These are called **well losses**, and they include the following items:

partial penetration losses

losses through the gravel pack

losses through the screen

partial perforation losses

turbulent losses within the screen and formation up to the pump inlet

A simple mathematical model that includes both formation losses and well losses is

$$s = BQ + CQ^m \qquad (3.7.1)$$

The first term in Equation 3.7.1 corresponds to those losses that occur as water moves through the geologic formation to the well. For a fully penetrating well, we have

$$B = \frac{1}{2\pi T} ln\left(\frac{R}{r_w}\right) \qquad (3.7.2)$$

while for a partially penetrating well, we may use

$$B = \frac{1}{2\pi T}\left(ln\left(\frac{R}{r_w}\right) + \left(\frac{1-p}{p}\right)ln\left(\frac{(1-p)L}{r_w}\right)\right) \qquad (3.7.3)$$

The second term in Equation 3.7.1 provides the additional losses associated with water entry into the well and within the well, pump, and intake piping. The early work of Jacob (1947) assumes that these losses are associated with turbulent flow and that $m = 2$. Rorabaugh (1953) presents a graphical method for well test analysis that leaves m as an unknown variable to be estimated. Field data generally show that m varies from 1.5 to 3.5. Here, we will use Jacob's model and assume that m = 2, in which case Equation 3.7.1 becomes

$$s = BQ + CQ^2 \qquad\qquad (3.7.4)$$

The well loss coefficient, C, may be estimated from a series of pumping tests run at increasing discharges, and separated by a long enough period of time for the water levels to re-equilibrate. An alternative test, and one which is more commonly performed in practice, is a **step-drawdown test**. In either case, one has a sequence of discharges, Q_i, and corresponding drawdown, s_i. With the assumption that $m = 2$, we can write Equation 3.7.4 as

$$\frac{s}{Q} = B + CQ \qquad\qquad (3.7.5)$$

According to Equation 3.7.5, a plot of s/Q versus Q should yield a straight line with the intercept providing the value of B, while the slope gives the coefficient C. This situation is shown schematically in Figure 3.7.1.

Walton (1962) has examined field data from a number of tests and characterized the state of a well by the value of its well loss coefficient, C. The data in Table 3.7.1 summarize his results (standard SI units for C are used, s²/m⁵).

If T and S are found from an aquifer test, and B is found from the step-drawdown test, then the value of B can be used to estimate the effective well radius, r_w. We have

$$r_w = \sqrt{\frac{2.25\,Tt}{S\,\exp(4\pi TB)}} \qquad\qquad (3.7.6)$$

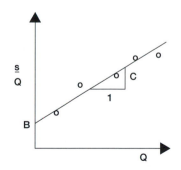

FIGURE 3.7.1   Analysis of step-drawdown data for estimation of well loss coefficient

TABLE 3.7.1   Well Loss Coefficient, C (s²/m⁵) after Walton (1962)

$C < 3800$	properly developed and designed well
$3800 < C < 7600$	mild deterioration
$C > 7600$	clogging is severe
$C > 15200$	difficult or impossible to restore well to original capacity

Well Efficiency

The magnitude of well losses determine the **well efficiency**, which is defined by

$$E = \frac{s_{\text{theory}}}{s_{\text{actual}}} \qquad (3.7.7)$$

With Equation 3.7.4, this defination gives

$$E = \frac{BQ}{BQ + CQ^2} = \frac{B}{B + CQ} \qquad (3.7.8)$$

Pump Selection

Pump selection is based on methods learned in hydraulics. An individual plots the system curve and the pump characteristic curves and identifies the potential operating points. The pump is selected which provides the desired discharge and has the greatest efficiency. The only difference from the standard hydraulics analysis is that the lift increases with time because of formation losses and the continually developing drawdown cone. This situation is shown schematically in Figure 3.7.2. The total losses include the following:

$$s = \underset{\text{formation}}{BQ} + \underset{\text{well}}{CQ^2} + \underset{\text{major}}{DQ^2} + \underset{\text{minor}}{EQ^2}$$

where the major losses are associated with uniform flow along the length of the intake piping and the minor losses are associated with bends, valves, and other obstructions along the intake pipe.

Specific Capacity

The **specific capacity** is the ratio of the discharge to the drawdown and is often taken as a characteristic constant for a well. With the well loss equation, the specific capacity may be written

$$\frac{Q}{s_w} = \frac{Q}{BQ + CQ^2} = \frac{1}{B + CQ} \qquad (3.7.9)$$

It is apparent that the specific capacity is smaller for larger discharges. This concept is shown in Figure 3.7.3. In addition, because B depends on time

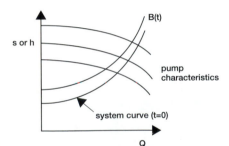

FIGURE 3.7.2   Typical pump characterist curve used for pump selection

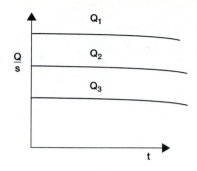

FIGURE 3.7.3   The specific capacity varies both with discharge and time

(through R), so does the specific capacity—although it is often treated as a constant. Figure 3.7.3 suggests that the specific capacity decreases slowly with time because of increasing R.

Power Costs During Pumping

The costs of pumping groundwater have become of major interest as water tables and piezometric surfaces continue to decline and the cost of energy continues to increase. From hydraulics, we know that the power usage may be calculated from

$$\text{Power} = \frac{Q\gamma h}{\eta} \tag{3.7.10}$$

where γ is the specific weight of water, η is the pump efficiency, and the head includes

$$h = h_{\text{static}} + h_{\text{dynamic}}$$

The static head is given by

$$h_{\text{static}} = \text{initial lift} + \text{well losses} + \text{partial penetration losses}$$

while the dynamic head is given by the formation losses

$$h_{\text{dynamic}} = s(r_w, t)$$

In terms of costs, one pays for energy use. The total energy use may be calculated from

$$\text{energy use} = \int \text{Power } dt$$

while the cost is calculated from the total energy use and the cost per kw-hr:

$$\text{cost} = \chi(\$/\text{kw} \cdot \text{hr}) \times \text{energy use}$$

Thus, in order to calculate the cost of pumping water, we want to evaluate the energy use. The static energy use is easily calculated from

$$\text{static energy use} = \frac{Q\gamma h_{\text{static}} t}{\eta}$$

For the dynamic energy use, we have

$$\frac{Q\gamma}{\eta}\int_0^t s\,dt = \frac{Q\gamma}{\eta}\frac{Q}{4\pi T}\int_0^t W(u)\,dt = \frac{Q\gamma}{\eta}\frac{Q}{4\pi T}t(W(u_t)-1) = \frac{Q\gamma}{\eta}t\left(s(r_w,t)-\frac{Q}{4\pi T}\right)$$

Thus, the *total energy use* (T.E.U.) is given by

$$\text{T.E.U.} = \frac{Q\gamma t}{\eta}\left(h_{\text{static}} + s(r_w,t) - \frac{Q}{4\pi T}\right) \qquad (3.7.10)$$

When evaluating the energy usage and power costs, the question of units comes into play. A consistent set of units is provided by $Q = m^3/d$, $\gamma = 9.81$ kN/m^3, $t =$ days, h, $s = $ m, and $T = $ m^2/d. Then, the T.E.U. is given in terms of kN $-$ m $= kJ$. The total cost is calculated from

$$\chi\left(\frac{\$}{\text{kW}\cdot\text{hr}}\right) \times \frac{1\text{ hour}}{3600\text{ sec}} \times \text{T.E.U. (kJ)} \qquad (3.7.11)$$

EXAMPLE PROBLEM   **3.7.1   Power Costs During Pumping**

A well with radius $r_w = 0.4$ m produces water from a confined aquifer at a rate of $Q = 1000$ m^3/d for a period of 6 months per year. What is the annual power cost if the unit rate is $\chi = 0.09\$/(\text{kW-hr})$. Assume that $\eta = 0.7$, that the water levels recover each year, and that the static lift is $h_{\text{static}} = 20$ m. For the aquifer, we have $T = 150$ m^2/d and $S = 0.0002$. For this data, the dynamic lift is

$$s = \frac{Q}{4\pi T}\ln\left(\frac{2.25Tt}{r_w^2 S}\right) = 11.3\text{ m}$$

Equation 3.7.10 gives a total energy use of $TEU = 7.77 \times 10^7$ kJ, while Equation 3.7.11 gives an annual cost of $1,940.

3.8   Multiple Well Problems

Problems involving a number of wells are of great interest, especially in design of dewatering systems for drainage engineering. The principle of superposition may always be used directly to calculate drawdown and required well discharge. For a confined aquifer, this principle gives

$$2\pi Ts = \sum_{i=1}^{N} Q_i \ln\left(\frac{R}{r_i}\right) \qquad (3.8.1)$$

and if the discharge from each well is the same, we have

$$\frac{2\pi Ts}{Q} = \sum_{i=1}^{N}\ln\left(\frac{R}{r_i}\right) \qquad (3.8.2)$$

where Q is the discharge from each well. Equation 3.8.2 is important, because it shows that with a multiple number of wells, all pumping at the same rate, the drawdown at any point depends only on the geometry of the system through the

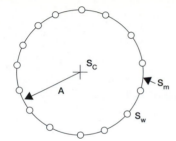

FIGURE 3.8.1 Circular arrangement of dewatering system of wells

expression on the right. Simplified formulas are found by working with this factor. In this section, we consider two simple cases. There are many others in the technical literature.

3.8.1   Circular Dewatering System

Jacob (1950) has analyzed the problem with a number of wells, N, arranged in the geometry of a circle, as shown in Figure 3.8.1. One assumes that each well pumps at the same rate. We are interested in the drawdown at the center of the set of wells, which might correspond to the center of an excavation for example, and the drawdown at each of the wells and at the midpoint between wells on the circle of the system.

For the drawdown at the center of the system of wells, the radii from each of the wells to the center is the same and is given by $r_i = A$. Thus, Equation 3.8.2 becomes

$$\frac{2\pi T s_c}{Q} = \sum_{i=1}^{N} ln\left(\frac{R}{A}\right) = N \, ln\left(\frac{R}{A}\right)$$

Because $NQ = Q_T$, however, which is the total discharge from the system of wells, this equation allows us to calculated the required total discharge without knowing the number of wells from

$$Q_T = \frac{2\pi T s_c}{ln\left(\frac{R}{A}\right)} \tag{3.8.3}$$

For the drawdown at each well, Equation 3.8.2 becomes

$$\frac{2\pi T s_w}{Q} = ln\left(\frac{R}{r_w}\right) + \sum_{i=1}^{N-1} ln\left(\frac{R}{r_i}\right)$$

Jacob uses a number of trigonometric identities to show that this reduces to

$$s_w = s_c + \frac{Q}{2\pi T} ln\left(\frac{A}{Nr_w}\right) \tag{3.8.4}$$

For the drawdown midway between wells, Equation 3.8.2 becomes

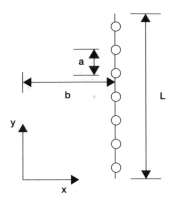

FIGURE 3.8.2   Geometry for a line of wells

$$\frac{2\pi T s_m}{Q} = \sum_{i=1}^{N} ln\left(\frac{R}{r_i}\right)$$

and again, using trigonometric identities, one can show that this equation is equal to

$$s_m = s_c - \frac{Q}{2\pi T} ln(2) \qquad\qquad \textbf{(3.8.5)}$$

Note that $s_w > s_c > s_m$.

3.8.2   Linear Well System

As a second example, we consider a line of wells such as shown in Figure 3.8.2. The constant well spacing is a, and the number of wells in the line is N. The number of wells, well spacing, and length of the line are related through $N = L/a$, so that the length of the line is considered to extend a distance $a/2$ beyond the last well at each end. We are interested in being able to calculate the drawdown at an arbitrary point away from the line of wells. The results from Charbeneau and Wright (1983) and Irmay (1984) are followed.

The drawdown at an arbitrary distance b from one end of the row (chosen as the origin) satisfies Equation 3.8.2, and we have the following equation:

$$\frac{2\pi T s}{Q} = \sum_{i=1}^{N} ln\left(\frac{R}{r_i}\right) = \frac{1}{2}\sum_{i=1}^{N} ln\left(\frac{R^2}{r_i^2}\right) = \frac{1}{2}\sum_{i=1}^{N} ln\left(\frac{R^2}{b^2 + y_i^2}\right)$$

The summation can be approximated by the integral (because both equal N),

$$\sum_{i=1}^{N} \cong \int_0^L \frac{dy}{a}$$

and we have

$$\frac{2\pi T s}{Q} \cong \frac{1}{2}\int_0^L ln\left(\frac{R^2}{b^2 + y^2}\right)\frac{dy}{a} = \frac{L}{2a}\left(ln\left(\frac{R^2}{L^2 + b^2}\right) + 2\left(1 - \frac{b}{L}\tan^{-1}\left(\frac{L}{b}\right)\right)\right)$$

Because $L/a = N$ and $NQ = Q_T$, however,

$$\frac{4\pi Ts}{Q_T} = ln\left(\frac{R^2}{L^2 + b^2}\right) + 2\left(1 - \frac{b}{L}\tan^{-1}\left(\frac{L}{b}\right)\right) \qquad (3.8.6)$$

Likewise, for the unconfined aquifer,

$$\frac{2\pi K(H_R^2 - H^2)}{Q_T} = ln\left(\frac{R^2}{L^2 + b^2}\right) + 2\left(l - \frac{b}{L}\tan^{-1}\left(\frac{L}{b}\right)\right) \qquad (3.8.7)$$

Equation 3.8.7 takes the same form as Equation 3.8.6 if we use the corrected drawdown and set $T = KH_R$.

EXAMPLE PROBLEM

3.8.1   Evaluation of the Approximation

A simple calculation shows that the approximation of replacing a summation by an integral is quite good. Figure 3.8.3 shows two pumping wells and an observation well. One can assume that each well produces at a rate of $Q = 550$ m³/d, that the radius of influence is $R = 1000$ m, and that the aquifer transmissivity is $T = 100$ m²/d. The distance between wells is $r = \sqrt{50^2 + 75^2} = 90.14$ m, so the drawdown at the observation well is

$$s = \frac{2Q}{2\pi T}ln\left(\frac{R}{r}\right) = 4.21 \text{ m}$$

With the line-of-wells approximation, this figure represents two rows of wells, each of length 100 m, situated 75 m from the observation point. The geometry factor from Equation 3.8.6 gives

$$ln\left(\frac{R^2}{L^2 + b^2}\right) + 2\left(1 - \frac{b}{L}\tan^{-1}\left(\frac{L}{b}\right)\right) = 4.16 + 0.61 = 4.77$$

and so Equation 3.8.6 becomes

$$\frac{4\pi Ts}{2Q} = 4.77$$

With this result, one finds that $s = 4.18$ m. The line-of-wells approximation is off by 3 cm.

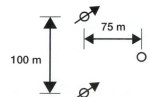

100 m

75 m

FIGURE 3.8.3   Well arrangement of Example 3.8.1

EXAMPLE **3.8.2 Design of a Dewatering System**
PROBLEM This second example considers a more typical application. Two lines of wells are used to dewater an excavation in an unconfined aquifer. The aquifer has an initial saturated thickness of $H_R = 110$ ft, a hydraulic conductivity of $K = 15$ ft/d, and a specific yield of $S_y = 0.1$. The layout is shown in Figure 3.8.4. We want a drawdown at the center of the excavation of 40 ft within one year. If each well can produce up to 10,000 ft³/d (50 gpm), how many wells are necessary?

After one year of pumping, the radius of influence is calculated from

$$R = 1.5\sqrt{\frac{KH_Rt}{S_y}} = 3{,}680 \text{ ft}$$

For drawdown at the center, $L = 750$ ft and $b = 200$ ft. The geometry factor is

$$ln\left(\frac{(3680 \text{ ft})^2}{(750 \text{ ft})^2 + (200 \text{ ft})^2}\right) + 2\left(1 - \frac{200 \text{ ft}}{750 \text{ ft}} \tan^{-1}\left(\frac{750 \text{ ft}}{200 \text{ ft}}\right)\right) = 3.11 + 1.30 = 4.41$$

Four half-rows cause the drawdown at the center. The drawdown is the sum of that from each half-row, and the total discharge is the sum of that for each half-row. Thus, the term

$$\frac{4\pi KH_Rs'}{Q_T} = 4.41$$

may be used with s', the total corrected drawdown, and Q_T, the total discharge. The corrected drawdown corresponding to an actual drawdown of 40 ft is

$$s' = s - \frac{s^2}{2H_R} = 40 \text{ ft} - \frac{(40 \text{ ft})^2}{2 \times 110 \text{ ft}} = 32.7 \text{ ft}$$

This equation gives

$$Q_T = \frac{4\pi KH_Rs'}{4.41} = 154{,}000 \text{ ft}^3/\text{d}$$

Because each well can produce 10,000 ft³/d, 16 wells are required—8 on each side. The well spacing would then be $a = 1500/8 = 187.5$ ft.

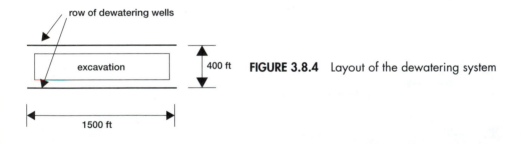

FIGURE 3.8.4 Layout of the dewatering system

3.9   Potential Theory for Stratified Aquifers

For the analysis of groundwater hydraulics near a well, the assumption of effective parameter values (T, K, S, etc.) is often adequate for engineering calculations. For large-scale projects such as dewatering of deep excavations, however, these approximations may not be as useful. For alluvial systems, one usually finds that variations in lithology associated with facies changes are much more pronounced in the vertical direction than along the (usually horizontal) bedding plane. This concept is a natural characteristic of a depositional environment that changes over (geologic) time. When significant groundwater level variations are involved in calculations, one might not be able to find constant effective parameter values. Methods are available, however, for conditions where vertical changes in lithology cause vertically changing hydraulic parameters that remain uniform along the horizontal plane. These methods provide an alternative that is sometimes useful compared to the general reliance on numerical simulation methods.

Figure 3.9.1 shows the cross-section for unconfined flow in a stratified aquifer. Girinskii (1946) has shown that so long as the flow is essentially horizontal, then a **potential function**[2] can be defined for stratified aquifers, which generalizes the saturated thickness used for homogeneous unconfined aquifers (also see Bear, 1972). The primary assumption is that the energy losses associated with vertical flow are negligible, which implies that the Dupuit-Forchheimer assumptions apply.

In order to develop this potential function, we want to calculate the aquifer flux. Because the hydraulic conductivity varies with elevation and the water table elevation is not constant, however, the flux may be found by integrating the horizontal component of the Darcy velocity across the saturated thickness of the aquifer. Application of the Dupuit assumption $\{H = H(x,y)\}$ gives

$$\tilde{U}(x,y) = \int_0^{H(x,y)} \tilde{q}(z)dz = \int_0^{H(x,y)} - K(z)\nabla H(x,y)dz$$

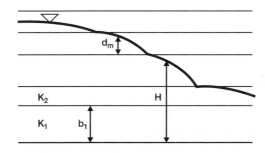

FIGURE 3.9.1   Cross-section for unconfined flow in a stratified aquifer

[2]For our present purposes, a potential function is one whose gradient gives the aquifer flux. Application of potential theory to groundwater flow is presented more fully in Chapter 6.

and because $\nabla H(x,y)$ is independent of z,

$$\tilde{U} = -\int_0^H K(z)dz\nabla H \tag{3.9.1}$$

Girinskii noted that if a potential function is defined by

$$\Phi_g \equiv \int_0^H (H - z)K(z)dz \tag{3.9.2}$$

then

$$\nabla\Phi_g = \nabla H((H - z)K(z))\big|_{z=H} + H\int_0^H \nabla HK(z)dz = \int_0^H \nabla HK(z)dz = -\tilde{U}$$

so that Φ_g provides a potential function for the stratified medium and

$$\tilde{U}(x,y) = -\nabla\Phi_g(x,y) \tag{3.9.3}$$

The potential function defined by Equation 3.9.2, $\Phi_g(x,y)$, is called the **Girinskii potential**.

The general form of the continuity equation for steady flow in an unconfined aquifer is

$$\nabla \cdot \tilde{U} = W + Q\delta(\tilde{x})$$

With Equation 3.9.3, the continuity equation for the unconfined stratified medium may be written

$$\nabla^2(\Phi_g) + W + Q\delta(\tilde{x}) = 0 \tag{3.9.4}$$

For one-dimensional linear flow, Equation 3.9.4 becomes

$$\frac{d}{dx}\left(\frac{d\Phi_g}{dx}\right) + W = 0 \tag{3.9.5}$$

Equation 3.9.5 integrates to give

$$\Phi_g + \frac{Wx^2}{2} = Ax + B \tag{3.9.6}$$

which is the same as Equation 2.3.11.

In order to apply this theory, we must be able to calculate Φ_g for a given set of layers making up the aquifer. With reference to Figure 3.9.1, Φ_g may be evaluated for a layered system with layer thicknesses $b_1, b_2, \ldots,$ and hydraulic conductivities $K_1, \ldots,$ numbered from the bottom-up. The definition of Φ_g gives

$$\Phi_g \equiv \int_0^H (H - z)K(z)dz = H\int_0^H K(z)dz - \int_0^H zK(z)dz \tag{3.9.7}$$

The first of these integrals on the right-hand-side gives

$$H\left(\sum_{i=1}^{m-1} K_i b_i + K_m d_m\right)$$

where d_m is the saturated thickness of the *mth* layer. The second integral on the right-hand-side may be evaluated piecewise across each layer. The final result is

$$\Phi_g = H\left(\sum_{i=1}^{m-1} K_i b_i + K_m d_m\right) - \frac{b_1^2}{2}(K_1 - K_2) - \frac{(b_1 + b_2)^2}{2}(K_2 - K_3) - \ldots \quad \text{(3.9.8)}$$

$$- \frac{(b_1 + \ldots + b_{m-1})^2}{2}(K_{m-1} - K_m) - \frac{K_m H^2}{2}$$

EXAMPLE
PROBLEM

3.9.1 Flow Through an Embankment

In this example, we consider the flow through the embankment shown in Figure 3.9.2. The layer thickness and hydraulic conductivity values are as follows: $b_1 = 10$ ft, $K_1 = 3$ ft/d; $b_2 = 15$ ft, $K_2 = 12$ ft/d; $b_3 = 20$ ft, $K_3 = 0.5$ ft/d. The upstream water depth is 35 ft, while the downstream depth is 20 ft. The length of the embankment is 150 ft. The infiltration rate is 0.833 ft/yr = 0.0023 ft/d. We want to find the discharge (per linear foot of length) through the embankment and the distance from the upstream end at which layer 3 becomes unsaturated.

Equation 3.9.6 is the general solution for the potential for linear flow through the embankment. If we take the upstream boundary as the x-axis origin, then the boundary conditions give

$$\Phi_g(0) = \Phi_u = B$$

$$\Phi_g(L) = \Phi_d = AL + \Phi_u - \frac{WL^2}{2}$$

These conditions give

$$\Phi_g(x) = (\Phi_d - \Phi_u)\frac{x}{L} + \Phi_u + \frac{Wx}{2}(L - x) \quad \text{(3.9.9)}$$

$$U(x) = -\frac{d\Phi_g}{dx} = \frac{\Phi_u - \Phi_d}{L} - \frac{WL}{2} + Wx \quad \text{(3.9.10)}$$

Now, the only problem is to evaluate the potential and find $\Phi_g(H)$. Using Equation 3.9.8, we have

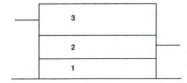

FIGURE 3.9.2 Flow through a stratified embankment

$$\Phi_g = \frac{K_1 H^2}{2} \text{ for } 0 \le H \le b_1 \tag{3.9.11}$$

$$\Phi_g = H(K_1 b_1 + K_2 d_2) - \frac{b_1^2}{2}(K_1 - K_2) - \frac{K_2 H^2}{2} \text{ for } b_1 \le H \le (b_1 + b_2) \le d_2 \le b_2 \tag{3.9.12}$$

For the third layer, we have

$$\Phi_g = H(K_1 b_1 + K_2 b_2 + K_3 d_3) - \frac{b_1^2}{2}(K_1 - K_2) - \frac{(b_1 + b_2)^2}{2}(K_2 - K_3) - \frac{K_3 H^2}{2} \tag{3.9.13}$$

for $(b_1 + b_2) \le H \le (b_1 + b_2 + b_3)$. At the upstream boundary, $H = 35$ ft and $d_3 = 10$ ft. This equation gives

$$\Phi_u = \Phi_g(35) = 4075 \text{ ft}^3/\text{d}$$

Similarly, at the downstream boundary, $H = 20$ ft and $d_2 = 10$ ft. Here, the potential is

$$\Phi_d = \Phi_g(20) = 1050 \text{ ft}^3/\text{d}$$

Thus, from Equation 3.9.10,

$$U(0) = \frac{4075 \text{ ft}^3/\text{d} - 1050 \text{ ft}^3/\text{d}}{150 \text{ ft}} - \frac{0.0023 \text{ ft/d} \times 150 \text{ ft}}{2} = 20.0 \text{ ft}^2/\text{d}$$

$$U(L) = \frac{4075 \text{ ft}^3/\text{d} - 1050 \text{ ft}^3/\text{d}}{150 \text{ ft}} - \frac{0.0023 \text{ ft/d} \times 150 \text{ ft}}{2} + 0.0023 \text{ ft/d} \times 150 \text{ ft} = 20.34 \text{ ft}^2/\text{d}$$

The difference between these discharges is $WL = 0.34$ ft^2/d, which is the infiltration recharge to the embankment.

In order to find the location at which the upper layer becomes dewatered, we need the location where $\Phi_g(x) = 1950$ ft^3/d, which is the potential corresponding to $z = 25$ ft. Using this value in Equation 3.9.9, we find

$$1950 \text{ ft}^3/\text{d} = \frac{1050 \text{ ft}^3/\text{d} - 4075 \text{ ft}^3/\text{d}}{150 \text{ ft}} x + 4075 \text{ ft}^3/\text{d} + \frac{0.0023 \text{ ft/d} \times 150 \text{ ft}}{2} x - \frac{0.0023 \text{ ft/d}}{2} x^2$$

Solution of this quadratic equation gives $x = 105.6$ ft.

3.10   The Interface in the Coastal Aquifer

The interaction between fresh water and seawater in a coastal aquifer is of great interest because communities and industries located near the coast have developed these fresh water resources for use. Overproduction of fresh water supplies has lead, in some cases, to encroachment of saline water into the aquifers and ultimate loss of the resource, at least from wells located near the coastline. A typical cross-section of a coastal aquifer is shown in Figure 3.10.1. Fresh water that has been recharged to the aquifer further inland moves toward the coast under the hydraulic gradient that is present. Seawater, because of its greater

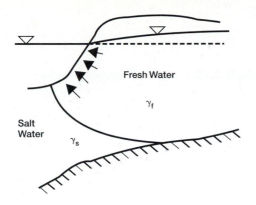

Fresh Water

γ_f

Salt
Water

γ_s

FIGURE 3.10.1 Typical cross-section of a coastal aquifer

density, migrates inland beneath the fresh water—forming a salt-water lens similar to that found in estuaries. The interface between the salt water and fresh water shown in Figure 3.10.1 is abrupt. The actual interface may be somewhat diffuse due to diffusion processes and lateral migration of the interface over time. The abrupt interface approximation, however, leads to simple and useful models of seawater and fresh water interactions.

3.10.1 Law of Ghyben and Herzberg

Beginning with Badon-Ghyben (1888) and Herzberg (1901), investigations of the coastal interface between fresh and saline water have been aimed at determining the relationship between its shape and position and the various hydrological components of a groundwater balance in the region near the coast. Both Badon-Ghyben and Herzberg found that in wells near the coast, salt water was not encountered at sea-level, but it was encountered at a depth below sea-level of about 40 times the height of fresh water above sea-level. The explanation suggested was that the fluids were in hydrostatic equilibrium with stationary seawater. Figure 3.10.2 shows the idealized Ghyben-Herzberg model of an interface in a coastal phreatic aquifer.

 With reference to Figure 3.10.2, consider a point on the fresh water-salt water interface. The pressure at this point is the same, whether approached from the fresh water side or from the salt-water side. Thus,

$$\gamma_s h_s = \gamma_f(h_s + h_f)$$

where γ_s and γ_f are the salt and fresh water specific weights. Solving for h_s, we find

$$h_s = \frac{\gamma_f}{\gamma_s - \gamma_f} h_f = \frac{\rho_f}{\rho_s - \rho_f} \equiv \alpha h_f \tag{3.10.1}$$

With $\rho_f = 1.0$ g/cm^3 and $\rho_s \cong 1.025$ g/cm^3, Equation 3.10.1 gives

$$\alpha = \frac{\rho_f}{\rho_s - \rho_f} \cong \frac{1.0}{0.025} = 40 \tag{3.10.2}$$

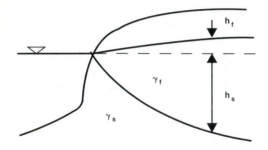

FIGURE 3.10.2   Ghyben-Herzberg interface model

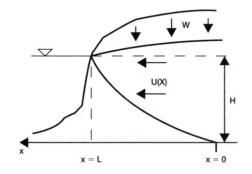

FIGURE 3.10.3   Hydrologic balance in a coastal phreatic aquifer

so

$$h_s = 40\, h_f \qquad\qquad (3.10.3)$$

which confirms the observations. This relation given by Equation 3.10.1 is called the **Law of Ghyben and Herzberg**.

3.10.2   Application of Dupuit-Forchheimer Theory

The Ghyben-Herzberg model provides a useful tool for estimating the depth of the saline interface under conditions of vertical equilibrium. When combined with the Dupuit assumption of horizontal flow, we can also examine the role played by the hydrologic balance in the coastal aquifer. Figure 3.10.3 shows an idealized cross-section of a coastal phreatic aquifer. H is the water level at the coastline, which is located at $x = L$. The fresh water aquifer flux toward the coast at the extent of penetration of the saline water lens ($x = 0$) is $U(0)$, and this flux increases because of diffuse recharge at a rate W. The objective is to determine the penetration of the lens as a function of the hydrologic fluxes.

For the situation shown in Figure 3.10.3 with $x > 0$, continuity gives

$$U(x) = U(0) + Wx \qquad\qquad (3.10.4)$$

while Darcy's Law gives

$$U = -K(h_f + h_s)\frac{dh_f}{dx} = -\frac{K}{2}(1 + \alpha)\frac{dh_f^2}{dx} \qquad\qquad (3.10.5)$$

where h_f and h_s are defined as in Figure 3.10.2. Combining these, we have

$$U(0) + Wx + \frac{K}{2}(1 + \alpha)\frac{dh_f^2}{dx} = 0 \qquad (3.10.6)$$

Equation 3.10.6 may be integrated to give

$$U(0)x + \frac{Wx^2}{2} + \frac{K}{2}(1 + \alpha)h_f^2 = A$$

where A is the constant of integration. The appropriate boundary condition is

$$h_f^2(0) = \frac{H^2}{\alpha^2} \qquad (3.10.7)$$

which specifies the fresh water head above mean sea level at the toe of the saline lens in terms of the Ghyben-Herzberg model. Using this boundary condition, we have for the general solution

$$U(0)x + \frac{Wx^2}{2} + \frac{K}{2}(1 + \alpha)h_f^2 = \frac{K}{2}(1 + \alpha)\frac{H^2}{\alpha^2} \qquad (3.10.8)$$

Equation 3.10.8 provides h_f as a function of x. If we use the condition that $h_f^2(L) = 0$, then we have

$$U(0)L + \frac{WL^2}{2} = \frac{K}{2}(1 + \alpha)\frac{H^2}{\alpha^2}$$

This equation is a quadratic equation in L that may be used to find

$$\frac{LW}{U(0)} = -1 + \sqrt{1 + \frac{KWH^2(1 + \alpha)}{U(0)^2\alpha^2}} \qquad (3.10.9)$$

Equation 3.10.9 is not appropriate if the diffuse recharge rate is small ($W \cong 0$). We can, however, return to Equation 3.10.8 with $W = 0$ and set $h_f^2(L) = 0$ to find

$$L = \frac{K(1 + \alpha)H^2}{2U(0)\alpha^2} \qquad (3.10.10)$$

EXAMPLE PROBLEM **3.10.1 Depth of Fresh Water on an Island**

A fresh water lens beneath an island receives recharge at a rate of 15 cm/yr. If the island can be considered to be roughly circular with an effective radius of 4 km, estimate the total thickness of the lens as well as the volume of fresh water that it contains. Assume that the porosity is 0.35, that the hydraulic conductivity is 5 m/d, and that the density of seawater is 1.25 g/cm³.

In radial coordinates, the continuity equation corresponding to Equation 3.10.4 is

$$2\pi r U_r(r) = \pi r^2 W$$

where $U_r(0) = 0$. The combination of Darcy's Law, the Ghyben-Herzberg model, and the Dupuit assumption brings about the model corresponding to Equation 3.10.5:

$$U_r = -\frac{K}{2}(1 + \alpha)\frac{dh_f^2}{dr}$$

Combining these and integrating with the boundary condition $h_f^2(R) = 0$, we have

$$h_f = \sqrt{\frac{W(R^2 - r^2)}{2K(1 + \alpha)}}$$

With the Ghyben-Herzberg model, the total fresh water thickness at any location r is

$$h_f + h_s = (1 + \alpha)h_f = \sqrt{\frac{W(1 + \alpha)(R^2 - r^2)}{2K}} \qquad \textbf{(3.10.11)}$$

The maximum thickness that occurs at $r = 0$ is

$$\sqrt{\frac{0.15 \text{ m/yr} \times 41 \times (4000 \text{ m})^2}{2 \times 5 \text{ m/d} \times 365 \text{ d/yr}}} = 164 \text{ m}$$

The volume of fresh water in storage is given by

$$V_{\text{storage}} = \int_0^R 2\pi r n(1 + \alpha)h_f dr = \pi n\sqrt{\frac{W(1+\alpha)}{2K}}\int_0^{R^2}\sqrt{R^2 - r^2}\ dr^2 = \frac{2}{3}\pi n R^3 \sqrt{\frac{W(1+\alpha)}{2K}} \qquad \textbf{(3.10.12)}$$

For the given conditions, the volume in storage is $1.926\ (10^9)$ m^3. By comparison, the annual recharge is 0.15 m/yr $\times \pi R^2 = 7.54(10^6)$ m^3. Thus, there is more than 250 years worth of recharge in storage.

3.10.3   General Considerations for Nonuniform Density Flows

In this section, we want to look at the difficulties involved in a rigorous investigation of flows with nonuniform density fields. In Chapter 2, we have seen that in general the driving force for flow is given by

$$\tilde{F} = -\nabla p - \rho g \hat{k} \qquad \textbf{(3.10.13)}$$

The curl of the driving force field is calculated from

$$\text{curl}(\tilde{F}) = -(\text{curl}(\nabla p) + \text{curl}(\rho g \hat{k}))$$

From vector analysis, we know that the curl of a gradient always vanishes, so we have

$$\frac{1}{g}\text{curl}(\tilde{F}) = -\frac{\partial \rho}{\partial y}\hat{i} + \frac{\partial \rho}{\partial x}\hat{j} \qquad \textbf{(3.10.14)}$$

Thus, as long as ρ is constant, then the force field has zero curl: $curl(\tilde{F}) = 0$. This in turn leads to the conclusion that the flow field can be derived from the gradient of a potential function. On the other hand, if the density is not constant and it varies on the horizontal plane, then the curl of the force field does not vanish. This concept has important implications for the resulting flow field.

The general form of Darcy's Law is

$$\tilde{q} = -\frac{k}{\mu}\left(\text{grad}(p) + \rho g \hat{k}\right)$$

When compared with Equation 3.10.13, this equation gives

$$\tilde{q} = -\frac{k}{\mu}\tilde{F} \qquad\qquad \textbf{(3.10.15)}$$

Thus, if the curl of the force field does not vanish, then the same is true of the resulting flow field. The curl of the velocity field is given by

$$\text{curl}(\tilde{q}) = \frac{kg}{\mu}\left(\frac{\partial\rho}{\partial y}\hat{i} - \frac{\partial\rho}{\partial x}\hat{j}\right) \qquad\qquad \textbf{(3.10.16)}$$

Such a flow field is not irrotational and cannot be described by a potential function (such as our velocity potential). Nevertheless, the general form of Darcy's Law still applies, as does the general continuity equation. This discussion shows that when the density field is nonuniform, the resulting groundwater flow field has large-scale vorticity. Our abrupt interface model considered in section 3.10.2 actually corresponds to a vortex sheet.

3.11   Other Transient Flow Problems

There are a wide variety of solutions to transient flow problems available in the literature. In large part, this situation is because the continuity equation takes a particularly simple form under approximations that may normally be accepted, at least provisionally. For a confined aquifer the continuity equation takes the form

$$\frac{\partial h}{\partial t} = \frac{T}{S}\nabla^2(h) \qquad\qquad \textbf{(3.11.1)}$$

which is a parabolic partial differential equation of the same type as appears in problems of diffusion and heat conduction. There is a vast literature available in these fields which is directly applicable to problems of groundwater flow (see, for example, Carslaw and Jaeger, 1959, and Crank, 1975). For flow in unconfined aquifers, one finds that so long as the drawdown is small compared to the saturated thickness, then the same solutions apply with the approximate relations $T \rightarrow K\,H_R$ and $S \rightarrow S_y$. In this section, we will take a brief look at two representative solutions to problems of bank storage during flood flow in a river and propagation of waves in aquifers.

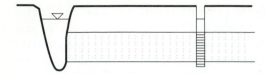

FIGURE 3.11.1   Cross-section of an aquifer in direct communication with a surface water body

3.11.1   Bank Storage Along Rivers

Consider an aquifer that is in direct communication with a river or other surface water body, as shown in Figure 3.11.1. It is of great interest to know how the aquifer will respond to rapid changes in water level in the river. For example, with a water level increase, such as might occur during a flooding event, water will be lost from the channel as groundwater recharge. The water will later return to the channel after the flood wave has past, and the losses are often referred to as **bank storage** because it is assumed that the water never moves far into the aquifer. This problem has been discussed by Ferris et al. (1962).

Assume that the initial water level in the river is the same as the piezometric surface in the aquifer at a level h_o and that the water level in the river increases to h_r abruptly at time zero. Then, the problem may be stated mathematically as one of solving

$$\frac{\partial h}{\partial t} = \frac{T}{S}\frac{\partial^2 h}{\partial x^2} \tag{3.11.2}$$

$$h(x,0) = h(\infty,t) = h_0 \; ; \; h(0,t) = h_r$$

The solution to this problem may be written

$$\frac{h(x,t) - h_o}{h_r - h_o} = \left(1 - \frac{2}{\sqrt{\pi}}\int_0^\zeta e^{-w^2}dw\right) \tag{3.11.3}$$

where

$$\zeta = \sqrt{\frac{Sx^2}{4Tt}} \tag{3.11.4}$$

The function appearing on the right of Equation 3.11.3 appears many times in problems dealing with diffusion-type phenomena, and it is given the name "complementary error function" and denoted erfc(ζ). Thus, Equation 3.11.3 may be written

$$\frac{h(x,t) - h_o}{h_r - h_o} = \text{erfc}\left(\sqrt{\frac{Sx^2}{4Tt}}\right) \tag{3.11.5}$$

The error function and complementary error function are tabulated (Appendix G) and discussed in Chapter 7, "Solute Transport by Diffusion," because they are important to diffusion-type problems.

To calculate the rate of water loss per unit stream length to *one side* of the channel, we need to calculate

$$U(0,t) = -T\frac{\partial h(0,t)}{\partial x} = (h_r - h_o)\sqrt{\frac{ST}{\pi t}} \qquad \text{(3.11.6)}$$

while the cumulative loss through time t is given by

$$\int_0^t U(0,t)dt = (h_r - h_o)\sqrt{\frac{4STt}{\pi}} \qquad \text{(3.11.7)}$$

EXAMPLE PROBLEM

3.11.1   Aquifer Response to Change in Reservoir Stage

The water level in a reservoir increases abruptly by 6 ft and is held at this new level for 3 days, after which it is returned to its previous level. Determine the water level response in an observation well located 500 ft from the reservoir at times of 3 days and 5 days. The aquifer has $T = 50$ ft^2/d and $S = 0.001$.

For this problem, we may take $h_o = 0$. At a time of 3 days, we have

$$\zeta = \sqrt{\frac{0.001 \times (500 \text{ ft})^2}{4 \times 50 \text{ ft/d} \times 3}} = 0.65$$

From Table G.2, we have $erfc(0.65) = 0.358$. Thus, $h(500 \text{ ft}, 3 \text{ days}) = 6 \times 0.358 = 2.15$ ft. For a time of 5 days, we can use the principle of superposition to write

$$h(x,t) = h_r(\text{erfc}(\zeta) - \text{erfc}(\zeta'))$$

where ζ' differs from ζ in that its time origin is 3 days later. We have

$$\zeta' = \sqrt{\frac{0.001 \times (500 \text{ ft})^2}{4 \times 50 \text{ ft}^2/\text{d} \times 5 \text{ d}}} = 0.50 \; ; \; \zeta' = \sqrt{\frac{0.001 \times (500 \text{ ft})^2}{4 \times 50 \text{ ft}^2/\text{d} \times 2 \text{ d}}} = 0.79$$

From Table G.2, $erfc(0.50) = 0.48$ and $erfc(0.79) = 0.26$. Thus, $h(500 \text{ ft}, 5 \text{ d}) = 6$ ft$(0.48 - 0.26) = 1.32$ ft.

3.11.2   Wave Propagation in Aquifers

Waves in surface water bodies will propagate into aquifers that are in direct communication with them. The problem has been investigated by Ferris (1951), who considered sinusoidal stage changes in the surface water body. If s is the stage change in the aquifer (departure from the equilibrium piezometric head value), then the problem is to solve Equation 3.11.2 subject to

$$s(0,t) = s_r \sin(\omega t) \qquad \text{(3.11.8)}$$

where s_r is the amplitude or half-range of the stage change in the surface water body and ω is its frequency (radians per unit time). The frequency and period, τ, of the stage change are related through $\omega = 2\pi/\tau$. The result found by Ferris (1951) is

$$s(x,t) = s_r \exp\left(-\sqrt{\frac{\omega S x^2}{2T}}\right) \sin\left(\omega t - \sqrt{\frac{\omega S x^2}{2T}}\right) \qquad \text{(3.11.9)}$$

Equation 3.11.9 is interesting. It shows that the wave amplitude decreases as

$$s_r \exp\left(-\sqrt{\frac{\omega S x^2}{2T}}\right) \tag{3.11.10}$$

so that short period (large frequency) waves die out quickly. This reason is why the usual surface waves on a beach are not propagated inland to any extent. On the other hand, long period (small frequency) waves are propagated with little change in amplitude. Equation 3.11.9 also shows that the time lag of occurrence of a given maximum or minimum groundwater stage is given by

$$t_{\text{lag}} = x\sqrt{\frac{\tau S}{4\pi T}} \tag{3.11.11}$$

The apparent wave celerity is

$$c = \frac{x}{t_{\text{lag}}} = \sqrt{\frac{4\pi T}{\tau S}} \tag{3.11.12}$$

where again, τ is the wave period. This says that for the aquifer considered in Example 3.11.1, the speed of a tidal wave with a 12-hour period would be 1,100 ft/day = 0.78 ft/minute.

Problems

3.2.1. A well is producing from a confined aquifer. An observation well 100 ft from the pumping well shows a drawdown of 5 ft, while an observation well 20 ft from the pumping well shows a drawdown of 16 ft. What is the drawdown at the pumping well of radius 8 inches?

3.2.2. If the well in Problem 3.2.1 produces 20 ft³/min, what is the transmissivity of the aquifer?

3.2.3. Repeat Example 3.2.2 for an unconfined aquifer, and develop a criterion for the water level in the cell that corresponds to zero saturated thickness in the well (and thus the well should pump-off).

3.2.4. A non-pumping well is perforated over three aquifers with transmissivities T_1, T_2 and T_3. If the aquifers have the same radius of influence and effective well radius, show that the head in the well is related to the potentiometric surface elevations far from the well, h_1, h_2, and h_3 by (Sokol, 1963).

$$h_w = \frac{T_1 h_1 + T_2 h_2 + T_3 h_3}{T_1 + T_2 + T_3}$$

3.2.5. For the conditions of problem 3.2.4, show that the water level fluctuations in the well Δh_w are related to the fluctuations of the potentiometric surface in aquifer 1 of Δh_1 by (Sokol, 1963)

$$\Delta h_w = \frac{T_1 \Delta h_1}{T_1 + T_2 + T_3}$$

3.2.6. Leakage occurs along the annulus of a well that penetrates two aquifers. If K_a, A_a, and L_a are the effective annulus hydraulic conductivity, cross-section area and length between the two aquifers, respectively, show that the leakage between aquifer 1 with head h_1 and aquifer 2 with head $h_2(<h_1)$ is

$$Q = \frac{h_1 - h_2}{\frac{\ln(R_1/r_w)}{2\pi T_1} + \frac{L_a}{K_a A_a} + \frac{\ln(R_2/r_w)}{2\pi T_2}}$$

3.2.7. Pumping wells are used to contain petroleum free-product (oil) that has been released. The aquifer is unconfined with $K = 5$ ft/d and the saturated thickness is 30 ft. It is assumed that the effective radius of influence of a well is 1000 ft, and that the well radius is 0.25 ft. What is the discharge per well if the drawdown is limited to 15 percent of the saturated thickness to control smearing of the oil over the aquifer?

3.2.8. A leaky aquifer has a transmissivity of 400 ft^2/d. The aquitard has a conductivity of 0.02 ft/d and a thickness of 20 ft. A well of radius 10 inches pumps at a rate of 9,600 ft^3/d. What is the drawdown at the well and its apparent radius of influence?

3.2.9. A well of radius 0.5 ft produces water from an unconfined aquifer with $K = 10$ ft/d. The well is located 900 ft from a large water body that acts as a constant head boundary. If the saturated thickness of the aquifer at the water body is 80 ft, what is the maximum production rate of the well?

3.2.10. A river crosses an unconfined alluvial aquifer and enters a large reservoir. The saturated thickness of the aquifer at the location where the stream enters the reservoir is 65 ft and its hydraulic conductivity is 12 ft/d. If a production well with radius 0.5 ft is placed at a distance of 'a' = 200 ft from the reservoir and 'b' = 250 ft from the river, what is the maximum possible long-term production rate? Develop an expression for the effective radius of influence of the well for this situation, similar to Equation 3.2.25.

3.3.1. A confined aquifer has a transmissivity of 50 ft^2/d and a storage coefficient of 0.001. A well of radius 1 ft produces water at a rate of 5,000 ft^3/d. Use the Theis equation to determine the drawdown at the production well and at an observation well 1,000 ft from the production well after 5 days.

3.3.2. Repeat problem 3.3.1 using the Jacob equation.

3.3.3. What is the apparent radius of influence of the production well in problem 3.3.1 after 5 days of pumping? Within what radius is the Jacob approximation valid?

3.3.4. A well produces water from a confined aquifer at a rate of 550 m^3/d for a period of 30 days, and then the pumping rate decreases to 300 m^3/d for another 20 days. What is the resulting drawdown at the well at the end of this 50-day pumping period if $T = 50$ m^2/d, $S = 0.001$, and the well radius is 0.2 m?

3.3.5. A confined aquifer has a transmissivity of 30 m^2/d and a storage coefficient of 0.0008. A production well of radius 0.2 m is pumped at 200 L/min for 96 hours and shut off. Determine the maximum drawdown and the drawdown 5 days after the start of pumping.

3.3.6. Four wells are used to depressurize a confined aquifer underlying an excavation for a building project. The wells are located 60 m from the center of the excavation at corners of a square. The radius of each well is 0.3 m, the transmissivity of the aquifer is 50 m^2/d, and the storage coefficient is 0.001. (a) What discharge from each well is required if the piezometric head is to be lowered 20 m at the center of the excavation within 30 days? (b) What will be the drawdown at each well? (c) What fraction of the drawdown at each well is associated with interference from each of the other wells present?

3.3.7. A well in a confined aquifer is in hydraulic communication with a reservoir located 1500 m from the well. If the transmissivity of the aquifer is 40 m^2/d and the storage coefficient is 0.001, how long after the start of production will it take for the water level in the well to stabilize if the well produces at a rate of 300 m^3/d?

3.4.1. The saturated thickness of an extensive confined aquifer is 100 ft. A production well fully penetrating the aquifer was continuously pumped at a constant rate of 150 gpm (28,900 ft^3/d) for a period of one day. The drawdown values listed below were measured in an observation well located 300 ft from the production well. Compute the transmissivity, hydraulic conductivity, and coefficient of storage for the aquifer using the Theis method with the spreadsheet described in Example 3.4.1.

Time (min)	Drawdown (ft)	Time (min)	Drawdown (ft)
1	.45	100	2.67
2	.74	200	2.96
4	1.04	300	3.11
7	1.28	400	3.25
10	1.45	600	3.41
21	1.79	800	3.50
40	2.17	1000	3.60
70	2.41	1440	3.81

3.4.2. Repeat problem 3.4.1 using the Jacob Method for analysis of pumping test data.

3.4.3. A well in a confined aquifer is located due north from the location of an inferred fault, and a pumping test is performed to determine whether the fault acts as a barrier or whether there is leakage across the fault. During the test the water level is observed in a monitoring well that is located 80 m due south from the pumping well. During the test the well produces at a rate 208 L/min, and the drawdown values listed below were measured in the monitoring well. Calculate T and S for this aquifer and estimate the effective distance from the pumping well to the fault if it acts as a barrier boundary. What drawdown would you predict in the monitoring well if the pumping well were to produce water at a rate or 800 m^3/d for a period of 30 days?

Time (min)	Drawdown (m)	Time (min)	Drawdown (m)	Time (min)	Drawdown (m)	Time (min)	Drawdown (m)
5	0.01	32	0.15	120	.34	620	.75
8	0.03	45	0.19	180	.43	760	.81
15	0.07	60	0.23	240	.49	840	.84
23	0.11	90	0.29	360	.60	960	.88

3.4.4. The drawdown data shown below is from a pumping test of a confined aquifer. The monitoring well is located a distance 80 ft to the west from the pumping well that is pumped at a rate of 5 ft^3/min. It is known that facies change (change in soil texture) cuts-off the aquifer to the west (that is, it acts as a no-flow boundary). Determine the transmissivity and storage coefficient of the aquifer, and the effective distance to the facies change?

Time (min)	Drawdown (ft)	Time (min)	Drawdown (ft)	Time (min)	Drawdown (ft)
2	0.01	60	4.38	660	9.92
5	0.62	85	5.01	844	10.69
8	1.11	121	5.71	1080	11.50
12	1.64	152	6.19	1440	12.49
17	2.17	180	6.54	1825	13.22
21	2.51	244	7.24	2560	14.55
32	3.22	365	8.24	3210	15.40
45	3.84	480	9.00	3620	15.84

3.4.5. An aquifer test was conducted in a shallow aquifer that is confined above by about 4.5 m of clay. A small creek is located about 60 m from the pumping well. The well was pumped at a rate of 95 L/min for 20 hours, and drawdown was measured in a monitoring well that is located 29 m from the pumping well. The data is shown below (from Walton, 1970). Using the Hantush leaky aquifer well function, estimate *T*, *S*, and *K'/b'*.

Time (min)	Drawdown (m)	Time (min)	Drawdown (m)	Time (min)	Drawdown (m)
5	0.23	75	1.34	958	1.91
28	1.01	244	1.67	1129	1.95
41	1.09	493	1.82	1185	1.96
60	1.24	669	1.86		

3.4.6. Using the parameters determined in Problem 3.4.5, evaluate whether the data could also be fit to a model that treats the creek as a recharge boundary for a confined aquifer.

3.4.7. During the pumping test of Problem 3.4.5, water levels were also measured in two other wells. For the three monitoring wells at distances from the pumped wells of 71.3 m, 28 m, and 29 m, the steady-state drawdown at the end of the test had values of 0.99 m, 2.04 m, and 1.96 m, respectively (from Walton, 1970). The wells are located in opposite directions from the pumping well. Are these data more consistent with a leaky aquifer or with a recharge boundary?

3.4.8. The drawdown data shown below is from a monitoring well located a distance 25 m from the pumping well in the direction of a surface water reservoir. The data shows that the confined aquifer is in hydraulic communication with the reservoir. If the average pumping rate is 550 m^3/d, estimate the transmissivity and storage coefficient of the aquifer and the effective distance to the surface water reservoir. (25 points)

Time (days)	Drawdown (m)	Time (days)	Drawdown (m)	Time (days)	Drawdown (m)
0.012	1.28	0.2	3.25	5	4.52
0.020	1.62	0.3	3.55	15	4.60
0.050	2.25	0.6	3.95	30	4.61
0.09	2.70	0.9	4.12	45	4.62
0.12	2.90	1.5	4.33	60	4.63
0.14	3.00	3	3.47	90	4.63

3.5.1. A well pumps from a confined aquifer at 450 L/min. The transmissivity and co-efficient of storage are 30 m^2/d and 0.00018, respectively. The screen radius is 15 cm while the casing radius is 20 cm. Using Figure 3.5.3, estimate the percent error is made in calculating the drawdown at the well screen using the

Theis or Jacob equation (which neglects well storage) at times of 30 sec, 1 min, 10 min, 100 min, and 1000 min after the start of pumping?

3.5.2. A well pumps from an unconfined aquifer at a rate of 6.7 ft^3/min. The aquifer has an initial saturated thickness of 80 ft, a horizontal hydraulic conductivity of 30 ft/d, a vertical conductivity of 3.0 ft/d, a coefficient of storage of 0.00015, and a drainage porosity (specific yield) of 0.15. At an observation well located 25 ft from the pumped well, use Figure 3.5.4 to estimate the drawdown after 1 hr and after 9 days of pumping?

3.5.3. A well produces 500 m^3/d from an unconfined aquifer with K_r = 10 m/d, an initial saturated thickness of 25 m, and S_y = 0.15. Estimate the drawdown at the well after two months (60 days) of pumping if the effective well radius is 0.2 m.

3.5.4. A groundwater well with an effective radius of 0.3 meters is used to supply irrigation water at a rate of 1500 m^3/d during the agricultural growing season. What is the drawdown at the well at the end of the 4 month (120 day) period of pumping, and does this well's drawdown interfere with the performance of a well located at a distance of 2 km? The aquifer is unconfined with hydraulic conductivity K = 8 m/d, coefficient of storage S = 0.0005, specific yield S_y = 0.15, and initial saturated thickness of H_R = 0.1530 meters.

3.5.5. A well with an effective radius of 1 ft is situated 1,000 ft from a recharge boundary and produces groundwater at a rate of 3.75 ft^3/min. The well penetrates the upper 40-ft of a 150-ft thick confined aquifer and has a steady-state drawdown of 27 ft. If this well is replaced by one with an effective radius of 2 ft penetrating the upper 80 ft of the aquifer and pumping at a rate of 4.75 ft^3/min, calculate the steady-state drawdown using both the Kozeny equation and the equation referenced by Huismann.

3.6.1. A slug test was performed on a cased well in an unconfined aquifer with a saturated thickness of 80 ft. The static water level was at a depth of 15 ft, the bottom of the screen at a depth of 30 ft, and the screen length is 10 ft. The casing radius is 3 inches and the effective well radius at the screened section, including a gravel pack, is 6 inches. A solid cylinder with a volume equivalent to a 3-ft change in water level was placed below the water table. After equilibrium levels were again achieved, the cylinder was quickly removed and the water level measurements were as follows:

Time (min.)	H — h (ft)
0	3
1	2.60
2	2.20
3	1.91
5	1.40
10	0.65
15	0.30

Estimate the hydraulic conductivity of the aquifer using the method of Bouwer and Rice, and Hvorslev's method.

3.6.2. A monitoring well is screened over a 2.6 m thick interval that is confined above and below by lower permeability units. The well has a screen radius of 6 cm and a casing radius of 8 cm. The initial head measured by a pressure transducer submerged within the water column is 2.4 m. Rapid removal of a submerged weight results in a drop in the head of 1.5 m. Using the data shown below, use Cooper's method to estimate the transmissivity, storage coefficient and hydraulic conductivity of the screened interval.

Time (sec)	Head (m)	Time (sec)	Head (m)
17	0.98	295	1.56
30	1.02	420	1.68
42	1.05	540	1.78
95	1.2	725	1.92
130	2.05	1015	2.09
205	1.39	1450	2.25
260	1.47	2450	2.34

3.7.1. The data from a step-drawdown test show that the drawdown reaches 5.48 ft when pumping at 13.4 ft^3/min, 7.13 ft when pumping at 17.1 ft^3/min, and 7.98 ft when pumping at 18.7 ft^3/min. From this data, calculate the specific capacity, efficiency, well losses, and well condition for a pumping rate of 15 ft^3/min.

3.8.1. Six wells are used to depressurize a confined aquifer that lies beneath a planned excavation. If the aquifer head is to be reduced by 20 meters at the center of the excavation over a 45-day period, what is the total groundwater-pumping rate required if all wells pump at the same rate and the aquifer transmissivity is 40 m^2/d and storage coefficient is 0.0006?

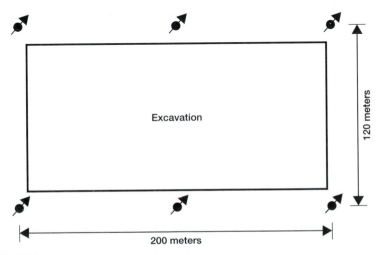

FIGURE 3.8.1 P

3.8.2. An excavation requires dewatering of an unconfined sandy aquifer. The excavation is to be 360 m in length and 90 m wide. Dewatering will be accomplished by two rows of wells along the longer sides of the excavation. The following hydraulic characteristics are estimated for the site: $K = 8$ m/d, $H_R = 42$ m, and $S_y = 0.1$. Dewatering is to occur during a 60 day time period. The required drawdown in the center of the excavation must exceed 18 m while the drawdown at the ends of the excavation must exceed 15 m. Assume that each well can pump at a rate of up to 380 L/min. Design an efficient dewatering system giving the number of wells and their locations and pumping rates.

3.8.3. Show that the following function satisfies the continuity equation for steady flow in a homogeneous aquifer (i.e., show that $f(x,y)$ satisfies Laplace's equation), and give a physical interpretation of the parameters m, a and b (Muskat, 1940). What physical condition does this function represent?

$$f(x,y) = m \ln\left[\cosh\left(\frac{2\pi(y - b)}{a}\right) - \cos\left(\frac{2\pi x}{a}\right)\right]$$

The following contour graph shows the function $f(x,y)$ = 'constant' for different equally spaced 'constant' values for parameter values $m = 1$, $a = 100$, $b = 50$.

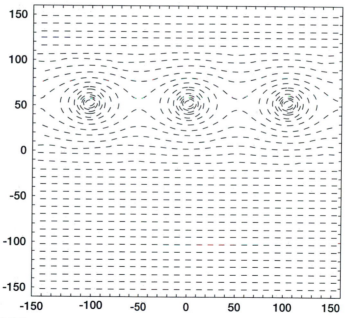

FIGURE 3.8.3 P

3.8.4. With the potential function from Problem 3.8.3, determine the parameter 'm' so that each well pumps at a rate Q_w from a confined aquifer with uniform transmissivity T.

3.8.5. Use the function from problem 3.8.3 to write the potential (head) function for a staggered line drive which consists of a line of injection wells locate along $y = -b$ with wells at $x = (j + \frac{1}{2})a, j = \ldots -2, -1, 0, 1, 2, \ldots$; and a line of production wells located along $y = b$ with wells at $x = j\,a$, $j = \ldots -2, -1,$ 0, 1, 2, . . .

3.9.1. Flow passes through a stratified embankment from a lake with a water level of 10 m on the upstream side to a water level of 3.4 m on the downstream side. The layers and their hydraulic conductivities are shown below. The length of the embankment between the upstream and downstream boundaries is 30 m and the infiltration rate is 25 cm/yr. Find the flow into the lower reservoir through the embankment and the distance from the upstream end at which layer 3 becomes unsaturated.

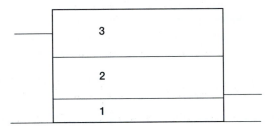

FIGURE 3.9.1 P

$$b_1 = 2.4 \text{ m} \; ; \; K_1 = 4 \text{ m/d} \; ; \; b_2 = 5.5 \text{ m} \; ;$$

$$K_2 = 0.5 \text{ m/d} \; ; \; b_3 = 6 \text{ m} \; ; \; K_3 = 2 \text{ m/d}$$

3.10.1. Estimate the volume of fresh water in storage beneath an island with a land surface area of 21 km^2 that receives an annual groundwater recharge of 25 cm. The hydraulic conductivity and porosity of the aquifer are 3 m/d and 0.35, respectively. Assume that the density of seawater is 1.025 g/cm^3. How many years worth of recharge is in storage?

Chapter Four
The Vadose Zone and Groundwater Recharge

The unsaturated zone, or **vadose zone**, extends from the ground surface down to the water table. The shallow part of the vadose zone is called the **zone of soil water**, and the processes that occur in the zone of soil water are of utmost importance to all terrestrial life. Soil scientists have studied the balance of water, nutrients, and salts within the upper few meters of the soil profile in relation to agriculture. Hydrologists have also investigated the processes in the vadose zone, because it is through this region that water must move in order to reach the water table and the zone of saturation. An understanding of the hydrologic processes that control the water balance in the unsaturated zone is of fundamental importance both to surface water and groundwater hydrology—and ultimately ties together these components of the hydrologic cycle.

For engineering applications, there are at least two major issues that are of interest in studying the hydrology of the unsaturated zone. The first of these concerns the calculation of **abstractions** during rainfall. Abstractions account for the difference between total rainfall and total runoff from a rainfall event over a watershed (Chow et al., 1988). They include interception (wetting of surfaces), depression storage (water accumulating in puddles), and infiltration. A standard and widely used method for calculating abstractions is based on the *Soil Conservation Service* (SCS) curve number (Soil Conservation Service, 1972). While this method is easy to apply, its foundations are obscure. There is considerable interest in developing physically based and practical models for estimating infiltration and direct runoff from rainfall events, and a number of rainfall-runoff and stormwater management models offer alternatives to the SCS method. A related issue in rainfall-runoff modeling is estimation of **antecedent moisture** conditions. For the same amount of rainfall, there is less runoff from a watershed when it is initially dry, compared to conditions immediately following a rainfall event. The practical question concerns how quickly a watershed will dry following rainfall.

The second major issue in studying the hydrology of the vadose zone concerns calculation of the **water balance**. The focus is on estimating the cumulative infiltration that may eventually become groundwater recharge. Estimation of the distribution of net recharge is necessary for understanding the water balance for a groundwater flow system. Estimates of net recharge are also essential for understanding the transport of contaminants from regions near the ground surface to the underlying aquifer, if one is present. Water balance estimates are at the heart of hydrology, and experience has shown that they are particularly difficult in arid environments.

There are a number of terms that must be understood when discussing water flow and the hydrologic budget for the unsaturated zone. The process of water entering from the ground surface into the vadose zone is called **infiltration**. Not all of the water that infiltrates the unsaturated zone actually reaches the water table. A significant amount is returned to the atmosphere through evaporation and transpiration from plants. **Percolation** refers to the passage of water through the various soil layers or rock (Ritzema, 1994). Internal drainage beyond the root zone is sometimes referred to as **deep percolation** (Hillel, 1982). The infiltrating water that actually reaches the water table is called **recharge**. The terms deep percolation and recharge are often used interchangeably.

There are several reasons why the processes controlling the flow of water in the vadose zone are more complex than in the saturated zone of an aquifer. In the vadose zone, only part of the void space is filled with water. The remainder is filled with soil air. The amount of water retained against gravity varies significantly with soil texture. Coarse-textured, sandy soils may hold as little as 10–20 percent water saturation after a long period of drainage. Fine-textured silts or clays may hold as much as 90 percent saturation. The permeability or hydraulic conductivity of the soil is a nonlinear function of the water content. This situation contrasts with conditions in an aquifer where all of the pore space is filled with water and the hydraulic conductivity at a given location is constant. The average hydraulic gradient in the vadose zone is often near unity, which is 100 to 1000 times larger than typical gradients in aquifers. Near the ground surface, the temperature varies in response to cyclic inputs of radiant energy, and water vapor can move because of both temperature and vapor pressure gradients. Finally, soils near the ground surface are very active chemically and biologically.

To describe the flow of water in the vadose zone, two phenomenological relationships are required. The first of these is the relationship between water pressure and volumetric water content (or water saturation). In the vadose zone, liquid water usually has a negative gauge pressure—that is, its pressure is below atmospheric pressure.[1] The drier the soil, the more negative the pressure. This first relationship of decreasing pressure with decreasing water content reflects the capacity of a soil mass to pull water in (imbibe water) as a function of the amount of water present. The second phenomenological relationship gives the hydraulic conductivity of the soil as either a function of water content or water

---

[1]An exception occurs within perched regions with saturated conditions, where the water pressure is positive.

pressure. This second relationship specifies the ability of the soil mass to transmit flow under a given head gradient. Both of these relationships are nonlinear and soil-dependent.

In this chapter, we first discuss the nature of soil water in the vadose zone and the phenomenal relationships between soil water saturation, pressure, and hydraulic conductivity. The generalized form of Darcy's Law that applies for unsaturated flow is examined. Methods for measuring soil properties are reviewed. Following this discussion, models that describe hydrologic processes including infiltration of water during rainfall events, redistribution of water following rainfall, and water loss to the atmosphere through evapotranspiration are discussed. The problem of evaporation from a shallow water table is briefly considered. Finally, methods for estimating the water balance for a soil profile are discussed.

4.1   Soil Water in the Vadose Zone

The water table is defined as the locus of points at atmospheric pressure. Below the water table the pressure is positive, while above the water table the pressure is negative (less than atmospheric pressure). The water pressure in the vadose zone is often called either the **capillary pressure** or **suction pressure (tension)**, and by either name it is treated as a positive quantity. Thus, large capillary pressures correspond to large negative water phase pressures, and soils under such conditions have small water saturations usually.

For unsaturated soil above the water table, part of the pore space is filled with water while the rest is filled with air. The sum of the volumetric water and air contents is equal to the total porosity.

$$\theta_w + \theta_a = n \tag{4.1.1}$$

Equation 4.1.1 may also be written in terms of saturation rather than volumetric water content. The equivalent form of Equation 4.1.1 is

$$S_w + S_a = 1 \tag{4.1.2}$$

Both the volumetric water content and the water saturation are macroscopic parameters. They apply to a mass of soil but not to individual pores on the microscopic scale.

It is important to understand the distribution of water at the pore scale within a mass of soil. Consider a soil mass that is initially dry. Upon addition of water, the water is first adsorbed as a film on the surface of the soil grains. The thin skin of **adsorbed water** covering the grains is called **pellicular water** (Bear, 1972). Pellicular water is held strongly by van der Waal's forces and can have a suction pressure corresponding to many tens of atmospheres (bars). This idea means that a suction pressure of many tens of bars would have to be applied to remove this water from the soil. The suction pressure is a thermodynamic measure of the energy state of the water within the soil. Free water would cavitate before the suction pressure reached one bar. The water within a distance of about 100 A (approximately 30 molecular layers) is strongly held by these forces.

With further addition of water, the water starts to accumulate at the contact points between grains that represent the smallest pore-space openings in the soil. This accumulating water is referred to as **pendular water**. The pendular water (pendular rings) is held at the contact points by capillary forces. Capillary forces are caused by the presence of surface tension between the water and air phases within the soil pore space, and these capillary forces cause water to move into the smallest pores and pore openings first. Pellicular and pendular water are shown schematically in Figure 4.1.1.

At small volumetric water content conditions, water movement in soil is slow, even under large gradients, because the water must move across the thin films of adsorbed water as it passes from one pendular ring to the next. As the water content increases, the pendular rings grow, and the thickness of the pellicular water film increases. With a film thickness in excess of 500 to 1000 A, the water is free to move under imposed energy gradients. The saturation at which a continuous wetting phase forms that is to move is called the **equilibrium wetting phase saturation**. Above this critical saturation, the saturation is called **funicular**, and flow of water is possible. Funicular saturation is shown in Figure 4.1.2.

As the water content continues to increase, eventually the air becomes isolated in individual pockets in the larger pores, and flow of the air phase is no longer possible. The saturation at which this process happens is called **insular** saturation and is shown in Figure 4.1.3. If the air is under positive gauge pressure, these air pockets will eventually dissolve, and complete water saturation will be achieved. Under normal conditions in the unsaturated zone, however, full water saturation is not achieved because of residual or entrapped air.

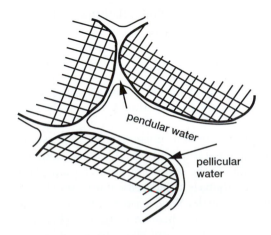

pendular water

pellicular water

FIGURE 4.1.1 Distribution of water as pellicular and pendular water at small soil water contents

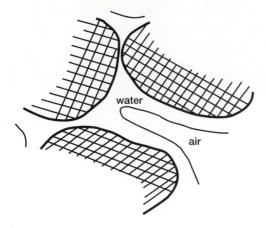

FIGURE 4.1.2 Intermediate water content showing funicular saturation

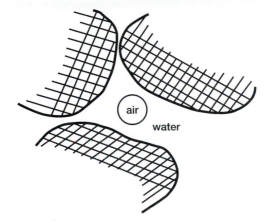

FIGURE 4.1.3 High water content showing insular saturation

There are two issues that arise when considering flow processes in unsaturated porous media. These issues are associated with the presence of the interface separating the water and air phase and the presence of the solid phase. The first of the issues concerns the media **wettability**. For mineral soils, there is more interfacial energy associated with air-solid phase interfaces than with water-solid phase interfaces. Thus, minimizing the total amount of air-solid interface minimizes the overall system energy, and the soil surface is preferentially covered with water. For this system, water is the wetting phase and air is the nonwetting phase, with respect to the soil. This situation is generally assumed to be the case for natural soils and leads to the description given earlier of the distribution of water that is introduced to dry soil.

The second issue concerns the pressure difference across an interface that is curved. Consider an interface separating the water and air phases, as shown

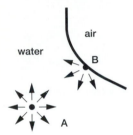

FIGURE 4.1.4   Cohesive forces between molecules within the interior of a fluid and at the interface

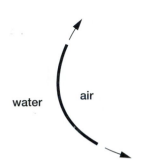

FIGURE 4.1.5   Curved interface separating water and air phases

in Figure 4.1.4. For a molecule at point A within the interior of the water phase, the cohesive forces between molecules pull equally in all directions, and there is no net force. For a molecule at point B near the surface, the cohesive forces result in a net force toward the interior of the water phase. To move a molecule from A to B, we must break the neighbor bonds and move against this force field. The result is that molecules at the interface have more energy than molecules within the bulk phase. The excess surface energy, σ, is called the **interfacial energy** (ergs/cm^2), or **surface tension** (dynes/cm).

Figure 4.1.5 shows water and air phases separated by a curved air-water interface. The interface acts as a membrane under tension, and a force balance shows that the pressure in the air phase is greater than that in the water phase, given the curvature of the interface shown (see Adamson, 1978). This pressure difference is given by

$$p_a - p_w = \sigma\left(\frac{1}{r_1} + \frac{1}{r_2}\right) = \frac{2\sigma}{r_c} \tag{4.1.3}$$

where r_1 and r_2 are two radii of curvature of the interface taken along perpendiculars within the surface through the normal to the interface, and r_c is the average radii of curvature. A theorem due to Euler shows that r_c is an invariant for the surface. This idea means that the value of r_c does not depend upon which set of perpendiculars within the surface that r_1 and r_2 are taken along. Within the theory of capillarity, Equation 4.1.3 is called the **Laplace equation**.

In the vadose zone, the air pressure is often close to atmospheric pressure. Under such conditions, Equation 4.1.3 gives

$$p_w = -\frac{2\sigma}{r_c} \qquad (4.1.4)$$

Equation 4.1.4 shows that the water (gauge) pressure is negative, and that it becomes more negative with smaller values of the average radii of curvature of the air-water interface. The value of r_c becomes smaller as the interface is pulled into smaller and smaller pores, and the soil loses water.

Under hydrostatic conditions, the water pressure change with elevation is given by the following hydrostatic pressure equation:

$$p_w = -\rho_w g z \qquad (4.1.5)$$

where z is the elevation above the water table (recall that the water table is defined as the locus of points where the water pressure is equal to the atmospheric pressure), and it is assumed that the density of water is constant. Equation 4.1.5 also verifies that the water pressure is negative in the vadose zone—that is, the water is under a state of tension. Together, Equations 4.1.4 and 4.1.5 confirm that the datum for elevation measurement in Equation 4.1.5 is the water table, because this is the elevation that one measures in an observation well, which is simply a pore space with a large (nearly infinite) radius of curvature.

The **capillary pressure**, p_c, is defined as the pressure difference between the air (nonwetting phase) and the water (wetting phase) in a porous medium. This definition gives

$$p_c = p_a - p_w \qquad (4.1.6)$$

It is often convenient to measure the capillary pressure in terms of an equivalent height of a water column. The **capillary head** (or suction head) is defined by

$$\Psi = \frac{p_c}{\rho_w g} \qquad (4.1.7)$$

According to Equations 4.1.5 to 4.1.7, $\Psi > 0$ within the vadose zone.

From Equations 4.1.5 and 4.1.6, it is noted that the capillary pressure increases with elevation above the water table (assuming $p_a = 0$). The Laplace Equation 4.1.4 shows that the mean radius of curvature of the interfaces separating the water and air phases decreases with increasing capillary pressure, and thus with elevation above the water table. This idea suggests that the interfaces are present in smaller and smaller pores with increasing elevation. Thus, the water phase saturation decreases while the air phase saturation increases with elevation above the water table. These arguments suggest that at higher capillary pressures, the air-water interface is pulled out of the larger pores and remains present only in the smaller pores.

Therefore, at higher elevations above the water table, water is retained in the smaller pores while the larger pores contain air. A schematic view of the macroscopic pore water distribution for a soil mass is shown in Figure 4.1.6, where the bounding curve is the **pore size density function** (fraction of pores at a given size).

4.2   The Soil Water Characteristic Curve

The **soil water characteristic curve** provides the relationship between the capillary pressure and water content for a particular soil. As suggested by Equation 4.1.3, an increase in the capillary pressure results in the interface between the water and air phases moving into smaller pores, with a corresponding dewatering of the porous medium. The macroscopic distribution shown in Figure 4.1.6 of water and air phases within the pores of a soil mass corresponds to a particular capillary pressure. The integral of the water-filled pore fraction corresponds to the fraction of the total porosity that is filled with water. This quantity is the water saturation, S_w (see Equation 1.3.7). The relationship between capillary pressure and water content (or S_w) is a function of the pore size distribution.

Typical characteristic curves for sand and clay soils are shown in Figure 4.2.1. These curves correspond to soil samples that are initially saturated and that are drained by increasing the capillary pressure. At first, a relatively small number of the soil pores are drained with increasing p_c because there are a limited number of large pores (which are drained first, according to Equation 4.1.3). For the sand soil, there is a rapid decrease in water content with increasing p_c through the range of most plentiful pore size. The clay soil, on the other hand, has a wide range of pore sizes, so that there is not a range of rapid decrease in water content with small changes in suction. At large p_c, there is little change in water content with increasing capillary pressure. Ultimately, increasing the capillary pressure cannot remove more water. The saturation at which this condition occurs is called the **irreducible water saturation**.

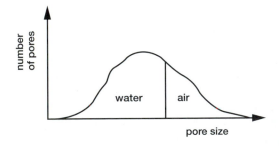

FIGURE 4.1.6 Water and air occupancy of different pore sizes for a soil mass

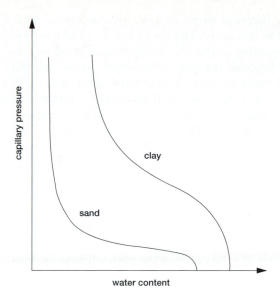

FIGURE 4.2.1 Typical characteristic curves for sand and clay soils

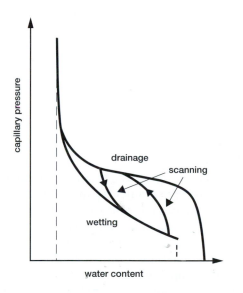

FIGURE 4.2.2 Soil water characteristic curve showing drainage, wetting and scanning curves

Hysteresis

The soil water characteristic for most soils shows **hysteresis**. This characteristic means that the $p_c - \theta_w$ relationship depends on the saturation history, as well as the existing water content. Figure 4.2.2 shows the soil water characteristic curve for drainage and wetting conditions. The upper curve corresponds to a soil sample that is initially saturated and is drained by increasing the capillary pressure. This curve is called the **drainage curve**, and the parameters for most soil characteristic models are fit to this curve. The lower curve, called the **wetting curve**, corresponds to a re-wetting of the soil. Ultimately, insular saturation is reached, and the air phase becomes trapped in the largest pores—and the wetting process ceases. From this point onward, there is no unique relationship between the water pressure and the water content. If the wetting or drainage process had stopped anywhere between the endpoint limits and had reversed, then the **scanning curves** would be followed. What actually happens in the field during a rainfall event is that the soil initially becomes wetted, following a wetting scanning curve, and then starts to drain following a drainage scanning curve. If wetting or drainage occurs for a long enough time period, then the primary wetting or drainage curves are reached.

There are two primary reasons why the wetting and drying curves are different. First, the pores of the soil are not of uniform size, and variably sized pores drain and fill differently. The capillary pressure needed to drain a pore of variable radius is controlled by the minimum radius of the pore, whereas the capillary pressure needed to fill an empty pore with water is controlled by the maximum radius. The second reason is that when an unsaturated soil is wetted, air bubbles are inevitably trapped in the pores of the soil, which causes θ_w to be less than n at $\Psi = 0$. Even wetting a column of soil with the base of the column immersed in a pool of water and the top subjected to a vacuum does not remove all the air bubbles from a soil. More complete discussion of hysteresis may be found in articles by Topp and Miller (1966) and Mualem (1986), and a good summary of nomenclature is provided by Klute (1986).

4.3   Darcy's Law and Richards' Equation

In Chapter 2, we saw that Darcy's Law is interpreted as a balance between the forces that drive the flow (gravity and pressure gradients) and viscous resistance forces. Darcy's Law results when the viscous forces are proportional to the flow rate (Darcy velocity) to the first power. The same conditions are true for flow in the vadose zone when the soil matrix is not saturated with water. Buckingham (1907) proposed a model for describing the flow of water though unsaturated soil. The two primary assumptions upon which his model is based are that 1) the driving force for water flow is the sum of the **matric** and gravitational potential gradients, and 2) that the hydraulic conductivity of the unsaturated soil is a function of the water saturation or matric potential. The term "matric" is used to denote the thermodynamic pressure of water within unsaturated soil. We have already seen that in the vadose zone, the water pressure is negative (less than atmospheric) because of capillary forces. For dry soils, adsorption of water to the soil particles (hydration envelopes are formed over the particle surfaces) also becomes important. The matric potential (or matric suction) is used to denote both the effects of capillarity and adsorption (Hillel,

1980). For our discussions, we will use the capillary head, Ψ, to denote the effects of both capillarity and adsorption. Two decades after Buckingham published his model, it was recognized as a form of Darcy's Law (Ritzema, 1994). For this reason, the model is sometimes called the Buckingham-Darcy Law (Jury et al., 1991).

For the flow of water in unsaturated porous media, Darcy's law may be written as

$$\tilde{q}_w = - K_w(\theta_w)\nabla h_w \qquad (4.3.1)$$

In Equation 4.3.1, $K_w(\theta_w)$ is the hydraulic conductivity to water flow for a soil at a water content θ_w. In an unsaturated porous medium, more than one fluid phase is present and more than one phase can flow, so an additional subscript is needed to denote which fluid phase is being referenced. We will use the subscripts "w" for water, "a" for air, and in later chapters, "o" for oil, which is a nonaqueous phase liquid (such as gasoline).

Because the soil pore space contains both water and air, the value of $K_w(\theta_w)$ is less than the water saturated hydraulic conductivity K_{ws}. The water **relative permeability** is defined as the ratio of the hydraulic conductivity of a porous medium at a specified water content to its conductivity under fully saturated conditions.

$$k_{rw}(\theta_w) = \frac{K_w(\theta_w)}{K_{ws}} \qquad (4.3.2)$$

The relative permeability for water varies from 1 under water-saturated conditions to 0 at the irreducible water content. The relative permeability may be expressed either as a function of water content or capillary head. Measurements of the $k_{rw}(\theta_w)$ relation do not show much hysteresis upon wetting and drying of the soil. The $k_{rw}(\Psi)$ function, however, does show considerable hysteresis. Where appropriate, we prefer to work with the water content (or water saturation) relationship. For heterogeneous media, however, where abrupt changes in water content are possible, then the capillary head formulation is more appropriate.

For unsaturated flow of water, the water head is

$$h_w = z - \Psi \qquad (4.3.3)$$

With Equations 4.3.1, 4.3.2, and 4.3.3, Darcy's Law may be written

$$\tilde{q}_w = - K_{ws}k_{rw}(\theta_w)\hat{k} + K_{ws}k_{rw}(\theta_w)\nabla\Psi \qquad (4.3.4)$$

Equation 4.3.4 states that water will move downward under the force of gravity at a rate determined by the soil hydraulic conductivity and wetness and will additionally move in response to capillary suction gradients at a rate that is also dependent on soil wetness.

Continuity Relations

The general water continuity equation for unsaturated flow is

$$\frac{\partial \theta_w}{\partial t} + \text{div}(\tilde{q}_w) = S' \tag{4.3.5}$$

which should be compared with Equation 2.3.32. In Equation 4.3.5, S' is the source strength, which in the root zone is negative and accounts for plant uptake of soil water. Beneath the root zone, there usually are no sources or sinks and the term vanishes, which is the case considered next. With $S' = 0$, the continuity equation states that the rate of increase in water content at a point plus the net volumetric flux of water away from that point is equal to zero. Substitution of Darcy's Law in the form of Equation 4.3.1 or 4.3.4 yields

$$\frac{\partial \theta_w}{\partial t} = \text{div}(K_w(\theta_w) \, \text{grad}(h_w)) = \text{div}(K_w(\theta_w)\hat{k} - K_w(\theta_w) \, \text{grad}(\Psi)) \tag{4.3.6}$$

The problem with this formulation is that both θ_w and h_w (or Ψ) appear, and we have one equation with two unknowns. Recall, however, that the soil water characteristic relates θ_w and Ψ. The simplest way to proceed is to either write the equation in terms of Ψ or write it in terms of θ_w. The former gives

$$C(\Psi)\frac{\partial \Psi}{\partial t} = \text{div}(K_w(\Psi) \, \text{grad}(\Psi)) - \frac{\partial K_w(\Psi)}{\partial z} \tag{4.3.7}$$

In Equation 4.3.7,

$$C(\Psi) = -\frac{d\theta_w}{d\Psi}$$

is the **capillary (soil water) capacity**, determined from the characteristic curve for the soil. The alternative formulation is

$$\frac{\partial \theta_w}{\partial t} = \text{div}(D(\theta_w) \, \text{grad}(\theta_w)) + \frac{\partial K_w(\theta_w)}{\partial z} \tag{4.3.8}$$

where

$$D(\theta_w) = -K_w(\theta_w)\frac{d\Psi}{d\theta_w}$$

is the **capillary (soil water) diffusivity**. There are advantages and disadvantages to both formulations. The formulation with Ψ as the unknown variable, however, is more general and versatile in its application. As an example, for a soil profile with an abrupt change in texture, the water content changes discontinuously while the pressure remains continuous. This situation means that if Equation 4.3.8 is applied, then it must be applied piecewise with a flux balance condition at the boundary between regions of homogeneous texture. Furthermore, Equation 4.3.7 is applicable for both saturated and unsaturated conditions, while Equation 4.3.8 applies to only unsaturated conditions (the capillary diffusivity becomes infinite under saturated conditions). Both formulations (most often the former) are called **Richards' equation**, after their development by Richards (1931). You should recognize that Richards' equation neglects the resistance to water flow caused by the flow of air. Under most applications, this characteristic is not a problem. Various exceptions arise, however, including that for infiltration under ponded conditions with a shallow water table, wherein this resistance cannot be ignored. In addressing these issues, Morel-Seytoux (1973) and Vauclin (1989) present alternative formulations based on fractional flow theory.

There are two limiting cases that are also of interest. For large water contents, the product $D(\theta_w)\nabla\theta_w = -K_w\nabla\Psi$ is small, and Equation 4.3.8 takes the approximate form

$$\frac{\partial\theta_w}{\partial t} - \frac{\partial K_w(\theta_w)}{\partial z} = 0 \qquad \textbf{(4.3.9)}$$

Equation 4.3.9 leads to the **kinematic theory** of modeling of vertical unsaturated flow, where capillary pressure gradients are neglected. The kinematic theory is also applicable if $\Psi \cong constant$ within the profile. For this theory, Darcy's Law for vertical flow takes the particularly simple form

$$q_z = -K_w(\theta_w) \qquad \textbf{(4.3.10)}$$

That is, the flow is downward under a "unit gradient."

The second limiting case occurs when gravitational forces are negligible with respect to capillary pressure forces. This theory is the theory of absorption or desorption of soil water, and Equations 4.3.7 and 4.3.8 become

$$C(\Psi)\frac{\partial\Psi}{\partial t} = \mathrm{div}(K_w(\Psi)\,\mathrm{grad}(\Psi)) \qquad \textbf{(4.3.11)}$$

$$\frac{\partial\theta_w}{\partial t} = \mathrm{div}(D(\theta_w)\,\mathrm{grad}(\theta_w)) \qquad \textbf{(4.3.12)}$$

These are forms of the **nonlinear diffusion equation**, and they are especially useful for modeling evaporation processes—as will be seen later in this chapter.

4.4   Measurement of Soil Properties

Analysis of most problems dealing with flow and contaminant transport in unsaturated soil begins with an understanding of liquid fluxes. If one cannot accurately predict the direction of water movement or the flux, there is little hope of accurately predicting the water balance or rates and patterns of contaminant movement. Also, the accuracy of predictions of flow and contaminant transport in unsaturated soil is limited by one's ability to describe accurately the subsurface conditions and to quantify soil properties. Imperfect characterization of subsurface stratigraphy and inaccurate description of the properties of the subsoils (including the spatial distribution of those properties) usually are the most important factors that limit the overall accuracy of modeling.

Two critical relationships must be defined for all unsaturated, subsurface units that are to be assessed. These are 1) the soil water characteristic (retention) curve and 2) the relationship between hydraulic conductivity and either pressure head or water content, which for convenience is expressed through the relative permeability function such that

$$K_w(\Psi) = k_{rw}(\Psi)K_{ws} \; ; \; K_w(\theta_w) = k_{rw}(\theta_w)K_{ws} \qquad \textbf{(4.4.1)}$$

The hydraulic conductivity at full saturation, K_{ws}, is often evaluated independently from the other parameters. Typically, the moisture characteristic curve and relative permeability function are determined in the laboratory, whereas the hydraulic conductivity at saturation is assessed with a combination of laboratory and in situ measurements. To define initial soil conditions, information is required on the initial capillary pressure heads that exist in the field soil; this information can be obtained by direct measurement of Ψ (laboratory or field) or by measuring θ_w of the soil and estimating Ψ from the moisture characteristic curve.

4.4.1   Measurement of Soil Water Content and Capillary Pressure Head

A number of methods are available for measuring the soil water content (Gardner, 1986). The simplest and most widely used is the **gravimetric method**. A moist soil sample is weighed, dried in an oven (usually at 105 degrees C for 24 hours), and then reweighed. The gravimetric water content is the mass of water per mass of dry soil. The gravimetric method is simple but destructive of the soil sample in that its state is modified through measurement.

Nondestructive methods for measuring the soil water content include neutron scattering, gamma ray attenuation, and time-domain reflectometry. The **neutron scattering** method uses a radiation source emitting high-energy neutrons (~5MeV) and a detector that are both lowered into the soil profile through an aluminum access tube. The neutrons collide with nuclei of atoms in the surrounding soil and are slowed substantially by collisions with nuclei of similar mass (hydrogen in water). The detector will count only slowed-down (thermal) neutrons, and thus the neutron count may be calibrated with the amount of water present. The measurement volume varies from about 15 cm in wet soil to 50–70 cm in dry soil (Van Bavel et al., 1956; Gardner, 1986). Neutron attenuation is often used in the field because repeated measurements of the water content profile can be made, and the instrument has a rapid response time. Measurements near the soil surface are inaccurate, however, and interference from other light nuclei is possible. In addition, the radiation source (usually americium-beryllium or radium-beryllium) can pose a health risk unless appropriate care is taken.

The **gamma ray attenuation** method uses a narrow beam of gamma rays emitted from a cesium−137 source that are sent through a soil sample of known thickness and measured by a detector located on the other side of the sample. The amount of attenuation depends on the wet bulk density of the sample, so changes in attenuation may be calibrated with changes in water content. This method is most commonly used in laboratory experiments with columns or sandboxes with small thickness (width) but arbitrary depth and length. The soil water content can be obtained with good resolution, although the equipment is costly and can pose a health hazard unless proper shielding precautions are taken. If two sources with different energies are used, then the method can be applied with swelling soils or for multiphase flow experiments.

Time domain reflectometry (TDR) is a recent method for nondestructive measurement of the soil water content without using a radiation source. The measurement uses the dielectric properties of the soil and how these change with water content. A probe with two or three arms (approximately 30 cm length or less) that act as a wave guide is placed in the soil. A step pulse of electromagnetic radiation is sent along the wave-guides and its propagation travel time is measured. This travel time depends on the dielectric properties of the soil surrounding the guides, and these properties may be calibrated with water content. The relationship between the apparent dielectric constant and volumetric water content is only weakly dependent on soil type, soil density, soil temperature, and salt content (Topp and Davis, 1985). Thus, TDR may be used for many soils without calibration. Highly accurate measurements can be made. Topp et al. (1980) reported measured volumetric water content with an accuracy of +/- 0.02 (m^3/m^3).

Capillary Pressure Measurement

The capillary pressure head can be measured in two ways: (1) an undisturbed sample of soil can be obtained and Ψ measured in the laboratory on that sample, or (2) the measurement can be obtained directly in the field. The most commonly used device for measuring Ψ in both the laboratory and field is the tensiometer, such as that shown in Figure 4.4.1. The **tensiometer** consists of a porous element that is inserted into the soil and a pressure-sensing device. The tensiometer is initially saturated with a liquid (usually water). When the saturated porous element is brought into contact with the soil, the soil will try to suck water out of the tensiometer. A negative pressure develops in the tensiometer, however. In a short period of time, equilibrium is established between

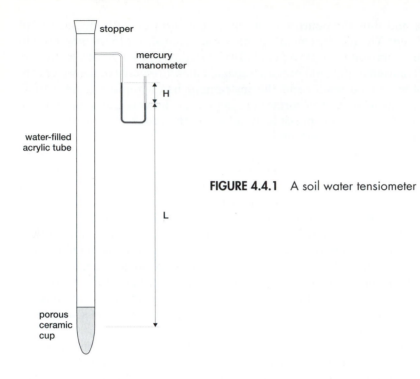

stopper

mercury
manometer

H

water-filled
acrylic tube

L

FIGURE 4.4.1   A soil water tensiometer

porous
ceramic
cup

the soil water and the water in the tensiometer. The negative pressure reading indicated on the tensiometer is equal to Ψ, although a correction for the elevation difference between the pressure gauge and the porous element must be made. Tensiometry is described in detail by Cassell and Klute (1986) and Stannard (1990).

One limitation of the tensiometer is that it cannot read Ψs more negative than approximately 0.9 bars due to problems with cavitation of water. The **thermocouple psychrometer** (Rawlins and Campbell, 1986) is a convenient device for laboratory measurement of water potentials between 1 and 75 bars (100 to 7500 kPa). The thermocouple psychrometer works as follows. A tiny drop of water is formed on the measuring junction of a miniature thermocouple. As the drop evaporates, the junction cools to the dew point. The thermocouple generates an electromotive force that is proportional to the difference between wet and dry bulb temperatures, which can be related to the relative humidity of the soil air. A **microvoltmeter** is used to measure the output from the thermocouple psychrometer. Provided that equilibrium exists, the relative humidity of the soil air can be related to the total water potential. One problem with the thermocouple psychrometer is that it measures the total water potential, which includes capillary, osmotic, and adsorptive effects. Also, the thermocouple is sensitive to temperature fluctuations and is less accurate near the upper end of its range—in the range of about 1 to 3 bars (100 to 300 kPa).

4.4.2   **Soil Water Characteristic Curve**

An important problem encountered when attempting to measure the moisture characteristic curve is the fact that different curves are measured for wetting and drying conditions. This situation is the problem of hysteresis described in Section 4.2 and shown schematically in Figure 4.2.2. To develop a moisture characteristic curve for a soil, the first step is to determine whether or not hysteresis will be taken into account. Hysteresis is usually ignored to simplify the analysis. Even if hysteresis is ignored, however, it is important to recognize which condition (wetting or drying) is applicable to a particular problem to be modeled and to develop the moisture characteristic curve for that condition.

Three approaches can be taken to determine the moisture characteristic curve for a particular soil. The first technique is to estimate the curve from published data for similar soils. Information on the soil type can be used in conjunction with catalogs of published soil moisture characteristic curves, e.g., Mualem (1976a), to estimate the moisture characteristic curve for a particular soil. Gupta and Larson (1979) describe the use of grain-size distribution data, organic content, and bulk density to estimate the moisture characteristic curve. Rawls and Branensiek (1982) collected and analyzed data from 500 soils. Ahuja et al (1985) compared predicted soil moisture characteristic curves with measured Ψ-θ_w relationships for 189 soil cores; the mean error in predicted water content (based on measured Ψ) was 10–30 percent.

Analytic Models of the Soil Characteristic Curve

The second and probably most common technique is to assume an analytic function for $\Psi(\theta_w)$. An advantage of using these parametric models is that there has been a large amount of field data collected and fit to the models, and in this respect, this approach to estimating the soil characteristic curve is closely aligned with the first. The literature tabulations and data models provide an estimate of the parameters and their expected range based on the soil texture. This information is particularly helpful, because the type of information that is initially available for a field site or regional investigation is often limited. In this way, one can often make rough estimates of the parameters from texture information without requiring additional laboratory or field-testing. Because developers of closed-form analytical solutions and computer codes for flow problems in unsaturated soils often assume that the $\Psi(\theta_w)$ function follows one of the commonly used relationships, the engineer or scientist is usually compelled to force the assumed $\Psi(\theta_w)$ relationship to fit the available data as best as possible. There are a number of parametric models that have been suggested in the literature. The two most widely used models are the power-law models of Brooks and Corey (1964) and Campbell (1974) and the model suggested by van Genuchten (1980). Most of the data tabulations for these models are for fitting the drainage curve, such as those shown in Figure 4.2.1.

Brooks and Corey Model

The Brooks and Corey (1964) model takes the form of a power-law relating the effective or **reduced saturation**, Θ, to the *capillary pressure head*, Ψ. The mathematical form of this model is

$$\Theta = 1 \text{ for } \Psi \leq \Psi_b \ ; \ \Theta = \left(\frac{\Psi_b}{\Psi}\right)^{\lambda} \text{ for } \Psi > \Psi_b \qquad \textbf{(4.4.2)}$$

where the reduced saturation is defined by

$$\Theta = \frac{\theta_w - \theta_{wr}}{n - \theta_{wr}} = \frac{S_w - S_{wr}}{1 - S_{wr}} \qquad \textbf{(4.4.3)}$$

The parameters θ_{wr} and S_{wr} are the irreducible water content and **irreducible saturation**, respectively. The parameter Ψ_b is the **displacement pressure head** (also called the bubbling or air entry capillary pressure head). The parameter λ is called the **pore size distribution index**.

With reference to Figure 4.4.2, the displacement pressure head has the physical interpretation as the height of capillary rise, such as the thickness of the saturated region of the capillary fringe above the water table. It provides a measure of the size of the largest soil pores; that is, those that would be dewatered first under an applied suction. The value of Ψ_b is smaller for the sand than for the clay because the size of the largest pore within the soil matrix is greater for the sand, and thus, this largest pore may be dewatered at a smaller suction.

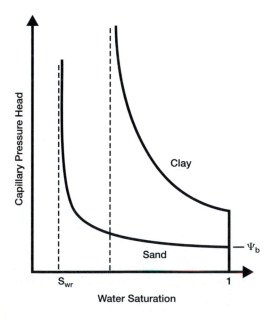

FIGURE 4.4.2 Brooks and Corey model representations of the soil water characteristic curve

The parameter λ characterizes the range of pore sizes within the soil, with larger values corresponding to a narrow size range and small values corresponding to a wide distribution of pore sizes. A soil characterized by a large value of λ would have the size of many of its pores fall within a narrow range. These would all be dewatered through a relatively small change in the capillary pressure. Such a soil would have a sharp, well-defined capillary fringe, such as suggested for the sand in Figure 4.4.2. A soil with a small λ, on the other hand, would experience only a small change in water saturation with height above the capillary fringe—at least under equilibrium conditions.

The parameters θ_{wr} and S_{wr} represent the water content of the soil that cannot be removed through application of a large suction pressure. They correspond to the adsorbed film of water and the water held tightly at points of contact between grains.

The model of Campbell (1974) is essentially identical to Equation 4.4.2, except that the irreducible water content is assumed to equal zero, and Campbell uses a coefficient $b = 1/\lambda$ as the power.

van Genuchten Model

The form of the model suggested by van Genuchten (1980) is

$$\Theta = \left(\frac{1}{1 + (\alpha\Psi)^N}\right)^M \tag{4.4.4}$$

for $\Psi \geq 0$. The parameters in the **van Genuchten model** are α, N, and M. With Mualem's (1976) relative permeability model (see the next section), the parameters N and M are related through $M = 1 - 1/N$ or $N = 1/(1 - M)$. This model results in a continuous soil characteristic curve throughout the range $\Psi > 0$. For this reason, it is sometimes preferred to the Brooks and Corey model for use in numerical modeling.

For large capillary heads, the Brooks and Corey and the van Genuchten models become identical if

$$\lambda = N - l \ ; \ \Psi_b = \frac{1}{\alpha} \tag{4.4.5}$$

Equation 4.4.5 is useful for relating the parameters of the two models. Alternately, Lenhard, et al. (1989) suggest that throughout the range of water contents, a better relationship is found from

$$\lambda = \frac{M}{1 - M}(1 - 0.5^{1/M}) \ ; \ \Psi_b = \frac{\Theta_*^{1/\lambda}}{\alpha}(\Theta_*^{1/M} - 1)^{1-M} \tag{4.4.6}$$

where

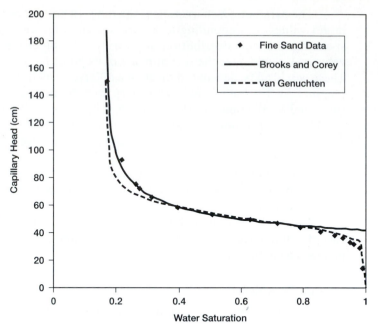

FIGURE 4.4.3   Comparison of fitted Brooks and Corey ($\lambda = 3.7$, $\Psi_b = 4\ 1$ cm) and van Genuchten ($M = 0.871, \alpha = 0.0202$) models ($S_r$-0.167) with data from Brooks and Cory

$$\Theta_* = 0.72 - 0.35 \exp(-N^4)$$

Figure 4.4.3 compares the fit of the Brooks and Corey and van Genuchten models to measured data for a fine sand (Brooks and Corey, 1964).

Parameters and Soil Texture Data

Measurements of the characteristic curve for a soil provide paired values of water content and capillary pressure. Such data may be used directly, but for many applications it is convenient to fit the parametric models of the soil water characteristic described above to laboratory or field data. Fitting of parametric models may be done manually. Alternatively, there are computer programs such as the *Retention Curve* (RETC) model (van Genuchten et al., 1991) that may be used to optimally fit parametric models to measured data.

Extensive studies have been undertaken by the *United States Department of Agriculture* (USDA) and the United States *Environmental Protection Agency* (EPA) to determine the soil water retention parameters for the Brooks and Corey model and the van Genuchten model, respectively. One of the objectives of these studies is to provide estimates of the model parameters and their variation based on readily obtainable soil texture information. Rawls and Brakensiek (1989) and Rawls et al. (1993) summarize much of the USDA data, while

TABLE 4.4.1 Mean (Standard Deviation) Values of Soil Water Retention Parameters (after Carsel and Parrish, 1988)

Soil Texture	Porosity	Hydraulic Conductivity (m/d)	Irreducible Water Content	Capillary Pressure Head* (m)	Pore Size Distribution Index
Clay	0.38 (0.09)	0.048 (0.10)	0.068 (0.034)	1.25 (1.88)	0.09 (0.09)
Clay loam	0.41 (0.09)	0.062 (0.17)	0.095 (0.010)	0.53 (0.42)	0.31 (0.09)
Loam	0.43 (0.10)	0.25 (0.44)	0.078 (0.013)	0.28 (0.16)	0.56 (0.11)
Loamy sand	0.41 (0.09)	3.5 (2.7)	0.057 (0.015)	0.081 (0.028)	1.28 (0.27)
Silt	0.46 (0.11)	0.060 (0.079)	0.034 (0.010)	0.62 (0.27)	0.37 (0.05)
Silty loam	0.45 (0.08)	0.11 (0.30)	0.067 (0.015)	0.50 (0.30)	0.41 (0.12)
Silty clay	0.36 (0.07)	0.0048 (0.0260)	0.070 (0.023)	2.0 (2.0)	0.09 (0.06)
Silty clay loam	0.43 (0.07)	0.017 (0.046)	0.089 (0.009)	1.0 (0.6)	0.23 (0.06)
Sand	0.43 (0.07)	7.1 (3.7)	0.045 (0.010)	0.069 (0.014)	1.68 (0.29)
Sandy clay	0.38 (0.05)	0.029 (0.067)	0.100 (0.013)	0.37 (0.23)	0.23 (0.10)
Sandy clay loam	0.39 (0.07)	0.31 (0.66)	0.100 (0.006)	0.17 (0.11)	0.48 (0.13)
Sandy loam	0.41 (0.09)	1.1 (1.4)	0.065 (0.017)	0.13 (0.066)	0.89 (0.17)

*Carsel and Parrish (1988) report the mean and standard deviation of α. Using Equation 4.4.4 and a first-order expansion, the standard deviation of Ψ_b is approximated by $\sigma_{\Psi_b} \cong \sigma_\alpha / \bar{\alpha}^2$.

Carsel and Parrish (1988) present the EPA data. Carsel and Parrish used a database for the 12 *Soil Conservation Service* (SCS) textural classifications obtained from measurements for all soils reported in SCS Soil Survey Information Reports. These reports (published by the state) generally contain soil data for the predominant soil series within a state. A total of 42 books representing 42 states were used to develop the database. They used a multiple regression equation developed by Rawls and Brakensiek (1985) to estimate the retention parameters for the Brooks and Corey model and estimated the corresponding van Genuchten model parameters from these using Equation 4.4.5. Table 4.4.1 shows the average Brooks and Corey model parameters determined from their study. Clapp and Hornberger (1978) present results from fitting of the Campbell model to soil water retention data from a number of different soil textures.

Measurement of the Soil Moisture Characteristic Curve

The third approach for developing a soil moisture characteristic curve is to measure it directly. Several methods of measurement may be used. The methods may be divided into two categories: 1) incremental equilibrium methods and 2) dynamic methods.

With incremental equilibrium methods, the soil is allowed to come to equilibrium at some θ_w (or Ψ), and then Ψ (or θ_w) is measured. Next, θ_w (or Ψ) is changed, enough time is allowed for equilibrium to be established, and finally Ψ (or θ_w) is measured. The process is repeated until a sufficient number of $\Psi - \theta_w$ points have been measured to define the moisture characteristic curve. With dynamic methods, the soil is continuously wetted or dried, and both Ψ and θ_w are measured directly. Incremental equilibrium methods are much more widely used than dynamic methods because of greater simplicity. Dynamic methods yield a moisture characteristic curve much more quickly than incremental equilibrium methods (Charbeneau and Daniel, 1993).

Among the incremental equilibrium methods, by far the most commonly used method of measurement is the **pressure plate**. The *American Society for Testing and Materials* (ASTM) (1992a) and Klute (1986) describe the procedures. Soil samples are placed in rings, soaked with water, and then placed in contact with a high-air-entry-value, saturated porous disk. A chamber is placed around the assembly, and a positive air pressure is applied to the chamber. The air pressure forces water out of the soil samples until the strength of the soil water tension holding water in the pores is equal to the applied air pressure; the outflow may be monitored to confirm equilibrium, which usually takes a few days to establish. The applied air pressure and any negative pressure in the porous disk are measured and are used to calculate Ψ. After equilibrium is established, the chamber is disassembled, and the soil samples are oven dried to determine θ_w. The technique is used primarily for development of drying curves but can also be used to measure a wetting curve. The applicable suction range is from 0–15 bars. The pressure membrane is a similar apparatus, but the high air entry value of the membrane allows testing at even higher pressure (suction range from 0–70 bars). Richards (1941), Coleman and Marsh (1961), and ASTM (1992b) describe pressure membrane methods.

Another incremental equilibrium technique involves sealing the soil in a cell and measuring Ψ with tensiometers or thermocouple psychrometers. After equilibrium is established, the soil is wetted or dried by adding a small amount of water or allowing some evaporation, and the cell is sealed. Equilibrium is established after a few days, Ψ is measured, and the cell is weighed to track changes in θ_w. After completion of several wetting or drying increments, the soil is oven dried to determine the final θ_w. Daniel (1983) gives examples of this technique, which is particularly convenient for determining the wetting curve. Campbell and Gee (1986) and Klute (1986) describe other techniques, including vapor equilibrium methods and the null method.

With dynamic methods, θ_w must be measured nondestructively, and Ψ must be measured with instruments with a fast response time, because Ψ is continuously changing. Water content is typically measured with nuclear methods or flux integration, and Ψ is measured with tensiometry (Topp et al., 1967, Perroux et al., 1982).

Moisture characteristic curves can also be measured in situ using techniques similar in principle to those described for dynamic measurements in the laboratory. Bruce and Luxmoore (1986) summarize the techniques.

4.4.3 Hydraulic Conductivity

Hydraulic conductivity in unsaturated soils is determined from the hydraulic conductivity at saturation, K_{ws}, and the relative permeability, k_{rw} (Equation 4.4.1). The relative permeability can be related to Ψ, θ_w, or the saturation S_w. The $k_{rw}(\theta_w)$ or $k_{rw}(S_w)$ relationship, however, is usually preferred over $k_{rw}(\Psi)$, because the $k_{rw}(\Psi)$ relationship is hysteretic; in contrast, $k_{rw}(\theta_w)$ and $k_{rw}(S_w)$ exhibit virtually no hysteresis.

Estimating Relative Permeability

There are many examples in the literature of procedures to estimate $k_{rw}(\theta_w)$ or $k_{rw}(S_w)$ from other properties of the soil, including the moisture characteristic curve (e.g., Childs and Collis-George, 1950; Marshall, 1958; Millington and Quirk, 1961; Green and Corey, 1971; Elzeftawy and Mansell, 1975; Mualem, 1976a; Ahuja et al., 1985; and Mualem, 1986). Estimating k_{rw} is far easier than measuring this function.

Methods for estimating the relative permeability function are generally based on the pore size distribution of the soil. Such an approach is conceptually based on Figure 4.1.6. This figure shows that the smaller pores are filled with water (the wetting fluid), while the larger pores are filled with air (the nonwetting fluid). If an effective permeability can be assigned to pores of a given size, through Poiseuille's law for example, then one may integrate across the range of pore sizes that are filled with a given fluid to assign permeability for the porous medium to that fluid. The Poiseuille formula (Equation 2.1.8) shows that the fluid velocity in a capillary tube is proportional to its radius squared. The Laplace equation shows that the pore radius is inversely proportional to the capillary head. The permeability associated with a pore size range dr is proportional to

$$r^2 f(r) dr = \frac{1}{\Psi^2} d\Theta \qquad (4.4.7)$$

where $f(r)$ is the relative frequency of pore sizes of radius r. Furthermore, the tortuosity increases with decreasing water saturation, and Burdine (1953) found that

$$\frac{\tau(\Theta = 1)}{\tau(\Theta)} = \Theta^2 \qquad (4.4.8)$$

Combining these relations leads to the **Burdine equation** for the water (wetting phase) relative permeability (Wyllie and Gardner, 1958; and Brooks and Corey, 1964, 1966). An equivalent formulation may be developed for the flow of air through unsaturated soil. The resulting relationships are

$$k_{rw} = \Theta^2 \left(\frac{\int_0^\Theta \frac{d\Theta'}{\Psi^2}}{\int_0^1 \frac{d\Theta'}{\Psi^2}} \right) \quad ; \quad k_{ra} = (1 - \Theta)^2 \left(\frac{\int_\Theta^1 \frac{d\Theta'}{\Psi^2}}{\int_0^1 \frac{d\Theta'}{\Psi^2}} \right) \tag{4.4.9}$$

The leading terms in Equation 4.4.9 account for the relative tortuosity.

Brooks and Corey (1964, 1966) used their power law characteristic model of Equation 4.4.2 in the Burdine equations to derive the water and air relative permeabilities as

$$k_{rw} = \Theta^{(3+2/\lambda)} = \Theta^\varepsilon \tag{4.4.10}$$

$$k_{ra} = (1 - \Theta)^2 (1 - \Theta^{(1+2/\lambda)}) \tag{4.4.11}$$

In Equation 4.4.10, $\varepsilon = 3 + 2/\lambda$.

The Burdine equations are only one of many relations that have been suggested to estimate the medium permeability from its soil water characteristic curve (Brutsaert, 1967). Mualem (1976) developed a second model that is of interest here. Mualem's model takes the following form:

$$k_{rw} = \sqrt{\Theta} \left(\frac{\int_0^\Theta \frac{d\Theta'}{\Psi}}{\int_0^1 \frac{d\Theta'}{\Psi}} \right)^2 \quad ; \quad k_{ra} = \sqrt{1 - \Theta} \left(\frac{\int_\Theta^1 \frac{d\Theta'}{\Psi}}{\int_0^1 \frac{d\Theta'}{\Psi}} \right)^2 \tag{4.4.12}$$

van Genuchten (1980) used his model for the soil water characteristic curve, Equation 4.4.4, in the **Mualem** model to find for the water and air relative permeabilities:

$$k_{rw} = \sqrt{\Theta} (1 - [1 - \Theta^{1/M}]^M)^2 \tag{4.4.13}$$

$$k_{ra} = \sqrt{1 - \Theta} (1 - \Theta^{1/M})^{2M} \tag{4.4.14}$$

At least for the wetting phase, Equation 4.4.10 leads to the far simpler model, and for this reason it has been widely used in practice.

Brooks and Corey (1964) also present experimental data reporting the water and air relative permeability curves. This data, along with the theoretical curves of Equations 4.4.10, 4.4.11, 4.4.13, and 4.4.14 with parameters estimated from the fit of the characteristic curve from Figure 4.4.3 are shown in Figure 4.4.4. A significant point of interest is that the permeability data was not fit to the models. Rather, the model parameters that are derived from fitting the soil water characteristic data are used in the theoretical permeability models. This idea suggests that one may be able to use laboratory measurement of the soil water characteristic in lieu of the much more difficult relative permeability measurements. The Brooks and Corey model appears to better fit the permeability data, while the van Genuchten model does a better job with the characteristic data. Based on the experience to date, there is no "best" model to use in all

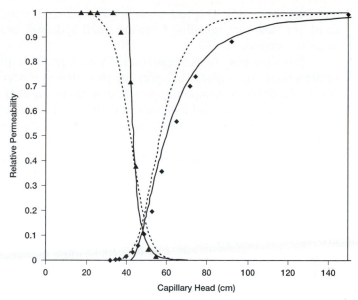

FIGURE 4.4.4 Comparison of Brooks and Corey (solid line) and van Genuchten (dashed line) relative permeability models for parameters estimated from the capillary pressure curves with data measured for fine sand (triangles-water; diamonds-air)

cases, and there is much experimental data that is poorly fit by both models. The advantage of the Brooks and Corey model is its simplicity, although it provides a less satisfactory fit to measured data of capillary pressure at high water saturation.

Measuring Relative Permeability

A number of techniques have been developed in the laboratory to measure k_{rw}. One of the main problems in measuring k_{rw} is that to induce flow of water through a soil column, a hydraulic gradient must exist; if a gradient exists, either (1) flow must take place vertically under gravity drainage at unit gradient, or (2) Ψ must vary along the soil column (which means that θ_w and k_{rw}, which is the parameter to be measured, must also vary).

Gravity drainage through a vertical soil column in unsaturated soils is accomplished by placing a high-resistance porous plate on top of the soil column. Water is applied to the top of the plate. Gravity drainage through the plate drives a downward flux of water through the plate and soil column. The hydraulic resistance of the porous plate limits the flux. Once steady conditions are established, the flux is measured, and either θ_w or Ψ (or both) is measured in the soil. The high-resistance porous plate is replaced with a plate having a different resistance, and the experiment is repeated. Several repetitions lead to development of a k_{rw} curve. Ahuja (1974) describes the procedure more fully.

Steady-state methods induce steady flow through a soil column with a controlled Ψ at both ends of the soil column and a small hydraulic gradient to drive flow. Procedures are described by Nielsen and Biggar (1961) and Klute and Dirksen (1986). This technique is not practical for soils with low K_{ws}, because the flow rate resulting from the small hydraulic gradient is too low to measure accurately.

Transient tests may be performed with a pressure plate (**pressure plate outflow test**). Soil is placed in a pressure plate device, a step increase in air pressure in the chamber is imposed, the rate of water flow out of the soil is measured as a function of time, and hydraulic conductivity is computed from the resulting data. Gardner (1956) and Olson and Daniel (1981) report procedures and methods of calculation.

Another method of transient testing is called the **instantaneous profile method**. Instantaneous profiling provides a powerful means for determining K_w in the laboratory or field. The soil is continuously wetted; by monitoring the rate of change of Ψ and/or θ_w with position (z) and time (t) along a soil column, K_w can be calculated. The Ψ-z curve is differentiated to obtain the hydraulic gradient, and the θ_w-z curve is integrated from a given point in the soil column to the downstream end of the column to determine the volume of water that exists downstream of a point, which is used to compute the rate of flow. Richards and Weeks (1953), Weeks and Richards (1967), and Daniel (1983) give details of the method.

Klute and Dirksen (1986) summarize several other methods of measurement of unsaturated hydraulic conductivity in the laboratory. Clothier and White (1981), Green et al (1986), Perroux and White (1988), and Clothier and Smettem (1990) provide examples of field applications.

Daniel notes that there are no simple guidelines as to which methods are best to use for determining relative permeability (Charbeneau and Daniel, 1993). Many testing techniques are available, and these types of tests (particularly for contaminant transport studies) tend to be performed by experienced, specialized laboratories. The main points are (1) simple analytic functions can be assumed for relative permeability, and in many cases no further effort is needed; (2) for important or sensitive projects, the relative permeability function can be measured using one of several available techniques, and the measurements can be used to confirm an assumed relationship or to determine appropriate curve-fitting parameters for an analytic function; and (3) the relative permeability is multiplied by the hydraulic conductivity at saturation—there is no point in going to extremes to define the relative permeability function accurately unless the hydraulic conductivity at saturation (K_{ws}) is well characterized. In nearly all practical situations, uncertainties in K_{ws} will far outweigh uncertainties in k_{rw}, particularly for values of soil water saturation (S_w) greater than about 70 percent.

Hydraulic Conductivity at Saturation

The hydraulic conductivity at saturation (K_{ws}) is a parameter than can vary over many orders of magnitude. Although much has been published about K_{ws}'s of many soils, it is inappropriate to guess a value of K_{ws} in detailed modeling work. Although other properties of the soil may be used to aid in estimating K_{ws}, e.g., grain-size distribution data or the moisture characteristic curve (e.g. Messing, 1989), it is best to measure directly the hydraulic conductivity at saturation for each significant soil unit at the site. One reason why it is best to measure K_{ws} directly is that unpredictable secondary porosity features, e.g., cracks, fractures, or worm holes, can play a dominant role in controlling K_{ws}—and yet are not reflected in parameters such as grain-size distribution.

Laboratory permeameters for measuring K_{ws} fall into two categories: rigid-wall permeameters and flexible-wall permeameters. With rigid-wall devices, the soil is contained in a rigid cylinder. One problem with rigid-wall cells is that spurious side-wall flow may occur with soils having low K_{ws}. The problem of side-wall flow is overcome by confining the soil column with a flexible latex membrane in what is known as a **flexible-wall permeameter**. Hydraulic conductivity tests may be performed with a constant head, falling head, or constant flux. Use of constant head and falling head permeameters are described in Section 2.2.6. Recommended sources of information for testing details include Olson and Daniel (1981), Daniel et al (1984), and Klute and Dirksen (1986).

Laboratory and field hydraulic conductivities at saturation often correlate poorly, because the structural features that control flow in the field exist at a scale that is too large to be accurately reflected in small laboratory test specimens. Olson and Daniel (1981) compared laboratory- and field-measured K_{ws}'s for 72 data sets involving clay soils and found that the ratio of $(K_{ws})_{field}/(K_{ws})_{laboratory}$ ranged from 0.3 to 46,000. In 52 of the 72 data sets, the field-measured K_{ws} was greater than the lab-measured value. The $(K_{ws})_{field}$ was less than $(K_{ws})_{laboratory}$ in only 13 of 72 cases. These trends are typical: One usually finds that the hydraulic conductivity increases with increasing scale of measurement due to structural features of a soil that usually are measurable in a reliable manner only on a large scale (see, for example, Bradbury and Muldoon, 1990).

Numerous methods of measurement of K_{ws} in the field have been described in the literature. Most of the literature describes testing of soils below the water table, however. The interest in this chapter is on methods of measurement of K_{ws} in unsaturated soils. For unsaturated soils, there are essentially two methods of measurement: **infiltration testing** and **borehole testing**.

Infiltration tests are performed with single- or double-ring infiltrometers. The rate of infiltration and the depth of wetting front are measured as a function of time. The Green-Ampt method typically is used to calculate K_{ws}; the wetting-front pressure head is calculated, assumed, or measured with an air-entry permeameter. Bouwer (1966), Amoozegar and Warrick (1986), Amoozegar (1989), and Reynolds and Elrick (1991) provide more information on the infiltration test.

The second type of test is a borehole test. A hole is drilled into the soil, the hole is filled with water, and the rate of inflow is measured, typically while a constant head is maintained in the borehole. Stephens and Neuman (1982), Philip (1985), Reynolds, Elrick, and Clothier (1985), Reynolds and Elrick (1987), and Herzog and Morse (1990) review methods of calculating K_{ws}.

4.4.4   Field Monitoring

Field monitoring of unsaturated soils uses many of the techniques that have already been described. Neutron scattering remains the most commonly used technique for precise measurements of water content. There are various types of porous blocks (gypsum or nylon) and heat dissipation sensors that provide a low-cost alternative for matric potential measurement, although these require calibration for each soil, and the probes deteriorate over time (Gardner, 1986). Advances in technology associated with time domain reflectometry and nuclear magnetic resonance (Paetzold et al., 1987) and the recent development of commercial devices for these techniques is likely to cause a shift toward these non-nuclear-source techniques.

Capillary pressure measurements in the field are made using tensiometers and thermocouple psychrometers, as described in Section 4.4.1.

For soil water sampling from unsaturated soil, the most common device used is the pressure-vacuum lysimeter (Parizek and Lane, 1970; Everett et al., 1988). The device consists of a porous element (preferably with a high air entry pressure) attached to the end of a tube. The device is placed at the base of a borehole, and the space around the porous element is backfilled with fine silica flour. A vacuum is drawn inside the lysimeter. If the pressure inside the lysimeter is less than the capillary pressure in the soil water, water will be drawn into the lysimeter. The lysimeter may be evacuated several times to capture a large sample of water. The device is pressurized to lift the water sample to the surface via tubing.

There are many potential sources of error in sampling with lysimeters. The most significant errors are loss of volatiles and sorption of solutes onto the porous element. In addition, the lysimeter will draw water most easily from the largest saturated pores, and these may not provide a representative sample of the soil water quality. See Section 6.11 for a further discussion.

Pan and trench lysimeters (Everett, 1990) have also been used to sample pore fluids from unsaturated soil. Holder et al (1991) describes a capillary-wick pan lysimeter. Pan lysimeters are particularly useful for sampling soils located near the surface, e.g., for soil treatment systems.

4.5   Infiltration Models

Infiltration is the important hydrologic process of water entry into the soil, which has been extensively studied by hydrologists and soil scientists. There are many empirical and physically based infiltration models. A classic review is presented by Philip (1969), who himself has contributed significantly to our current level of understanding of infiltration processes. This section looks at a number of the more familiar infiltration equations and shows that all of them may adequately describe observed and simulated data. This information leads to the conclusion that the important criterion in choosing one model form over another must be based on ease of use in estimating the model parameters prior to the experiment or observation. A more detailed look is then taken at the Green and Ampt (1911) model, which is physically based, easy to use, and one of the earliest models. The Green and Ampt model continues to experience broad application, and in fact, it is probably the most popular infiltration model in use at the present time.

In this section and the following sections of this chapters, θ is the volumetric water content. Subscripts refer to various conditions placed upon the volumetric water content. Unless it can lead to confusion, the subscript 'w' referring to water will be dropped.

In order to compare the infiltration models, consider a problem of infiltration into a soil with a uniform initial water content θ_a, which is the **antecedent water content**. The **cumulative infiltration** at a given time, $I(t)$, expressed per unit area of soil surface, is calculated as the integral over depth of the water content profile in excess of the antecedent water content:

$$I(t) = \int_0^\infty [\theta(z,t) - \theta_a]dz \qquad (4.5.1)$$

The **infiltration rate** at the soil surface is defined by

$$i(t) = \frac{dI}{dt} \qquad (4.5.2)$$

After a long period of infiltration with a sufficient supply of water at the ground surface, the infiltration rate becomes limited by the rate at which the soil can transmit flow under a unit gradient (i.e., the capillary pressure gradient becomes negligible and only the gravitational gradient remains). Because entrapped air cannot be displaced, the soil does not achieve its full saturated hydraulic conductivity, and the resulting maximum hydraulic conductivity that can be achieved under field conditions is called the hydraulic conductivity at **natural (field) saturation**, θ_{ns}. Thus, the long-term rate of infiltration is given by

$$i(t \to \infty) = K(\theta_{ns}) \qquad (4.5.3)$$

The long-term behavior of the infiltration models should obey this relationship. For the short-term, if water is readily available at the ground surface, then the capillary pressure gradients are initially infinite, and the initial infiltration rate must approach $i(t\to 0) = \infty$. An appropriate infiltration model should also satisfy this relationship.

Parlange and Haverkamp (1989) discuss four conventional infiltration models: Green and Ampt (1911), Kostiakov (1932), Horton (1940), and Philip (1957). The Green and Ampt (1911) infiltration model may be written

$$I(t) = K_{ns}t + (\theta_{ns} - \theta_a)(H + \Psi_f)ln\left(1 + \frac{I(t)}{(\theta_{ns} - \theta_a)(H + \Psi_f)}\right) \qquad (4.5.4)$$

where K_{ns} is the hydraulic conductivity at natural saturation, H is the ponding depth at the ground surface, and Ψ_f is the constant effective suction head at the wetting front. If Equation 4.5.4 is differentiated, one finds

$$i(t) = K_{ns}\left(1 + \frac{(\theta_{ns} - \theta_a)(H + \Psi_f)}{I(t)}\right) \tag{4.5.5}$$

It is apparent that Equation 4.5.5 does satisfy the requirement of Equation 4.5.3. As shown below, it is also possible to give a firm physical interpretation to the other parameters appearing in the Green and Ampt model.

The Kostiakov (1932) equation is given by

$$i(t) = \alpha_1 t^{-\alpha_2} \tag{4.5.6}$$

where α_1 and α_2 are two empirical parameters with $\alpha_1 > 0$ and $0 < \alpha_2 < 1$. Integration of Equation (6) yields

$$I(t) = \frac{\alpha_1}{1 - \alpha_2} t^{1-\alpha_2} \tag{4.5.7}$$

A look shows that Equation 4.5.6 does not satisfy Equation 4.5.3. Aware of this problem, Kostiakov proposed a maximum time range of application given by

$$t_{max} = \left(\frac{\alpha_1}{K_{ns}}\right)^{1/\alpha_2}$$

The empirical parameters α_1 and α_2 may be found by regression analysis over the experimental data $I(t)$.

Horton (1940) proposed an exponential form of the infiltration equation, where the infiltration rate is given by

$$i(t) = \beta_1 + \beta_2 e^{-\beta_3 t} \tag{4.5.8}$$

and by integration, the cumulative infiltration is

$$I(t) = \beta_1 t + \frac{\beta_2}{\beta_3}(1 - e^{\beta_3 t}) \tag{4.5.9}$$

Comparison with Equation 4.5.3 shows that β_1 must equal K_{ns}. Equation 4.5.8, however, shows that the initial infiltration rate is $i(0) = \beta_1 + \beta_2$, so that the model does not satisfy physical constraints.

Philip (1957a) has developed a quasi-analytical time series solution to infiltration expressed by

$$I(t) = K_a t + \sum_{m=1}^{M} S_m t^{m/2} \tag{4.5.10}$$

where K_a is the hydraulic conductivity corresponding to the initial water content θ_a, and S_m are a series of coefficients that may be calculated from the soil water characteristic curve and the boundary conditions for the infiltration problem. In particular, $S_1 = S$ is known as the **sorptivity** (Philip, 1957b). The two-term infiltration equation of Philip (1957b) is given by

$$I(t) = St^{1/2} + At \qquad (4.5.11)$$

while the corresponding infiltration rate is given by

$$i(t) = \frac{S}{\sqrt{4t}} + A \qquad (4.5.12)$$

Equation 4.5.12 does not satisfy the long-time requirement of Equation 4.5.3 unless $K_a = K_{ns}$. Empirical evidence using Equation 4.5.10, however, has shown that the theoretical value of $A = K_a + S_2 \neq K_{ns}$. Accordingly, as the only model to satisfy the physical constraint of Equation 4.5.3, the approach of Green-Ampt (1911) continues in popularity.

One can use the Green and Ampt model to obtain an estimate of the sorptivity, S, which appears in Philip's model and in other applications in soil physics. The series expansion of the logarithm function is

$$ln(1 + x) = x - \frac{x^2}{2} + \dots$$

Using this expansion, the short-time expansion of Equation 4.5.4 gives

$$I(t) = \sqrt{2K_{ns}(\theta_{ns} - \theta_a)(H + \Psi_f)t}$$

which suggests that the sorptivity may be estimated from

$$S = \sqrt{2K_{ns}(\theta_{ns} - \theta_a)(H + \Psi_f)} \qquad (4.5.13)$$

The sorptivity may also be viewed as a fitting parameter.

Parlange and Haverkamp (1989) examined the four infiltration models specified by Equations 4.5.4, 4.5.7, 4.5.9, and 4.5.11, for comparison with experimental and simulated infiltration data. They used two different soil types: clay soil and coarse sand. For the coarse sand, experimental data was used with $H = 2.25cm$, $\theta_a = 0.0816$, $\theta_{ns} = 0.312$, and $K_{ns} = 15.32$ cm/hr. For each of the four models, the parameters were treated as fitting parameters to obtain the best match with the experimental data. For the coarse sand, the following results were obtained:

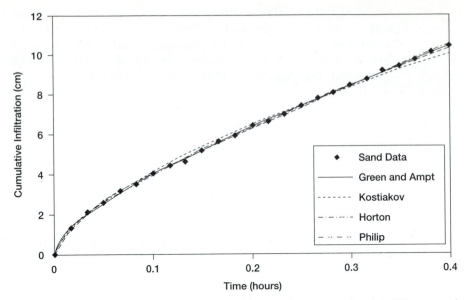

FIGURE 4.5.1   Comparison of cumulative infiltration for sand soil with 2.25 cm ponding depth

Green and Ampt: $(\theta_{ns} - \theta_a)(H + \Psi_f) = 2.845$ cm, $K_{ns} = 15.012$ cm/hr
Kostiakov: $\alpha_1 = 11.519$, $\alpha_2 = 0.361$
Horton: $\beta_1 = 21.317$ cm/hr, $\beta_2 = 73.284$ cm/hr, $\beta_3 = 36.9641$/hr
Philip: $A = 11.125$ cm/hr, $S = 9.246$ cm/$\sqrt{hr}$

Figure 4.5.1 compares the experimental data with the model results based on these parameters.

For the clay soil, no experimental data was available, so Parlange and Haverkamp used an accurate numerical simulation model with conditions specified by $H = 10$ cm, $\theta_a = 0.2376$, $\theta_{ns} = 0.4950$, and $K_{ns} = 0.04428$ cm/hr for model comparisons. In fitting the model parameters to the simulated results they find

Green and Ampt: $(\theta_{ns} - \theta_a)(H + \Psi_f) = 9.077$cm, $K_{ns} = 0.0394$ cm/hr
Kostiakov: $\alpha_1 = 0.406$, $\alpha_2 = 0.380$
Horton: $\beta_1 = 0.072$ cm/hr, $\beta_2 = 0.268$ cm/hr, $\beta_3 = 0.0701$ /hr
Philip: $A = 0.0350$ cm/hr, $S = 0.780$ cm/$\sqrt{hr}$

Figure 4.5.2 compares the experimental data with the model results based on these parameters.

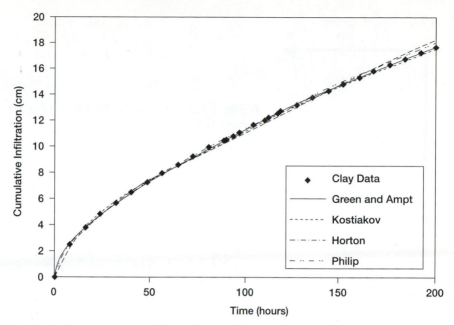

FIGURE 4.5.2 Comparison of cumulative infiltration for clay soil with 10 cm ponding depth

Figures 4.5.1 and 4.5.2 make it clear that all of the models are capable fitting the experimental and simulated infiltration data. Then, how does one choose between them? The answer must lie with the capabilities of the models to predict infiltration before or independently of the observations. One must be able to estimate the model parameters from readily obtainable information. For this reason, physically based models are greatly preferred over empirical models. In this light, there is a strong preference for the Green and Ampt model, and for the Philip model to a lesser degree. There are a number of other physically based infiltration models available in the literature, including those of Morel-Seytoux and Khanji (1974), Smith and Parlange (1978), Parlange et al. (1982), and Parlange and Haverkamp (1989). All of these models share attributes of the Green and Ampt model. Nevertheless, only the Green and Ampt model is presented here.

Green and Ampt Infiltration Model

The Green and Ampt (1911) infiltration model assumes that the initial or antecedent water content, θ_a, is uniform; that there is an abrupt wetting front; and that the water content behind the wetting front is uniform. The form of the water content profile is shown in Figure 4.5.3, where H is the ponded depth of

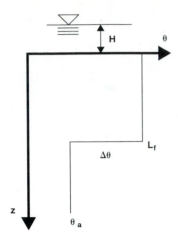

FIGURE 4.5.3   Assumed profile for the Green and Ampt infiltration model

water on top of the soil profile, $\Delta\theta$ is the jump in water content across the wetting front, and L_f is the depth of the wetting front. The development of the model shown below follows Neuman (1976).

To develop the Green and Ampt model, it is simplest to take the z-direction as positive downward and write Darcy's Law following Figure 4.5.3 as

$$q = K\frac{\partial\Psi}{\partial z} + K \qquad (4.5.14)$$

Integrating Equation 4.5.14 over the region from the ground surface down to the wetting front gives

$$\int_0^{L_f} q\,dz = \int_{-H}^{\Psi_a} K\,d\Psi + \int_0^{L_f} K\,dz \qquad (4.5.15)$$

At any given time within the region, from the ground surface to $z = L_f$, both q and K are constant. In addition, if the soil under antecedent conditions is fairly dry, Ψ_a is large. In terms of the relative hydraulic conductivity, one may approximate this limit with $\Psi_a \rightarrow \infty$. Under these conditions, Equation 4.5.15 becomes

$$qL_f = KH + K\int_0^{\infty} k_r d\Psi + KL_f \qquad (4.5.16)$$

The wetting front suction head is defined by

$$\Psi_f = \int_0^{\infty} k_r d\Psi \qquad (4.5.17)$$

With Equation 4.5.16, this expression gives

$$q = K\left(\frac{H + \Psi_f + L_f}{L_f}\right) \tag{4.5.18}$$

Equation 4.5.18 is a direct form of Darcy's Law with important physical interpretations. Consider the three terms adding to the head gradient. The first, H/L_f, corresponds to the driving force due to ponded water at the ground surface. The second term, Ψ_f /L_f, corresponds to the suction gradient caused by the negative pressures below the wetting front. The last term has a unit gradient and corresponds to the gravitational component.

To further develop the theory, consider the volume balance at the wetting front. During a time interval, dt, the front moves downward an increment, dL_f. The volume of fluid added to this increment is $\Delta\theta \, dL_f$. This volume comes from the Darcy flux shown above and is given by $q \, dt$. Equating these, we obtain

$$q = \Delta\theta\frac{dL_f}{dt} \tag{4.5.19}$$

By combining Equations 4.5.18 and 4.5.19 and separating variables, we have

$$\frac{K}{\Delta\theta}\,dt = \frac{L_f}{H + \Psi_f + L_f}\,dL_f = \left(1 - \frac{H + \Psi_f}{H + \Psi_f + L_f}\right)dL_f \tag{4.5.20}$$

Equation 4.5.20 is integrated with $L_f = 0$ at $t = 0$ to give

$$\frac{Kt}{\Delta\theta} = L_f + (H + \Psi_f)ln\left(\frac{H + \Psi_f}{H + \Psi_f + L_f}\right)$$

The cumulative infiltration, I(t), however, is equal to

$$I(t) = \Delta\theta L_f(t) \tag{4.5.21}$$

Thus, the model may be written

$$I(t) = Kt + \Delta\theta(H + \Psi_f)ln\left(1 + \frac{I(t)}{\Delta\theta(H + \Psi_f)}\right) \tag{4.5.22}$$

Equation 4.5.22 is perhaps the most important form of the Green and Ampt model. The flux or infiltration rate, i, may be written as

$$i(t) = q = K\left(1 + \frac{\Delta\theta(H + \Psi_f)}{I(t)}\right) \tag{4.5.23}$$

Equation 4.5.23 shows that the potential infiltration rate is a function of the cumulative infiltration, and both of these are functions of time. Time is not an explicit variable in Equation 4.5.23, which represents the **time condensation** form of the Green and Ampt model. Accordingly, Equation 4.5.23 is especially important for problems with time-varying rainfall rates.

Application to Rainfall and Runoff Modeling

For rainfall and runoff applications, $H = 0$. In addition, the Green and Ampt model predicts that the infiltration rate is infinite at the beginning of a rainfall event (because $I = 0$ in Equation 4.5.23). In this case, one may call this rate the potential infiltration rate, i_p, or **infiltration capacity**, because this rate is the rate at which the soil profile can absorb water. The actual rainfall rate is finite, however, so that the actual infiltration rate for early times is equal to the rainfall rate. As long as the rainfall rate is less than the potential infiltration rate, then all of the rainfall will infiltrate. It is only when the rainfall intensity exceeds the potential infiltration rate that surface runoff can occur. The objective here is to find the time at which the soil profile can no longer absorb all of the rainfall. This time is called the **time to ponding**, or t_p.

If P is the precipitation rate (assumed piecewise constant), then at ponding time the infiltration rate equals the rainfall rate: $i = P$. In addition, if P has been constant, then the cumulative infiltration is equal to

$$I = Pt_p$$

Substitution of these into Equation 4.5.23 and solving for the ponding time gives

$$t_p = \frac{K\Delta\theta\Psi_f}{P(P - K)} \tag{4.5.24}$$

This result is presented by Mein and Larson (1973).

In order to calculate the cumulative infiltration after ponding the following method is useful (Chow et al., 1988). The Green and Ampt model is assumed to be valid at and beyond ponding time. If one writes Equation 4.5.22 for time t and for time $t + \Delta t$ and eliminates t between these two expressions, one finds

$$I_{t+\Delta t} = I_t + K\Delta t + \Delta\theta\Psi_f \, ln\left(\frac{I_{t+\Delta t} + \Delta\theta\Psi_f}{I_t + \Delta\theta\Psi_f}\right) \tag{4.5.25}$$

This equation is generally true for the Green and Ampt model. Moreover, it is especially helpful when t is taken as t_p and $I_t = Pt_p$.

The cumulative *runoff*, R, is equal to the difference between the rainfall and the infiltration:

$$R_t = Pt - I_t \tag{4.5.26}$$

Generalizations for a time variable rainfall rate are available (Chow et al., 1988).

Parameters for Rainfall-Runoff Modeling

The Green and Ampt infiltration model has received much attention as an alternative to the Soil Conservation Service curve number method for calculating rainfall abstractions, and there has been a significant amount of work done in the area of associated parameter estimation. The required parameters are the wetting front capillary suction, Ψ_f, the hydraulic conductivity of the wetting region, and the water content jump across the wetting front. According to Equation 4.5.17, the wetting front suction head may be calculated from the relative permeability relationship for a soil. Using the Brooks and Corey power law model with $k_r = \Theta^\varepsilon$ for $\Psi > \Psi_b^*$ and $k_r = 1$ for $\Psi < \Psi_b^*$, the integral in Equation 4.5.17 gives

$$\Psi_f = \frac{2 + 3\lambda}{1 + 3\lambda} \, \Psi_b^* \qquad (4.5.27)$$

In addition, since infiltration follows the wetting soil characteristic curve while the model parameters are estimated for the drainage curve, Brakensiek (1977) suggests using the following equation:

$$\Psi_b^* = \frac{\Psi_b}{2} \qquad (4.5.28)$$

During infiltration, water does not completely displace all the air from the soil. Because of entrapped air, the effective permeability is less than the saturated one. Bouwer (1966) suggests calculating the hydraulic conductivity from the condition

$$K = K_{ns} = \frac{K_{ws}}{2} \qquad (4.5.29)$$

This condition is considered to be the condition at **natural saturation** of the soil. Thus, according to this model, at natural saturation $k_r = \frac{1}{2}$, and using the Brooks and Corey model,

$$\theta_{ns} = \theta_{wr} + (n - \theta_{wr})\left(\frac{1}{2}\right)^{1/\varepsilon} \qquad (4.5.30)$$

This equation represents the largest water content expected in the soil profile under field conditions.

The last parameter needed is the antecedent water content. In order to find this quantity, one needs to know how the water content changes with time following a rainfall event. The antecedent water content generally lies somewhere between **field capacity** and **wilting point**. Field capacity refers to the water content remaining in the soil profile after gravitational drainage has slowed significantly or has ceased. It is a measure of the water content that is held by capillary forces and is available to plants. Water in excess of field capacity readily drains from the profile. One pragmatic definition from the literature considers field capacity to be the water content remaining after two days of drainage following irrigation or significant rainfall (for a discussion, see Hillel, 1982). More recently, the measure of field capacity has been operationally defined as the water content corresponding to a capillary suction of $1/3$ *bar*. Wilting point corresponds to the water content that exists when the water is held so strongly by capillary and hygroscopic forces that it is no longer available to plants for transpiration. At this point, a plant starts to wilt. The current measure of wilting point is the water content corresponding to a capillary suction of *15 bars* (Brady, 1974). Thus, one generally finds

$$\theta_{wp} < \theta_a < \theta_{fc}$$

where θ_{wp} and θ_{fc} correspond to the water contents at wilting point and field capacity, respectively. As a default, one may assume that $\theta_a = \theta_{fc}$. (Also see Example 4.7.1.)

With θ_a known from this model, one has

$$\Delta\theta = \theta_{ns} - \theta_a \tag{4.5.31}$$

Equations 4.5.27 through 4.5.31 provide the parameters for the Green and Ampt infiltration model.

EXAMPLE PROBLEM

4.5.1   Estimation of Field Capacity

We want to estimate field capacity for a fine sand soil using the Brooks and Corey model. The soil has the following properties: $S_{wr} = 0.167$, $\Psi_b = 41$ cm, $\lambda = 3.74$, $n = 0.36$, and $K_{ws} = 2.80 \times 10^{-5}$ m/s. The residual water content is given by $\theta_{wr} = S_{wr} \times n = 0.060$.

One bar corresponds to a pressure of 101.33 kPa, which in turn corresponds to a suction head of $101.33/9.81 = 10.33$ m. Thus, a suction head of $1/3$ bar corresponds to $\Psi = 344$ cm, and from Equation 4.4.2, the reduced saturation and field capacity are

$$\Theta = \left(\frac{41 \text{ cm}}{344 \text{ cm}}\right)^{3.74} = 3.5 \times 10^{-4}$$

$$\theta_{fc} = \theta_{wr} + (n - \theta_{wr})\Theta = 0.0601$$

Note that for this soil, θ_{fc} and θ_{wr} are essentially the same, and for this reason, field capacity and irreducible water content are often taken to be identical concepts, especially in hydrologic literature. For example, Bear (1972, 1979) comments that "the notion of field capacity of unsaturated flow is identical to the notion of irreducible wetting fluid saturation." This usage is contrary to the notion of field capacity in soil science. To see the distinction, consider the data for a clay loam soil from Rawls and Brakensiek (1989). From a sample of 366 soils, they found the following mean values: $n = 0.464$, $\theta_{wr} = 0.075$, $\Psi_b = 26$ cm, and $\lambda = 0.194$. For the clay loam, field capacity is estimated to be

$$\theta_{fc} = 0.0075 + 0.389\left(\frac{26 \text{ cm}}{344 \text{ cm}}\right)^{0.194} = 0.311$$

while for its wilting point, we have

$$\theta_{wp} = 0.075 + 0.389\left(\frac{26 \text{ cm}}{15 \times 1033 \text{ cm}}\right)^{0.194} = 0.188$$

Both of these values are considerably larger than the irreducible water content.

EXAMPLE PROBLEM

4.5.2 The Green and Ampt Infiltration Model

A design storm has a rainfall of 11 cm occurring over a 2-hour period. Estimate the amount of infiltration and runoff assuming that the soil has the following characteristics: $\lambda = 0.60$ ($\varepsilon = 3 + 2/0.6 = 6.33$), $\theta_{wr} = 0.06$, $n = 0.44$, $\Psi_b = 30$ cm, and $K_{ws} = 0.8$ m/d. Assume that the antecedent water content is equal to field capacity.

The antecedent water content and water content at natural saturation are calculated as follows.

$$\theta_a = \theta_{fc} = 0.06 + 0.38\left(\frac{30 \text{ cm}}{344 \text{ cm}}\right)^{0.6} = 0.148$$

$$\theta_{ns} = 0.06 + 0.38\left(\frac{1}{2}\right)^{1/6.33} = 0.401$$

Thus, the increase in water content across the wetting front is $\Delta\theta = 0.401 - 0.148 = 0.253$. The wetting front suction head is given by Equations 4.5.27 and 4.5.28.

$$\Psi_f = \frac{2 + 3 \times 0.6}{1 + 3 \times 0.6}\left(\frac{30 \text{ cm}}{2}\right) = 20.4 \text{ cm} = 0.204 \text{ m}$$

Finally, the hydraulic conductivity at natural saturation is $K_{ns} = (0.8 \text{ m/d})/2 = 0.4$ m/d, and the mean precipitation rate is $P = (0.11 \text{ m})/(2 \text{ hr}) = 0.055$ m/hr $= 1.32$ m/d. With these values, the time to ponding is calculated from Equation 4.5.24.

$$t_p = \frac{0.4 \frac{\text{m}}{\text{d}} \times 0.253 \times 0.204 \text{ m}}{1.32 \frac{\text{m}}{\text{d}} \times \left(1.32 \frac{\text{m}}{\text{d}} - 0.4 \frac{\text{m}}{\text{d}}\right)} = 0.017 \text{ d} = 0.408 \text{ hr}$$

The cumulative infiltration at ponding time $t = t_p$ is $I_t = 0.055$ m/hr $\times 0.408$ hr $= 0.0224$ m. With I_t and $\Delta t = 2.0$ hr $- 0.408$ hr $= 1.59$ hr $= 0.0663$ d, we can use Equation 4.5.25 to find

$$I_{t+\Delta t} = 0.0224 \text{ m} + 0.4\tfrac{\text{m}}{\text{d}} \times 0.0663 \text{ d} + 0.0516 \text{ m} \times ln\left(\frac{I_{t+\Delta t} + 0.0516 \text{ m}}{0.0224 \text{ m} + 0.0516 \text{ m}}\right)$$

This equation is solved by iteration for $I_{t+\Delta t}$. For the first trial, use $I_{t+\Delta t} = 0.0224$ m + 0.4 m/d × 0.0663d = 0.0489 m within the logarithm to give $I_{t+\Delta t} = 0.0647$ m. Using this value inside of the logarithm again gives $I_{t+\Delta t} = 0.0722$ m. Repeating this procedure, we obtain successively $I_{t+\Delta t} = 0.0755$ m, 0.0768 m, 0.0773 m, 0.0776 m, and 0.0777 m. Thus, we have for the cumulative infiltration $I_{t+\Delta t} = 0.0777$ m, and the amount of runoff is given by $R_{t+\Delta t} = 0.11$ m − 0.0777 m = 0.0323 m. The depth of the wetting front, L_f, may be estimated from Equation 4.5.21 to find

$$L_f = \frac{I_{t+\Delta t}}{\Delta \theta} = 0.307 \text{ m}$$

Thus, the upper 30 cm of the soil profile have reached natural saturation.

4.6   Redistribution of Soil Water

When rainfall ceases and surface storage is depleted, the infiltration process comes to an end. The movement of soil water that has been added to the profile during infiltration, however, does not cease immediately and may continue for a long period of time as it redistributes within the profile. Water drains downward, primarily under the force of gravity, and may eventually recharge the water table. At the same time, a significant amount of moisture may be lost to evaporation and transpiration by plants. The processes of liquid moisture redistribution are important in determining the amount of water retained at various times by different depths within the profile. In turn, these quantities define the amount and time sequence of groundwater recharge, and the antecedent moisture content for subsequent rainfall events. A schematic view of the soil water content profile at different time epochs is shown in Figure 4.6.1.

Following a significant rainfall or wetting event, the upper soil layers are wet to natural saturation. As water drains from these upper layers, it drains first from the larger pores and only later from the smaller pores. As the larger pores are drained, there is a significant decrease in the hydraulic conductivity of the soil. As the soil hydraulic conductivity decreases, so does the rate of drainage. Thus, the overall rate of drainage from a soil will depend on its hydraulic conductivity at natural saturation and its pore size distribution. This idea means that a soil with a narrow pore size range, such as uniform sand, will have most of its pores drain at the same time, so that its period of drainage is brief. On the other hand, a soil with a wide range of pore sizes, such as a loam, would have only a small fraction of pores draining at any time, and the corresponding decrease in effective hydraulic conductivity would be small—with the overall drainage time being much longer.

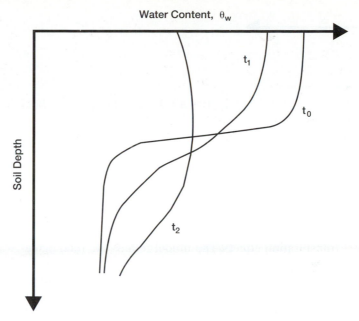

FIGURE 4.6.1   Soil water redistribution following a rainfall event

The most direct way to obtain quantitative information on both infiltration and redistribution is to solve Richards' equation (Equation 4.3.7 or 4.3.8) for specified inflow boundary conditions during infiltration, and either no-flow or an estimated evaporation loss rate as the boundary condition during redistribution. As might be expected, such an approach requires application of numerical simulation methods. For an example, see Rubin (1967) who investigated redistribution of soil moisture using an implicit finite-difference equation. More complex models include the effects of thermal gradients and both liquid and vapor flow. While such approaches are of scientific interest, in applications it is sometimes useful to look towards simple models.

In this section, we will examine two simple models for soil water redistribution. The first of these is a rectangular profile model that retains the profile form of the Green and Ampt infiltration model. This model is the simplest and perhaps most useful. The second model is based on kinematic theory. While the mathematical formulation of the kinematic model is more difficult, it does provide a good interface for combining redistribution and evaporation process models. Charbeneau (1989) discusses features of both models in greater detail.

4.6.1   Rectangular Profile Model

The rectangular profile model of soil water redistribution is the simplest model available. Here, we follow Morel-Seytoux (1984, 1985) in the development of a model for the downward movement of soil water, neglecting evaporation and

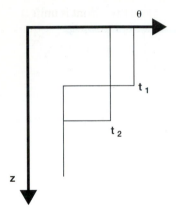

FIGURE 4.6.2 Resitribution Profile Model for Two Different Times

transpiration effects. The model retains the rectangular water content profile form of the Green and Ampt infiltration model. With increasing time, the profile moves downward and the maximum water content decreases. The basis of the model is application of the continuity principle to the profile as a whole and for displacement of the wetting front. The resulting profile is shown at two different times in Figure 4.6.2.

The amount of water within the profile (in excess of the antecedent water content, which is assumed to be field capacity) is equal to the infiltration volume:

$$W = I = (\theta - \theta_{fc})z_f \qquad (4.6.1)$$

In Equation 4.6.1, W is the amount of water available for drainage from the soil, I is the cumulative infiltration from the wetting event, θ_{fc} is the water content at field capacity, and z_f is the depth to the wetting front. We assume that the soil permeability at field capacity is essentially zero and normalize the water content with respect to θ_{fc} rather than θ_{wr}. This means that the **reduced water content** may be written as

$$\Theta^* = \frac{\theta - \theta_{fc}}{n - \theta_{fc}} \qquad (4.6.2)$$

For the simplest case we neglect evaporation and transpiration. The continuity principle applied to the entire profile gives $W = \Theta^*(n - \theta_{fc})z_f = constant$. Differentiation with respect to time gives the following relation:

$$\frac{dW}{dt} = (n - \theta_{fc})\,z_f\,\frac{d\Theta^*}{dt} + (n - \theta_{fc})\,\Theta^*\,\frac{dz_f}{dt} = 0 \qquad (4.6.3)$$

Equation 4.6.3 expresses continuity for the rectangular soil water profile.

Consider the displacement of the wetting front. During a time increment dt, the front will have moved downward a distance dz_f, and the volume of pore space filled up would be $(n - \theta_{fc})\Theta^* dz_f$. This volume, however, comes from the

Darcy flux from above. Because the gradient behind the wetting front is uniform, this volume is given by $qdt = K(\Theta^*)dt$ (pressure gradients are absent). Equating these volumes gives

$$K_w(\Theta^*)dt = (n - \theta_{fc})\Theta^* dz_f \qquad (4.6.4)$$

Equation 4.6.4 expresses continuity for the wetting front.

Combining Equations 4.6.3 and 4.6.4, the redistribution soil water profile gives

$$(n - \theta_{fc})z_f \frac{d\Theta^*}{dt} + K_w(\Theta^*) = \frac{W}{\Theta^*}\frac{d\Theta^*}{dt} + K_w(\Theta^*) = 0 \qquad (4.6.5)$$

With the Brooks and Corey power-law permeability model written as $K_w(\Theta^*) = K_{ws}(\Theta^*)^\epsilon$ and with the initial condition $\Theta^*(t = 0) = \Theta_i^*$, Equation 4.6.5 may be integrated to give

$$\Theta^* = \frac{1}{\left(\left(\frac{1}{\Theta_i^*}\right)^\epsilon + \frac{\epsilon K_{ws}t}{W}\right)^{1/\epsilon}} \qquad (4.6.6)$$

Equation 4.6.6 provides the water content within the profile as a function of time. Knowing this quantity and the infiltration volume, one can readily find the depth of the wetting front at any later time from Equation 4.6.1, such that

$$z_f = \frac{W}{(n - \theta_{fc})\Theta^*} \qquad (4.6.7)$$

with $\Theta^*(t)$ taken from Equation 4.6.6. Furthermore, Equations 4.6.6 and 4.6.7 may be used to find the time at which the wetting front reaches the water table by setting $z_f(t) = z_{wt}$ (depth of water table) and solving for the time.

An interesting characteristic of the rectangular profile model is that while the water content is uniform over depth at any time, the Darcy velocity varies with depth. This characteristic follows because the water content changes uniformly with time, and the Darcy flux at any depth must equal the rate of change in water volume above that depth (because evaporation is neglected). This concept means that at any time t, the Darcy flux must increase linearly with depth. To calculate the Darcy flux, we use

$$q(z,t) = -z(n - \theta_{fc})\frac{d\Theta^*}{dt} = \frac{(n - \theta_{fc})K_{ws}z}{W\left(\left(\frac{1}{\Theta_i^*}\right)^\epsilon + \frac{\epsilon K_{ws}t}{W}\right)^{1+1/\epsilon}} \qquad (4.6.8)$$

Equation 4.6.8 is valid for $z < z_f$, the depth to the wetting front. The cumulative groundwater recharge, G, is found by integrating Equation 4.6.8 from the time at which the wetting front first reaches the water table, t_{wt}, to the time of interest. This information gives

$$G(t) = \int_{t_{wt}}^{t} q(z_{wt},\tau)d\tau = (n - \theta_{fc})z_{wt}(\Theta^*(t_{wt}) - \Theta^*(t)) \qquad \textbf{(4.6.9)}$$

Equation 4.6.9 could be developed intuitively, because $(n - \theta_{fc})z_{wt}\theta^*(t)$ is the amount of water available for drainage above the water table at time t.

4.6.2   Kinematic Model

The second class of simple redistribution models is the gravity drainage or **kinematic wave** models (Sisson et al. (1980), Smith (1983), Charbeneau (1984), and Morel-Seytoux (1987)). An individual soil water wave propagating downward through the soil is shown in Figure 4.6.3. Figure 4.6.3 (a) shows the assumed wave immediately after the end of the period of infiltration. The wave consists of a rectangular portion added onto the profile for antecedent water content, which is assumed to be field capacity. The horizontal line indicating a drop in water content from the wave above to the unwetted soil below is referred to as the **wetting front** of the wave. Figure 4.6.3 (b) shows the wave after a short period of drainage. The wave here consists of a draining part with increasing water content with depth (the curved portion of the wave), a constant or plateau part, and the wetting front. The draining part of the wave has not reached the wetting front and is separated from the wetting front by the vertical plateau. Finally, Figure 4.6.3 (c) shows the wave at a later time after the drainage profile has caught up with the wetting front, and the plateau region is no longer present.

The basic assumption in application of kinematic models is that pressure gradients are negligible, and thus Equation 4.3.9 is applicable and may be written as follows:

$$(n - \theta_{fc})\frac{\partial \Theta^*}{\partial t} + \frac{dK_w(\Theta^*)}{d\Theta^*}\frac{\partial \Theta^*}{\partial z} = 0 \qquad \textbf{(4.6.10)}$$

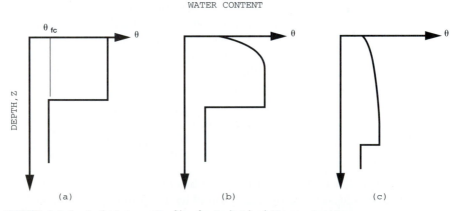

FIGURE 4.6.3   Soil Moisture Profiles for Individual Kinematic Wave

Because we are now working with z positive downward, the sign of the second term in Equation 4.6.10 has changed (compare with Equation 4.3.9). We want to solve Equation 4.6.10 using the **method of characteristics** (MOC). Accordingly, we recognize that the reduced effective saturation of Equation 4.6.2 is a function of z and t, $\Theta^*(z,t)$, and that its total differential may be expressed by

$$d\Theta^* = \frac{\partial \Theta^*}{\partial t} dt + \frac{\partial \Theta^*}{\partial z} dz \qquad (4.6.11)$$

Equation 4.6.11 may be thought of as expressing a vector displacement in z-t-Θ^* space. The basic idea of the MOC is that Equation 4.6.10 may also be interpreted as specifying a vector displacement in z-t-Θ^* space. Comparison of Equations 4.6.10 and 4.6.11 shows that these displacements are parallel so long as the following ratios are the same:

$$\frac{dt}{n - \theta_{fc}} = \frac{dz}{dK_w/d\Theta^*} = \frac{d\Theta^*}{0} \qquad (4.6.12)$$

Equation 4.6.12 expresses the MOC solution to Equation 4.6.10. The first-order partial differential equation has been reduced to a set of two ordinary differential equations. These are the **characteristic equations**. The first expresses the slope of the characteristic in the **base characteristic** z-t plane as

$$\frac{dz}{dt} = \frac{1}{n - \theta_{fc}} \frac{dK_w(\Theta^*)}{d\Theta^*} \qquad (4.6.13)$$

Because $dK/d\Theta^*$ is a function only of Θ^*, however, and according to the last of Equation 4.6.12, Θ^* is constant along such paths, the image of these paths must be straight lines in the z-t plane. Sisson et al. (1980) present a number of specific formulations for different K-Θ^* models. In particular, if the water content at the ground surface changes abruptly from Θ^*_i during infiltration to $\Theta^* = 0$ during redistribution, one may integrate Equation 4.6.13 for Θ^* and determine directly how the drainage profile evolves. For the power law model of Equation 4.4.7, Equation 4.6.13 gives

$$\frac{z}{t} = \frac{1}{n - \theta_{fc}} \frac{dK_w}{d\Theta^*} = \frac{\varepsilon K_{ws}(\Theta^*)^{\varepsilon-1}}{n - \theta_{fc}} \qquad (4.6.14)$$

and one finds (Sisson et al. (1980), Charbeneau (1984)),

$$\Theta^* = \left(\frac{(n - \theta_{fc})z}{\varepsilon K_{ws}t} \right)^{1/(\varepsilon-1)} \qquad (4.6.15)$$

where t is measured from the time at which the infiltration process ended. According to the kinematic model, within the drainage profile, the Darcy velocity is given by Equation 4.3.13 so with the power law relative permeability model,

$$q(z,t) = \left(\frac{1}{K_{ws}}\left(\frac{(n - \theta_{fc})z}{\varepsilon t}\right)^{\varepsilon}\right)^{1/(\varepsilon-1)} \tag{4.6.16}$$

The amount of water available for redistribution between the ground surface and the depth Z is given by

$$W_Z = (n - \theta_{fc})\int_0^Z \Theta^* dz = \frac{(\varepsilon - 1)Z(n - \theta_{fc})}{\varepsilon}\Theta^*_Z \tag{4.6.17}$$

where Θ^*_Z is the normalized water content at depth $z = Z$.

Analysis of the Wetting Front

Equations 4.6.10 through 4.6.17 are appropriate as long as the profile remains continuous and no wetting fronts are encountered. At the **wetting front,** the water content gradient becomes large, and in the kinematic model it becomes infinite. Thus, Equation 4.6.10 is no longer appropriate, and the usual way around this problem is to replace the continuous diffusive profile with an equivalent kinematic profile that is discontinuous. This process is shown schematically in Figure 4.6.4.

To analyze this problem, we recall that continuity for the volume of liquid soil water in one-dimensional vertical flow is most generally given in its integrated or control volume formulation as

$$\frac{d}{dt}\int_{z_1}^{z_2}\theta dz + q(z_2,t) - q(z_1,t) = 0 \tag{4.6.18}$$

where z_1 and z_2 are the depths shown in Figure 4.6.4. Equation 4.6.18 states that the time rate of increase in water content between elevations z_1 and z_2 must equal the volume flux into this region at depth z_1 minus the flux out of the region at depth z_2. This concept assumes that there are no local volumetric sources or sinks for soil moisture. If the $\theta(z,t)$ and $q(z,t)$ fields remain continuous then the size of the control volume can be made infinitesimal and in this limit Equation 4.6.18 is equivalent to a one-dimensional version of Equation 4.3.5 with

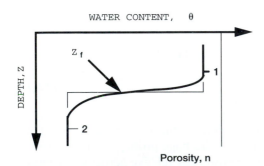

FIGURE 4.6.4 Approximate Wetting Front Profile

$S' = 0$. Physically, capillary pressure gradients dictate that the θ-field remains continuous except at an interface of two different media. For some applications, however, it is appropriate to approximate the smooth but sharply varying function θ in the limit with a discontinuous function that shows jumps in θ-values at certain locations. This situation is shown in Figure 4.6.4, where the continuous wetting front is approximated by the step function between points 1 and 2 respectively at depths z_1 and z_2. The location of the sharp front, z_f, is specified by the requirement of (volume) conservation:

$$\int_{z_1}^{z_2} \theta dz = \theta_1(z_f - z_1) + \theta_2(z_2 - z_f)$$

Because θ_1, θ_2, z_1, and z_2 are constant and z_f is a function of time, one has the exact result

$$\frac{d}{dt}\int_{z}^{z_2} \theta dz = (\theta_1 - \theta_2)\frac{dz_f}{dt}$$

Substituting this result in Equation 4.6.18 and noting that according to the kinematic model, $q = K(\theta)$ at points 1 and 2, the physical law that the speed or celerity of a sharp wetting front must satisfy is

$$\frac{dz_f}{dt} = \frac{K_w(\theta_1) - K_w(\theta_2)}{\theta_1 - \theta_2} \tag{4.6.19}$$

Equation 4.6.19 is a jump or step equation, which is similar to the jump or shock equations found in gas dynamics, nonlinear chromatography, open channel hydraulics, and other fields (see Landau and Lifshitz (1959) or Whitham (1974)). This relationship is also used in the Buckley and Leverett displacement analysis of petroleum hydrocarbons (Marle, 1981).

In the present application of Equation 4.6.19, we have $\theta_2 = \theta_{fc}$ and $K(\theta_2) \cong 0$, so Equation 4.6.19 becomes

$$(n - \theta_{fc})\Theta_1^* \frac{dz_f}{dt} = K_{ws}(\Theta_1^*)^\varepsilon \tag{4.6.20}$$

Equation 4.6.20 governs the displacement of the wetting front while Equation 4.6.10 governs the displacement of the liquid water profile so long as it remains continuous.

With reference to Figure 4.6.3, two separate cases describing the draining profile can be identified. In Figure 4.6.3(b), there exists a plateau region between the draining part of the profile and the wetting front. This fact implies that $\Theta_1^* = \Theta_i^*$ is constant in Equation 4.6.20 and is equal to the reduced saturation corresponding to the rainfall rate ($\Theta_i^* = \Theta_{ns}^*$ if $P > K_{ns}$) and that the celerity of the wetting front is constant. Thus, the image of the wetting front is also a straight line in the base characteristic plane. This fact is shown in Figure

4.6.5 as the lower boundary of the plateau. At time t_{dp}, the plateau has disappeared from the profile, and only the drainage wave remains. In this case, the water content arriving at the wetting front decreases with time, and the celerity of the wetting front also decreases. Combination of Equations 4.6.14 and 4.6.20 provides

$$\frac{dz_f}{dt} = \frac{K_{ws}}{n - \theta_{fc}}(\Theta^*)^{\varepsilon-1} = \frac{z_f}{\varepsilon t} \tag{4.6.21}$$

Equation 4.6.21 is integrated from the initial point (z_{fdp}, t_{dp}) to give

$$\frac{t}{t_{dp}} = \left(\frac{z_f}{z_{fdp}}\right)^{\varepsilon} \tag{4.6.22}$$

To complete the model, the location of the initial point (z_{fdp}, t_{dp}) must be found. This point lies at the intersection of the wetting front path originating from $z = z_{fi}$ at time zero and the characteristic with water content Θ_i^* originating from the ground surface at time zero. From Equations 4.6.20 and 4.6.14, the intersection is found to be

$$z_{fdp} = \frac{\varepsilon}{\varepsilon - 1} z_{fi} \; ; \; t_{dp} = \frac{(n - \theta_{fc})z_{fi}}{(\varepsilon - 1)K_{ws}(\Theta_i^*)^{\varepsilon-1}} \tag{4.6.23}$$

where the initial depth of the wetting front, z_{fi}, is still given by Equations 4.6.1 and 4.6.2 with $\Theta^* = \Theta_i^*$. The image of the drainage wave in the base characteristic plane consists of a plateau at water content Θ_i^* and a centered simple wave spanning the characteristics carrying $\Theta^* = \Theta_i^*$ to $\Theta^* = 0$, as shown in Figure 4.6.5.

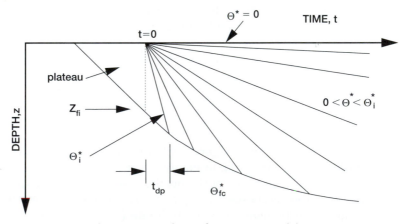

FIGURE 4.6.5   Base Characteristic Plane of Kinematic Model

EXAMPLE **4.6.1   Estimation of Soil Water Redistribution and Recharge**
PROBLEM A rainfall event has a rainfall of 5 inches over a 5-hour period on a fine sand soil
with the following characteristics: $\lambda = 2.0$ ($\varepsilon = 3 + 2/\lambda = 4$); $\theta_{wr} = 0.05$; $n = 0.36$;
$\Psi_b = 5$ inches; and $K_{ws} = 5$ ft/d. If the soil profile is initially at field capacity, estimate the time for recharge to reach a shallow water table at a depth of 8 ft—
and the recharge rate thereafter.

For this rainstorm event the average rainfall rate is

$$P = \frac{(5 \text{ in})/(12 \frac{\text{in}}{\text{ft}})}{(5 \text{ hr})/(24 \frac{\text{hr}}{\text{d}})} = 2 \text{ ft/d}$$

while the hydraulic conductivity at natural saturation is $K_{ns} = (5 \text{ ft/d})/2 = 2.5$ ft/d.
Because $P < K_{ns}$, ponding will never occur, and all of the rainfall will infiltrate.
To estimate the water content corresponding to this precipitation rate, we use
Equation 4.3.10 with 4.4.10 to write

$$q_w = P = K_{ws}(\Theta^*)^\varepsilon \tag{4.6.24}$$

which gives $\Theta_i^* = (P/K_{ws})^{1/\varepsilon} = 0.795$. For this soil, we have

$$\theta_{fc} = 0.05 + 0.31\left(\frac{5 \text{ in} \times 2.54 \frac{\text{cm}}{\text{in}}}{344 \text{ cm}}\right)^2 = 0.0504$$

If we use the rectangular profile model, then Equation 4.6.7 gives for the effective reduced water content at the time the wetting front reaches the water
table

$$\Theta_{wt}^* = \frac{(5 \text{ in})/(12 \frac{\text{in}}{\text{ft}})}{(0.36 - 0.0504) \times 8 \text{ ft}} = 0.168$$

Using this value, Equation 4.6.6 may be written to find the time of initial
recharge.

$$t_{wt} = \frac{W}{\varepsilon K_{ws}}((\Theta_{wt}^*)^{-\varepsilon} - (\Theta_i^*)^{-\varepsilon}) = 26.1 \text{ days} \tag{4.6.25}$$

Equation 4.6.8 provides the infiltration rate

$$q_{wt} = \frac{0.234}{(0.0521 + t)^{5/4}} \text{ ft/d} \; ; \; t \text{ in days}$$

for $t > 26.1$ days.

For comparison, we also consider the kinematic model for soil water redistribution. For the base characteristic plane shown in Figure 4.6.5, the reduced

effective water content in the plateau that originates at time $t = -5/24$ days at the ground surface is again equal to 0.795. The time after the end of the storm event and corresponding depth at which this plateau disappears are given by Equation 4.6.20: $z_{fdp} = 2.26$ ft and $t_{dp} = 0.0695$ d, where z_{fi} is given by Equation 4.6.5 with $\Theta^*_i = 0.795$ as $z_{fi} = 1.69$ ft. With Equation 4.6.22, we then have for the time at which the wetting front reaches the water table, $t = 10.3$ d, and the recharge rate is given by Equation 4.6.16 as

$$q_{wt} = 0.309 \, t^{-4/3} \text{ ft/d} \; ; \; t \text{ in days}$$

for $t > 10.3$ days.

The groundwater recharge rates predicted by these two models are compared in Figure 4.6.6. While the kinematic profile model predicts a much earlier time to reach the water table and larger initial recharge rates, after a time of 26 days the recharge rate estimates between the models are similar. One may understand the conditions that lead to this conclusion by considering the water content profiles at the time that the rectangular profile reaches the water table, such as shown in Figure 4.6.7. The area of the kinematic profile located below the water table corresponds to the predicted recharge at this time. The water content at the water table is nearly the same for both models, and investigation of Equation 4.6.8 shows that it predicts that the Darcy flux at the wetting front is the same as for a kinematic model. The kinematic model predicts a much more rapid movement of water away from the ground surface. With increasing time, the recharge rate estimate of the rectangular profile model exceeds that of the kinematic model, and the ultimate area under both curves shown in Figure 4.6.6 is the same, as it must be.

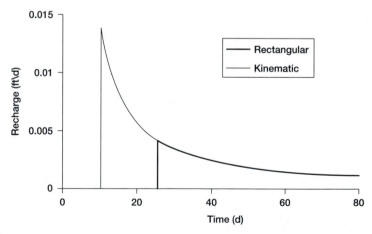

FIGURE 4.6.6 Groundwater recharge rates calculated by the rectangular profile and the kinematic models

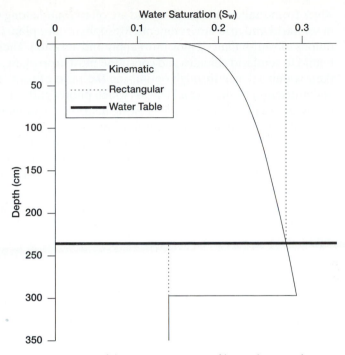

FIGURE 4.6.7 Comparison of the water content profiles at the time the rectangular profile reaches the water table

In summary, the kinematic model retains the largest water content at the wetting front, which in turn means that the wetting front moves downward more rapidly than for the rectangular profile model, and that the predicted recharge rate is initially larger. For later times, however, the recharge rate predicted by the rectangular profile model is larger. With the sandy soil considered in this example, this difference is significant. For a soil with a finer texture (smaller λ), however, the difference is not as great.

4.7   Evaporation and Desorption Models

Evaporation in the field can take place directly from the soil surface or from plants that cover the surface. Plants uptake water through their roots and transpire this water to the atmosphere. Together, the processes of **evaporation** and **transpiration** are called **evapotranspiration**. A significant fraction of the water that falls on the surface of the earth is returned to the atmosphere through evapotranspiration.

Climate and environmental conditions determine the type and number of plants present in a given location. Phreatophytes are adapted to drawing water from shallow water tables and are often found along the banks of streams in semiarid and arid environments. Mesophytes are plants growing under conditions of well-balanced moisture supply and aeration. These are found in semihumid to semiarid climates and constitute most crop plants. Mesophytes control their water balance through optimizing the ratio of roots (which supply water and nutrients) to shoots and leaves (which photosynthesize and transpire). Xerophytes are plants adapted to growing in arid environments through development of features that minimize water loss.

Plants transpire because of the vapor pressure difference between the surfaces of shoots and leaves, and that of the atmosphere. Only a very small fraction (generally less than 1 percent) of the water absorbed by plants is used in photosynthesis, while most (often more than 98 percent) is lost through transpiration (Hillel, 1982). The processes involved are complex. The flow path includes liquid water movement in the soil towards the plant roots, transport as liquid and vapor across the soil-to-plant contact zone, absorption into the roots and vascular tubes (the plants internal piping system), transfer up the stem to the leaves, evaporation within the intercellular spaces of the leaves, diffusion from within the leaf structure to the atmospheric boundary layer, and finally transport to the turbulent atmosphere that carries away the water.

The concepts of field capacity and wilting point were operationally defined in Section 4.5. According to the conventional concepts of soil-water availability to plants, a plant could function and grow without detriment if the soil water content corresponded to conditions within these limits. Newer concepts recognize that there is a continuum of conditions that affect plant functioning and growth, which includes properties of the plant, properties of the soil, and meteorological conditions (Hillel, 1982). Nevertheless, the concepts of field capacity and wilting point remain of value in hydrologic as well as agricultural applications.

Plants can control the hydrologic balance of a location. In semihumid to semiarid climates, there will be little groundwater recharge beneath healthy stands of vegetation. Gee, et al. (1994) show that even for desert sites, the presence of plants can control the soil water content and recharge potential. The physics of soil water movement through dry soils and evaporation is best studied under conditions where plant influences may be neglected. In the following discussion, we focus on evaporation from bare-surface soils.

Three conditions are necessary if evaporation from a soil is to persist. First, there must be a supply of thermal energy to meet the latent heat requirements. It requires 590 calories to transform a gram of liquid water to water vapor at a temperature of 15 degrees C. This source of thermal energy is mostly supplied from solar radiant energy, although advected energy and the soil itself can supply some of this energy (if the soil supplies the energy, then the soil cools). The second requirement for evaporation is that there must exist a vapor pressure gradient between the soil surface and the atmosphere. Water vapor diffuses upwards across the gradient into the atmosphere. These two conditions for evaporation depend on exterior influences and are independent of the soil.

They determine the **evaporative capacity** of the atmosphere. The third requirement for continued evaporation is that water must be supplied from the soil interior to the surface. The rate at which water can be supplied to the surface depends on both the wetness of the soil and on its hydraulic conductivity. As the soil dries, the surface layers are dried first, creating a partial barrier to upward water movement from deeper in the soil profile. The upward movement of liquid water to the soil surface occurs under induced capillary pressure gradient. When the soil surface is wet the soil can readily supply water to the atmosphere, and the evaporation rate is equal to the evaporative capacity of the atmosphere. Later, as the soil dries, the rate of upward water movement falls below the atmospheric evaporative capacity, and the soil profile and its hydraulic characteristics determine the rate of evaporation.

4.7.1   Absorption/Desorption Models for Capillary Flow

Absorption refers to the movement of water into or out (**desorption**) of a soil mass under the force of capillary pressure gradients. This process represents the second limiting case of Richards' equation considered in Section 4.3, and for vertical flow Equation 4.3.12 is written

$$\frac{\partial \theta}{\partial t} = \frac{\partial}{\partial z}\left(D(\theta)\frac{\partial \theta}{\partial z}\right) \tag{4.7.1}$$

For analysis of the absorption/desorption problem we assume that the soil profile has a uniform initial water content $\theta(z > 0, t = 0) = \theta_1$ and a constant different water content at the ground surface $\theta(z = 0, t > 0) = \theta_0$. If the capillary diffusivity D is constant, then Equation 4.7.1 is linear in θ and the solution is easily obtained (Gardner, 1959; Crank, 1978; also see Chapter 7, "Solute Transport by Diffusion").

$$\frac{\theta - \theta_0}{\theta_1 - \theta_0} = \text{erf}\left(\frac{z}{\sqrt{4Dt}}\right) \tag{4.7.2}$$

The function erf( ) is the error function, and it is described in Appendix G. The flux at the ground surface is found from

$$-q = D\left.\frac{\partial \theta}{\partial z}\right|_{z=0} = (\theta_1 - \theta_0)\sqrt{\frac{D}{\pi t}} \tag{4.7.3}$$

The cumulative flux is calculated from

$$\int_0^t -q\,d\tau = (\theta_1 - \theta_0)\sqrt{\frac{4Dt}{\pi}} \tag{4.7.4}$$

Gardner (1959) retains the form of Equations 4.7.3 and 4.7.4 for the general case where D is not a constant by using a weighted mean diffusivity. Even without the use of a weighted constant diffusivity, one can show that the basic behavior of the solutions given by Equations 4.7.3 and 4.7.4 holds for general $D(\theta)$.

For the general problem where D is a function of the water content, $D(\theta)$, an exact solution for the water content profile cannot be determined. Important results, however, can still be obtained as follows. For this problem, there are no characteristic length or time scales, which means that z and t cannot appear in the solution as independent variables. Instead they must combine to form a single variable which combines both length and time. Such a variable is called a **self-similar variable**. For the nonlinear diffusion equation, the self-similar variable is given by the **Boltzmann transformation** and is represented by

$$\phi = \frac{z}{\sqrt{t}} \tag{4.7.5}$$

With this transformation, we have

$$\frac{\partial}{\partial t} = \frac{\partial \phi}{\partial t}\frac{d}{d\phi} = -\frac{\phi}{2t}\frac{d}{d\phi}$$

$$\frac{\partial}{\partial z} = \frac{\partial \phi}{\partial z}\frac{d}{d\phi} = \frac{1}{\sqrt{t}}\frac{d}{d\phi}$$

Application of these transformation rules to the diffusion Equation 4.7.1 gives

$$\frac{\phi}{2} + \frac{d}{d\theta}\left(D(\theta)\frac{d\theta}{d\phi}\right) = 0 \tag{4.7.6}$$

The initial and boundary conditions transform to $\theta(\phi \to 0) = \theta_1, \theta(\phi \to \infty) = \theta_0$. The solution to this problem may be written either as $\theta = \theta(\phi)$ or as $\phi = \phi(\theta)$. Analysis of this problem is presented by Philip (1957d). The most important feature is that a number of fundamental results follow directly from this formulation without actually trying to solve the problem.

4.7.2   Application to Evaporation Modeling

As an application of the desorption model, consider the water loss from a profile that is initially at water content equal to field capacity, θ_{fc}. The soil surface is brought to the air-dry water content θ_0 and maintained at that value. This situation is the problem of desorption of water from the soil profile and evaporation. After some period of evaporation the water content profile is as shown in Figure 4.7.1.

The shaded area in Figure 4.7.1 represents the **cumulative evaporative loss**, E. To calculate this loss, we write

$$E = \int_0^\infty (\theta_{fc} - \theta(z))dz = \int_{\theta_0}^{\theta_{fc}} z(\theta)d\theta \tag{4.7.7}$$

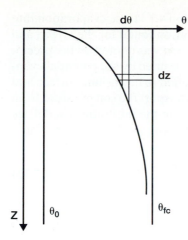

FIGURE 4.7.1 Computation of cumulative evaporation from an initially uniform saturation

Both of these integrals cover the shaded area in Figure 4.7.1. With the Boltzmann transformation of Equation 4.7.5, however, we have $z = \phi \sqrt{t}$ and

$$E = \int_{\theta_0}^{\theta_{fc}} \phi(\theta) d\theta \sqrt{t} \qquad (4.7.8)$$

where the integral represents a characteristic of the soil:

$$S_E = \int_{\theta_0}^{\theta_{fc}} \phi(\theta) d\theta \qquad (4.7.9)$$

The name given to this characteristic, S_E, is the **desorptivity**.

We have arrived at the important result that the cumulative evaporation is given by

$$E = S_E \sqrt{t} \qquad (4.7.10)$$

The evaporation rate, e, is given by

$$e = \frac{dE}{dt} = \frac{S_E}{\sqrt{4t}} \qquad (4.7.11)$$

Equations 4.7.10 and 4.7.11 are exact results. Comparison with Equations 4.7.3 and 4.7.4 shows that the desorptivity and effective mean capillary diffusivity are related through

$$S_E = (\theta_1 - \theta_0) \sqrt{\frac{4D}{\pi}} \qquad (4.7.12)$$

Philip (1969) presented Equation 4.7.12.

It should be noted that according to Equation 4.7.11, the evaporation rate is initially infinite. This fact is similar to the situation found for the Green and Ampt infiltration model. There, the infiltration rate was limited by the precipitation rate. In the case of evaporation, the rate is limited by the potential evaporation rate that is controlled by the energy supplied to the ground surface and by the ability of the atmosphere to absorb and transport the soil moisture from the ground surface. Following the approach used in the infiltration model, we eliminate time from Equations 4.7.10 and 4.7.11 (time condensation) to find

$$e = \frac{S_E^{\;2}}{2E} \tag{4.7.13}$$

Equation 4.7.13 gives the potential evaporation rate e as a function of the cumulative evaporation, E. As noted above, however, the maximum potential evaporation rate is determined by external atmospheric conditions. This potential evaporation rate, e_p, may be estimated from the Penman (1948) equation or from some other method. The time of transition from atmospheric limiting conditions to those controlled by soil water profile evaporation is called the **time of evaporative capacity**, t_{ec}. The cumulative evaporation to this time is $E_{ec} = e_p t_{ec}$. Substituting this value in the time condensation equation with $e = e_p$ and solving for t_{ec} gives

$$t_{ec} = \frac{S_E^{\;2}}{2e_p^{\;2}} \tag{4.7.14}$$

The cumulative evaporation varies linearly with time until t_{ec} is reached. Thereafter, the cumulative evaporation follows the square-root law of Equation 4.7.10. Writing this relation for times t and $t + \Delta t$, and combining the results to eliminate t gives

$$E_{t+\Delta t} = E_t \sqrt{1 + \Delta t \left(\frac{S_E}{E_t}\right)^2}$$

This result is most conveniently written

$$E(t) = E_{ec} \sqrt{1 + (t - t_{ec})\left(\frac{S_E}{E_{ec}}\right)^2} \tag{4.7.15}$$

This model is outlined in the Figure 4.7.2. For any time $t > t_{ec}$, the cumulative evaporation is given by Equation 4.7.15 and is represented as the shaded area shown in Figure 4.7.2 up to the time t.

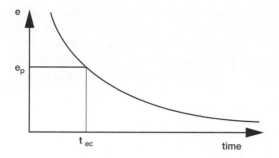

FIGURE 4.7.2 Evaporation model showing time to evaporation capacity

One issue that remains is to estimate the desorptivity. Gardner (1959) uses an effective mean diffusivity along with Equation 4.7.12 to model evaporation. With the Brooks and Corey power-law model, Charbeneau (1989) compares estimates of the desorptivity obtained using the numerical model of Philip (1955) to solve Equation 4.7.1 with estimates obtained using the weighted mean capillary diffusivity suggested by Parlange et al. (1985) and found that the results compare favorably. Use of the weighted mean capillary diffusivity and the power-law model provides the following estimate of the desorptivity, where θ_{ep} is the initial water content in the evaporative profile.

$$S_E = \sqrt{\frac{16(\varepsilon - 3)K_{ws}\Psi_b(n - \theta_r)}{3(\varepsilon + 3)(\varepsilon + 5)}} \left(\frac{\theta_{ep} - \theta_r}{n - \theta_r}\right)^{(\varepsilon + 5)/4} \qquad \textbf{(4.7.16)}$$

Equation 4.7.16 provides an estimate of the desorptivity that may be used to calculate evaporation from the soil profile. For calculation of evaporation following a period of internal drainage, one can take $\theta_{ep} = \theta_{fc}$ in Equation 4.7.16. For calculations immediately following a wetting event, however, a larger value should be used. Jury et al. (1991) report field experiments on two sandy loam plots where evaporation from bare soil was measured. These field experiments give $S_E = 0.54$ and 0.66 cm/d$^{1/2}$ for the two plots. Using mean sandy loam parameter values from Table 4.4.1 in Equation 4.7.16, the corresponding values of θ_{ep} are 0.18 and 0.19, which compares with an estimated field capacity of $\theta_{fc} = 0.083$.

Charbeneau (1989) discusses models for combining soil water redistribution and evaporation processes. While the resulting models are more complex than those presented earlier, they still provide useful tools for estimation of hydrologic processes. Gardner, et al (1970) suggest that the simple evaporation model of Gardner (1959) (Equations 4.7.10 and 4.7.12) is still appropriate during redistribution with most of the drainage occurring during the first day or two. Moreover, Charbeneau (1989) notes that the evaporation water content profile resulting from numerical solution of Equation 4.7.4 is similar in shape to the kinematic profile resulting from soil water redistribution, and suggests that these profiles be combined for modeling redistribution with evaporation. Here, evaporation is included as *excess drainage*.

The relatively simple models reviewed in this chapter may be combined to develop a continuous time water balance model. Charbeneau and Asgian

(1991) have shown that a model which combines the Green and Ampt model for infiltration, the kinematic model for redistribution, and the evaporation model from Equation 4.7.15 yields comparable results to standard daily time step water balance models. Appendix F outlines a simple spreadsheet model for a single rainfall event that combines infiltration, redistribution and evaporation processes.

EXAMPLE PROBLEM

4.8.1   Calculation of Antecedent Water Content

Charbeneau (1989) shows that the water content profile calculated from numerical solution of the nonlinear desorption problem has the same shape as the kinematic profile and that evaporation could be modeled as an equivalent amount of drainage. Use these concepts to develop a model for predicting the antecedent water content θ_a (average water content over a depth L of the soil profile).

It is assumed that the water content is initially θ_{ep} and that it decreases from θ_{ep} to 0. If z_p is the depth below ground surface to the base of the kinematic evaporation profile and W_{z_p} is the water quantity above the depth z_p, then the following relationships hold:

$$\theta = \theta_{ep}\left(\frac{z}{z_p}\right)^{1/(\varepsilon-1)} \; ; \; W_{z_p} = \frac{(\varepsilon-1)z_p\theta_{ep}}{\varepsilon} \; ; \; \theta_{ep}z_p = W_{z_p} + E \quad \textbf{(4.7.17)}$$

The first of these equations is the kinematic profile (compare with Equation 4.6.12), the second is the water quantity remaining between the ground surface and depth z_p that is available for evaporation (compare with Equation 4.6.14), and the third is the continuity equation for the profile (compare with Figure 4.7.1). Combining the second and third of Equation 4.7.17 gives $z_p = \varepsilon E/\theta_{ep}$, which may be substituted in the first to find

$$\theta(z) = \theta_{ep}\left(\frac{\theta_{ep}z}{\varepsilon E}\right)^{1/(\varepsilon-1)}$$

For $L \le z_p$, the average water content over depth L is

$$\theta_a(L) = \frac{1}{L}\int_0^L \theta(z)dz = \frac{\varepsilon-1}{\varepsilon}\theta(L) = \frac{\varepsilon-1}{\varepsilon}\theta_{ep}\left(\frac{2u_{ep}L}{\varepsilon E}\right)^{1/(\varepsilon-1)} \quad \textbf{(4.7.18)}$$

Equation 4.7.18 may be used to estimate the antecedent water content over a depth L (10 to 15 cm), where the cumulative evaporation may be calculated using Equation 4.7.15.

4.8   Evaporation from a Shallow Water Table

A shallow water table may serve as a discharge point for a groundwater flow system. As water leaves the saturated domain, its salts remain behind and the regions of water loss from a shallow water table can develop higher levels of salinity, making the water unusable for many applications. Moore (1939) first

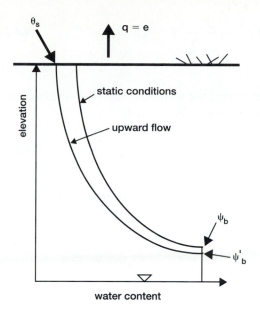

FIGURE 4.8.1 Soil Water Content Profile above a Shallow Water Table Including Evaporation from the Water Table

studied the steady-state upward flow of water from a water table through the soil profile to an evaporation zone at the soil surface. Several workers including Philip (1957c), Gardner (1958), and Anat et al (1965) have undertaken theoretical investigations of this process.

In this section, we follow Corey (1977) to estimate the potential evaporative loss as a function of depth to the water table. If we use the power-law or Brooks and Corey model for the soil water characteristic, then under static conditions the profile remains saturated up to an elevation Ψ_b above the water table. If there is upward flow, the level of water saturation drops (because of an added pressure decline associated with the flow). This process is shown in Figure 4.8.1, where the dynamic bubbling pressure head is designated as Ψ_b'.

To find the saturated depth of the capillary fringe, Ψ_b', under conditions of upward flow, we consider Darcy's equation for steady conditions. This equation gives

$$q = -K_{wsv}\frac{h_{cf} - h_{wt}}{\Psi_b'}$$

where K_{wsv} is the saturated vertical water hydraulic conductivity; h_{cf} is the head at the top of the capillary fringe; h_{wt} is the head at the water table; and Ψ_b' is the thickness of the capillary fringe. Choice of the water table as the datum dictates that $h_{wt} = 0$. The head at the top of the capillary fringe is $h_{cf} = z - \Psi = \Psi_b' - \Psi_b$. Note that the capillary pressure at the top of the capillary fringe is still the bubbling pressure, even though it occurs at a lower elevation than Ψ_b above the water table. Thus,

$$\Psi_b' = \frac{\Psi_b}{1 + \dfrac{q}{K_{wsv}}} \tag{4.8.1}$$

From the top of the capillary fringe, the water content decreases with elevation. Within this region, Darcy's Law may be written

$$q = -D(\theta)\frac{\partial\theta}{\partial z} - K(\theta) \tag{4.8.2}$$

For steady upward flow, q is fixed, and the terms in Equation 4.8.2 may be separated and integrated to give

$$z_s - \Psi_b' = \int_{\theta_s}^{n} \frac{D}{q + K} d\theta \tag{4.8.2}$$

where z_s is the elevation of the ground surface above the water table and θ_s is the water content at the ground surface. Alternatively, we can write the same equation in terms of the capillary pressure head to find

$$z_s - \Psi_b' = \int_{\Psi_b}^{\Psi_s} \frac{K}{q + K} d\Psi \tag{4.8.3}$$

In Equation 4.8.3, Ψ_s is the capillary pressure head at the ground surface. Equations 4.8.2 and 4.8.3 are important because they relate the water content or capillary pressure at the ground surface to the distance from the top of the capillary fringe to the ground surface and to the upward flow rate. With the Brooks and Corey model for the soil water characteristic function, we may write Equation 4.8.3 as

$$\frac{z_s - \Psi_b'}{\Psi_s - \Psi_b} = \int_0^1 \frac{d\varphi}{1 + \dfrac{q}{K_s}\left(1 + \left(\dfrac{\Psi_s}{\Psi_b} - 1\right)\varphi\right)^{3\lambda+2}} \tag{4.8.4}$$

where

$$\varphi = \frac{\Psi - \Psi_b}{\Psi_s - \Psi_b}$$

Equation 4.8.4 may be integrated for differing values of the pore size distribution function and other parameters. Typical results for a sandy loam soil are shown in Figure 4.8.2. This figure suggests that with a water table at a depth of 5 m, the evaporation loss could be as large as 3.7 cm/yr.

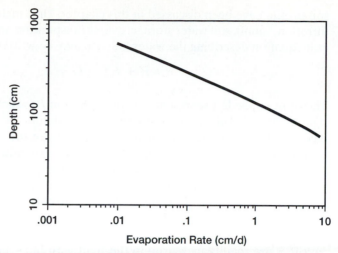

FIGURE 4.8.2 Evaporation rate from a shallow water table in a sandy loam soil

4.9   Water Balance and Groundwater Recharge

For both hydrologic and groundwater pollution investigations, one of the most important variables is the average or net annual groundwater recharge rate. The average recharge rate provides one of the significant groundwater source terms for regional hydrogeologic models. The average recharge rate is also used to determine the mean residence time (travel time) for pollutants in the vadose zone. There are a number of methods that may be used to estimate the water balance for a soil profile. These generally include the application of physical measurements, chemical measurements, or mathematical models.

The types of groundwater recharge may be distinguished by location or distribution. On one hand, diffuse recharge occurs over large areas, and is usually equated with deep percolation. Diffuse recharge probably dominates in humid environments. Local recharge, on the other hand, occurs where there is a prolonged period of ponding of water, such as along losing streams or playa lakes. Local recharge is probably most important in arid and semiarid locations. For calculation of pollutant travel times in the vadose zone and the contaminant source terms for subsurface investigations, the diffuse recharge is most important, and diffuse recharge is the focus of this section.

Physical Measurement of the Net Groundwater Recharge Rate

The **water balance** provides one method for calculating groundwater recharge. This balance represents an indirect method in that other components of the water budget are measured and groundwater recharge is then calculated from the difference. The hydrologic processes that control the water balance for a soil profile have been discussed in this chapter. They include precipitation, infiltration, runoff, soil water storage, evapotranspiration and recharge. The general equation describing the water balance may be written as

$$P = R + ET + G + \Delta S \qquad (4.9.1)$$

The term on the left represents the precipitation P (which includes applied irrigation water). This term represents the net addition of water to the soil surface. The terms on the right of Equation 4.9.1 are the surface runoff R, the evapotranspiration ET, the deep percolation or groundwater recharge G, and the water storage change (increase) ΔS of the soil profile. Each of the terms in Equation 4.9.1 represent depths of water flows or storage changes over some arbitrary time interval. To estimate the net groundwater recharge, Equation 4.9.1 is applied over a long time interval (e.g., months to years), and the water storage change is generally neglected. Both the precipitation P and surface runoff R are readily measured in the field with fairly high accuracy. Evapotranspiration ET is not directly measured. Instead, it is calculated or estimated from other measured quantities. Estimates of potential ET are most readily made. These estimates require measurements of micrometeorological variables including incoming or net radiation, air temperature, relative humidity, and wind speed. Some methods require an estimate of the turbulent transfer of water vapor from the ground surface. These methods are generally described in hydrology or micrometeorology textbooks (see for example, Chow et al., 1988, Ritzema, 1994, Chapter 5).

It is important to note that micrometeorological methods are actually used to calculate the potential ET. The actual ET may be much less the potential ET when a supply of water is not readily available near the ground surface. Thus, the micrometeorological methods for estimating actual ET are much better in humid climates rather than arid or semi-arid climates. Nevertheless, errors in ET estimates can exceed recharge estimates, making meaningful calculations of recharge using this method questionable.

The only method available for directly measuring recharge is **lysimetry**. Drainage and weighing lysimeters are soil-filled containers in which changes in water content can be directly measured. They are often used to calculate actual ET, because deep percolation is directly measured. Meyer et al. (1996) and Stephens et al. (1995) describe other physical measurement based methods. These methods include the zero-flux plane, unit hydraulic gradient, groundwater basin outflow and stream gauging. They generally have limited range of application, except for the basin outflow and stream gauging methods that are useful in humid environments with shallow water tables.

Chemical Measurement Methods

Various chemical, radioactive and environmental tracers can be used to estimate natural groundwater recharge rates in arid and semi-arid climates. The most common tracers used include tritium $H-3$, chlorine-36, and chloride.

With these tracers, two different methods are used to estimate natural ground-water recharge. **Tritium** (H−3) is a radioactive isotope of hydrogen with a half-life of about 12.4 years. It was introduced into the atmosphere during the 1950s through atmospheric testing of nuclear weapons. Its atmospheric concentration increased substantially over a relatively short time period. Before 1953, tritium concentrations in precipitation were less than 10 *tritium units* (TU). Tritium concentrations increased culminating in the period 1963–1964 (Phillips et al., 1988), when the concentration exceeded 1000 TU. The tritium concentration decreased during the subsequent decade to concentrations less than 100 TU. Tritium, as part of water vapor, fell to the earth as precipitation and infiltrated the soil profile. The net infiltration may be estimated from the depth of the peak tritium concentration in the soil profile. If L is the depth to the center of mass of the tritium pulse and if θ is the average water content through depth L, then the net infiltration I_n may be estimated from

$$I_n = \theta \frac{L}{\Delta t} \qquad (4.9.2)$$

where Δt is the time increment. If the tritium peak is still within the root zone, then the estimated net infiltration from Equation 4.9.2 exceeds the true recharge rate, because water is still being removed from the zone by plant transpiration. In addition, tritium can also move as water vapor if strong temperature gradients are present.

    Chlorine−36 is another tracer that can be used to estimate recharge. Cl−36 is radioactive with a half-life of about 300,000 years. It was produced as a byproduct of thermonuclear testing near the oceans in the 1950s. Chlorine is very stable, soluble and nonvolatile, so it makes an excellent tracer for liquid-phase water movement. The depth of the Cl−36 bomb pulse in the soil profile may be used to estimate net infiltration using Equation 4.9.2.

    The **chloride mass balance** may also be used to estimate recharge in arid environments. This method uses the concentration of natural chloride that slowly accumulates in the soil profile from infiltration of precipitation. Chloride naturally dissolves in precipitation, and when water is evapotranspired back to the atmosphere, the dissolved chloride remains in the soil profile. If C_p is the chloride concentration in precipitation and C_s is the chloride concentration in the soil profile, then the net infiltration may be estimated from

$$I_n = P \frac{C_p}{C_s} \qquad (4.9.3)$$

where P is the average precipitation rate. Depending on the location, C_p generally ranges from about 0.2 to less than 5.0 mg/L. In arid environments, the chloride concentration within the soil profile may exceed 1,000 mg/L. The chloride mass balance may be used to date water at different depths within the soil profile (Scanlon, 1991).

Mathematical Models for Estimating Recharge

When extensive site characterization data are available for a location, then numerical models may be applied to estimate the recharge rate. Such models may be based on Richards' equation, groundwater flow equations, or simplified water balance models. These models typically require site-specific climatic data for precipitation, temperature and solar radiation. Soil characteristic data includes porosity, soil water retention characteristics, and saturated hydraulic conductivity. Parameters such as field capacity and wilting point may also be used. Other factors such as the vegetation cover and rooting characteristics may also be included. For specified climatic data, the models simulate the hydrologic processes of surface runoff, infiltration, evapotranspiration and snowmelt. Infiltrated water that cannot be held in storage or transpired by plants becomes available for deep percolation.

The range of recharge rates for different locations in the United States may be estimated using water balance models such as the *Hydrologic Evaluation of Landfill Performance* (HELP) model (Schroeder et al., 1989). The HELP model has a climatological database and sub-model that can generate synthetic precipitation, temperature and solar radiation data for many U.S. cities based on their location and historical records. To obtain conservative estimates of deep percolation and recharge, the sites are modeled as consisting of bare ground with no vegetative cover, eliminating rainfall interception and transpiration of soil water. Further, the shallowest evaporative depth recommended for each city is used, reducing soil evaporative losses, and sites are modeled with a mild slope of 1 percent to limit runoff of surface water. Default soil parameters for different soil textures were uniformly applied to a 10-foot deep soil column, and 20-year simulations were performed (Hemstreet, 1996). Table 4.9.1 provides a listing of simulation results for a few cities. Johnson et al. (1998) presents a more comprehensive listing.

Travel Time through the Vadose Zone

The average travel time for water to move through the vadose zone from the ground surface to the water table may be estimated if both the net infiltration rate (or deep percolation rate) and average water content are known or estimated. The average infiltration rate is not the same as the average velocity at which a water particle moves. Deep percolation is an estimate of the Darcy velocity below the root zone. The Darcy velocity is the discharge crossing a unit cross-section area assuming that water can move through the entire cross-section area. In a porous medium, water can only move through the saturated part of the cross-section area. If the average volumetric water content is known, then the average water displacement velocity, **pore water velocity**, or **seepage velocity** v is given by

$$v = \frac{q}{\theta} = \frac{G}{\theta} \qquad (4.9.4)$$

where the **Darcy velocity** or **specific discharge** q is equal to the net groundwater recharge rate G. The travel time to the water table at a depth L, t_L, is given by

TABLE 4.9.1 Prediction of Deep Percolation (cm/yr) for Various U.S. Cities Using the HELP Model

Location	Annual Precipitation (cm/yr)	Sand	Sandy Loam	Loam	Silty Loam	Clay Loam	Clay
Bakersfield, CA	14.1	1.2	1.4	1.4	1.5	1.0	1.1
San Francisco, CA	48.8	20.6	17.2	14.6	13.0	9.1	4.7
Denver, CO	36.1	2.5	1.5	1.0	1.1	0.8	0.6
Miami, FL	129.0	48.1	32.2	22.8	17.7	10.5	3.1
Atlanta, GA	123.0	49.7	35.9	27.0	22.5	14.6	5.9
Chicago, IL	83.5	21.4	14.5	10.9	8.7	6.1	2.8
Boston, MA	110.0	33.4	25.3	19.8	16.9	11.8	5.4
Detroit, MI	77.7	17.9	12.5	9.7	8.2	6.2	3.3
Albuquerque, NM	20.8	0.9	0.8	0.9	0.9	0.7	0.6
New York, NY	106.7	36.0	26.9	21.1	17.9	12.6	6.0
Austin, TX	74.9	23.2	14.0	9.5	7.1	4.6	2.0
El Paso, TX	18.4	1.3	1.1	0.9	1.0	0.6	0.6
Houston, TX	105.8	41.3	28.3	20.0	16.1	10.0	3.4
Cheyenne, WY	32.6	1.9	1.9	1.5	1.4	1.1	1.1

$$t_L = \frac{L}{v} = \frac{L\bar{\theta}}{G} \qquad (4.9.5)$$

Volumetric water contents are usually measured in the field using neutron attenuation or in the laboratory using gravimetric methods on soil samples (Gardner, 1986). When field data are not available, models may be used to estimate average soil water content. Near the ground surface, the water content is quite variable over time—and both gravitational and capillary pressure forces are important. Below the upper one or two meters, however, the water content does not vary significantly with either depth or time for a homogeneous soil. A nearly uniform water content dictates a nearly constant capillary pressure, so that capillary pressure is nearly constant below the upper soil zone. If capillary pressure gradients are assumed to be small, one may use the kinematic form of Darcy's Law, Equation 4.3.10, which gives for flow positive downward

$$q = K_w(\theta) \qquad (4.9.6)$$

In Equation 4.9.6, the hydraulic conductivity is taken as a function of the volumetric water content, θ. The soil water content adjusts itself so that the hydraulic conductivity is sufficient to pass the volumetric flux $\bar{q}$ in the downward direction under the force of gravity, and the water content θ can thus be inferred from knowledge of q and $K(\theta)$. With the Brooks and Corey (1964) **power law model**, Equation 4.9.4 gives

TABLE 4.9.2   Average Solute Displacement Velocities (cm/yr) (after Charbeneau and Daniel, 1993)

Average Annual Recharge (cm)				
Soil Type	**5**	**10**	**25**	**50**
Clay	16	31	75	148
Clay Loam	19	37	86	164
Loam	26	49	113	211
Loamy Sand	53	99	225	416
Silt	21	39	88	164
Silt Loam	22	41	93	174
Silty Clay	16	30	74	145
Silty Clay Loam	16	30	72	137
Sand	68	127	286	527
Sandy Clay	18	35	82	158
Sandy Clay Loam	25	48	112	212
Sandy Loam	39	73	167	308

$$v = \frac{G}{\theta_r + (n - \theta_r)\left(\frac{G}{K_s}\right)^{\frac{\lambda}{3\lambda+2}}} \qquad (4.9.7)$$

where the denominator is $\overline{\theta}$, which is found by inverting Equation 4.9.6 for the power law model.

Table 4.9.2 provides estimates of the average seepage velocity v from Equation 4.9.7 for different soil textures and recharge rates. For each soil texture the average parameters determined by Carsel and Parrish (1988) were used. These values are presented in Table 4.4.1. It is important to note that the velocities reported in Table 4.9.2 are quite small. This fact suggests that the average residence time for water (and thus chemical pollutants) in the unsaturated zone can be long, especially at locations where the water table is found at great depths. It also should be noted that Equation 4.9.7 and Table 4.9.2 provide a rough guide only. They are based on the assumption of steady recharge and uniform flow in a homogeneous soil profile. The effects of solute mixing or dispersion, macropores, and spatial variability are not included, and these effects can be dramatic.

Problems

4.1.1. Estimate the size of a water molecule given that one mole of water weighs 18 grams and contains $6.02(10^{23})$ molecules. The density of water may be taken as 1.0 g/cm^3.

4.1.2. A pore with a diameter greater than 1000 A contains water that is free of adsorptive forces. What capillary pressure head must be developed to dewater such a pore, assuming that $\sigma = 60$ dynes/cm?

4.5.1. A clay-loam soil has the following Brooks and Corey model characteristics: $n = 0.46$, $\theta_{wr} = 0.08$, $\Psi_b = 35$ cm, $\lambda = 0.2$, and $K_{ws} = 0.2$ cm/hr. Estimate the water content and saturation at natural saturation, field capacity, and wilting point.

4.5.2. A sandy-loam soil has the following van Genuchten model characteristics: $\alpha = 0.05 cm^{-1}$, $M = 0.25$, $\theta_{wr} = 0.04$, n(porosity) $= 0.45$, and $K_{ws} = 0.8$ m/d. Estimate the water saturation at natural saturation, field capacity, and wilting point.

4.5.3. The 25-year return period, one-hour design storm for Travis County, Texas, is 3.62 inches. For the clay-loam soil from Problem 4.5.1, apply the Green and Ampt model to estimate the infiltration and runoff resulting from this storm. Assume that the soil has moisture content of 0.20 at the beginning of the storm.

4.5.4. Repeat the arguments leading to Equation 4.5.24 to determine the time to ponding for Philip's infiltration model. That is, eliminate time between Equations 4.5.11 and 4.5.12, and then use $I = Pt_p$ and $i = P$ to solve for the ponding time. Compare the amounts of infiltration and runoff calculated from Philip's model and the Green and Ampt model for the case of a uniform rainfall at a rate of 2 cm/hr for a period of 2-hours over a loam soil with an initial water content of $\theta_a = 0.12$. Assume that $K_{ns} = 0.10$ m/d, $\theta_{ns} = 0.40$, and $\Psi_f = 0.20$ m, and use $A = K_{ns}$ in Philip's infiltration model. (*Hint:* In order to apply Philip's model after the time of ponding, one needs to solve for the new apparent time origin by finding t_o from

$$Pt_p = S\sqrt{t_p - t_o} + K_{ns}(t_p - t_o).$$

The result is that $t_o = t_p - \dfrac{(\sqrt{S^2 + 4K_{ns}Pt_p} - S)^2}{4K_{ns}^2}$. One may then calculate the cumulative infiltration for $t > tp$ from $I_t = S\sqrt{t - t_o} + K_{ns}(t - t_o)$.

4.6.1. Groundwater is recharged by diverting river water during periods of peak flow to a spreading basin, where it is ponded and infiltrates to the water table. The surface area of the spreading basin is 4 hectares. During one period of recharge, water is ponded to an average depth of 25 centimeters for 1.5 days. The subsurface material is a sandy loam with $n = 0.40$, $K_{ns} = 1$ m/d, $\theta_{ns} = 0.38$, $\theta_{fc} = 0.13$, $\Psi_b = 10 cm$, and $\lambda = 0.85$. If the underlying aquifer is at a depth of 15 m, determine the cumulative aquifer recharge over the 30-day period following the termination of the streamflow diversion. Use the kinematic redistribution model.

4.6.2. During an infiltration experiment 40 cm of water is infiltrated and the ground surface is covered. The following characteristics are estimated: the depth to the water table is $L_{wt} = 3$ m, $K_{ws} = 2$ m/d, $\epsilon = 5$, $W = 0.4$ m, $(n - \theta_{fc}) = 0.2$, and $\theta_i^* = 0.9$. Calculate the groundwater recharge through the 10-day period following the infiltration event using the rectangular profile model.

4.7.1. A loam soil has been wetted by rainfall to initial water content of 0.30. Estimate the cumulative evaporation from the soil during the subsequent 10 days if the potential evaporation rate is 6 mm per day. Use Table 4.4.1 for the soil parameters.

4.9.1. An arid environment receives an average of 20 cm of precipitation per year. If the average rainfall Cl concentration is 2 mg/L and the soil Cl concentration is 2000 mg/L, estimate the average annual net infiltration rate.

4.9.2. Use the spreadsheet described in Appendix F to investigate the parameter sensitivity for a loam soil to the following event. A rainfall of $P = 5$ cm/hr occurs for 2 hours. The potential evaporation rate is 0.5 cm/day. Calculate the infiltration, runoff, and evaporation during the subsequent 30 days using average parameter values from Table 4.4.1, as well as values one standard deviation above and below the mean. Assume antecedent water saturation, S_a, and effective evaporation profile water saturation, S_{ep}, values equal to field capacity, wilting point, and twice field capacity. Vary only one parameter at a time, leaving others at their mean values. Present your results in a table expressing the percent change in the variable value and discuss.

Chapter Five
Groundwater Contamination

Groundwater is a valuable natural resource. In the United States, nearly half of the population relies upon groundwater as a source for domestic uses, either through public or domestic self-supply (Solley et al., 1998). For potable and other uses, this water is often provided with no treatment or only disinfection before use. Not all groundwater is of sufficient quality for direct potable use, however. Even under natural conditions, this water may contain chemical constituents that make it unfit for consumption without some form of treatment. When groundwater has been impacted by human activities (industrial discharges, agricultural return flows, urban runoff, etc.), then pretreatment before use is probably a necessity. In this chapter, we review the sources of groundwater contamination and the nature of its impacts. The general mechanisms of chemical mass transport are introduced, and solute partitioning and biodegradation are assessed.

5.1   Sources of Subsurface Contamination

The quality of subsurface waters may be impacted by naturally occurring processes as well as by actions directly attributable to human activities. Four general ways in which the chemical composition of groundwater may be changed include natural processes, nonpoint agricultural and urban runoff, waste-disposal practices, and spills, leaks, and other unintentional releases. The magnitude of environmental impact associated with each of these processes may be far different.

The first general way in which groundwater quality may be impacted is due to **natural processes**. Leaching of natural chemical deposits can result in increased concentrations of chlorides, sulfates, nitrates, iron, and other inorganic chemicals. In arid and semiarid areas, evapotranspiration from shallow water tables can further concentrate salts, because the water is lost to the atmosphere while the salts remain.

The most significant and widespread source of subsurface contaminants is **runoff** from agricultural and urban watersheds (EPA, 1983; NRC, 1994). These waters can have high concentrations of nutrients, metals, pesticides, microorganisms and other organic chemicals. Because they are widespread and are not associated with a single identifiable source area, they are classified as nonpoint source contaminants.

The third general source of groundwater contamination is activities associated with **waste-disposal practices**. In the United States, a comprehensive review of waste-disposal effects on groundwater quality was undertaken as called for in passage of the Safe Drinking Water Act in 1974. This act directed the EPA to conduct a survey of waste-disposal practices (including residential waste) which could endanger underground water supplies of public water systems, and means of control of such waste disposal. The resulting report to Congress (U.S. EPA, 1977) noted that waste-disposal practices had affected the safety and availability of groundwater, but the overall usefulness had not been diminished on a national basis. Importantly, this report also noted that instances of groundwater contamination are usually not discovered until after a drinking water source had been affected. This statement points to the long-term problem of groundwater contamination: Groundwater moves slowly, so it takes a long time for contaminants to appear at potential receptor locations. It follows that it will take a correspondingly long time to remediate the initial release of contaminants. The waste-disposal practices identified by the EPA survey include those listed in Table 5.1.1. This list is fairly comprehensive and includes most groundwater contaminant sources from waste-disposal practices.

The fourth general source of groundwater contamination is also a direct result of human activities but is unrelated to waste-disposal practices (Pye and Kelley, 1984). These include such things as **accidental spills and leaks**, agricultural activities, mining, salt-water intrusion, and others. Of these, perhaps the

TABLE 5.1.1   Groundwater Contamination from Waste-Disposal Practices (EPA, 1977)

(1) industrial waste water that is contained in surface impoundments (lagoons, ponds, pits, and basins);
(2) municipal and industrial solid refuse and sludge that are disposed of on land;
(3) sewage wastes from homes and industries that are discharged to septic tanks and cesspools;
(4) municipal sewage and storm-water runoff that are collected, treated, and discharged to the land;
(5) municipal and industrial sludge that is land-spread;
(6) brine from petroleum exploration and development that is injected into the ground or stored in evaporation pits;
(7) solid and liquid wastes from mining operations that are disposed of in tailing piles, lagoons, or discharged to land;
(8) domestic, industrial, agricultural, and municipal waste-water that is disposed of in wells; and
(9) animal feedlot waste that is disposed of on land and in lagoons.

most significant problem is associated with leaking underground storage tanks. There are about 2 million underground storage tanks located at more than 700,000 facilities in the United States which contain petroleum products, and it is estimated that approximately 25 percent of these tanks are non-tight, potentially leaking, and polluting the subsurface environment (EPA, 1988). The majority of these tanks store petroleum products, principally heating and motor fuels, while others contain inorganic products such as acids and caustic products. In appreciation of the magnitude of this problem, EPA estimates that the cost for cleanup of leaking underground tanks will approach $60 billion over the next 30 years (Plehm, 1985).

Nature of Groundwater Contamination Impacts

The nature of groundwater contamination impacts may be assessed in terms of the characteristics of the chemicals that are released and in terms of their distribution and difficulty of restoration or containment. In addition, impacts are directly related to potential human and ecological exposures and risk.

The primary chemicals of concern are classified as **organic chemicals**, **metals**, and **radionuclides**. Other inorganic chemicals such as chlorides, sulfates, nitrates, and sodium from irrigation return flows and other sources are also significant. Table 5.1.2 lists the 25 most frequently detected groundwater contaminants at hazardous waste sites. This list was generated using groundwater data from the National Priorities List of sites to be cleaned up under *Comprehensive Environmental Response, Compensation, and Liability Act* (CERCLA) and includes 9 metals and 16 organic chemical species. These chemicals pose a wide range of behavior in terms of transport and fate characteristics. When multiple chemicals are present at a site, selection of the most appropriate remediation technologies is especially difficult. Important characteristics of these chemicals are discussed in the following sections.

Difficulty of remediation of groundwater contaminants depends on their mobility and time duration since release to the subsurface environment. Some metals and radionuclides, such as lead, nickel, zinc, cesium, radium and thorium, are immobile within the subsurface and are most significant as soil contaminants, rather than groundwater contaminants. For organic contaminants, the most significant chemical characteristics are the chemical's solubility (directly related to its polarity), and whether the chemical is released as an aqueous solution or as a *non-aqueous phase liquid* (NAPL). NAPLs include petroleum products such as gasoline and oils, and solvents such as *trichloroethylene* (TCE). NAPLs are often released through spills during handling and through leaking storage tanks.

Figure 5.1.1 shows a case of a release of a NAPL that is less dense than water (LNAPL). The release is from a leaking underground storage tank. The hydrocarbon moves downward, pulled by the force of gravity, until it reaches the water table. There, because the hydrocarbon is less dense than water, it ponds at the water table and within the capillary fringe. If a sufficient quantity of LNAPL is released, then LNAPL heads will build up, which will cause lateral spreading along the water table.

TABLE 5.1.2 Contaminants Found at Hazardous Waste Sites (NRC, 1994)

Rank	Compound	Common Sources
1	Trichloroethylene	Dry cleaning; metal degreasing
2	Lead	Gasoline; mining; construction materials
3	Tetrachloroethylene	Dry cleaning; metal degreasing
4	Benzene	Gasoline; manufacturing
5	Toluene	Gasoline; manufacturing
6	Chromium	Metal plating
7	Methylene chloride	Degreasing; solvents; paint removal
8	Zinc	Manufacturing; mining
9	1,1,1-Trichlorethane	Metal and plastic cleaning
10	Arsenic	Mining; manufacturing
11	Chloroform	Solvents
12	1,1-Dichloroethane	Degreasing; solvents
13	1,2-Dichloroethene, trans-	Transformation product of 1,1,1-trichloroethane
14	Cadmium	Mining; plating
15	Manganese	Manufacturing; mining; natural oxide
16	Copper	Manufacturing; mining
17	1,1-Dichlorethene	Manufacturing
18	Vinyl chloride	Plastic and record manufacturing
19	Barium	Manufacturing; energy production
20	1,2-Dichloroethane	Metal degreasing; paint removal
21	Ethylbenzene	Gasoline; styrene and asphalt manufacturing
22	Nickel	Manufacturing; mining
23	Di(2-ethylhexyl)phthalate	Plastics manufacturing
24	Xylenes	Solvents; gasoline
25	Phenol	Wood treating; medicines

As the LNAPL spreads through the subsurface, some of it is left behind, trapped by capillary forces. Other volatile constituents may move into the subsurface vapor phase and spread therein. Soluble constituents will partition into the groundwater flowing beneath the floating lens, forming a dissolved phase plume. The result is that the contaminant slowly spreads throughout the subsurface environment.

When there is control of the release and the source is cut off, the contamination potential still exists. Much of the hydrocarbon phase may remain trapped within the subsurface porous media by capillary forces and act as a continual source of pollutants through phase partitioning. Figure 5.1.2 shows the potential migration patterns from NAPL left at residual saturation.

The situation may be far different if the release involves a NAPL that is denser than water (DNAPL), such as a chlorinated solvent. The DNAPL will tend to penetrate the water table upon reaching it, though it too must overcome capillary pressure forces. The major distinction between a DNAPL and an LNAPL release is shown in Figure 5.1.3. Because of density and capillarity effects, the DNAPL migration becomes unstable within the saturated zone of the aquifer, and the migration occurs through fingers of NAPL. The NAPL will

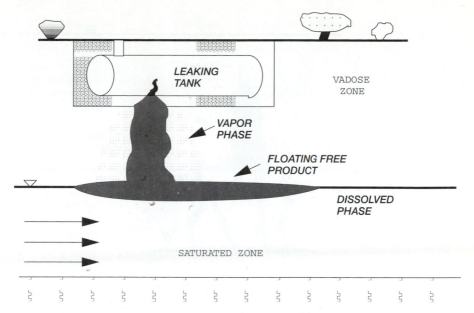

FIGURE 5.1.1   LNAPL migration pattern at the water table

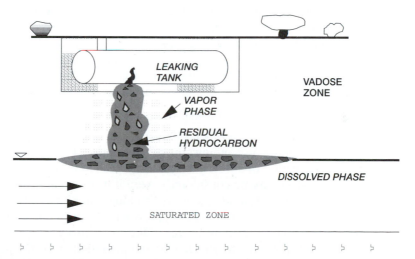

FIGURE 5.1.2   LNAPL migration from residual hydrocarbon

form pools both at the water table (unless there are heterogeneities, such as root holes, which offer an easier way into the water saturated domain) and at the base of the aquifer. DNAPL will continue to dissolve from these pools and from residual hydrocarbon trapped during its downward migration. Volatile constituents will also move into the gas phase and be transported therein by diffusion and bulk air movement.

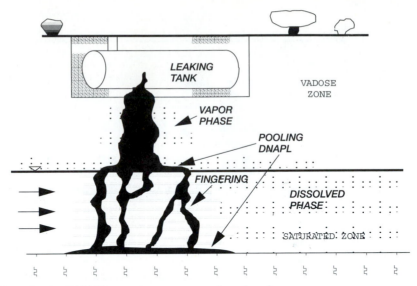

FIGURE 5.1.3   DNAPL migration pattern in the subsurface

The potential human health and environmental impacts from chemicals that are soil contaminants are through dermal contact, inhalation of suspended sediment, and mass transfer to the food chain. For aqueous phase liquids and dissolution of NAPLs, the primary route of impact is through ingesting of drinking water and use of groundwater for livestock and irrigation, with subsequent ingestion exposures. Also, groundwater contamination can lead to surface water contamination if groundwater serves as a significant source to a surface water body.

Magnitude of Groundwater Contamination Impacts

There are a number of problems that arise when one attempts to determine the overall magnitude of groundwater contamination. Pye and Kelley (1984) note that the definition of severity may be approached in several ways. A contamination incident may be considered severe if it results in groundwater concentrations that exceed standards set for drinking water if the intended use is water supply. If the groundwater is not appropriate for drinking, then the same incident might not be considered severe. One could also attempt to assess severity by the number of persons affected by contamination, or by the percentage of an aquifer that is contaminated, or, on a nationwide basis, by the percentage of the available groundwater that is affected. Still other measures of severity would arise if one includes the volume of contaminant released, its toxicity and persistence, and its mobility within the subsurface.

Pertinent measures of the magnitude of groundwater contamination impacts are the number of sites where groundwater is contaminated and the expected remediation costs. EPA estimates that there are approximately 330,000 hazardous waste sites where groundwater may be contaminated (EPA, 1993). Other estimates range from 300,000 to 400,000 sites (NRC, 1994). By far, most of these are sites contaminated by leaking underground storage tanks. The cost of cleaning up underground storage tank leaks varies widely and may be as low as $2,000 for some sites and as high as $1 million for others. According to EPA data, the average cost of cleaning up a leaking underground storage tank is $100,000, while the average cost of cleaning up a Superfund site is $27 million (EPA, 1993). Estimates are that the United States as a whole will spend $480 billion to $1 trillion (with a best guess of $750 billion) over the next 30 years cleaning up 300,000 to 400,000 sites where groundwater may be contaminated. With 90 million households in the nation, this number represents a cost of $8,000 per household (NRC, 1994).

Estimates of the amount of groundwater that has been contaminated are difficult. Experience has shown that it takes considerable effort and expense to establish the size and concentration distribution within a contaminant plume at a single site. For a national analysis, all one can do is extrapolate from the limited data available from the contaminated sites that have been properly characterized. Using available data, the EPA (1980) obtained an order-of-magnitude estimate of the extent of groundwater contamination from landfills and surface impoundments, which were considered to be the primary sources of subsurface contamination at that time. Using information on whether the facilities are situated over usable groundwater, the length of time that the facilities have been operating, and the amount of groundwater in storage, they estimated that between 0.1 and 0.4 percent of the usable shallow aquifers were contaminated by industrial impoundments and landfill sites. The EPA (1980) also investigated secondary sources such as septic tanks and petroleum exploration and mining. Extrapolating from the available data, they concluded that such sources had contaminated about 1 percent and 0.1 percent, respectively, of the nation's usable shallow aquifers.

In another study, Lehr (1982) assumed that for a worst-case scenario, there are 200,000 sources of contamination which have operated for longer durations —and that groundwater moves at faster rates than assumed in the EPA assessment. For this worst-case scenario, he estimates that between 0.2 and 2 percent of the shallow groundwater has become contaminated, and he concludes that the fraction of polluted groundwater is small.

The conclusion from these and other assessments is that the total fraction of subsurface water that has been contaminated from point sources is small on a national scale. Primarily, the shallow aquifers are susceptible to contamination, and there are vast quantities of uncontaminated groundwater within deeper aquifers throughout the world. This statement does not imply that the problem of groundwater contamination is insignificant. At numerous locations, subsurface releases of chemicals and radioactive constituents have impacted human health and the environment, and these releases must be remediated (or at least contained). The estimated costs of cleanup are high. Regulations have been enacted, however, that will limit and control future contaminant releases to the environment from industrial and municipal facilities. The *Resource Conservation and Recovery Act* (RCRA) and its amendments (1976, 1984) establish regulations on the generation, transport and disposal of hazardous substances. These will manage future releases of contamination from waste storage and disposal facilities. CERCLA (or Superfund) and its amendments (1980, 1986) provide the basis for cleanup of existing and abandoned contaminated sites. With these and other pieces of legislation, it appears that the problem of groundwater contamination from point sources has been controled and that the magnitude of the problem will decrease over time. Agriculture and urban watersheds remain significant nonpoint sources that impact groundwater quality on a regional scale, and regulatory agencies and laws do not address them.

Regulatory Requirements for Groundwater Cleanup

Table 5.1.3 lists the most significant laws that are applicable to groundwater and soil cleanup. The most important laws are RCRA and CERCLA. RCRA, enacted in 1976 and significantly amended in 1984, addresses the treatment, storage, and disposal of hazardous waste at operating facilities. The 1984 amendments brought *solid waste management units* (SWMUs), which are currently inactive but formerly used hazardous waste disposal sites within the boundary of an operating facility, under the umbrella of RCRA, as well. CERCLA, enacted in 1980 and amended in 1986, governs the cleanup of soil and groundwater at inactive facilities. Facilities that began cleanup under RCRA but later became listed on the National Priorities List can have cleanup regulated under both RCRA and CERCLA.

TABLE 5.1.3   Federal Regulations that Guide Groundwater Cleanup

Federal Laws Remediation	General Description	Significance to Groundwater
Resource Conservation and Recovery Act (RCRA), 1976; and Hazardous and Solid Waste Amendments (HSWA), 1984	Establishes regulations controlling generation, transportation, and treatment, storage, and disposal of hazardous materials for active industrial facilities.	RCRA contains requirements for groundwater monitoring, and if groundwater contamination is identified, then the RCRA corrective action process is implemented (see text for description). In addition, HSWA identifies "solid waste management units" (SWMUs) for monitoring and cleanup.
Comprehensive Environmental Response, Compensation, and Liabilities Act (CERCLA), 1980; and Superfund Amendment and Reauthorization Act (SARA), 1986	Establishes regulations for cleanup of inactive hazardous waste sites and determines the distribution of cleanup costs among the parties who generated and handled the hazardous substances disposed at these sites.	Requires cleanup of inactive facilities and establishes procedures and requirements for site characterization and remedy selection. SARA provides additional requirements on documentation and distribution of information on releases of pollutants from facilities.

TABLE 5.1.3 Continued

Federal Laws Remediation	General Description	Significance to Groundwater
Federal Facility Compliance Act (FFCA), 1992	Establishes a definition of "mixed waste," requires development of plans for its management, and waives the DOE, DOD (and other federal facilities) immunity for EPA and state hazardous waste regulations and sanctions.	Establishes that federal facilities are subject to and liable under federal and state waste management regulations, including CERCLA and RCRA.
Safe Drinking Water Act (SDWA), 1974	Develops drinking water standards that govern the quality of water delivered to the consumer and establishes the underground injection control program that classifies and regulates types of injection well practices.	If groundwater beneath a facility can be used as drinking water, then the SDWA establishes that cleanup levels must satisfy maximum contaminant levels (MCLs) for regulated chemicals.
Clean Water Act (CWA), 1972, and amendments	Establishes requirements controlling discharge of pollutants to surface waters.	If groundwater beneath a facility discharges to surface water bodies, then the CWA may be used to establish cleanup levels.
Toxic Substances Control Act (TSCA), 1976, and amendments	Establishes responsibility of manufacturers to provide data on health and environmental effects of chemical substances and provides the EPA with authority to regulate manufacture, use, distribution, and disposal of chemical substances.	TSCA includes special management provisions for handling and cleaning up material containing **polychlorinated biphenyls** (PCBs), which are present in environmental media at facilities.

RCRA established a manifest program to track hazardous waste at active facilities from the point of generation through transport to treatment, storage, and disposal facilities. The program is administered through a system of permits. To obtain a permit to operate a facility that is subject to RCRA, the facility owner must monitor the groundwater beneath the facility and downgradient of the operation to determine whether statistically significant increases in contaminant concentrations (higher than that occurring upgradient of the site) exist. If the monitoring program indicates that contaminant concentrations are increasing, then the facility owner must determine whether concentrations exist at levels above predetermined groundwater protection standards. Where such standards are exceeded, the site owner must implement a "corrective action" program to clean up the contaminated groundwater and soil. Figure 5.1.4 shows the steps in the RCRA corrective action process.

CERCLA authorized the federal government to require cleanup of abandoned or inactive facilities where groundwater and soil are contaminated. Under

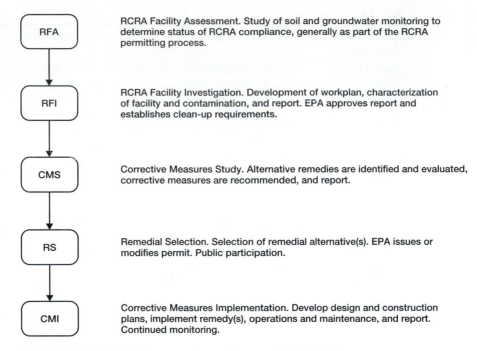

RFA — RCRA Facility Assessment. Study of soil and groundwater monitoring to determine status of RCRA compliance, generally as part of the RCRA permitting process.

RFI — RCRA Facility Investigation. Development of workplan, characterization of facility and contamination, and report. EPA approves report and establishes clean-up requirements.

CMS — Corrective Measures Study. Alternative remedies are identified and evaluated, corrective measures are recommended, and report.

RS — Remedial Selection. Selection of remedial alternative(s). EPA issues or modifies permit. Public participation.

CMI — Corrective Measures Implementation. Develop design and construction plans, implement remedy(s), operations and maintenance, and report. Continued monitoring.

FIGURE 5.1.4   RCRA Corrective Action Process (NRC, 1999)

CERCLA, current and former site owners can be held liable for cleanup costs. CERCLA established a federal fund, the Superfund, to pay for cleanup of sites where responsible parties cannot be identified. Because CERCLA facilities are no longer active, the program, unlike RCRA, is not operated through a permit system. Rather, the federal government is charged with identifying the nation's most highly contaminated sites, listing them in a database known as the National Priorities List, and ensuring that the sites are cleaned up. Figure 5.1.5 outlines the CERCLA remedial process.

Although the regulatory mechanisms under RCRA and CERCLA differ, the processes for selecting cleanup remedies under the two programs are similar. For example, CERCLA's remedial investigation/feasibility study corresponds to the RCRA facility investigation/corrective measures study (see Figures 5.1.4 and 5.1.5). Development of the CERCLA *record of decision* (ROD) corresponds to the RCRA remedy selection step. The two programs also require consideration of similar criteria when selecting cleanup remedies. CERCLA requires consideration of the nine evaluation criteria listed in Table 5.1.4, and nearly the same set of criteria are used in remedy selection under RCRA. Not all criteria listed in Table 5.1.4 receive equal weight. The first two are threshold requirements. Any remedy selected must be protective of human health and the environment, and "*applicable or relevant and appropriate requirements*" (ARARs—other federal or state/tribal laws that apply to a particular site

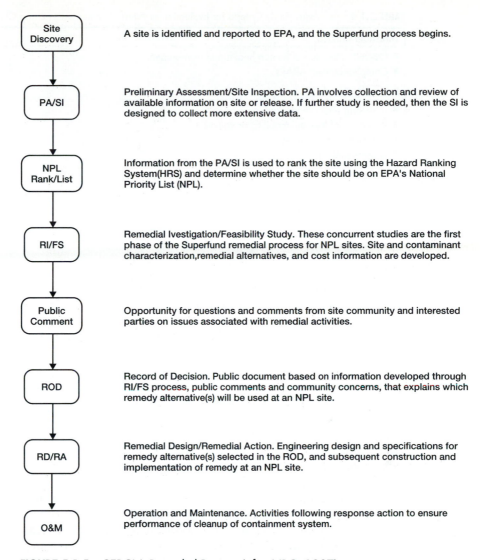

FIGURE 5.1.5   CERCLA Remedial Process (after NRC, 1997)

cleanup) must be followed. The next five criteria are considered balancing criteria. The selected remedy must be cost effective and use permanent solutions and treatment to the maximum extent practicable. The last two criteria are modifying criteria (NRC, 1999).

The Federal Facilities Compliance Act, shown third in Table 5.1.3, is the key law underlying all DOE, DOD, and other federal facilities cleanups because it established for the first time that federal facilities must comply with existing environmental statutes. The act makes federally owned and operated

TABLE 5.1.4 Feasibility Study Criteria for Evaluation of Alternative Remedies under CERCLA

Threshold Criteria
■ Overall protection of human health and environment
■ Compliance with ARARs
Balancing Criteria
■ Long-term effectiveness and permanence
■ Reduction of toxicity, mobility, or volume
■ Short-term effectiveness
■ Implementability
■ Cost-effectiveness
Modifying Criteria
■ Regulatory acceptance
■ Community acceptance

facilities subject to state-imposed fines and penalties for violation of hazardous waste requirements. The act also authorizes environmental regulators to treat federal facilities like privately owned industrial facilities and to subject them to the same rules and liabilities.

The Safe Drinking Water Act and Clean Water Act, listed fourth and fifth in Table 5.1.3, do not directly regulate cleanup of contaminated groundwater and soil; rather, they provide the basis for setting groundwater and soil cleanup goals. In particular, if groundwater beneath a facility can be used for drinking water, then the SDWA established that cleanup levels must satisfy **maximum contaminant levels** (MCLs) for regulated chemicals. The MCL levels for selected chemicals are listed in Table 5.1.5. The Toxic Substances Control Act, listed last in Table 5.1.3, includes special management provisions for cleanup of material containing *polychlorinated biphenyls* (PCBs), which are found at many facilities.

For contaminated soils, there are no ARARs equivalents to the drinking water MCLs. Until recently, soil cleanup goals were negotiated on a case-by-case basis, which increased the time to develop cleanup goals and costs and resulted in cleanup requirements that varied with location. Recognizing this limitation, EPA (1996) has developed soil-screening guidance for establishment of cleanup levels. The soil screening guidance provides a tiered approach to estimate **soil-screening levels** (SSLs) that may serve as preliminary remediation goals under certain conditions. The soil-screening framework considers potential exposures from ingestion of soil, inhalation of volatiles and fugitive dusts, and ingestion of contaminated groundwater caused by migration of chemicals through soil to an underlying drinking water aquifer. SSLs are generally based on health-based limits at the point of potential exposure. For ground water pathways, the SSLs are back calculated using drinking water standards at the site boundary.

An alternative strategy for evaluating and selecting cleanup remedies for sites has recently been developed. This is the Risk-Based Corrective Action

TABLE 5.1.5 Selected EPA Maximum Contaminant Levels (MCLs) for Drinking Water

Chemical	MCL (μg/L)
Alachlor	2
Aldicarb	3
Atrazine	3
Benzene	5
Benzo[a]pyrene	0.2
Carbontetrachloride	5
Chlorodane	2
1,2-Dichloroethane	5
1,1-Dichloroethylene	7
cis−1,2-Dichloroethylene	70
trans−1,2-Dichloroethylene	100
Di(ethylhexyl)phthalate	6
Ethylbenzene	700
Ethylene dibromide (EDB)	0.05
Lindane	0.2
Methylene chloride	5
PCBs	0.5
Pentachlorophenol	1
Tetrachloroethylene	5
Toluene	3
1,1,1-Trichloroethane	200
Trichloroethylene (TCE)	5
Vinyl chloride	2
Xylenes	10,000
Barium	2,000
Cadmium	5
Chromium (total)	100
Copper	1,300
Lead	15
Mercury	2
Nitrate (as N)	10,000
Selenium	50
Uranium	20

(RBCA, pronounced "Rebecca") program, which was originally developed for sites contaminated with petroleum hydrocarbons and later expanded to other chemicals (ASTM, 1995). RBCA integrates site assessment, remedy selection, and site monitoring through a tiered approach involving increasingly sophisticated levels of data collection and analysis. The initial site assessment identifies source areas of chemicals of concern, potential human and environmental receptors, and potentially significant transport pathways. Sites are then classified and initial response actions identified based on the urgency of need. Based on the information obtained during the initial site assessment, project managers perform a "tier 1" evaluation (according to steps specifically outlined in the

RBCA standard) to determine whether the site qualifies for quick regulatory closure or warrants a more site-specific evaluation. In determining risk, the tier 1 evaluation uses standard exposure scenarios with current reasonable maximum exposure assumptions and toxicological parameters. When the tier 1 evaluation indicates the possible presence of risk to human health, project managers can decide to clean up the site or proceed to a more detailed site risk evaluation, known as tier 2. At the end of tier 2, project managers again have the option of closing the site (if the more detailed evaluation shows that there is no risk), cleaning up the site to protect against risks as computed in tier 2, or proceeding to a final level of highly detailed evaluation, known as tier 3. Tier 3 provides the flexibility for more complex calculations to establish cleanup levels and may include additional site assessment, probabilistic evaluations, and sophisticated chemical fate and transport models. (NRC, 1999)

The general framework for the RBCA process relies upon three pieces of information, as shown in Figure 5.1.6. First, one must know whether there has been a release of contamination to the environment. In general, one would need to know the location and timing of the release, as well as the amount. However, such detailed information is rarely available. The second piece of information that is required concerns the potential transport pathways from the point of release to the point of exposure. Here we are generally concerned with subsurface transport pathways, though for problems with contaminated soil we may also look at suspension of the soil and atmospheric transport. The third requirement for deciding whether remediation is necessary at a site is a determination of whether there is a potentially exposed individual or population. If one examines the transport pathways and finds that there are no potentially exposed individuals or ecosystems at the various terminal points with the biosphere, then there is no risk and the need for remediation may be questioned. The middle step in the process shown in Figure 5.1.6 may be represented through a factor that relates the source concentration to the exposure concentration. Such a factor is called a **dilution-attenuation factor** (DAF), *natural-attenuation factor* (NAF), or similar name. Fundamentals for calculating such factors are a major topic of the following chapters.

5.2   Mass Transport Processes

In order to determine the environmental consequences of groundwater contamination, one must know where the contaminant will interface with the biosphere, when it will arrive, and what are the potential exposure concentrations. Assessment of subsurface fate and transport must address questions of source charac-

FIGURE 5.1.6   Assessment of contamination impacts

terization (what is released, where, when, how much, etc.), vadose zone transport and processes, groundwater transport, and exposure and dose assessment.

There are three basic physical mechanisms by which miscible and immiscible pollutants are transported in the subsurface environment: **advection**, **diffusion**, and **mechanical dispersion**. Emphasis in this chapter, as well as in Chapter 6, "Solute Transport by Advection," and Chapter 8, "Advection-Dispersion Transport and Models," is placed on water-soluble species, and only single-phase flow is considered. Issues associated with multiphase transport of pollutants are considered in Chapter 7, "Solute Transport by Diffusion," and Chapter 9, "Multiphase Flow and Hydrocarbon Recovery."

Advection refers to the soluble species being carried along with the flow of subsurface water. The volumetric flux of water, or the Darcy velocity, in general is given by

$$\tilde{q} = -\frac{k}{\mu}(\text{grad}(p) + \rho g \hat{k}) \tag{5.2.1}$$

where k is the intrinsic permeability of the medium and $\hat{k}$ is the upward unit vector. For water treated as a homogeneous and incompressible fluid, this relationship simplifies to

$$\tilde{q} = -K \, \text{grad}(h) = K\tilde{I} \tag{5.2.2}$$

where K is the hydraulic conductivity, h is the hydraulic head, and $\tilde{I}$ is the hydraulic gradient. For unsaturated flow, both k and K are functions of the water content.

The pore water velocity or **seepage velocity**, $\tilde{v}$, gives the average speed of the water movement. For saturated flow conditions, the seepage and Darcy velocity are related through

$$\tilde{v} = \frac{q}{n_e} \tag{5.2.3}$$

where n_e is the kinematic or **effective porosity**. For unsaturated flow the corresponding relation is

$$\tilde{v} = \frac{q}{\theta} \tag{5.2.4}$$

where θ is the volumetric water content. For most soils under saturated conditions, it is assumed that $n_e \cong n$, and no distinction is made between the effective porosity and the total porosity. An exception is drawn for clay soils and clay covers and liners for waste disposal facilities, where the difference between the effective and total porosity may be significant.

For a chemical that is present in an aqueous solution at a solute concentration c, its **advection mass flux vector** gives the total mass of the chemical which is carried across a unit area oriented normal to the bulk fluid motion. The advection mass flux vector is

$$\tilde{J}_{\text{advection}} = \tilde{q}c \qquad\qquad (5.2.5)$$

With Darcy's Law, it is seen that the advection flux equals $\tilde{K}\tilde{I}c$. Thus, the bulk transport is proportional to the medium hydraulic conductivity, the energy gradient, and the local concentration.

While advection is associated with the bulk macroscopic groundwater movement, **diffusion** is a molecular-based phenomenon. If one could see individual molecules, one would note the continual movement of each molecule and of one molecule relative to another. When these random molecular movements occur in a field with a concentration gradient, there is a net movement of the species toward regions of lower concentration. This process is known as diffusion. According to Fick's law of diffusion (Fick's first law), the **diffusive mass flux vector** for a saturated porous medium is

$$\tilde{J}_{\text{diffusion}} = -nD_s\,\text{grad}(c) \qquad\qquad (5.2.6)$$

where D_s is the apparent diffusion coefficient in soil for the chemical species. D_s is smaller than the molecular diffusion coefficient because the solute is confined to moving along a tortuous path through the pore space. For unsaturated flow, the tortuosity increases with decreasing water content and D_s is smaller still. Values of the apparent diffusion coefficient in a saturated soil are on the order of 10^{-4} m²/d (10^{-5} cm²/s).

The third mechanism of solute transport is associated both with bulk fluid movement and with the presence of the porous medium with its complex, intertwining pore space. Fluid particles that are at one time close together tend to move apart because of at least four physical mechanisms. First, the particles nearest the walls of the pore channel move more slowly than those near the channel center. Second, the variations of pore dimensions along the pore axes cause the particles to move at different relative speeds. Third, adjacent particles in one channel can follow different streamlines that lead to different channels. These particles may later come together in another channel or they may continue to move farther apart. The fourth mechanism is associated with heterogeneities in the hydraulic conductivity field, allowing solute molecules to move at different speeds—even when the hydraulic gradient is uniform. When these mechanisms occur in the presence of a concentration gradient, the resulting transport relative to the bulk water movement is referred to as **mechanical dispersion**. In a definite sense, dispersion occurs because of our inability to follow the details of pore-to-pore scale groundwater movement.

Statistically, advection refers to the average rate of movement while mechanical dispersion refers to the deviation from the mean. Also statistically, dispersion will be scale dependent. The farther a particle moves in the subsurface, the greater the range of heterogeneities of hydraulic conductivity it will experience. For example, the particle may either start off in sand or in clay. For short distances of movement, it will remain in the same type of material it started off in and the dispersion coefficient will be characteristic of that material. As it moves farther from its initial point, however, it may move from sand to clay to sand, etc., with each unit having its own characteristic velocity. Considering two particles, it is apparent that the expected deviation of their locations from the mean position will increase more through this actual heterogeneous system than it would through an idealized homogeneous system.

The mechanical dispersion mass flux is usually modeled as a diffusion or Fickian type of process. Field and laboratory experience, however, shows that the rate of mixing is greater in the direction of flow than transverse to this direction, and that the dispersion coefficient (mixing coefficient) is proportional to the flow rate. For uniform flow in the x-direction, the dispersion flux in the direction of flow is specified by the longitudinal dispersion as

$$- a_L q \frac{\partial c}{\partial x} \qquad (5.2.7)$$

while the flux in the direction transverse to the mean flow is given by

$$- a_T q \frac{\partial c}{\partial y} \qquad (5.2.8)$$

One of the goals of the theory of dispersion is to generalize these relationships for nonuniform flow fields. The coefficients a_L and a_T are referred to as the **longitudinal and transverse dispersivities** (units of length). In laboratory experiments, the longitudinal dispersivity is usually found to be 5 to 20 times larger than the transverse dispersivity. In the laboratory, a_L has been found to vary from 0.1 to 10 mm. In the field, the dispersivities are sometimes measured through single and multiple well tracer tests. More often, however, what is usually done is that measured field concentrations are simulated with mathematical models and the coefficients adjusted to get an adequate match. The values found in this fashion are usually much larger than laboratory values. Recent literature has shown field values of a_L to vary from 1 to 100 m or larger. These values of a_L are larger than laboratory values by a factor of up to 10^5, suggesting that dispersion play a different role in the field than in the laboratory.

In practice, one usually combines the coefficients of diffusion and mechanical dispersion into a single **hydrodynamic dispersion** coefficient (see Chapter 8). Because of mechanical dispersion, this new coefficient will depend upon direction (mixing is greater along the direction of flow as compared with the transverse direction), and the hydrodynamic dispersion coefficient is actually a second order symmetric tensor. The dispersion process and its characterization are discussed in detail in Chapter 8.

5.3 General Continuity Equation

A **solute** is a chemical substance dissolved in a given solution. In soil, several solutions may coexist, such as water, air, and an immiscible oil, each one being a separate **phase** (a phase is a separate, homogeneous part of a heterogeneous system). A physical interface exists between each of the phases in contact, which is a dividing surface between the phases that compounds can migrate across. Investigations of fate and transport of chemicals in both the saturated and unsaturated zones must inherently deal with a multiphase system consisting of water, air, and soil. In addition, for certain applications, such as spills, leaking tanks, or land treatment of petroleum hydrocarbons, there also is a separate liquid hydrocarbon phase that is immiscible with water.

The fundamental equation of pollutant transport is the **conservation of mass equation**. This law states that for an arbitrary region of the aquifer,

> *the rate of mass increase within the region*
> *is equal to the net mass flux into the region*   **(5.3.1)**
> *plus the increase in mass within the region*
> *due to biotic and abiotic processes*

The mass increase term represents the total mass per bulk volume, including the mass in all of the separate phases that are present. The pore space is filled by the sum of the fluids, so

$$n = \theta_w + \theta_a + \theta_o \qquad (5.3.2)$$

In Equation 5.3.2, n is the porosity, θ_w is the volumetric water content, θ_a is the volumetric air content, and θ_o is the volumetric content of the *nonaqueous phase liquid* (NAPL) or alternatively, *organic immiscible liquid* (OIL), which may be present. Individual chemical constituents partition themselves among the various phases according to thermodynamic equilibrium principles and mass transfer kinetic factors. The concentration of a constituent in the water, air, and NAPL phases are designated c_w, c_a, and c_o, respectively, all on the basis of mass per unit phase volume. The solute adsorbed to soil particles (**soil phase concentration**) is specified as mass of chemical sorbed per mass of soil, and is designated c_s. For the general case, the constituent mass density or **bulk concentration**, m, which is the mass of constituent per bulk volume (mg/L, or pCi/L for radionuclides), may be represented as

$$m = \theta_w c_w + \theta_a c_a + \theta_0 c_0 + \rho_b c_s \qquad (5.3.3)$$

where ρ_b is the soil bulk density (mass of soil per unit bulk volume) which is related to the soil density ρ_s through $\rho_b = (1 - n)\rho_s$. The soil density ρ_s generally has a value that ranges from 2.6 to 2.7 g/cm^3.

The net flux term in Equation 5.3.1 includes advection, diffusion, and dispersion in all phases. The **general mass flux vector**, $\tilde{J}$, is the mass crossing a unit surface area per unit time in the direction normal to the surface. The reaction term includes radioactive decay, biodegradation of organic pollutants, precipitation and redox chemical reactions that may immobilize a pollutant, and others. They may all be considered local source terms and represented by the symbol S^+. S^+ is the source strength and has units of mass of the species generated per unit volume per unit time.

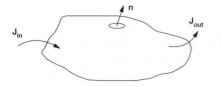

FIGURE 5.3.1   Control Volume for Mass Balance

For the arbitrary control volume shown in Figure 5.3.1, the conservation of mass equation takes the form

$$\frac{d}{dt}\iiint m dV = -\iint \tilde{J} \cdot \hat{n} dA + \iiint S^+ dV \tag{5.3.4}$$

where the first term represents the time rate of increase in the total mass of the chemical of interest within the control volume, the second term is the net flux of mass into the volume across the control surface with $\hat{n}$ the *outward* unit normal vector to the control surface, and the last term is the mass increase due to sources located within the volume. According to the divergence theorem, for any vector field $\tilde{J}$,

$$\iint \tilde{J} \cdot \hat{n} dA = \iiint \mathrm{div}(\tilde{J}) dV \tag{5.3.5}$$

In addition, with the volume fixed and nondeformable, the time derivative in the first term may be taken within the volume integral. This equation gives

$$\iiint \left\{ \frac{\partial m}{\partial t} + \mathrm{div}(\tilde{J}) - S^+ \right\} dV = 0 \tag{5.3.6}$$

Because the control volume is arbitrary, the integrand must vanish identically, and one has the differential form of the **continuity equation**

$$\frac{\partial m}{\partial t} + \mathrm{div}(\tilde{J}) = S^+ \tag{5.3.7}$$

Equations 5.3.4 and 5.3.7 are the most general form of the continuity equation and serve as the starting point for all further investigations of the subsurface transport and fate of chemical constituents.

5.4   Solute Partitioning

For most applications, the general continuity equation takes a form that is simplified in one fashion or another. These simplifications involve making further assumptions as to how to model the various processes that are of interest. Phenomenological models are needed to describe the mass transport processes, and solute partitioning among the various phases that are present must be quantified. This section and the following sections provide an introductory discussion of solute partitioning in multiphase systems and simple models for constituent loss through biodegradation.

Fluid Mixtures

Development of quantitative models to estimate the environmental impact of releases of gasoline, fuel oil, chlorinated solvents or other organic-phase liquids requires an understanding of their physical-chemical characteristics. Most petroleum products are mixtures of many individual constituents. The physical characteristics of a mixture may be estimated from the characteristics of the individual constituents that form the mixture. This process is most easily done if the mixture behaves 'ideally'. For an **ideal mixture**, the properties of the mixture result from a summation of the partial molar properties of the constituents, weighted by the mole fraction of the constituents in the mixture. There is no volume change upon mixing of different constituents within a phase.

The mole fraction, X_i, of constituent 'i' in a mixture of N constituents is related to the concentrations of the components of the mixture through

$$X_i = \frac{\dfrac{c_i}{\omega_i}}{\displaystyle\sum_{j=1}^{N} \frac{c_j}{\omega_j}} \tag{5.4.1}$$

where ω_i is the molecular weight of the ith constituent (g/mol). The molecular weight (g/mol) of a mixture is given by

$$\omega = \sum_i \omega_i X_i = \frac{\displaystyle\sum_i c_i}{\displaystyle\sum_{j=1}^{N} \frac{c_j}{\omega_j}} \tag{5.4.2}$$

ω is the mass per mole of solution. The molar volume (volume/mol) of an ideal mixture is equal to the sum of the product of the partial molar volumes of the constituents, v_i, and their mole fractions:

$$v = \sum_i v_i X_i \tag{5.4.3}$$

The relationship between the partial molar volume and the density of a constituent is

$$v_i = \frac{\omega_i}{\rho_i}$$

The density of the mixture is given by

$$\rho = \frac{\omega}{v} = \frac{\displaystyle\sum_i c_j}{\displaystyle\sum_j \frac{c_j}{\rho_j}} \tag{5.4.4}$$

EXAMPLE **Example 5.4.1   Density and Molecular Weight of an Ideal Mixture**
PROBLEM A hydrocarbon mixture that is to be used for laboratory experiments contains
benzene ($\rho_i = 0.874$ g/cm^3), ethylbenzene ($\rho_i = 0.867$ g/cm^3), toluene ($\rho_i = 0.862$
g/cm^3) and xylene ($\rho_i = 0.880$ g/cm^3), each at one-quarter fraction by volume.
What is the density and molecular weight of the mixture?

One liter of solution contains 250 mL of each of the constituents. The concentration of each constituent is $c_{\text{benzene}} = 0.874$ g/mL $\times$ 250 mL/L $= 218.5$ g/L,
$c_{\text{ethylbenzene}} = 216.8$ g/L, $c_{\text{toluene}} = 215.5$ g/L, and $c_{\text{xylene}} = 220.0$ g/L. For each constituent in this mixture, $c_i/\rho_i = \frac{1}{4}$. Thus, for this case,

$$\rho = \sum_{i=1}^{4} c_i = 871 \text{ g/L}$$

The molecular weights of benzene, ethyl benzene, toluene and xylene are 78.1,
106.2, 92.1 and 106.2, respectively. Using Equation 5.4.2, the molecular weight
of the mixture is

$$\omega = \frac{871}{\frac{218}{78.1} + \frac{217}{106.2} + \frac{216}{92.1} + \frac{220}{106.2}} = 94.2 \text{ g/mol}$$

Multiphase Equilibrium

Figure 5.4.1 shows a schematic representation of a hydrocarbon-contaminat-
ed soil. There are four phases present: air, soil, water, and the hydrocarbon or
NAPL phase. A petroleum hydrocarbon such as gasoline consists of more than

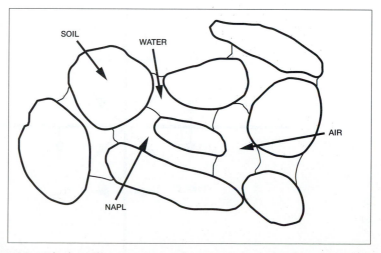

FIGURE 5.4.1   A hydrocarbon contaminated unsaturated zone has four physcial phases (air,
soil, water, NAPL). A chemical constituent may be present in any one, or all four phases

one hundred chemical constituents. These constituents may dissolve in or attach to any or all of the phases present. When considering the transport of the constituent within the multiphase system, a fundamental question concerns how the concentrations of constituents within the various phases relate to each other. The simplest and most common approach assumes that the rate of mass transport through the soil within a phase is slow compared with the rate of mass transfer between phases in contact locally. Then the concentrations remain in thermodynamic equilibrium, which is called the **local equilibrium assumption**.

The local equilibrium assumption assumes that the problem is *separable*, that is, even though a solute can exist in any one of four phases: air, soil, water, and NAPL, at any point where two of these phases touch each other, the equilibrium set up at that interface is assumed to hold independent of the presence of the other phases. Thus, it is being assumed that the presence of NAPL does not affect the water-soil partitioning properties of a medium; the total amount of material just gets shared. The equilibrium relationship for a multiphase system is shown schematically in Figure 5.4.2. If the constituent of interest is lost from one phase, for example by leaching of the water phase, then the other phases serve as a reservoir of contaminant that resupplies the phase that is losing mass while maintaining equilibrium partitioning.

Miller et al. (1990) have experimentally investigated the dissolution characteristics of trapped NAPL and found that equilibrium between the water and NAPL phases is achieved rapidly, over a wide range of NAPL saturation values and aqueous phase velocities. Jennings and Kirkner (1983), Valocchi (1985) and Parker and Valocchi (1986) discuss the limitations of the local equilibrium assumption. Karickhoff (1980) discusses use of kinetic models for sorption.

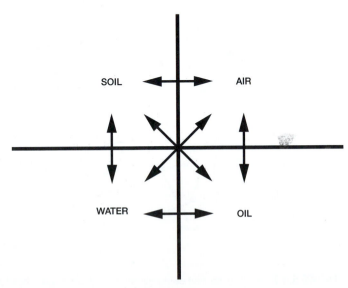

FIGURE 5.4.2   Partitioning in a multiphase system

Linear Partitioning Relations

For analysis and computation of solute transport in a multiphase system, it is convenient to be able to reference concentrations in any one phase to concentrations in another phase, or to the bulk concentration. For example, in the analysis of solute leaching it is convenient to reference concentrations in other phases to the concentration in water, while for analysis of volatilization (vaporization of a solute with subsequent transport to the atmosphere), reference to the air phase concentration is more appropriate. The local equilibrium assumption allows one to express the bulk concentration in terms of the concentration in a single phase. The water phase serves as the reference phase, and one has

$$c_a = K_H c_w \tag{5.4.5}$$

$$c_s = K_d c_w \tag{5.4.6}$$

$$c_o = K_o c_w \tag{5.4.7}$$

In these equations, K_H is the Henry's law constant, while K_d and K_o are the soil-water and NAPL-water partitioning coefficients, respectively. Both K_H and K_o are dimensionless, while K_d has units of volume per mass. Substituting Equations 5.4.5 through 5.4.7 in Equation 5.3.3 gives

$$m = (\theta_w + \theta_a K_H + \theta_0 K_0 + \rho_b K_d)c_w = B_w c_w \tag{5.4.8}$$

where B_w is the **bulk water partition coefficient**. There are similar bulk partition coefficient definitions for the other phases. For example, one has

$$m = B_a c_a = \frac{B_w}{K_H} c_a \tag{5.4.9}$$

$$m = B_s c_s = \frac{B_w}{K_d} c_s \tag{5.4.10}$$

$$m = B_0 c_0 = \frac{B_w}{K_0} c_0 \tag{5.4.11}$$

for the air, soil, and hydrocarbon (NAPL) phases, respectively.

Henry's Law

Henry's law states that water-vapor partitioning is described by a linear relationship under equilibrium conditions. This relationship is shown in Equation 5.4.5. The **Henry's law constant** in Equation 5.4.5 is dimensionless. This constant is often expressed in the literature as the ratio of the vapor pressure (P_{vp}) to the solubility (S) for the constituent with units of $\dfrac{atm - m^3}{mol}$:

$$K_H{}' = \frac{P_{vp}}{S} \tag{5.4.12}$$

The Henry's law constants in Equations 5.4.5 and 5.4.12 are related through

$$K_H = \frac{K_H'}{RT} \tag{5.4.13}$$

where R is the gas constant $8.2 \times 10^{-5} \dfrac{\text{atm} - \text{m}^3}{\text{mol} - \text{K}}$ and T is the temperature in

degrees Kelvin. The vapor pressure, solubility, and Henry's law constant for selected chemicals are presented in Table 5.4.1. Alternative units that are used for vapor pressure are mm mercury (or torr), where 1 atmosphere = 760 mm Hg. The values in this table were selected from an extensive tabulation presented by Mercer et al (1990), and Table 5.4.1 includes many of the chemical species that are identified at hazardous waste sites.

Sorption on Soil Organic Carbon: Hydrophobic Theory

Nonpolar organic compounds sorb onto the solid organic matter component of the soil matrix primarily because they do not like being in the water phase (Karickhoff, 1984). Such compounds are **hydrophobic**, and the resulting chemical interactions are weak and nonspecific (strong chemical bonds are not formed). When the organic compounds are present in trace concentrations, linear partitioning relationships (linear sorption isotherms) are often observed. A **sorption isotherm** is the relationship between the sorbed and solution concentrations. A linear isotherm plots as a straight line with a slope equal to the **distribution coefficient** K_d in Equation 5.4.6. The distribution coefficient is a function of the hydrophobic character of the organic compound and the amount of organic matter present in the soil, and may be written (Karickhoff, et al, 1979; Schwarzenbach and Westall, 1981; Brown and Flagg, 1981)

$$K_d = K_{oc} f_{oc} \tag{5.4.14}$$

where K_{oc} is the organic carbon partition coefficient and f_{oc} is the fraction of organic carbon within the soil matrix. The fraction of organic carbon f_{oc} must be measured for a particular soil. Sorption partition coefficients, indexed to organic carbon (K_{oc}) are relatively invariant for natural sorbents, and K_{oc}'s can be estimated from other physical properties of pollutants such as aqueous solubility or octanol/water partition coefficients (K_{ow}). For various organics, Karickhoff (1981) suggests

$$K_{oc} = 0.411 K_{ow} \tag{5.4.15}$$

while for a number of pesticides, Rao and Davidson (1980) find

$$\log(K_{oc}) = 1.029 \log(K_{ow}) - 0.18 \tag{5.4.16}$$

TABLE 5.4.1 Partitioning Characteristics for Selected Chemicals

	Formula	Molecular Weight	Density (g/cm³)	Water Solubility (mg/L)	Vapor Pressure (atm)	Henry's Law K_H (atm.m³/mol)	Organic Carbon K_{oc} (L/Kg)
Acetone(dimethyl ketone)	CH_3COCH_3	58.1	0.79	infinite	3.55E−01	2.06E−05	2.20E+00
Anthracene	$C_{14}H_{10}$	178.2		4.50E−02	2.56E−07	1.02E−03	1.40E+04
Benzene	C_6H_6	78.1	0.87	1.75E+03	1.25E−01	5.59E−03	8.30E+01
Benzo(a)pyrene	$C_{20}H_{12}$	252.3	1.35	1.20E−03	7.37E−12	1.55E−06	1.15E+06
Bromoform (tribromomethane)	$CHBr_3$	282.8	2.90	3.00E+03	8.17E−03	5.30E−04	1.16E+02
Bis-(2-ethylhexyl)phthalate	$C_6H_4(COOC_8H_{17})_2$	390.6		2.85E−01	2.63E−10	3.61E−07	5.90E+03
Carbon tetrachloride (CTC)	Ccl_4	153.8	1.59	8.25E+02	1.43E−01	2.98E−02	4.39E+02
Chlorobenzene	C_6H_5Cl	112.6	1.11	4.66E+02	1.54E−02	3.72E−03	3.30E+02
Chloroethane	CH_3CH_2Cl	64.5		5.74E+03	1.32E+00	6.15E−04	1.70E+01
Chloroform (trichloromethane)	$CHCl_3$	119.4	1.49	8.20E+03	1.99E−01	2.87E−03	4.70E+01
Chloromethane (methyl chloride)	CH_3Cl	50.5		6.50E+03	5.67E+00	4.40E−02	3.50E+01
o-Chlorotoluene	$C_6H_4CH_3Cl$	126.6	1.08	7.20E+1	3.55E−03	6.25E−03	1.60E+03
Chrysene	$C_{18}H_{12}$	228.2		1.80E−03	8.29E−12	1.05E−06	2.00E+05
Dibutyl phthalate	$C_6H_4(COOC_4H_9)_2$	278.4		1.30E+01	1.32E−08	2.82E−07	1.70E+05
1,2-Dichlorobenzene	$C_6H_4Cl_2$	147.0	1.28	1.00E+02	1.32E−03	1.93E−03	1.70E+03
1,1-Dichloroethane	CH_3CHCl_2	99.0	1.17	5.50E+03	2.39E−01	4.31E−03	3.00E+01
1,2-Dichloroethane	CH_2ClCH_2Cl	99.0	1.25	8.52E+03	8.42E−02	9.78E−04	1.40E+01
1,1-Dichloroethylene	CH_2CCl_2	97.0	1.22	2.25E+03	7.89E−01	3.40E−02	6.50E+01
1,2-Dichloroethylene (trans)	$CHClCHCl$	97.0	1.26	6.30E+03	4.26E−01	6.56E−03	5.90E+01
Ethyl benzene	$C_6H_5C_2H_5$	106.2	0.87	1.52E+02	9.00E−03	6.43E−03	1.10E+03
Ethylene dibromide (EDB)	$C_2H_4Br_2$	187.9	2.25	4.30E+03	1.54E−02	6.73E−04	4.40E+01
Hexachlorobenzene	C_6Cl_6	284.8	2.04	6.00E−03	1.43E−08	6.81E−04	3.90E+03
Hexachloroethane	C_2Cl_6	236.7	2.09	5.00E+01	5.26E−04	2.49E−03	2.00E+04
Methylene chloride (dichloromethane)	CH_2Cl_2	84.9		2.00E+04	4.76E−01	2.03E−03	8.80E+00
Methyl ethyl ketone	$(C_2H_5)(CH_3)CO$	72.1	0.80	2.68E+05	1.02E−01	2.74E−05	4.50E+00

TABLE 5.4.1 Continued

Formula	Molecular Weight	Density (g/cm³)	Water Solubility (mg/L)	Vapor Pressure (atm)	Henry's Law K_H (atm.m³/mol)	Organic Carbon K_{oc} (L/Kg)
Methyl *tert*-butyl ether (MTBE) $CH_3OC(CH_3)_3$	88.2		4.80E+04	3.23E−01	9.90E−04	1.12E+01
Napthalene $C_{10}H_8$	128.2	1.14	3.17E+01	3.03E−04	1.15E−03	1.30E+03
Pentachlorophenol C_6Cl_5OH	266.3	1.98	1.40E+01	1.45E−07	2.75E−06	5.30E+04
Phenol C_6H_5OH	94.1	1.05	9.30E+04	4.49E−04	4.54E−07	1.42E+01
Pyrene $C_{16}H_{10}$	202.3		1.32E−01	3.29E−09	5.04E−06	3.80E+04
1,2,3,4-Tetrachlorobenzene $C_6H_2Cl_4$	215.9		3.50E+00	4.00E−02		1.80E+04
Tetrachloroethylene (PCE) CCl_2CCl_2	165.8	1.63	1.50E+02	2.30E−02	2.59E−02	3.64E+02
Toluene $C_6H_5CH_3$	92.1	0.86	5.35E+02	3.70E−02	6.37E−03	3.00E+02
1,2,3-Trichlorobenzene $C_6H_3Cl_3$	181.4	1.45	1.20E+01	2.76E−04	4.23E−03	7.40E+03
1,1,1-Trichloroethane CH_3CCl_3	133.4	1.35	1.50E+03	1.62E−01	1.44E−02	1.52E+02
Trichloroethylene (TCE) $CHClCCl_2$	131.5	1.47	1.10E+03	7.60E−02	9.10E−03	1.26E+02
Vinyl chloride (chloroethylene) CH_2CHCl	62.5		2.67E+03	3.50E+00	8.19E−02	5.70E+01
o-Xylene $C_6H_4CH_3CH_3$	106.2	0.88	1.75E+02	9.00E−03	5.10E−03	8.30E+02

(from Mercer et al. (1990), Cohen and Mercer (1993), Pankow and Cherry (1996), Squillace et al. (1996)).

Many similar relations are available in the literature for various classes of organic compounds (see Lyman et al., 1982). Values of K_{oc} are also presented in Table 5.4.1 for selected chemicals. Equation 5.4.14 is valid only for $f_{oc} > 0.001$. Otherwise, sorption of organic compounds on non-organic solids (clay and mineral surfaces) can become significant. Also, the linear isotherm model is valid only if the solute concentration remains below one half of the solubility limit of the compound. Use of hydrophobic theory to estimate the distribution coefficient for modeling subsurface pollutant transport assumes that the sorbed concentration is in equilibrium with the concentration in solution.

For hydrophobic chemicals there is a strong inverse correlation between the chemical's solubility in water and its tendency to associate with the organic carbon matter of the soil, as measured by K_{oc}. Figure 5.4.3 shows this aqueous solubility—organic carbon partitioning relationship for a series of aromatic compounds using the data from Table 5.4.1. This figure is worth considerable attention. First, it is noted that the compound aqueous solubilities may range over many orders of magnitude, as may their organic carbon partition coefficients. A number of significant chemical series are included in this table. Representing constituents contained in gasoline, the BTEX chemicals (benzene, toluene, ethylbenzene, and xylene) are included. The influence of chlorinated aromatics is shown through benzene, chlorobenzene, dichlorobenzene, trichlorobenzene, and tetrachlorobenzene. Addition of the -OH functional group to make phenol

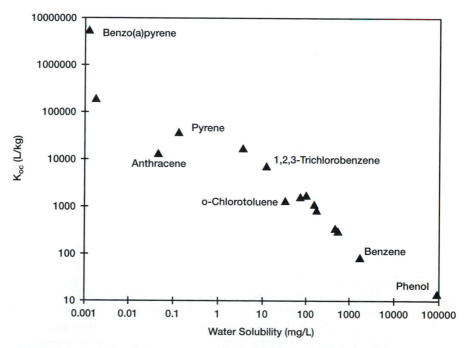

FIGURE 5.4.3 The K_{oc} and solubility relationships for a series of aromatic compounds including PAH's

from benzene increases the molecule's polarity and thus its solubility in water. Its corresponding K_{oc} decreases as compared with benzene. In contrast, addition of a halide or another organic functional group decreases the molecules solubility and increases its K_{oc} value. The K_{oc} values range over 5 orders of magnitude. The values are largest for the multiple ringed compounds (polycyclic aromatic hydrocarbons, PAHs), which suggests that they are very immobile in the subsurface environment.

Soil-Water Partitioning for Metals and Radionuclides

Analysis of transport of metals and radionuclides in soil is more difficult than for hydrophobic organic chemicals because the interaction between the soil and aqueous solution is more difficult to predict. The metals in the first two columns of the periodic table usually exist as cations in the soil solution. These include the common elements Na^+, K^+, Mg^{++}, and Ca^{++}, as well as Ba^{++} (which is used in manufacturing and energy production and is found at many hazardous waste sites) and the radionuclides Sr^{++}, Cs^+ and Ra^{++}. The degree to which these chemicals partition onto the soil is determined by the soil's cation exchange capacity and the presence of other cations that compete for the exchange sites. The cation exchange capacity is greatest for soils with high clay content, especially montmorillonite, and organic matter. Many of the other metals can exist in multiple oxidation states, and they are often complexed with ligands that are present in the soil solution. Thus the mobility of these constituents depends on both the oxidation state and speciation of the metals. For example, chromium is used in metal plating and is found at many hazardous waste sites. It exists in two oxidation states: Cr(III) and Cr(VI). Trivalent chromium is easily precipitated and is not very mobile in the subsurface, while hexavalent chromium forms stable complexes and is quite mobile and toxic.

Although the transport characteristics of metals and radionuclides is difficult to predict, estimates of effective soil-water **distribution coefficients** are useful for predicting the relative mobility of different species, and for making screening calculations in environmental assessments. Thibault et al. (1990) performed an extensive literature review for default soil solid/liquid partition coefficients for use in environmental assessments. They estimated K_d values for soils based on the soil texture, where soils that contained greater than 70 percent sand-sized particles were classed as sand soils, those containing greater than 35 percent clay-sized particles were classed as clay soils, loam soils had an even distribution of sand-, clay-, and silt-sized particles or consisted of up to 80 percent silt-sized particles, and organic soils contained greater than 30 percent organic matter. If no data existed in the literature for a given element, then the soil-to-plant concentration ratio was used as an indicator of the element's bioavailability and a means to predict a default K_d value. The geometric mean values were reported for each soil type and element (because K_d values are lognormally distributed). Table 5.4.2 provides the geometric mean K_d values for selected elements.

TABLE 5.4.2 Summary of K_d Values (L/kg) for Selected Elements*

Element	Sand	Silt	Clay	Organic
Am	1,900	9,600	8,400	112,000
C	5	20	1	70
Cd	80	40	560	800
Co	60	1,300	550	1,000
Cr	70	30	1,500	270
Cs	280	4,600	1,900	270
I	1	5	1	25
Mn	50	750	180	150
Mo	10	125	90	25
Ni	400	300	650	1,100
Np	5	25	55	1,200
Pb	270	16,000	550	22,000
Pu	550	1,200	5,100	1,900
Ra	500	36,000	9,100	2,400
Se	150	500	740	1,800
Sr	15	20	110	150
Tc	0.1	0.1	1	1
Th	3,200	3,300	5,800	89,000
U	35	15	1,600	410
Zn	200	1,300	2,400	1,600

*After Thibault et al. (1990)

NAPL-Water Partitioning

The NAPL-**water partition coefficient**, K_o, is not a constant, but instead depends on the composition of the NAPL phase. Because this composition changes with time as the pollutant ages, one should expect that K_o will also change with time. **Compositional models**, which simulate the behavior of the many components of the hydrocarbon phase, are used regularly in petroleum engineering, though their use in environmental studies has been limited (Corapcioglu and Baehr , 1987; Baehr and Corapcioglu, 1987; Baehr, 1987). **Raoult's Law** is used to describe the partitioning between the NAPL and water phases. This law states that *the aqueous phase concentration is equal to the aqueous phase solubility of the constituent in equilibrium with the pure constituent phase, multiplied by the mole fraction of the constituent in the OIL phase.* This law may be written for species k as

$$c_{wk} = S_k X_k \tag{5.4.17}$$

Using the definition of the mole fraction from Equation 5.4.1 and solving for the NAPL-water partition coefficient $K_o = c_o/c_w$ gives the following equation:

$$K_o = \frac{\omega_k \displaystyle\sum_{j=1}^{N} \frac{c_{oj}}{\omega_j}}{S_k \gamma_k} \tag{5.4.18}$$

TABLE 5.4.3   Pseudo-Gasoline Mixture (Baehr and Corapcioglu, 1987)

Constituent	Constituent Concentration (g/L)	Molecular Weight ω_j (g/mole)
benzene	8.2	78
toluene	43.6	92
xylene	71.8	106
1-hexene	15.9	84
cyclohexane	2.1	84
n-hexane	20.4	86
other aromatics	74.0	106
other paraffins (C_4–C_8)	336.7	97.2
heavy ends (> C_8)	145.1	128
	717.8	

TABLE 5.4.4   Fuel/Water Partition Coefficients Measured by Cline et al. (1991) Compared with K_0 Values Calculated from Corapacioglou and Baehr (1987) in Parentheses

Constituent	Average K_0	Coefficient of Variation (% dev.)
MTBE	15.5	19
benzene	350 (310)	21
toluene	1,250 (1,200)	14
ethyl benzene	4,500	13
m-, p-xylene	4,350	12
o-xylene	3,630 (4,200)	12
n-propylbenzene	18,500	30
3-, 4-ethyltoluene	12,500	19
1,2,3-trimethylbenzene	13,800	20

Equation 5.4.18 is written for a species k which is one out of N species which make up the NAPL phase; ω_j is the molecular weight of the jth constituent (g/mol), c_{oj} is the concentration of the jth constituent in the NAPL phase (g/L), S_k is the solubility of species k in water (g/L), and γ_k is the activity coefficient of the kth species (which equals 1 for ideal solutions). Equation 5.4.18 makes it apparent that K_o changes as the composition of the NAPL phase changes, because of dissolution, volatilization, and degradation of constituents.

Baehr and Corapcioglu (1987) used a simplified mixture to represent gasoline that is shown in Table 5.4.3. From this composition, several K_o's are calculated from Equation 5.4.18 and are listed in Table 5.4.4. Note that benzene, toluene and xylene are all hydrophobic, but the degree of hydrophobicity varies widely. Included in the tables are data for methyl *tert*-butyl ether (MTBE), an octane enhancer which may occupy up to 15 percent of gasoline by volume (Cline et al., 1991). The values calculated by using the mixture of Baehr and Corapcioglu (1987) compare favorably with the values measured by Cline et al. (1991).

Assuming ideality, Cline et al. (1991) used a further approximation to Raoult's law, which can be stated as

$$K_o = \frac{1 \times 10^6 \omega_k \left(\dfrac{\rho_o}{\omega_o}\right)}{S_k} \qquad (5.4.19)$$

where ρ_o is the NAPL phase density (g/mL), ω_o is the average molecular weight of the NAPL phase (g/mol), ω_k is the molecular weight of constituent k (g/mol), and S_k is the solubility of the constituent of interest in water (mg/L). Cline et al. (1991) demonstrated that this approximation provided an adequate fit to the measured partition coefficients from their 31 samples of gasoline. Cline et al. used an average gasoline density of 0.74 g/mL and average gasoline molecular weight of 100–105 g/mol. The measured partition coefficients showed approximately 30 percent variation and the fitted Raoult's law relationship adequately represented the trend of the values on a log-log plot.

In addition to the partition coefficient, the composition of the NAPL is important in determining the constituent concentration in contaminated groundwater. Because the water phase concentration depends on the NAPL phase concentration, the composition of the NAPL dictates both the partition coefficient and the amount of constituent that is available for contamination of the water phase.

EXAMPLE PROBLEM

Example 5.4.2 Partitioning of BTX from Gasoline

A gasoline spill whose constituents are given in Table 5.4.3 occurs on a soil whose bulk density $\rho_b = 1.6$ kg/L and organic carbon fraction $f_{oc} = 0.01$. The porosity $n = 0.4$ is divided among the water, NAPL and air phases as $\theta_w = 0.15$, $\theta_o = 0.05$, and $\theta_a = 0.20$. Determine the partitioning among soil, water, NAPL and air phases of the BTX compounds (benzene, toluene and xylene).

(1) The molar concentration for the gasoline mixture is calculated by dividing the constituent concentration (in g/L) in Table 5.4.3 by the corresponding molecular weight (in g/mol) and summing over all constituents:

$$\sum_{j=1}^{9} \frac{c_{oj}}{\omega_j} = 7.0 \text{ mol/L}$$

(2) The bulk concentration for each constituent is calculated from $m_j = \theta c_{oj}$. For benzene, this value gives $m_{\text{benzene}} = (0.05)(8.2) = 0.41$ g/L. The other resulting bulk concentrations are $m_{\text{toluene}} = 2.18$ g/L and $m_{\text{xylene}} = 3.59$ g/L.

(3) Using Equations 5.4.13, 5.4.14, and 5.4.18, the partitioning characteristics from Table 5.4.1, and an assumed temperature of 25 degrees C (298 K), the resulting partitioning coefficients are shown here. The bulk water partitioning coefficients are calculated from Equation 5.4.8.

Partitioning Coefficients				
Constituent	K_H	K_d (L/kg)	K_o	B_w
benzene	0.23	0.83	310	17
toluene	0.26	3.0	1200	65
xylene	0.21	8.3	4200	220

(4) The BTX concentrations in the water, air, soil and NAPL phases are calculated from

$$c_w = \frac{m}{B_w}, \; c_a = \frac{m}{B_a} = \frac{K_H m}{B_w}, \; c_s = \frac{m}{B_s} = \frac{K_d m}{B_w}, \text{ and } c_o = \frac{m}{B_o} = \frac{K_o m}{B_w}.$$

The corresponding percentages of the constituent in the water, air, soil and NAPL phases are calculated from

$$\%w = 100\frac{\theta_w}{B_w}, \; \%a = 100\frac{\theta_a K_H}{B_w}, \; \%s = 100\frac{\rho_b K_d}{B_w} \text{ and } \%o = 100\frac{\theta_o K_o}{B_w}.$$

The resulting concentrations and percentages are shown here.

Phase Concentration (Percentage)				
Constituent	water (mg/L)	air (mg/L)	soil (mg/kg)	NAPL (g/L)
benzene	24 (0.88)	5.5 (0.27)	20 (7.8)	7.5 (91)
toluene	34 (0.23)	8.7 (0.08)	100 (7.4)	40 (92)
xylene	16 (0.07)	3.4 (0.02)	130 (5.9)	67 (94)

EXAMPLE PROBLEM

Example 5.4.3   Leaching Model for a Gasoline Spill

A gasoline spill of 5,000 L spreads over a surface area of 100 m². Determine the leaching characteristics of xylene, which is present as 5 percent of the initial gasoline mass. Assume the same gasoline composition and soil characteristics as in Example 5.4.2.

(1) The depth of penetration L_o, as shown in Figure 5.4.4, is calculated from

$$L_o = \frac{V_{\text{spill}}}{\theta_{or} A_{\text{spill}}} = \frac{5 \text{ m}^3}{0.05 \times 100 \text{ m}^2} = 1 \text{ m}$$

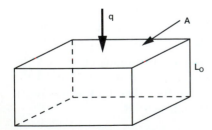

FIGURE 5.4.4   Leaching model for a surface spill

(2) Based on a gasoline density of 718 kg/m^3, the initial xylene concentration is 0.05×718 kg/m^3 = 35.9 kg/m^3. The partitioning coefficients are the same as in Example 5.4.2. Once the gasoline has spread through the soil profile the initial bulk concentration is found from $m_o = \theta_o c_o$(initial) = 0.05×35.9 kg/m^3 = 1.80 kg/m^3. With this bulk concentration and the bulk partition coefficients, the individual phase concentrations are calculated to be $c_w = 7.96$ mg/L, $c_o = 33{,}800$ mg/L, $c_a = 1.67$ mg/L, and $c_s = 0.0661$ mg/kg. Note that because of the presence of the other phases, the xylene concentration in gasoline has dropped from 35.9 to 33.8 kg/m^3. The percentage of xylene in the water, oil, air, and soil phases are 0.0664 percent, 94.03 percent, 0.0186 percent, and 5.88 percent, respectively.

(3) According to Equation 5.4.6, the bulk concentration is $m = B_w c_w$, and the total mass present is

$$M = AL_0 B_w c_w$$

The mass advection flux from the contaminated region is

$$JA = q c_w A$$

The continuity equation then says that

$$\frac{dM}{dt} = AL_0 B_w \frac{dc_w}{dt} = -qA c_w$$

which may be integrated to find

$$c_w = c_{\text{leachate}} = c_{wo} \exp\left(-\frac{qt}{L_o B_w}\right) \tag{5.4.20}$$

where c_{wo} is the initial aqueous phase concentration. Thus, the leachate concentration decreases exponentially, as does the total mass within the region. If the net infiltration rate is 0.25 m/yr, then the concentration of xylene in the leachate is given by

$$c_w(\text{xylene}) = 7.96 e^{-0.00111t} \text{ mg/L}$$

where t is the time in years. This value corresponds to a leaching "half-life" of 624 years, which is so large because of the small partitioning of the xylene into the mobile water phase. Note that we have neglected degradation and volatilization in this example, and we have not considered the change in K_o that occurs as other components are leached from the gasoline phase.

Compositional Models

It is not difficult to develop compositional models for chemical leaching from contaminated soils, though the computational effort is significantly greater than that presented earlier. For each constituent, the mass balance relation Equation 5.4.3 and Raoult's law may be written as follows:

$$m_i = (\theta_w + \theta_a K_H + \rho_b K_d)c_{wi} + \theta_o c_{oi} \; ; \; c_{wi} = S_i \frac{c_{oi}/\omega_i}{\sum_j (c_{oj}/\omega_i)} \qquad (5.4.21)$$

These relations may be combined to give

$$c_{oi} = \frac{m_i}{(\theta_w + \theta_a K_H + \rho_b K_d)\dfrac{S_i/\omega_i}{\sum_j (c_{oj}/\omega_j)} + \theta_0} \qquad (5.4.22)$$

If the initial soil concentrations are known, then at any later time the magnitude of m_i and the other parameters are known for all constituents, and Equation 5.4.22 may be solved iteratively for the set of values c_{oi}. The left side is only slightly sensitive to the assumed values in the denominator of the right side, so convergence is fast. Once the composition set ($c_{oi}, i = 1, N$) are found from Equation 5.4.22, the corresponding set of aqueous concentrations may be found from the first of Equation 5.4.21. Then, if the soil is contaminated over a depth L_o, mass conservation gives

$$L_0 \frac{dm_i}{dt} = J_i = q_w c_{wi} \; ; \; L_0 \frac{d\theta_0}{dt} = \sum_j \frac{J_j}{\rho_j} \qquad (5.4.23)$$

Together, Equations 5.4.21 to 5.4.23 provide the leachate and soil concentrations, and the NAPL content for a contaminated soil. One could also include mass loss through soil air diffusion for volatile chemicals.

EXAMPLE PROBLEM

Example 5.4.4   Compositional Leaching Model

A solvent solution to be used in a laboratory column leaching experiment consists of TCE, PCE, and CTC at initial concentrations 620, 465, and 465 g/L, respectively. The soil column is held vertically and has porosity 0.45, bulk density 1,460 g/L, fraction organic carbon 0.01, and average water saturation 0.5 during the test. The upper 5 cm section of the column retains residual solvent at an initial saturation 0.15. Water is introduced at the top of the column at volumetric flux (Darcy velocity) 4 cm/d. Estimate the leachate concentration and residual saturation as a function of time.

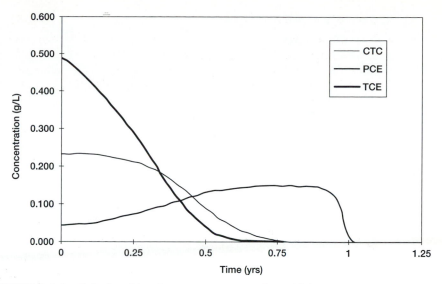

FIGURE 5.4.5 Calculated leachate concentrations from a laboratory investigation

This problem is similar to that of Example 5.4.3, except that the compositional model will be used instead of constant partition coefficients. The properties of the three solvents are provided in Table 5.4.1. The initial bulk concentrations are estimated from $m_i(0) = c_{oi}(0)\theta_o(0)$. The initial solution concentrations were then used in Equation 5.4.22 to find the initial NAPL concentrations in soil. After 2 iterations, the results are 604.9, 461.7, and 442.3 g/L for TCE, PCE, and CTC, respectively. A simple direct-step computation of Equation 5.4.23 was then performed to find the solute concentrations and NAPL content. All calculations were performed using a spreadsheet. Figure 5.4.5 shows the leachate concentration as a function of time. TCE concentrations are initially largest and decrease most rapidly because of its greater solubility and initial concentration. Note that even initially, the TCE leachate concentration is significantly less than its solubility of 1.1 g/L. The solubility of CTC is somewhat less than that of TCE but much larger than that of PCE. CTC disappears shortly following TCE. The constant partition coefficient model would not predict the behavior of PCE in leachate. Its concentration is initially small because of its small solubility. As the other constituents are leached from the NAPL solution, however, its mole fraction increases, as does its solubility. Eventually the PCE leachate concentration reaches its aqueous solubility, corresponding to when it is the only NAPL present. This behavior is followed in the NAPL saturation history shown in Figure 5.4.6. The rate of decrease in saturation is greatest initially when TCE and CTC are at their largest concentrations in leachate. At a time of about 0.4 years, PCE is the dominant constituent in leachate, and because of its smaller solubility, the rate of decrease in NAPL saturation also decreases. If a solution contains constituents all with the same solubility (or a single

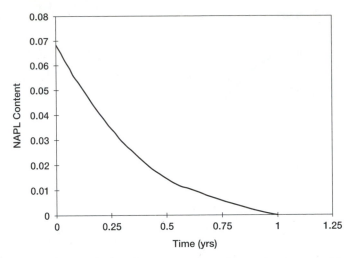

FIGURE 5.4.6   NAPL (residual) volumetric content during the computed leaching experiment

constituent), then the NAPL content magnitude should decrease at a constant rate until the phase is dissolved.

5.5   Degradation Losses of Soil and Groundwater Contaminants

Biodegradation is an important environmental process that causes the breakdown of organic chemicals in soil. The chemical transformations are associates with, or **mediated**, through the activities of microorganisms that are naturally present, or those that are introduced through engineered systems. The transformation of organic carbon to inorganic carbon (CO_2) is accomplished through enzymatic oxidation. The microorganisms produce enzymes that are small (10 to 100 nm) proteins. These enzymes catalyze the transfer of electrons from the chemical that is being degraded, the substrate, to another chemical that accepts the electrons. This process is an oxidation-reduction reaction, where the substrate is oxidized and the chemical that receives the electrons is reduced. The rate of the reaction will depend on the amount of both the electron-donor and electron-acceptor chemicals and on the amount of the enzyme present. Together, the electron-donor and electron-acceptor are the **primary substrates** that are essential to ensure the growth of the microorganisms.

Mineralization refers to the complete degradation of an organic chemical to inorganic products such as carbon dioxide, water, sulfate, nitrate, or ammonia. **Partial degradation** describes a level of degradation less than complete mineralization. The degradation products may be more or less toxic than their parent compounds. Chemical compounds that are not easily degraded are said to be **recalcitrant**, and they persist in the environment. (Alexander, 1965; Alexander, 1977; Scow, 1982; Valentine and Schnoor, 1986; Kuhn and Suflita, 1989).

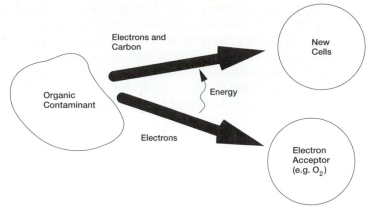

FIGURE 5.5.1   The process of microbial degradation of organic chemicals (after NRC, 1993)

Microorganisms participate in the biodegradation of organic chemicals because they can use the contaminants for their own growth and reproduction. This is shown schematically in Figure 5.5.1. The microbes receive energy through the process of breaking chemical bonds and (through their enzymes) transferring electrons from the contaminants to an electron acceptor. Under aerobic conditions the electron acceptor is usually oxygen (O_2). Under anaerobic conditions, the potential electron acceptors include nitrate (NO_3^-), sulfate (SO_{42}^-), iron (Fe_3^+), manganese (Mn_4^+), and other chemicals. Part of this energy, along with additional electrons and carbon, are used to generate new cells. During the degradation process, the microbe biomass utilizes the organic contaminant for growth and maintenance while transforming the contaminant to other chemical forms as a degradation byproduct. Certain organic contaminants are not able to serve as the primary substrate. These include various volatile organic chemicals such as chlorinated solvents. These chemicals may be degraded through a process called **cometabolism**, wherein another chemical serves as the primary substrate, and the produced enzymes are also able to serve to partially degrade the contaminant of concern, even though it is not able to serve as the primary energy source for the organisms.

The rate of biodegradation is determined by the number and type of microorganisms present, the toxicity of the parent compound or its daughter products to the microorganism population, the water content and temperature of the soil, the presence of electron acceptors and the oxidation-reduction potential, the soil pH, the availability of other nutrients for microbial metabolism, the water solubility of the chemical, and possibly other factors. Alexander and Scow (1989) review various models for the rate of biodegradation. We have considered only first-order decay in the transport and fate models. According to the first-order decay model, the rate of chemical loss is proportional to the amount of chemical present. For degradation of organic chemicals, first-order loss equations are only an approximation.

The basic equations for the kinetics of enzyme reactions were developed by Henri in 1903 and were extended by Michaelis and Menten in 1913 (see Moore, 1972). These were adapted by Monod (1942) to describe the growth of microorganisms and the chemical loss rate, and together they are referred to as Monod or Michaelis-Menten kinetics. As applied to biodegradation, these expressions model both the chemical substrate and the quantity of microorganisms. For each component in Monod kinetics, the rate-of-increase, μ, is given by

$$\mu = \mu_{max} \frac{c}{K_m + c} \tag{5.5.1}$$

where μ_{max} is the maximum rate-of-increase and K_m is known as the half-saturation constant and is defined as the rate-of-increase-limiting substrate concentration that allows the constituent to increase at half the maximum specific rate. Equation 5.5.1 has the same form as the Langmuir isotherm, except that the Langmuir isotherm refers to the sorption of a chemical, while the Monod expression refers to the rate-of-increase of a given constituent of concern. As applied to the substrate chemical loss, Equation 5.5.1 takes the form

$$\frac{dc}{dt} = -\frac{\mu_{max} M c}{Y(K_m + c)} \tag{5.5.2}$$

where M is the microbial mass in mg/L and Y is the yield coefficient, which is a measure of the organisms formed per unit substrate utilized. In turn, the maximum rate-of-increase is a function of the electron acceptor concentration. If O represents the electron acceptor concentration, then using the Monod model, one may write

$$\mu_{max} = \mu_o \frac{O}{K_o + o} \tag{5.5.3}$$

where μ_o and K_o represent analogous parameters to μ_{max} and K_m in Equation 5.5.1. An equation similar to 5.5.2 and 5.5.3 is written for the electron acceptor. The corresponding balance equation for the microbial mass is

$$\frac{dM}{dt} = \frac{\mu_{max} M Y c}{K_m + c} - bM \tag{5.5.4}$$

where b is a first-order decay coefficient that accounts for cell death. Borden and Bedient (1986) have used this Monod kinetics formulation in a groundwater fate and transport model.

Equations 5.5.1 and 5.5.2 show that the chemical substrate loss rate is proportional to the microbial mass present, the substrate concentration, and the electron acceptor concentration. Borden et al. (1986) suggest that microbial growth reaches equilibrium rapidly with respect to the rate of groundwater

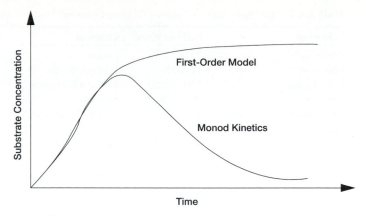

FIGURE 5.5.2   Effect of microbial growth on degradation of organic chemicals

flow, and therefore, the microbial concentration can be assumed constant under steady-state conditions in the subsurface environment. If both the microbial mass and electron acceptor concentrations are constant (or O is large compared with K_o), and the substrate concentration is small relative to K_m, then the first-order loss model is obtained. For higher substrate concentrations compared with K_m, however, the loss rate is zero-order (independent of the substrate concentration). In general, the loss rate is a nonlinear function of the concentrations of the primary substrates.

Equation 5.5.1 shows that the microorganism mass will grow at a rate that is proportional to the substrate concentration for low concentrations and at a constant rate for high concentrations. The new features that the Monod model introduces are shown in Figure 5.5.2. If a substrate is introduced to a porous media, the first-order model predicts that the concentration will increase to a level that depends on the substrate half-life, the inlet concentration, and the residence time. The Monod model, however, suggests that the introduction of substrate promotes the growth of microorganism mass, which increases the rate of substrate degradation. At early times, the predicted concentrations are similar to those from the first-order model. For later times, however, the substrate concentration decreases due to the increased microbial mass.

Although it is an approximation, for most investigations it is assumed that degradation can be described by first-order rate reactions in which the rate of loss of a chemical is proportional to the chemical concentration present in the soil. In terms of the bulk concentration, this model is written:

$$\frac{dm}{dt} = -\lambda m \qquad (5.5.5)$$

where λ (time-1) is the first-order rate constant which is related to the half-life $T_{1/2}$ by

TABLE 5.5.1   Half-Life in Soils for Selected Organic Chemicals*

Chemical	Half-Life (days)	Chemical	Half-Life (days)
Acetone	2–14	Lindane	5.9–240
Benzene	10–730	Methyl chloride	14–56
Benzo(a)pyrene	114–1,060	Methyl ethyl ketone	2–14
Benz(a)anthracene	200–1,350	Methanol	1–7
Bis-(2-ethylhexyl) Phthalate	10–389	Methyl t-butyl ether	56–365
Carbon tetrachloride	7–365	Napthalene	1–258
Chlordane	570–2,770	Pentachlorophenol	46–1,520
Chloroethane	14–56	Phenanthrene	32–400
Chloroform	56–1,800	Phenol	0.5–7
Chrysene	740–2,000	Pyrene	420–3,800
DDT	16–11,300	Tetraethyl lead	14–56
1,1-Dichloroethane	64–154	Toluene	7–28
1,2-Dichloroethane	100–365	1,1,1 Trichloroethane	140–546
1,1-Dichloroethylene	56–132	Trichloroethylene	321–1,650
1,2-Dichloroethylene	56–2,880	Vinyl chloride	56–2,880
Ethylbenzene	6–228	Xylenes	14–365

*From Howard et al. (1991), *Handbook of Environmental Degradation Rates.*

$$T_{1/2} = \frac{ln(2)}{\lambda} \cong \frac{0.693}{\lambda} \tag{5.5.6}$$

Measured values of $T_{1/2}$ are quite variable between the laboratory and the field, and from site to site in the field. A number of references provide typical values. See for example Jury et al. (1984); Wilkerson et al. (1984); American Petroleum Institute, (1984); and Rao et al. (1985). Representative values for selected chemicals are shown in Table 5.5.1. These values were taken from Howard et al. (1991).

The first-order rate equation of Equation 5.5.5 shows that the source term of Equation 5.4.7 may be represented with an apparent or effective rate constant, λ, using

$$S^+ = -\lambda m \tag{5.5.7}$$

If the actual loss rate is proportional to the aqueous concentration rather than the bulk concentration then Equation 5.5.7 is replaced by

$$S^+ = -\lambda_c \theta_w c_w \tag{5.5.8}$$

Equation 5.5.8 is put in a form appropriate for the continuity Equation 5.4.7 by combining Equations 5.4.8 and 5.5.8 to give

$$S^+ = -\frac{\theta_w \lambda_c}{B_w} m = -\lambda m \tag{5.5.9}$$

Equation 5.5.9 shows how to relate the two rate constants λ_c and λ. Furthermore, if there is a separate hydrocarbon phase present, then it may be toxic to aerobic soil microbes and there would be no degradation. Even under these conditions, however, biodegradation could continue through anaerobic bacteria.

Radioactive Decay

Certain naturally occurring and manmade chemicals are radioactive and are potentially harmful to human health. Radioactive chemicals transform through changes in the composition of their nucleus. These transformations result in release of alpha (essentially a helium nucleus), beta (an electron), and other types of particles. In addition, the nuclear transformations release gamma radiation in the form of photons. The rate of release of radiation is called the **activity**, and it is directly proportional to the amount of nuclide present. Radioactive decay is described through a first-order loss equation. If N represents the number of atoms of a radionuclide in a sample of material at any time, then the law of radioactive decay is

$$\frac{dN}{dt} = -\lambda N \tag{5.5.10}$$

where λ is again the first-order rate coefficient that is related to the nuclide half-life through Equation 5.5.6. Because the nuclide activity is the nuclide loss rate, Equation 5.5.10 shows that the nuclide activity, A, is related to the amount of atoms present through

$$A = -\frac{dN}{dt} = \lambda N \tag{5.5.11}$$

Equation 5.5.11 shows that the first-order loss equation also applies for the nuclide activity, A.

The standard unit for measure of the radionuclide activity is the curie (Ci), which is the decay rate of 1 g of radium−226. This value is equal to 3.7×10^{10} disintegrations/sec (sec^{-1}). The SI unit for activity is the becquerel (Bq), which is 1 disintegration/sec. Thus, $1 C_i = 3.7(10^{10})Bq$. The standard unit for measurement of radionuclide activity in groundwater is pCi/L, where $1pC_i = 10^{-12}C_i$.

Another unit of measurement of radioactivity is the **specific activity**, which is the activity per unit mass. Using Avogadro's number, 6.02×10^{23} atoms/mol, the relationship between specific activity, SA, and molecular weight is

$$SA(Bq/g) = \frac{6.02 \times 10^{23}\lambda}{\omega} \tag{5.5.12}$$

A simple calculation shows that the specific activity of radium−226, with a half-life of 1,600 years and a molecular weight of 226, is $SA = 3.7 \times 10^{10}Bq/g = 1C_i/g$.

If a radionuclide has a decay product that is also radioactive, then the decay product is called the daughter of the parent, or primary nuclide. If the parent has

TABLE 5.5.2 Half-Life and Specific Activity for Selected Radionuclides

Radionuclide	Half-Life (yrs)	Specific Activity (Ci/g)	Source
H−3 (tritium)	12.3	9,690	Cosmic radiation, nuclear power, nuclear weapons
C−14	5,730	4.46	Cosmic radiation, nuclear weapons
K−40	$1.28 \ (10^9)$	$6.98 \ (10^{-6})$	Naturally occurring primordial nuclide
Mn−54	0.855	7,740	Nuclear weapons, nuclear power
Fe−55	1.80	3,610	Nuclear weapons, nuclear power
Co−60	5.27	1,130	Nuclear weapons, nuclear power
Sr−90	28.8	138	Fission product
Tc−99	$2.1 \ (10^5)$	0.0172	Fission product
Ce−134	2.06	1,300	Nuclear power
Ce−137	30.2	86.4	Nuclear weapons, nuclear power
Pb−210	22.3	76.3	NORM, U−238 decay chain
Rn−222	$1.047 \ (10^{-2})$	$1.54 \ (10^5)$	NORM, U−238 decay chain
Ra−226	1,600	0.989	NORM, U−238 decay chain
Ra−228	5.8	270	NORM, Th−232 decay chain
Th−232	$1.41 \ (10^{10})$	$1.09 \ (10^{-7})$	NORM
U−235	$7.04 \ (10^8)$	$2.16 \ (10^{-6})$	NORM (0.72 percent U in Earth's crust)
U−238	$4.5 \ (10^9)$	$3.34 \ (10^{-7})$	NORM
Np−237	$2.1 \ (10^6)$	$7.18 \ (10^{-4})$	Fission rectors, decay of Am−241
Pu−239	24,100	0.0621	Thermal reactors, used in nuclear weapons
Am−241	432	3.43	Decay of Pu−241

a long half-life compared with the daughter, then the activity of the daughter is limited to that of the parent. When the parent and daughter have the same activity, they are said to be in **secular equilibrium**. Under conditions of secular equilibrium, the number of atoms of each nuclide present in a sample are related through

$$\lambda_1 N_1 = \lambda_2 N_2 \qquad \textbf{(5.5.13)}$$

If nuclide 1 is the parent with a long half-life and nuclide 2 is the daughter with a short half-life, Equation 5.5.13 shows that if they are in equilibrium, there is much more mass of nuclide 1 present in the sample than of nuclide 2. (Only nuclides with a high molecular weight have radioactive daughters, so the emission of a nuclear particle during decay does not change the atomic mass significantly.)

Table 5.5.2 lists the half-life and specific activity (Ci/g) for selected radionuclides. The **specific activity** is used to convert between mass concentration units and activity units. A number of the radionuclides listed in this table are *naturally occurring radioactive materials* (NORM). These include the long-lived nuclide potassium−40 and the parents of three chains of radioactive elements: the uranium series originating with U−238, the thorium series originating with Th−232, and the actinium series originating with U−235. Other naturally

occurring radionuclides include tritium and C−14, which are produced by cos-
mic rays. H−3 and C−14, and the other nuclides listed are also generated
through nuclear fission and capture reactions associated with nuclear power
applications and nuclear weapons testing.

The whole-body or specific tissue dose equivalent measures the potential
human health impacts from exposure to radioactivity. The standard unit for dose
equivalent is the **rem**, which is calculated from the adsorbed dose and a quality
factor that accounts for the biological effectiveness of different kinds of radiation. The SI dose-equivalent unit is called the *sievert* (*Sv*), where $1\ Sv = 100\ rem$.

EXAMPLE
PROBLEM

Example 5.5.1   Calculation of Radiological Dose

Table 5.1.5 shows that uranium has a drinking water standard (MCL) of 20
µg/L. Calculate the radiological dose associated with ingestion of groundwater
with this concentration. Use an ingestion dose (whole-body) conversion factor
for U−238 of 2.70 (10^{-4}) mrem/pCi (from EPA, 1988).

An ingestion rate of drinking water of 2 L/day is assumed, 365 days per
year. This value gives for the total yearly uranium ingestion: 20 µg/L × 2 L/d ×
365 d/yr = 1.46(10^4)µg = 0.0146gU. From Table 5.5.2, the specific activity of
U−238, which accounts for 99.28 percent of the naturally occurring U, is 3.34
(10^{-7}) *Ci*/g, which gives a yearly ingestion activity 0.0146g × 3.34(10^{-7}) *Ci*/g ×
10^{12} *pCi/Ci* = 4900 *pCi*. With the dose conversion factor, this value amounts to
a whole-body dose equivalent of 4900 *pCi* × 2.70(10^{-4}) mrem/*pCi* = 1.3 mrem.
For comparison, the average annual dose equivalent to persons in the US from
natural and artificial sources (excluding radon) is approximately 160 mrem,
while the average radon dose equivalent is 2400 mrem (NRC, 1990).

5.6   Simplified Forms of the Continuity Equation

For the transport of contaminates in aquifers in the absence of NAPL, Equation 5.5.8 takes the form

$$m = (n + \rho_b K_d)c \qquad (5.6.1)$$

The mass flux vector, including its advection and dispersion components, is expressed by

$$\tilde{J} = \tilde{q}c - n\tilde{\tilde{D}}_{hd} \cdot \text{grad}(c) \qquad (5.6.2)$$

where $\tilde{\tilde{D}}_{hd}$ is the hydrodynamic dispersion tensor. Substituting Equations 5.6.1,
5.5.7, and 5.6.2 in the general continuity Equation 5.4.7 gives

$$(n + \rho_b K_d)\frac{\partial c}{\partial t} + \text{div}(\tilde{q}c) - \text{div}(n\tilde{\tilde{D}}_{hd} \cdot \text{grad}(c)) + \lambda(n + \rho_b K_d)c = 0 \qquad (5.6.3)$$

If the flow field is steady and there are no volumetric sources or sinks (leakage, infiltration or evaporation, etc.) then the water volume continuity equation is

$$\text{div}(\tilde{q}) = 0 \tag{5.6.4}$$

Because

$$\text{div}(\tilde{q}c) = c\,\text{div}(\tilde{q}) + \tilde{q} \cdot \text{grad}(c)$$

the advection term in Equation 5.6.3 may be written

$$\tilde{q} \cdot \text{grad}(c) \tag{5.6.5}$$

Also, if the porosity is constant and approximately equal to the kinematic porosity, then one may introduce the **retardation factor** as

$$R = 1 + \frac{\rho_b K_d}{n} \tag{5.6.6}$$

The physical significance of the retardation factor is that it measures how much slower the solute migrates than water because the solute spends part of its time sorbed on the soil matrix and immobile. Thus, a retardation factor of 10 means the average speed of the solute is 10 times slower than that of the water.

Multiplying the continuity Equation 5.6.3 by n^{-1} gives

$$R\frac{\partial c}{\partial t} + \tilde{v} \cdot \text{grad}(c) - \text{div}(\tilde{\tilde{D}}_{hd} \cdot \text{grad}(c)) + \lambda Rc = 0 \tag{5.6.7}$$

Equation 5.6.7 is the form of the continuity equation that is used for most analytical models in one-, two-, and three-dimensions with the added assumption that the velocity field is uniform. In particular, the one-dimensional form of the continuity equation is

$$R\frac{\partial c}{\partial t} + v\frac{\partial c}{\partial x} - D\frac{\partial^2 c}{\partial x^2} + \lambda Rc = 0 \tag{5.6.8}$$

The first term in Equation 5.6.8 gives the concentration change at a given location (including the sorbed mass as well as the mass in solution). The second term accounts for the change in concentration associated with advection while the third term accounts for the concentration change associated with mixing or diffusion. The last term is the mass sink that is modeled as first-order decay. Solutions to Equations 5.6.7 and 5.6.8 for various boundary conditions will be discussed in Chapter 8.

EXAMPLE
PROBLEM

Example 5.6.1. Length of a Contaminant Plume

Dissolution of constituents from trapped residual NAPL results in a contaminant plume, the maximum length of which is determined by the interplay between advection and decay. Develop an expression for the maximum plume length if the constituent source-term concentration is c_o.

Retaining the second and fourth terms in Equation 5.6.8, the continuity equation gives

$$v\frac{dc}{dx} + \lambda Rc = 0 \rightarrow L = \frac{v}{\lambda R} \ln\left(\frac{c_0}{c}\right) = \frac{vT_{1/2}}{\ln(2)R} \ln\left(\frac{c_0}{c}\right) \qquad (5.6.9)$$

As an example, consider benzene dissolving from a gasoline release. We may take $c_o = 2.4$ mg/L, which follows from Example 5.4.2 with a reduction by a factor of 10 to account for mixing within the source zone. We may also use $c = 5\mu g/L$, which is the drinking water MCL from Table 5.1.5. Assume that $v = 1$ ft/d and that $T_{1/2} = 60$ days (Table 5.5.1) and $R = 2$. These values give $L = 270$ ft, which is similar to the length of benzene plumes that are often observed in the field.

Problems

5.2.1. Within a region of an aquifer, the hydraulic gradient and hydraulic conductivity are 0.003 and 4 m/d, respectively. If the porosity is 0.3, how far will a conservative solute move over the period of one year?

5.2.2. An aquifer has a porosity and hydraulic conductivity of 0.35 and 8 m/d. The local hydraulic gradient is 0.008. For a solute with apparent molecular diffusion coefficient $5(10^{-5})$ m^2/d and concentration 100 mg/L, what concentration gradient results in a diffusion flux equal to the advection mass flux?

5.2.3. Repeat Problem 5.2.2 to consider mechanical dispersion instead of molecular diffusion, when the longitudinal dispersivity has a value of 0.1 m.

5.4.1. A synthetic OIL contains benzene (density = 0.874g/cm^3), ethyl benzene (density = 0.867g/cm^3), toluene (density = 0.862g/cm^3), and xylene (density = 0.880g/cm^3), each at one-quarter fraction by mass. Assuming ideal mixing (that is, that the final volume is equal to the sum of the added volumes), answer the following questions:

i)   What is the density of the resulting mixture?

ii) What is the molar concentration of the mixture?

iii) What is the effective solubility for benzene in water in contact with this mixture?

iv) What is the mixture OIL-water partition coefficient for toluene?

5.4.2. Water in the vadose zone contains chlorobenzene at a concentration of 50 mg/L. What is the equilibrium concentration of soil air in contact with this solution?

5.4.3. An aquifer with porosity 0.35 has a distribution coefficient for neptunium (Np) of 5 L/kg. What fraction of Np is in aqueous solution and sorbed to the aquifer solids?

5.4.4. An aqueous solution containing chloroform is released to a surface soil. The soil has porosity, volumetric water content, and fraction organic carbon of 0.45, 0.25, and 0.03, respectively. What percent of chloroform will partition to the soil water, soil air, and sorbed on the soil solids? Assume a temperature of 25 C.

5.4.5. A gasoline fuel has a density of 760 g/L and a molecular weight of 102 g/mol. MTBE is present in the gasoline with a concentration $c_o = 60g/L$. What groundwater concentration is in equilibrium with this fuel?

5.4.6. What is the leaching "half-life" of strontium (Sr) in a sand soil with porosity 0.4 and water content 0.2 if the average infiltration rate is 0.25 m/yr, and if the soil is contaminated to a depth of 0.1 m?

5.5.1. Under batch conditions, an organic chemical with $K_d = 2.5L/kg$ has an apparent half-life of 40 days (based on the aqueous concentration loss rate) in a soil with a porosity of 0.4. Estimate its true half-life for the chemical in the aqueous phase. Also estimate the chemicals apparent or effective half-life in a similar soil with $K_d = 5L/kg$ and the loss rate (in mg per liter of soil per day) when the aqueous chemical concentration is 100 mg/L.

5.5.2. An organic chemical with $K_d = 4L/kg$ in a soil with porosity 0.4 has an aqueous phase half-life of 30 days for first-order biotic transformation losses, and a sorbed half-life of 300 days for abiotic transformation losses. What is the apparent loss half-life for the chemical including both processes?

5.5.3. Radium-226 is present within the minerals of natural soils at concentrations ranging from 0.1 to 0.3 pCi/g, and at higher levels in various rocks. Radon-222, which is a decay product of Ra-226, is released through alpha particle recoil, and the emission fraction gives the fraction of the decay product that ends up in the pore space. Emission fractions range from 0.1 to 0.5. What groundwater radon-222 concentration (pCi/L) is in secular equilibrium with soil (porosity 0.35) containing 0.2 pC_i/g Ra-226 if the emission fraction is 0.3? (*Hint:* the bulk concentrations should be the same when measured in pC_i/L.)

5.6.1. An aquifer has a porosity of 0.35. Estimate the retardation factor for the pesticides atrazine, dieldrin, and parathion. The values of K_d (in L/kg) for these three compounds are 0.82, 8.5, and 52, respectively.

5.6.2. Strontium (Sr-90) is present at a location in a sand aquifer (porosity 0.35) at a groundwater concentration of 500 pC_i/L. The local groundwater seepage velocity is 50 m/yr. Assuming steady-state conditions, what groundwater concentration is expected at a location 100-m downgradient?

5.6.3. Following Example 5.6.1, estimate the length of an MTBE plume originating from a gasoline spill. Assume that the petroleum product contained MTBE at a 0.2 mole-fraction. Use $v = 1$ ft/d and assume that in the dissolved plume $T_{1/2} = 365$ days and $R = 1.1$ for MTBE, and that a concentration limit $c = 20\mu g/L$ is used.

Chapter Six
Solute Transport by Advection

When groundwater moves, it carries with it dissolved constituents. This transport of chemicals with the bulk fluid movement is called **advection**. Advection is the most significant mass transport process. It results from large-scale gradients in fluid energy (head), although the resulting rates of mass transport are much less than those found for atmospheric or surface water transport. In this chapter, we discuss some of the tools that are used to analyze advective transport. The most important of these is the potential theory for groundwater flow, and this subject is presented in some detail. Residence time distribution theory, which provides a useful tool for many applications, is also discussed.

One simplification that can often be made when analyzing solute transport is that the groundwater flow field is steady. This situation implies that the velocity at a given location is constant, or that if the velocity does vary, the time period over which the velocity varies is short compared with the time period for solute transport so that a sequence of steady-state flow fields may be used. Justification for this simplification comes from the small velocities that are typically found in groundwater flow, and the relatively short duration of hydraulic transients that are caused by such things as changing the pumping rate of a well. The distances moved by groundwater solutes during the hydraulic transient time period are small and may usually be neglected.

6.1   Advective Transport

Advection transport is the movement of a chemical as it is carried along with the bulk fluid movement. The sorbed mass does not take part in this displacement. The most general form of the mass conservation equation for advection transport is

$$\frac{\partial m}{\partial t} + \text{div}(\widetilde{q}c) = 0 \qquad (6.1.1)$$

Equation 6.1.1 is the same as Equation 5.4.7 with no sources or sinks, and only advective transport (Equation 5.2.5, $\tilde{J} = \tilde{q}c$). This equation shows that advection transport is determined by the mass density of the chemical (both its bulk density and solute density) and by the volume flux vector (Darcy velocity). For steady groundwater flow and equilibrium linear sorption, Equation 6.1.1 simplifies to

$$R\frac{\partial c}{\partial t} + \tilde{v} \cdot \text{grad}(c) = 0 \qquad \textbf{(6.1.2)}$$

where R is the retardation factor and $\tilde{v}$ is the seepage velocity vector. Equation 6.1.2 specifies how the chemical concentration changes at any point in an aquifer as a function of time. It provides an **Eulerian description** of the transport field. A **Lagrangian representation** follows the chemical as it moves about. Mathematically, this process is given by the **characteristics** of Equation 6.1.2 and results in the following (see the discussion in Section 4.6.2):

$$\left.\frac{d\tilde{x}}{dt}\right|_c = \frac{\tilde{v}}{R} \qquad \textbf{(6.1.3)}$$

Equation 6.1.3 is mathematically equivalent to Equation 6.1.2. It states that the fluid **particle** with concentration c is transported with the bulk fluid movement at a velocity of magnitude v/R. In order to determine the advection mass flux, one must first determine the flow field and the seepage velocity, and then the chemical is transported at a velocity slower than this by a factor equal to the retardation factor, R.

Advective Transport in Aquifers

Some of the most important applications of groundwater transport theory are for flow in unconfined and confined aquifers. While the local flow field may be three-dimensional, transport through large distance may often be treated as two-dimensional. The usual approach to modeling these systems is to average the flow characteristics over the thickness of the aquifer and thus achieve a two-dimensional representation. Such a formulation was considered in Chapter 2 (see Equations 2.3.16 and 2.3.35, for example). These models include unsteady flow, diffuse recharge, and sources and sinks. For some applications, however, i.e., flow near injection and production wells of a pump-and-treat system, the diffuse recharge is small compared with the lateral flow rate in the aquifer, and the water table level changes rapidly with changes in well flow rates compared with the rate of solute transport. For such cases, the storage and source terms are negligible, and the continuity equation becomes

$$\text{div}(\tilde{U}) = 0 \qquad \textbf{(6.1.4)}$$

Equation 6.1.4 is the continuity equation for steady flow in an aquifer without sources or sinks and is appropriate for both confined and unconfined aquifers.

It is known from calculus that if the divergence of a vector field vanishes, then the field vector may be generated from the gradient of a potential function. This information leads us to consider potential theory and its application to advective transport.

6.2   Introduction to Potential Theory

Potential theory is one of the oldest and most widely studied areas of mathematical physics. It has applications in the theories of gravitational attraction, electrostatics, heat conduction and dynamics of ideal fluids. The theory applies to flow fields where the flux may be derived from the gradient of a scalar field called a potential field. Application of the continuity principle then yields the potential or Laplace equation.

   In our applications, we will be using the **aquifer flux** or discharge potential. We will find that in casting the problem in terms of the aquifer flux potential (rather than the velocity potential), the equations for both confined and unconfined aquifers are the same, so solutions for one aquifer type may be used for the other. In this way, the mathematical model formulation simplifies the applications. The assumptions that must generally be made are that constant effective values of the hydraulic conductivity or transmissivity can be found—and that the effective parameters are isotropic. Volume sources and sinks are absent except for wells. Furthermore, for an unconfined aquifer, the base of the aquifer is essentially flat and serves as a datum. Finally, we assume steady-state flow conditions.

6.2.1   Potential Functions

For steady flow in a **confined aquifer**, the continuity equation is Equation 6.1.4. In this equation, the aquifer flux, $\tilde{U}$, is equal to the product of the Darcy velocity and the aquifer thickness ($\tilde{U} = \tilde{q}b$), or with Darcy's Law written in terms of the aquifer transmissivity,

$$\tilde{U} = -T\nabla h \qquad (6.2.1)$$

If a constant effective value of the transmissivity can be identified, then we can define a **discharge potential** by

$$\Phi = Th \qquad (6.2.2)$$

and Equation 6.2.1 may be written

$$\tilde{U} = -\nabla\Phi \qquad (6.2.3)$$

Equation 6.2.3 is fundamental in that it says that the aquifer flux is given by the gradient of a potential, in this case the discharge potential. Substituting Equation 6.2.3, the continuity equation takes the form

$$\nabla^2 \Phi = 0 \qquad \textbf{(6.2.4)}$$

Equation 6.2.4 is the **Laplace equation**, which is one of the most important equations in mathematical physics. Solutions to the Laplace equation are called **potential functions**, and the equation is often called the **potential equation**. A potential equation arises whenever a vector field has zero divergence $(\nabla \cdot \tilde{U} = 0)$, and the flux may be written as the gradient of a potential (Equation 6.2.3). Its physical significance follows from the general continuity equation. At any point within the domain, there is no change in "storage", nor are there any sources or sinks. Thus, the flow of water into any point is equal to the flow of water away from the point. In this sense, the Laplace equation represents a physical condition of equilibrium or balance.

For steady flow in an **unconfined aquifer** with negligible vertical recharge the continuity equation also takes the form of Equation 6.1.4. If the base of the aquifer is nearly flat, and its elevation is chosen as the datum with H the saturated thickness of the aquifer, then the aquifer flux is given by

$$\tilde{U} = -KH\nabla H \qquad \textbf{(6.2.5)}$$

If a constant effective value of the hydraulic conductivity can be identified, then we can define a discharge potential by

$$\Phi = \frac{KH^2}{2} \qquad \textbf{(6.2.6)}$$

With Equation 6.2.6, the flux is again given by Equation 6.2.3, and the continuity equation again becomes Equation 6.2.4, the Laplace equation.

It is apparent from Equations 6.2.2 and 6.2.6 that the potential function is directly related to the head field. Pictures of the head field and potential field should look similar. The important difference is given by Equation 6.2.3; the gradient of the potential field gives the aquifer flux, which is the quantity of interest. Lines along which the potential takes the same constant value are called **equipotentials** and may be thought of as elevation contours on a map.

Equation 6.2.3 states that the aquifer flux vector is perpendicular to the equipotential lines. With the aquifer flux known, the Darcy velocity for a confined aquifer is found from

$$\tilde{q} = \frac{\tilde{U}}{b} \qquad \textbf{(6.2.7)}$$

where b is the aquifer thickness, and for an unconfined aquifer from

$$\tilde{q} = \frac{\tilde{U}}{H} = -\frac{\nabla\Phi}{\sqrt{\dfrac{2\Phi}{K}}} = -\sqrt{2K}\nabla\Phi^{1/2} = -\nabla\sqrt{2K\Phi} \qquad \textbf{(6.2.8)}$$

For both cases, the seepage or average linear velocity is given by

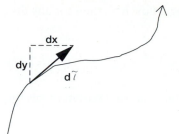

FIGURE 6.2.1   Flow Line for a Fluid Particle

$$\tilde{v} = \frac{\tilde{q}}{n} \qquad (6.2.9)$$

The common feature of these equations is that they represent parallel vectors:

$$\tilde{v} // \tilde{q} // \tilde{U} \qquad (6.2.10)$$

The common path of a fluid particle is called a **flow line** or a **streamline**. A streamline is shown in Figure 6.2.1 where the vector $\tilde{dl}$ is parallel with the velocities according to Equation 6.2.10.

For convenience, we have worked with the vector $\tilde{U}$ because its formulation in terms of the discharge potential is the same for confined and unconfined aquifers. Consider the displacement of a fluid particle along a flow line. If the displacement vector is given as

$$\tilde{dl} = dx\hat{i} + dy\hat{j}$$

then the condition that this vector is parallel with the discharge vector is that the cross product of the two vectors vanish:

$$\tilde{U} \times \tilde{dl} = (U_y\, dx - U_x\, dy)\hat{k} = \tilde{0}$$

Because the zero vector must have vanishing components, the condition that the displacement vector is parallel to the discharge vector is

$$\frac{dx}{U_x} = \frac{dy}{U_y} \qquad (6.2.11)$$

Equation 6.2.11 may be used to define the flow paths mathematically through use of a new function called a stream function.

6.2.2   The Stream Function

Equation 6.2.11 must be satisfied along a flow line or **streamline**. We wish to define such a line mathematically through use of a characteristic function called the stream function. The **stream function** is defined as the function $\Psi(x,y)$ which is constant along a flow line. If Ψ is constant along a flow line, then for any displacement along the flow line,

$$d\Psi = \frac{\partial \Psi}{\partial x}\, dx + \frac{\partial \Psi}{\partial y}\, dy = 0 \qquad (6.2.12)$$

Comparison with Equation 6.2.11, however, gives the following requirements for the stream function:

$$U_x = -\frac{\partial \Psi}{\partial y} \;\; ; \;\; U_y = \frac{\partial \Psi}{\partial x} \qquad (6.2.13)$$

Thus, if one can find the stream function, then the components of the discharge vector may be found directly through differentiation.

The stream function has a clear physical interpretation. Consider Figure 6.2.2, which shows two streamlines and an arbitrary line that crosses them. The unit tangent to the arbitrary line is given by

$$\hat{t} = \frac{dx}{dl}\, \hat{i} + \frac{dy}{dl}\, \hat{j}$$

while the unit normal is specified by

$$\hat{n} = \frac{dy}{dl}\, \hat{i} - \frac{dx}{dl}\, \hat{j}$$

Note that $\hat{t} \cdot \hat{n} = \tilde{0}$. The calculation of interest concerns the total discharge crossing the arbitrary line between the two flow lines. For each increment of the line, dl, this calculation is given by

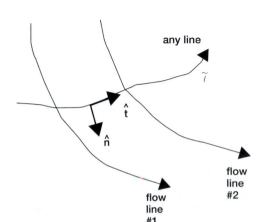

FIGURE 6.2.2 Calculation of the discharge between two streamlines

$$dQ = \tilde{U} \cdot \hat{n} dl = (U_x \hat{i} + U_y \hat{j}) \cdot \left(\frac{dy}{dl} \hat{i} - \frac{dx}{dl} \hat{j}\right) dl$$

$$= U_x dy - U_y dx = -\frac{\partial \Psi}{\partial y} dy - \frac{\partial \Psi}{\partial x} dx = -d\Psi$$

Thus, the discharge passing between flow lines #1 and #2 is given by the difference in their stream function values:

$$Q_{1-2} = \Psi_1 - \Psi_2 \tag{6.2.14}$$

It is of interest to note that this statement is true, even if T or K is nonhomogeneous.

Conjugate Functions

The discharge potential Φ is defined by the conditions

$$U_x = -\frac{\partial \Phi}{\partial x} \; ; \; U_y = -\frac{\partial \Phi}{\partial y} \tag{6.2.15}$$

For steady flow without recharge, these give

$$\nabla \cdot \tilde{U} = 0 = \nabla \cdot (-\nabla\Phi) = -\nabla^2\Phi$$

which is the Laplace equation. In order to calculate the **rotation** within the flow field, we calculate the curl of the field:

$$\nabla \times \tilde{U} = \left(\frac{\partial U_y}{\partial x} - \frac{\partial U_x}{\partial y}\right)\hat{k} = \left(\frac{\partial}{\partial x}\left(-\frac{\partial \Phi}{\partial y}\right) - \frac{\partial}{\partial y}\left(-\frac{\partial \Phi}{\partial x}\right)\right)\hat{k} = 0$$

This equation shows that the flow field is **irrotational**. In general, any field that can be derived from the gradient of a potential function is irrotational.

We now want to repeat the calculation of the rotation of the field using the stream function. We have

$$\nabla \times \tilde{U} = \left(\frac{\partial U_y}{\partial x} - \frac{\partial U_x}{\partial y}\right)\hat{k} = \left(\frac{\partial}{\partial x}\left(\frac{\partial \Psi}{\partial x}\right) - \frac{\partial}{\partial y}\left(-\frac{\partial \Psi}{\partial y}\right)\right)\hat{k} \equiv (\nabla^2\Psi)\hat{k}$$

Because we have already seen that the flow field is irrotational, however, we conclude that

$$\nabla^2\Psi = 0 \tag{6.2.16}$$

and the stream function also satisfies the Laplace equation.

The stream function and the potential function are said to be **conjugate functions**. By definition, the flow lines are parallel to the **streamlines** (lines of constant stream function value). On the other hand, the velocity vector is perpendicular to the lines of constant potential (this is the meaning of the gradient operator). Thus, the streamlines and potential lines are perpendicular.

In summary, the continuity equation for flow in either a confined or unconfined aquifer without leakage or recharge is given by

$$\nabla \cdot \tilde{U} = 0$$

A vector $\tilde{U}$ that satisfies this equation is called a **solenoidal vector**. An **irrotational flow** field satisfies

$$\nabla \times \tilde{U} = \tilde{0}$$

A field that is both solenoidal and irrotational may be derived from a pair of conjugate function, both of which satisfy the Laplace equation:

$$\nabla^2 \Phi = 0 \; ; \; \nabla^2 \Psi = 0$$

For each function Φ there is a corresponding function Ψ which is orthogonal to it, and vice versa. These functions may be interchanged, and we will still get a potential flow field. The physical interpretation of the stream function is that the difference in its magnitude between two streamlines corresponds to the total discharge between the streamlines ($Q_{12} = \Psi_1 - \Psi_2$). Thus, in regions where the streamlines are close, the velocity is fast, and where the streamlines are far apart, the velocity is slow.

For potential theory problems, the principle of **superposition** is valid. If each of the stream functions Ψ_1, Ψ_2, etc., or each of the potential functions Φ_1, Φ_2, satisfies the Laplace equation, then so does their superposition:

$$\Psi = \Psi^1 + \Psi^2 + \ldots \; ; \; \Phi = \Phi^1 + \Phi^2 + \ldots$$

This equation follows from the linearity of the Laplace equation. The flow fields of interest include 1) uniform flow, 2) point sources and sinks (wells), and 3) *doublet + uniform flow + source = leaking pond.* Strack (1989) presents many other examples.

6.2.3   Travel Time along Streamlines

One of the main interests in application of potential theory to groundwater transport problems is estimation of travel times along streamlines. If we know two points along a streamline, as shown in Figure 6.2.3, the travel time between them may be estimated from

$$\Delta t = \frac{\Delta s}{|\tilde{v}|} \tag{6.2.17}$$

where $\tilde{v}$ is the velocity calculated from Equation 6.2.9. In general, the travel time between any two points along a streamline may be estimated from

$$\text{time} = \int_{\tilde{x}_1}^{\tilde{x}_2} \frac{ds}{v} \tag{6.2.18}$$

where v is equal to the magnitude of $\tilde{v}$, and the displacements ds are taken along the streamline. For a few problems, we may actually evaluate Equation 6.2.18 analytically.

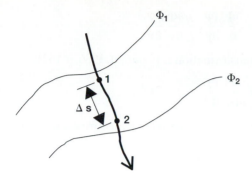

FIGURE 6.2.3 Calculation of travel time along a stremline

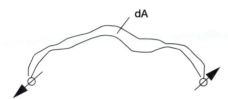

FIGURE 6.2.4 Stream tube leading from an injection well to a production well

We may also calculate the travel times along streamlines using the potential and stream functions directly. Figure 6.2.4 shows a stream tube leading from an injection well to a production well. The plan view area of the stream tube is dA, so that its fluid volume is $n\,bdA$ (where b is the average saturated thickness). According to Equation 6.2.14, the discharge through the stream tube is $d\Psi$. Thus, the travel time may be calculated from the ratio of the stream tube volume to discharge:

$$\text{time} = nb\frac{dA}{d\Psi} \qquad (6.2.19)$$

If the function $A(\Psi)$ is the area between the streamline Ψ and a reference streamline Ψ_o, and if $A(\Psi)$ may be found in a useful analytical form, then Equation 6.2.19 may be used to directly determine the travel times without having to first determine the velocity field.

Equation 6.2.19 can be written in another way. Each point in the flow domain may be specified by either its (x,y) coordinate or its corresponding (Φ,Ψ) coordinate. That is, we may write either $x = x(\Phi,\Psi)$ and $y = y(\Phi,\Psi)$, or $\Phi = \Phi(x,y)$ and $\Psi = \Psi(x,y)$. In this way, the (Φ,Ψ) coordinate system is a transformation of the (x,y) coordinate system, and vice versa. From calculus (see, for example, Courant, 1936, p. 368), we know that the Jacobian of the transformation is equal to the ratio of the area of the image region and the area of the original region, as evaluated at a given point where the transformation is continuous and invertible. For the transformation $\Phi = \Phi(x,y)$, $\Psi = \Psi(x,y)$, the ratio of areas is given by

$$|J| = \frac{\partial(\Phi,\Psi)}{\partial(x,y)} \equiv \frac{\partial\Phi}{\partial x}\frac{\partial\Psi}{\partial y} - \frac{\partial\Phi}{\partial y}\frac{\partial\Psi}{\partial x}$$

Likewise, the Jacobian of the inverse transformation is (Courant, 1936, p. 144):

$$\frac{1}{|J|} = \frac{\partial(x,y)}{\partial(\Phi,\Psi)}$$

With this transformation, the increment of area $dx\,dy$ is transformed into the area $d\Phi\,d\Psi$, where

$$dxdy = \frac{1}{|J|}d\Phi d\Psi$$

In Equation 6.2.19, the area dA of the stream tube is given by

$$dA = \left(\int_{-\infty}^{\infty}\frac{1}{|J|}d\Phi\right)d\Psi$$

so that Equation 6.2.19 may be written

$$\text{time} = nb\int_{-\infty}^{\infty}\frac{1}{|J|}d\Phi \qquad\qquad \textbf{(6.2.20)}$$

where the Jacobian is evaluated along a single streamline (Ψ fixed).

General Algorithm for Calculation of Streamline Travel Time

With Equations 6.2.18, 6.2.19, or 6.2.20, there are only a few problems for which one can find the travel time analytically. Because the potential and stream functions are easily found, however (see the next section), an alternative is to evaluate the travel time along a streamline numerically directly using Equation 6.2.18. The basic idea is to choose a step size and then use the potential and stream functions to take the step parallel with a streamline and determine the travel time for each step. An algorithm for doing so is shown in Figure 6.2.5 and is outlined as follows:

1. Start at point $\tilde{x}_1 = (x_1, y_1)$.
2. Calculate U_x and U_y at this point from $\tilde{U}(\tilde{x}_1) = -\nabla\Phi(\tilde{x}_1)$.

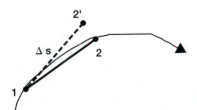

FIGURE 6.2.5   Algorithm for calculation of the travel time along a streamline

3. Calculate $\tilde{x}_2' = \tilde{x}_1 + \Delta\tilde{x}'$ from $\Delta s^2 = \Delta x^2 + \Delta y^2$ and $\dfrac{\Delta x}{U_x} = \dfrac{\Delta y}{U_y}$.

Combining these, we have

$$\Delta s^2 = \Delta x^2 \left(1 + \left(\frac{U_y}{U_x}\right)^2\right), \text{ or}$$

$$\Delta x' = \frac{\Delta s}{\sqrt{1 + \left(\dfrac{U_y}{U_x}\right)^2}} = \frac{\Delta s\, U_x}{\sqrt{U_x^2 + U_y^2}}; \ \Delta y' = \frac{\Delta s\, U_y}{\sqrt{U_x^2 + U_y^2}}$$

4. Calculate $\tilde{U}(\tilde{x}_2')$.

5. Find the average discharge from $\overline{U} = \dfrac{1}{2}(\tilde{U}(\tilde{x}_1) + \tilde{U}(\tilde{x}_2'))$.

6. Repeat step 3 using $\overline{U}$ in place of $\tilde{U}(\tilde{x}_1)$.

7. For an unconfined aquifer, find the midpoint

$$\tilde{x}_m = \tilde{x}_1 + \frac{\Delta\tilde{x}}{2}$$

and calculate the saturated thickness from

$$H_m = \sqrt{2\frac{\Phi(\tilde{x}_m)}{K}}$$

8. Calculate the average seepage velocity for a confined aquifer from

$$v = \frac{\overline{U}}{nb}$$

and for an unconfined aquifer from the same formula with H_m replacing b.

9. Calculate the time interval for the step from $\Delta t = \dfrac{\Delta s}{v}$ and $t_2 = t_1 + \Delta t$.

10. Repeat for the next step, etc.

Algorithms such as this are easy to program and have found fairly wide application in practice (for example, see Harpaz and Bear, 1963, or Bear, 1979; or Javandel, Doughty and Tsang, 1984, their RESSQ model in Chapter 3). Such models are used for front tracking and for estimation of pollutant outflow boundaries and travel times to these boundaries. Such assessments are very useful for estimation of the environmental consequences of groundwater contamination.

6.3   Calculation of Potential and Stream Functions

This section looks at some of the most important examples of potential and stream functions. These deal with uniform flow, flow to wells, and flow from larger water bodies such as ponds or lakes. The basic equations are the potential and stream function forms of Darcy's Law. In Cartesian coordinates, these equations are

$$U_x = -\frac{\partial \Phi}{\partial x} = -\frac{\partial \Psi}{\partial y} \qquad (6.3.1)$$

$$U_y = -\frac{\partial \Phi}{\partial y} = \frac{\partial \Psi}{\partial x} \qquad (6.3.2)$$

while the corresponding equations in cylindrical coordinates are

$$U_r = -\frac{\partial \Phi}{\partial r} = -\frac{1}{r}\frac{\partial \Psi}{\partial \theta} \qquad (6.3.3)$$

$$U_\theta = -\frac{1}{r}\frac{\partial \Phi}{\partial \theta} = \frac{\partial \Psi}{\partial r} \qquad (6.3.4)$$

Equations 6.3.1 and 6.3.2 are called the **Cauchy-Riemann equations**, and they are fundamental to application of complex variable theory to problems of groundwater flow. In particular, it may be shown that our potential and stream functions are analytic functions and the tools available through complex variable theory may be utilized. For a discussion and references, see Bear (1972).

Uniform Flow

Consider uniform flow at a rate U (L^2/T) directed at an angle α with respect to the x-axis. We have

$$U_x = U \cos(\alpha) \; ; \; U_y = U \sin(\alpha)$$

With Equations 6.3.1 and 6.3.2, these lead to

$$\Phi = \Phi_o - U(x \cos(\alpha) + y \sin(\alpha)) \; ; \; \Psi = \Psi_o - U(y \cos(\alpha) - x \sin(\alpha)) \quad (6.3.5)$$

The potential and stream functions for uniform flow are shown in Figure 6.3.1 for the case with $U = 1$ and $\alpha = \pi/6$. The solid lines are the streamlines, while the dashed lines are the potential lines.

Injection and Production Wells

For an injection well located at the origin of the coordinate system, continuity gives $Q = 2\pi r U_r$, or with Equation 6.3.3

$$U_r = \frac{Q}{2\pi r} = -\frac{\partial \Phi}{\partial r} = -\frac{1}{r}\frac{\partial \Psi}{\partial \theta}$$

Integrating, we have

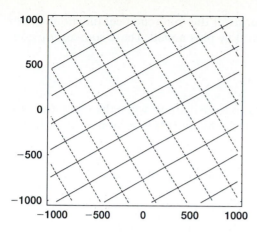

FIGURE 6.3.1 Potential and stream functions for uniform flow

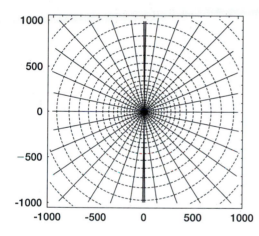

FIGURE 6.3.2 Potential and stream function for a well

$$\Phi = \Phi_o - \frac{Q}{2\pi} ln(r) \; ; \; \Psi = \Psi_o - \frac{Q}{2\pi} \theta \qquad \textbf{(6.3.6)}$$

Equation 6.3.6 shows that the lines of equal potential form a family of concentric circles while the streamlines form a family of rays emanating radially from the well. In Cartesian coordinates, these become

$$\Phi = \Phi_o - \frac{Q}{2\pi} ln(\sqrt{x^2 + y^2}) = \Phi_o - \frac{Q}{4\pi} ln(x^2 + y^2) \; ; \; \Psi = \Psi_o - \frac{Q}{2\pi} \tan^{-1}\left(\frac{y}{x}\right) \qquad \textbf{(6.3.7)}$$

Figure 6.3.2 shows the potential and streamlines for a well.

The potential and stream functions corresponding to uniform flow and wells are the most useful ones for many applications. There are, however, additional source terms which may be of interest, such as a pond, lagoon, or lake, and these may be represented in terms of the mathematical functions corresponding to a doublet plus uniform flow plus a well. These sources are considered next.

Doublet

The **doublet** consists of the limit of a source/sink pair when the separation distance (L) goes to zero while the well discharge (Q) goes to infinity in such a fashion that

$$2L\left(\frac{Q}{2\pi}\right) = \text{constant} = \lambda$$

With the injection (well 1) and production (well 2) well located along the x-axis at points (L,0) and ($-L$,0), respectively, the potential function for the pair of wells is

$$\Phi = \Phi_o + \frac{Q}{2\pi} \ln(r_2) - \frac{Q}{2\pi} \ln(r_1)$$

where r_1 is the distance from the injection well and r_2 is the distance from the production well. This equation may be written

$$\Phi = \Phi_o + 2L \frac{Q}{2\pi}\left(\frac{\ln(r_2) - \ln(r_1)}{2L}\right)$$

In the limit as $L \to 0$, the term inside the bracket corresponds to

$$\frac{\partial \ln(r)}{\partial x} = \frac{x}{r^2} = \frac{x}{x^2 + y^2}$$

and so we have

$$\Phi = \Phi_o + \frac{\lambda x}{x^2 + y^2} \tag{6.3.8}$$

In a similar fashion, for the stream function, one finds

$$\Psi = \Psi_o - \frac{\lambda y}{x^2 + y^2} \tag{6.3.9}$$

Figure 6.3.3 shows the potential and stream functions for the case with $\lambda = 10$. While the resulting picture is quite striking, it is difficult to see where this solution has much relevance to problems of groundwater flow. When we add uniform flow to a doublet, however, we get something quite useful.

Doublet Plus Uniform Flow

We can add uniform flow along the x-axis to the doublet solutions by adding ($-Ux$) to the potential function and ($-Uy$) to the stream function. If we further let $\lambda = UR^2$, then we find for the potential function

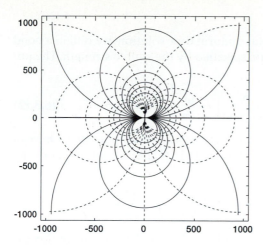

FIGURE 6.3.3 Potential and stream functions for a doublet

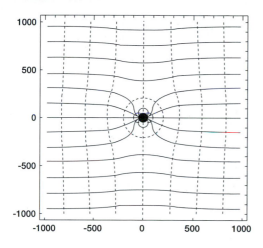

FIGURE 6.3.4 Potential and stream function for a doublet plus uniform flow (a pond)

$$\Phi = \Phi_o - Ux\left(1 - \frac{R^2}{x^2 + y^2}\right) \qquad (6.3.10)$$

This equation shows that the circle $R^2 = x^2 + y^2$ has a constant potential, and physically this may correspond to the boundary of a pond, lagoon, or other water body which is in direct contact with the aquifer. The same discharge enters the water body from the upstream side as leaves from the downgradient side. The corresponding stream function is

$$\Psi = \Psi_o - Uy\left(1 + \frac{R^2}{x^2 + y^2}\right) \qquad (6.3.11)$$

Figure 6.3.4 shows the potential and stream functions for the case with $U = 1$ and $R = 200$.

Doublet, Uniform Flow, Plus a Source

If we add a source to the doublet plus uniform flow solution, we obtain a pond or lagoon which is leaking at the rate specified by the source strength. The potential function is

$$\Phi = \Phi_o - Ux\left(1 - \frac{R^2}{x^2 + y^2}\right) - \frac{Q}{2\pi} ln(\sqrt{x^2 + y^2}) \qquad (6.3.12)$$

It is apparent that the circle $R^2 = x^2 + y^2$ still has a constant potential. One can show that the net volume flux across this circle is Q. The x-component of the aquifer flux is given by

$$U_x = -\frac{\partial \Phi}{\partial x} = U\left(1 - \frac{R^2}{x^2 + y^2} + \frac{2R^2x^2}{(x^2 + y^2)^2}\right) + \frac{Q}{2\pi}\frac{x}{x^2 + y^2}$$

To find the flow leaving the upstream side of the pond, we evaluate this expression at the point $(-R,0)$ to find

$$U_x = 2U - \frac{Q}{2\pi R} \qquad (6.3.13)$$

Thus, there will be outflow throughout the back of the pond if

$$Q > 4\pi RU \qquad (6.3.14)$$

Otherwise, there is some inflow to the pond from the upstream direction. Figure 6.3.5 shows the pond with recharge for conditions with $U = 1, R = 200$, and $Q = 1000\pi$. The image of the pond is also shown in the figure. The pond is still an equal potential surface, and one does not pay particular attention to the streamlines and potential lines within the boundary of the pond. It is clear that water is leaving the backside of the pond, and there is a stagnation point on the upstream side.

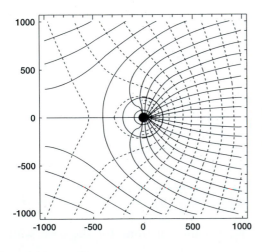

FIGURE 6.3.5 Potential and stream functions for a pond receiving recharge with water exiting the upstream boundary of the pond

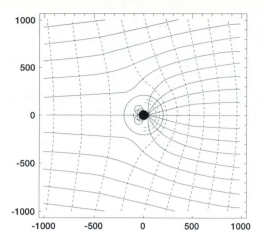

FIGURE 6.3.6 Potential and stream functions for a pond receiving recharge with water entering the upstream boundary of the pound

Figure 6.3.6 shows the potential and stream functions for the case with $U = 1$, $R = 200$, and $Q = 400\pi$. Because $Q < 4\pi RU$, there is water entering the pond from the upstream side. In addition, there is no stagnation point on the upstream side.

Image of a Well within the Pond

If there is an injection or production well in the neighborhood of the pond, then the addition of the well potential field results in the boundary of the pond no longer being a constant potential boundary. In order to retain the constant potential on the circle of radius R, we need to place an image well within the pond. The key to solving this problem comes from recognition that the potential field of a source/sink pair results in a family of circles (see Figure 3.2.4 and Section 6.4.2). If one of the circles represents the pond, the image point is the corresponding well within the circle. This image point is called the "inversion point" within the circle. The point is located on the same ray from the origin as the real well and at a distance that satisfies

$$r_r r_i = R^2 \qquad\qquad (6.3.15)$$

If we take the origin to be the center of the pond and the well located at point (x_w, y_w), then

$$r_r = \sqrt{x_w^2 + y_w^2}$$

while the corresponding ray is at the angle

$$\alpha = \tan^{-1}\left(\frac{y_w}{x_w}\right)$$

Then the image well is at

$$r_i = \frac{R^2}{r_r}$$

or

$$x_i = r_i \cos(\alpha) = \frac{R^2}{r_w^2} x_w = \frac{R^2 x_w}{x_w^2 + y_w^2} \; ; \; y_i = \frac{R^2 y_w}{x_w^2 + y_w^2} \qquad (6.3.16)$$

The equations developed in the last two sections provide tools for solution of a wide range of problems in groundwater hydraulics and contaminant transport. In the following sections, we will consider certain elementary problems dealing with flow and transport. We will also look at the travel time distributions along streamlines for multiple well problems, and at applications of fractional breakthrough curves.

Strack (1989) provides a general presentation of important and useful applications of potential theory.

6.4   Applications of Potential Theory

The continuity equations for steady flow are linear, which means that the principle of superposition of solutions is valid. According to this principle, if Φ_1 and Φ_2 are two solutions to the continuity equation, then so is the sum $\Phi = \Phi_1 + \Phi_2$:

$$\nabla^2 \Phi = \nabla^2 \Phi_1 + \nabla^2 \Phi_2 = 0$$

There have been a wide variety of applications of potential theory and the principle of superposition to groundwater flow and contamination problems in the literature. Here, we will take a somewhat in-depth look at a few of these. As a first example, we consider the problem of determination of the capture zone of a single well, or a series of wells. This problem has direct application to design of pump-and-treat systems for groundwater remediation.

6.4.1   Capture Zones of Production Wells

One of the most common methods of aquifer cleanup is to extract the contaminated groundwater and, after appropriate treatment to reduce pollutant levels, either inject the water back into the aquifer or release it to surface drainage. If one has characterized the site and knows the extent of the contaminant plume, major questions which must still be answered for the design of such remediation projects include the following (Javandel and Tsang, 1986):

1. What is the optimum number of pumping wells required?
2. Where should the wells be located so that no contaminated water can escape between the pumping wells?
3. What is the optimum pumping rate for each well?
4. Where should one inject the treated water back into the aquifer?

Finding general answers to these questions is difficult because of uncertainties in site data, pollutant extent, and other factors. Potential theory, however, does offer a simple guiding tool for some applications.

EXAMPLE
PROBLEM

6.4.1 Capture Zone of a Production Well

As part of a study to determine the zone of capture for a production well in an unconfined aquifer, it is of interest to identify the maximum downgradient extent to which the drawdown cone of the well will capture water. At the well, the saturated thickness is H_w, far from the well the regional aquifer flux is $U = 1$ ft²/d, and the well discharge is $Q = 10,000$ ft³/d (50 gpm).

This example involves a problem with a production well and uniform flow in an unconfined aquifer, and according to the principle of superposition, we can find the solution by combining flow to a well with the solution for linear flow. We take a coordinate system with the origin centered at the well and with the x-axis oriented in the direction of the regional flow. Superposition of the linear and radial flow solutions provides

$$\Phi = \Phi_o - Ux + \frac{Q}{2\pi} \ln(r)$$

where Φ_o is an unknown constant that determines the elevation of the $\Phi(x,y)$ surface. Evaluation of the potential at the well ($x = 0, r = r_w$) gives

$$\Phi_w = \frac{KH_w^2}{2} = \Phi_o + \frac{Q}{2\pi} \ln(r_w)$$

Using this last equation to solve for Φ_o, the potential function is given by

$$\Phi = \frac{Q}{2\pi} \ln\left(\frac{r}{r_w}\right) - Ux + \frac{KH_w^2}{2}$$

The aquifer flux at a point $\tilde{x}$ in the x-direction is given by

$$U_x(\tilde{x}) = -\frac{\partial \Phi}{\partial x} = -\frac{Q}{2\pi r} \cos(\theta) + U$$

where θ is the angle that a ray from the origin to point $\tilde{x}$ makes with the x-axis and r is the radius to point $\tilde{x}$ from the origin. At the downgradient extent of capture (which occurs along the x-axis so $\theta = 0$ and $r = x$), the x component of the flux vanishes, and we have

$$x = \frac{Q}{2\pi U} = \frac{10,000 \text{ ft}^3/\text{d}}{2\pi \times 1 \text{ ft}^2/\text{d}} = 1,600 \text{ ft}$$

Thus, the zone of capture extends more than 500 yards beyond the well in the downgradient direction.

The actual shape of the capture zone may be determined from the stream function. The stream function for this problem is given by superposition of Equations 6.3.5 and 6.3.6 to yield

$$\Psi = \frac{Q}{2\pi}\theta - Uy = \frac{Q}{2\pi}\tan^{-1}\left(\frac{y}{x}\right) - Uy$$

The downstream stagnation point along the dividing streamline is located at

$$x = \frac{Q}{2\pi U} \ , \ y = 0 \ , \ (\theta = 0)$$

This equation corresponds to the stream function value $\Psi = 0$. Because the stream function is constant along the dividing streamline, the capture zone satisfies

$$y = \frac{Q}{2\pi U}\tan^{-1}\left(\frac{y}{x}\right) \tag{6.4.1}$$

The only parameter appearing in this equation is the ratio of the well discharge to the regional aquifer flux, Q/U. The ultimate width of the capture zone is found by letting x go to negative infinity ($\theta \rightarrow \pm\pi$). This equation gives

$$y_{max} = \pm\frac{Q}{2U}$$

This result could have (and should have) been expected based on simple continuity considerations.

EXAMPLE
PROBLEM

6.4.2   Capture Zone of a Production Well (continued)

If changes in saturated thickness of the aquifer are small (or the aquifer is confined with a constant transmissivity) then one may go much further in the analysis of Example 6.4.1, following the work of Bear and Jacobs (1965). The issue of concern is "where did the water which is entering the well today originate from 20 years ago or 50 years ago?" This question is essentially the same question as "where will the front of injected fluid be 20 years from now, or 50 years from now, if injection continues at the same rate over this time period?" For either question, possible capture zones or injection displacement regions are shown in Figure 6.4.1. For a capture zone problem the regional flow is moving in the negative x-axis direction, while for an injection problem the regional flow is moving along the positive x-axis.

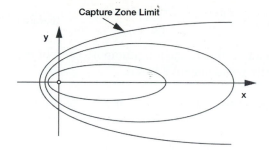

Capture Zone Limit

FIGURE 6.4.1 Injection fronts and capture zones at different times for a well in a uniform regional flow field

This problem may be solved analytically, because both the equations of the streamlines and the velocity along a streamline may be expressed in a sufficiently simple mathematical form. Using the equation for a streamline, one finds that along it the following relationship holds

$$x = y \cot\left(-\frac{2\pi}{Q}(\Psi + Uy)\right) = -y \cot\left(\frac{2\pi}{Q}(\Psi + Uy)\right)$$

which gives the path of the streamline Ψ. The y-component of the seepage or pore water velocity along this streamline is given by

$$v_y = -\frac{1}{nb}\frac{\partial\Phi}{\partial y} = \frac{Q}{2\pi nb}\frac{y}{x^2 + y^2}$$

Bear and Jacobs (1965) show that these two equations may be combined to eliminate x as a variable and integrated over the y-displacement region to give

$$\frac{2\pi U^2 t}{nbQ} = \ln\left(\frac{\sin\left(\tan^{-1}\left(\frac{y}{x}\right)\right)}{\sin\left(\tan^{-1}\left(\frac{y}{x}\right) + \frac{2\pi Uy}{Q}\right)}\right) + \frac{2\pi Ux}{Q} \tag{6.4.2}$$

This solution was presented by Bear and Jacobs (1965). It provides the travel time to a location (x,y) from a well located at the origin if the uniform regional flow rate U and pumping rate Q remain constant, and if the location is within the 'capture zone' of the well. Similarly, at a fixed time, Equation 6.4.2 gives the shape of the capture zone or injection front as an implicit function.

Equation 6.4.2 may also be written

$$\frac{2\pi U^2 t}{nbQ} = -\ln\left(\cos\left(\frac{2\pi Uy}{Q}\right) + \frac{x}{y}\sin\left(\frac{2\pi Uy}{Q}\right)\right) + \frac{2\pi Ux}{Q} \tag{6.4.2a}$$

which is evaluated along the x-axis to give the front location

$$\frac{2\pi U^2 t}{nbQ} = -\ln\left(1 + \frac{2\pi Ux}{Q}\right) + \frac{2\pi Ux}{Q} \qquad \text{(6.4.2b)}$$

This last equation may be used to estimate the longitudinal extent of the injection plume along both the positive and negative x-axis.

Equation 6.4.1 gives the shape of the capture zone for a single production well, while Equation 6.4.2 gives the travel time from any point $\tilde{x} = (x,y)$ within the capture zone to the production well. The case with two or more wells is more interesting, because depending on the well locations and production rates and on the magnitude of the regional flow, some of the contaminant may move either around or between the wells. Consider two production wells with discharge Q located a distance L apart at $(0,L/2)$ and $(0,-L/2)$ with regional flow of magnitude U along the x-axis. The potential and stream functions are

$$\Phi = \Phi_o + \frac{Q}{4\pi}\ln(x^2 + (y - L/2)^2) + \frac{Q}{4\pi}\ln(x^2 + (y + L/2)^2) - Ux$$

$$\Psi = \frac{Q}{2\pi}\tan^{-1}\left(\frac{y - L/2}{x}\right) + \frac{Q}{2\pi}\tan^{-1}\left(\frac{y + L/2}{x}\right) - Uy$$

where Φ_o is a constant. In the limiting case where no water passes between the wells, there will be a single stagnation point along the x-axis. Setting

$$U_x(x,0) = -\frac{\partial \Phi}{\partial x}(x,0) = 0$$

gives

$$x^2 - \frac{Q}{\pi U}x + \left(\frac{L}{2}\right)^2 = 0$$

The roots of this equation are

$$x = \frac{Q}{2\pi U} \pm \sqrt{\left(\frac{Q}{2\pi U}\right)^2 - \left(\frac{L}{2}\right)^2}$$

This equation provides a single root only if

$$L = \frac{Q}{\pi U} \qquad \text{(6.4.3)}$$

If L is less than or equal to the value given in Equation 6.4.3, then none of the contaminant can pass between the wells. If L is greater than this value, then there are no stagnation points along the x-axis, and the pollutant can pass directly between the wells.

For the limiting condition of Equation 6.4.3, the capture zone may be found as follows. We have just seen that the stagnation point is located at

$$\left(\frac{Q}{2\pi U},0\right)$$

Furthermore, Equation 6.4.3 shows that at the stagnation point, $x = L/2$ and the arguments of arctan in the stream function are ± 1. This fact gives

$$\Psi = \frac{Q}{2\pi}\left(\frac{\pi}{4}\right) + \frac{Q}{2\pi}\left(-\frac{\pi}{4}\right) = 0$$

Thus, the equation for the capture zone is

$$y = \frac{Q}{2\pi U}\left\{\tan^{-1}\left(\frac{y - L/2}{x}\right) + \tan^{-1}\left(\frac{y + L/2}{x}\right)\right\} \qquad \textbf{(6.4.4)}$$

Equation 6.4.4 shows, for example, that the width of the capture zone along the y-axis is Q/U while far upstream (where $\theta \to \pi$) the total width is $2\,Q/U$.

The condition with three or more wells is even more difficult to deal with analytically. For three wells, located at $(0,-L)$, $(0,0)$ and $(0,L)$, Javandel and Tsang (1986) show that the limiting well spacing where no fluid could escape between the wells is given by

$$L = (2)^{1/3}\frac{Q}{\pi U}$$

which is somewhat larger than the spacing required when two wells are used. The general result which is suggested is that the minimum spacing required between production wells, in order to prevent bypass of contaminated groundwater, is

$$L = 1.2\frac{Q}{\pi U} \qquad \textbf{(6.4.5)}$$

Evaluation of Plume Containment Using Numerical Models

In applications, the ability of a set of wells to contain and capture groundwater contaminants is often evaluated through use of numerical models. Containment is evaluated through examination of the distribution of drawdown that is created during the simulation. If a depression in the water table, sometimes called a **hinge line**, is developed at a location downgradient of the contaminant plume, and if the zone of depression is sufficiently wide, then this may be taken as strong but indirect modeling evidence of plume containment. Groundwater will flow towards the depression from all directions to eventually be captured.

More direct evidence of plume containment may be developed through use of models that predict the flow paths of fluid particles. The model MODPATH (Pollock, 1994) is a particle-tracking package that was developed by the USGS to compute three-dimensional flow paths using output from the MODFLOW model that was described in Section 2.5. MODPATH uses a particle tracking scheme that allows an analytical expression of the particle's flow path to be obtained within each finite-difference grid cell, and the paths are computed by tracking particles from one cell to the next until the particle reaches a boundary or an internal sink. Travel times are also computed. Plotting of the flow paths from the contaminant plume to the pumping well system provides direct evidence of plume containment.

A third approach for using numerical models to evaluate plume contain-ment is to simulate the contaminant plume migration using solute transport models. Such an approach is described in Section 8.10.

6.4.2   Flow between an Injection/Production Well Pair

In this section, we consider the case of a source/sink pair with the injection well located at $(-L,0)$ and the production well located at $(L,0)$, as shown in Figure 6.4.2. The potential and stream function are given by

$$\Phi = \Phi_o + \frac{Q}{4\pi} \ln\left(\frac{(x - L)^2 + y^2}{(x + L)^2 + y^2}\right) \tag{6.4.6}$$

$$\Psi = \frac{Q}{2\pi}\left(\tan^{-1}\left(\frac{y}{x - L}\right) - \tan^{-1}\left(\frac{y}{x + L}\right)\right) \tag{6.4.7}$$

It is not difficult to rearrange Equation 6.4.6, with

$$\alpha = \exp\left(\frac{4\pi(\Phi - \Phi_o)}{Q}\right)$$

to the form

$$\left(x - L\left(\frac{1 + \alpha}{1 - \alpha}\right)\right)^2 + y^2 = \left(\frac{2L\sqrt{\alpha}}{1 - \alpha}\right)^2 \tag{6.4.8}$$

Equation 6.4.8 is the equation for a circle with center location and radius

$$(x,y) = \left(L\left(\frac{1 + \alpha}{1 - \alpha}\right),0\right) \; ; \; r = \frac{2L\sqrt{\alpha}}{1 - \alpha}$$

Similarly, using the identity

$$\tan(a - b) = \frac{\tan(a) - \tan(b)}{1 + \tan(a) \tan(b)}$$

one may rearrange Equation 6.4.7, with

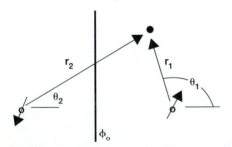

FIGURE 6.4.2   Notation for two-well pair travel time cal-culation

$$\beta = \tan\left(\frac{2\pi\Psi}{Q}\right)$$

to the form

$$x^2 + \left(y - \frac{L}{\beta}\right)^2 = \left(L\sqrt{1 + \frac{1}{\beta^2}}\right)^2 \tag{6.4.9}$$

Equation 6.4.9 is also the equation of a circle, showing that both the streamlines and equipotentials are circles.

For this example, one may calculate the travel time for a fluid particle along a streamline using direct integration of the velocity field according to Equation 6.2.18. Grove and Beetem (1971) follow this approach. It is more direct, however, to calculate the travel times using Equation 6.2.19 as shown next. With reference to Figure 6.4.3, the area between the central streamline along the *x*-axis and another arbitrary streamline is equal to the area of the sector minus the area of the triangle. This area is given by

$$A = r^2\theta - \frac{L^2}{\beta} = L^2\left(1 + \frac{1}{\beta^2}\right)\theta - \frac{L^2}{\beta}$$

Also, one can see from Figure 6.4.3 that $\theta = \tan^{-1}(\beta) = \frac{2\pi\Psi}{Q}$. Furthermore,

$$\frac{L}{r} = \frac{1}{\sqrt{1 + 1/\beta^2}} = \sin(\theta) = \sin\left(\frac{2\pi\Psi}{Q}\right) \; ; \; \frac{1}{\beta} = \cot\left(\frac{2\pi\Psi}{Q}\right)$$

With these results, the expression for the area becomes

$$A(\Psi) = \frac{2\pi\Psi L^2}{Q \sin^2\left(\frac{2\pi\Psi}{Q}\right)} - L^2 \cot\left(\frac{2\pi\Psi}{Q}\right)$$

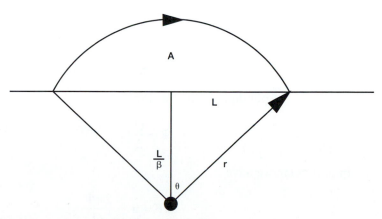

FIGURE 6.4.3   Area of the Segment of a Circle

Differentiation with Equation 6.2.19 leads to

$$t(\Psi) = nb\frac{dA}{d\Psi} = \frac{4\pi n b L^2}{Q \sin^2\left(\frac{2\pi\Psi}{Q}\right)}\left(1 - \frac{2\pi\Psi}{Q}\cot\left(\frac{2\pi\Psi}{Q}\right)\right) \qquad (6.4.10)$$

This result is essentially the same result found by Grove and Beetem (1971).

Rather than presenting the function $t(\Psi)$, it is convenient to determine the time for a given fraction of the injected fluid to appear in the production well effluent. If this fraction is designated as F, then the curve $t(F)$, or t_F, is called the **fractional breakthrough curve**. t_F gives the time at which a fraction F of a tracer which is continuously injected starting at time 0 has reached the production well. In order to develop the fractional breakthrough curve we note that the stream function is

$$\Psi = \frac{Q}{2\pi}(\theta_1 - \theta_2)$$

with the notation shown in Figure 6.4.2. The segment of the x-axis connecting the injection and production wells has $\theta_1 = \pi$ and $\theta_2 = 0$, so $\Psi = Q/2$. The x-axis beyond the production well has $\theta_1 = \theta_2 = 0$, so $\Psi = 0$. Similarly, for the streamline entering the production well directly from above, $\theta_1 = \pi/2$ and $\theta_2 = 0$ (at the well), so $\Psi = Q/4$. The streamline pattern is shown in Figure 6.4.4. At the production well, $\theta_2 = 0$, so the relation between the stream function and the angle at which a streamline enters the well is

$$\Psi = \frac{Q}{2\pi}\theta_1$$

The travel times for all streamlines with $\theta_1 < \theta < \pi$ are shorter than that for $\theta = \theta_1$. Thus, the fraction of the streamlines with travel times shorter than that with an entry angle θ_1 is

$$F = \frac{\pi - \theta_1}{\pi} = 1 - \frac{\theta_1}{\pi}$$

and the corresponding stream function is

$$\Psi = \frac{Q}{2}(1 - F)$$

In particular, we have

$$\pi F = \pi - \frac{2\pi\Psi}{Q}$$

With these results, the fractional breakthrough Equation 6.4.10 becomes

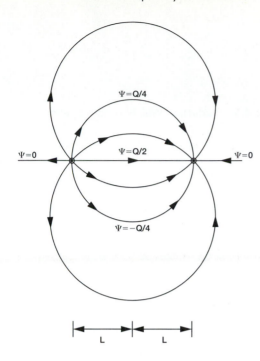

$\Psi=Q/4$

$\Psi=0$ $\Psi=Q/2$ $\Psi=0$

$\Psi=-Q/4$

FIGURE 6.4.4   Flow Pattern for a Source-Sink Pair

$|\longleftarrow L \longrightarrow|\longleftarrow L \longrightarrow|$

$$t_F = \frac{4\pi nbL^2}{Q}\left(\frac{1 - \pi F \cot(\pi F)}{\sin^2(\pi F)}\right) \qquad (6.4.11)$$

which gives the travel time t_F corresponding to fractional breakthrough F.

6.4.3   Water Quality for a Production Well near a River

The following example shows the application of potential theory to questions of water quality control. Consider a production well pumping from an unconfined aquifer adjacent to a river. In the absence of the well, there would be uniform groundwater flow to the river from the perpendicular direction. The well is located at a distance L from the river, as shown in Figure 6.4.5. The question of interest concerns how much of the well discharge comes from the aquifer and how much comes from the river.

Recall from image well theory that the way to introduce a constant potential boundary such as the river is to use an image injection well at the distance L from the other side of the river, as shown in Figure 6.4.6. With the coordinate system shown in Figure 6.4.5 we have

$$r_1 = \sqrt{(x - L)^2 + y^2} \;;\; r_2 = \sqrt{(x + L)^2 + y^2}$$

$$\theta_1 = \tan^{-1}\left(\frac{y}{x - L}\right) \;;\; \theta_2 = \tan^{-1}\left(\frac{y}{x + L}\right)$$

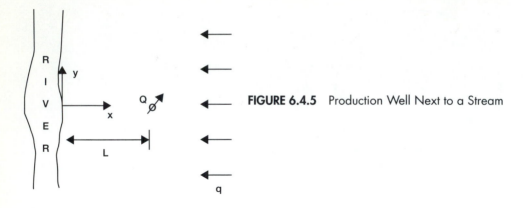

FIGURE 6.4.5   Production Well Next to a Stream

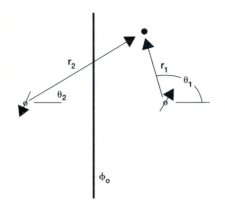

FIGURE 6.4.6   Representation of the Stream by an Image Well

Using superposition, the potential function is given by

$$\Phi = \Phi_o + Ux + \frac{Q}{2\pi} ln(r_1) - \frac{Q}{2\pi} ln(r_2)$$

We are interested in the possible location of points where $U_x = 0$ along the river ($x = 0$). These points will separate regions where the flow is into the river from regions where flow is out of the river. We find

$$U_x(0,y) = -\frac{\partial \Phi}{\partial x}(0,y) = -U - \frac{Q}{2\pi}\left(\frac{-L}{L^2 + y^2}\right) + \frac{Q}{2\pi}\left(\frac{L}{L^2 + y^2}\right) = 0$$

which simplifies to

$$\left(\frac{y}{L}\right)^2 = \frac{Q}{\pi LU} - 1$$

It is apparent that the group

$$\zeta = \frac{Q}{\pi LU} \tag{6.4.12}$$

is an important dimensionless parameter. The dividing points are found as solutions to

$$\left(\frac{y}{L}\right)^2 = \zeta - 1 \qquad\qquad (6.4.13)$$

There are three cases:

a. $\zeta < 1 \Rightarrow$ no real roots

b. $\zeta = 1 \Rightarrow$ one double root at $y = 0$

c. $\zeta > 1 \Rightarrow$ two roots at $\dfrac{y}{L} = \pm \sqrt{\zeta - 1}$

The fact that there are no real roots for $\zeta < 1$ implies that the flow is everywhere into the river, and thus the well receives only water from the aquifer. The limiting condition is found for $\zeta = 1$. Finally, for $\zeta > 1$, the water leaving the river between the points $y = -L\sqrt{\zeta - 1}$ and $y = L\sqrt{\zeta - 1}$ enters the well, as shown in Figure 6.4.7. In order to determine the well concentration, we need to know the fraction of water from the aquifer and from the river, and the corresponding aquifer and river concentrations. We could find this fraction by integrating $U_x(0,y)$ between the two limiting points for $\zeta > 1$. A much easier approach, however, comes from use of the stream function.

The stream function for this problem is again found through superposition to be

$$\Psi = Uy + \frac{Q}{2\pi}\theta_1 - \frac{Q}{2\pi}\theta_2$$

The positive x-axis beyond $x = L$ has the stream function value $\Psi = 0$. The streamline leading from the river to the well has $y = 0$, $\theta_1 = \pi$, and $\theta_2 = 0$. Thus, this streamline is given by $\Psi = Q/2$. If we are just interested in the stream function evaluated along the river, we may use the fact that along x=0, $\theta_1 = \pi - \theta_2$. Along the river, the stream function is

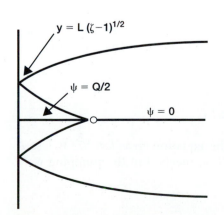

$y = L\,(\zeta-1)^{1/2}$

$\psi = Q/2$

$\psi = 0$

FIGURE 6.4.7   Streamline Pattern for Flow from River

$$\Psi(x = 0, y) = Uy + \frac{Q}{2} - \frac{Q}{\pi}\theta_2 = Uy + \frac{Q}{2} - \frac{Q}{\pi}\tan^{-1}\left(\frac{y}{L}\right)$$

The dividing streamline is located at $y = L\sqrt{\zeta - 1}$ along the positive y-axis, and the streamline evaluated at this location is designated Ψ^*. In order to determine the fraction of the well discharge coming from the river consider only the upper-half plane. The total discharge from this upper plane is $Q/2$. The amount from the river is equal to the difference in stream function values from the dividing point to the x-axis. Thus,

$$F_R = \frac{Q/2 - \Psi^*}{Q/2} = \frac{\frac{Q}{\pi}\tan^{-1}(\sqrt{\zeta-1}) - UL\sqrt{\zeta-1}}{Q/2} = \frac{2}{\pi}\left(\tan^{-1}(\sqrt{\zeta-1}) - \frac{\sqrt{\zeta-1}}{\zeta}\right) \qquad \textbf{(6.4.14)}$$

Equation 6.4.14 is the important result.

EXAMPLE PROBLEM

6.4.3   Well placement and water quality

A river has a *total dissolved solids* (TDS) concentration that averages $c_r = 1{,}200$ mg/L, while groundwater recharging the river has a TDS concentration $c_g = 300$ mg/L. We want to place a water well which will produce 20,000 ft³/d (100 gpm) as close to the river as possible and still meet a water quality limit of $c = 500$ mg/L TDS. The aquifer is unconfined and has a hydraulic conductivity of 15 ft/d. Before placement of the well, a monitoring well located 1,000 ft from the river has a saturated thickness of 65 ft, while the saturated thickness at the river is 60 ft.

The saturated thickness values at the monitoring well and river correspond to discharge potentials of $\Phi_m = 31{,}690$ ft³/d and $\Phi_r = 27{,}000$ ft³/d ($\Phi = KH^2/2$). These provide an estimate of the regional aquifer flux of

$$U = \frac{\Phi_m - \Phi_r}{\Delta x} = \frac{31{,}690 \text{ ft}^3/\text{d} - 27{,}000 \text{ ft}^3/\text{d}}{1{,}000 \text{ ft}} = 4.69 \text{ ft}^2/\text{d}$$

To find the required fraction from the river, we use $c = F_R c_r + (1 - F_R)c_g$, which gives $F_R = {}^2\!/_9$. Substituting this fraction into Equation 6.4.14 and solving by trial and error gives

$$\zeta = 2.28 = \frac{Q}{\pi L U}$$

With $Q = 20{,}000$ ft³/d and $U = 4.69$ ft²/d, this equation gives $L = 573$ ft. Use $L = 600$ ft. To find the water level that will be reached in the pumping well, assume a well radius of 1.5 ft (which includes the gravel pack). The potential function for this situation gives

$$\Phi_w = \Phi_o + UL + \frac{Q}{2\pi} ln\left(\frac{r_w}{2L}\right) = 8{,}540 \text{ ft}^3/\text{d} \;\rightarrow\; H_w = \sqrt{\frac{2\Phi_w}{K}} = 33.7 \text{ ft}$$

Finally, with $L = 600$ ft, $\zeta = 2.26$ so that $F_R = 0.22$ and $c = 498$ mg/L.

6.5 Residence Time Distribution Theory

Study of nonideal flow systems requires a computational method for relating the influent and effluent concentrations for a flow system when the pattern of flowlines (or streamlines) is complex—or even unknown. The **residence time distribution theory**, originally presented by Danckwerts (1953), provides an appropriate tool. Levenspiel (1972) presents an excellent discussion. Consider an experiment with a tracer migrating from a source to a sink. This experiment may be a release from a pond to a river, an injection well to a production well, the ground surface to the water table, etc. We are concerned with relating the sink concentration to that at the source. Two cases are of immediate interest. If the release at the source is instantaneous, then the sink concentration will depend on the travel time of impulse from the source to the sink along the various contributing flow paths. The effluent concentration represents the distribution of travel times, and it may be referred to as a residence time **density** function or impulse response function. On the other hand, if there is a step increase in source concentration, then the resulting sink concentration will rise from the initial concentration to the injected concentration in a fashion which is again determined by the distribution of travel times along the flow paths. The resulting effluent concentration curve is the residence time **distribution** function. The terms density and distribution correspond to their use in probability theory. In this section, we develop the theory of these two functions.

First, consider the slug release of mass M at the source. Then, the source representation is $M\,\delta(t)$, where $\delta(t)$ is the delta function. With the sink concentration given by $c(t)$, the source and sink concentrations are related through

$$Q\int_0^\infty c(t)dt = M \tag{6.5.1}$$

where Q is the sink discharge. If we want to normalize the effluent concentration curve so that it corresponds to a travel time 'density' function, then the area under the curve must equal unity. The **residence time density** curve, $E(t)$, can then be specified by

$$E(t) = \frac{Q}{M}c(t) \tag{6.5.2}$$

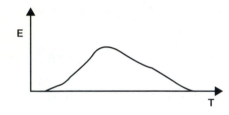

FIGURE 6.5.1   Residence Time Density Function

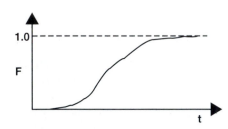

FIGURE 6.5.2   Residence Time Distribution Curve

The function $E(t)$ gives the fraction of flow lines leading from the source to the sink which have the travel time, t. An example density curve is shown in Figure 6.5.1.

The second kind of release is a step change in source concentration. If the initial concentration is 0 and the step change is to a concentration c_o, then the **residence time distribution** curve, or **fractional breakthrough curve**, is given by

$$F(t) = \frac{c(t)}{c_o} \tag{6.5.3}$$

An example residence time distribution curve (breakthrough curve) is shown in Figure 6.5.2.

If $\dot{m}$ is the mass release per unit time, then $c_o = \dot{m}/Q$, and the breakthrough curve may be calculated from

$$F(t) = \frac{Qc(t)}{\dot{m}} \tag{6.5.4}$$

Because $\dot{m} = \dfrac{dM}{dt}$, the density and distribution curves are related through

$$E = \frac{dF}{dt} \tag{6.5.5}$$

The **mean residence time**, $\bar{t}$, may be calculated from the area above the breakthrough curve, as shown in Figure 6.5.3. To see this result, note that

$$\bar{t} = \int_0^\infty tE(t)dt = \int_0^1 t\,dF(t) = \int_0^\infty (1 - F(t))dt \tag{6.5.6}$$

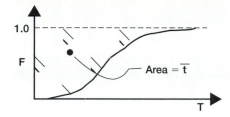

FIGURE 6.5.3   Mean Residence Time

In terms of the mean residence time, the displaced **pore volume**, V_p, is given by

$$V_p = Q\bar{t} \qquad (6.5.7)$$

Thus, knowledge of the effluent discharge and its breakthrough curve allows one to calculate the size of the displaced pore volume leading from the source to the effluent sink.

Convolution Integral

In this subsection, we are interested in predicting the concentration at a sink which receives flow from a number of stream tubes with differing travel times. To fix ideas, we may consider the flow from an injection well to a production well and simplify the problem by considering only advection and neglecting the influence of mixing and dispersion. Considering any streamline, the advection transport equation for the streamline is

$$R\frac{\partial c}{\partial t} + v\frac{\partial c}{\partial s} = -\lambda R c \qquad (6.5.8)$$

In Equation 6.5.8, the seepage velocity is a function of the distance along the streamline, $v = v(s)$, and also possibly a function of time, although we neglect this dependency for the present discussion. This situation means that we are considering steady flow conditions. A hypothetical flow pattern is shown in Figure 6.5.4.

The **travel time** along an arbitrary streamline from the source to a distance s is given by

$$\tau = \int_0^s \frac{ds'}{v(s')} \qquad (6.5.9)$$

so that the incremental travel time is

$$d\tau = \frac{ds}{v(s)} \qquad (6.5.10)$$

Use of Equation 6.5.10 in the advection transport Equation 6.5.8 gives

$$R\frac{\partial c}{\partial t} + \frac{\partial c}{\partial \tau} = -\lambda R c \qquad (6.5.11)$$

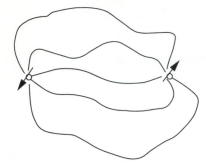

FIGURE 6.5.4 Streamline Pattern from an Injection Well to a Production Well

The *important* point to note is that while Equation 6.5.8 was valid only for one streamline (because of the dependency $v(s)$), Equation 6.5.11 is valid for any and all streamlines.

We now want to consider the advection transport Equation 6.5.11 that is valid for all streamlines. If its solution is considered as a function $c(\tau,t)$, the function $c(0,t)$ corresponds to the injection concentration at time t. Solution of the transport equation by the **method of characteristics** gives

$$\frac{dt}{R} = \frac{d\tau}{1} = \frac{dc}{-\lambda Rc} \rightarrow \left[\frac{dc}{dt} = -\lambda c \text{ or } \frac{dc}{d\tau} = -\lambda Rc\right] \text{ and } \frac{dt}{d\tau} = R \quad \textbf{(6.5.12)}$$

The last equation in Equation 6.5.12 gives the relationship for the **base characteristic** between time, t, and travel time along a streamline, τ, for a 'fluid particle' that enters the subsurface at time t_I:

$$t - t_I = R\tau \quad \textbf{(6.5.13)}$$

The time interval $(t - t_I)$ is the length of time that the 'fluid particle' has been in the subsurface. Because of the sorption interaction with the soil matrix, the 'fluid particle' will only move a fraction

$$\frac{\tau}{t - t_1} = \frac{1}{R}$$

of the distance along the streamline that a conservative (nonreactive) particle would move. Equation 6.5.12 then gives

$$c(t) = c_I(t_I)e^{-\lambda(t-t_I)} \; ; \; c(\tau) = c_I(\tau)e^{-\lambda R\tau}$$

where $c_I(t_I)$ is the influent concentration at time t_I. A look at the base characteristic solution 6.5.13, however, shows that these are actually the same.

$$c(\tau,t) = c_I(t - R\tau)e^{-\lambda(t-t_I)} = c_I(t - R\tau)e^{-\lambda R\tau} \quad \textbf{(6.5.14)}$$

Equation 6.5.14 states that the influent concentration is shifted along the streamline at a speed retarded by a factor R (in terms of the travel time) and decays at a rate determined by the time that the *fluid particle* remains in the subsurface. The projection of the solution onto the base characteristic plane is shown in

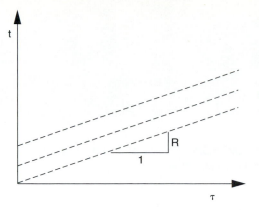

FIGURE 6.5.5 Base Characteristic Plane for Linear Retardation of Solutes

Figure 6.5.5. The slope of the characteristics is equal to the retardation factor, R.

To predict the effluent concentration in the production well at time t, $c_E(t)$, note that it is equal to the average concentration from all streamlines reaching the well at time t. Each streamline, however, has a different travel time. At a given time t, each stream tube supplies solute particles to the sink effluent which left the source at a time $t - R\tau = t_I$. Considering the fractional breakthrough curve, each increment dF corresponds to a stream tube with a given travel time, $dF(\tau)$. The concentration reaching the well from this stream tube is $c(t,\tau)$, and the fraction of the effluent concentration from this stream tube is $c(t,\tau)\,dF(\tau)$. The effluent concentration may be found from

$$c_E(t) = \int_0^1 c(\tau,t)dF(\tau) \qquad (6.5.15)$$

Equation 6.5.15 is the general result presented by Rainwater et al. (1987) and Charbeneau (1988) for applications with multicomponent exchange. For linear isotherms, the solution simplifies to

$$c_E(t) = \int_0^1 c_I(t - R\tau)e^{-\lambda R\tau}dF(\tau) = \int_0^\infty c_I(t - R\tau)e^{-\lambda R\tau}E(\tau)d\tau \quad (6.5.16)$$

which is the convolution integral presented by Levenspiel (1972).

The computation in Equation 6.5.16 is shown schematically in Figure 6.5.6. At a given time t', the stream tube with an increment ΔF has the influent concentration equal to that which entered the system at a time $R\tau$ earlier. This concentration is $c_I(t' - R\tau)$, and the resulting effluent concentration may be approximated by

$$c_E(t) \cong \sum c_I(t - R\tau)e^{-\lambda R\tau}\Delta F(\tau) \qquad (6.5.17)$$

A computational algorithm for Equation 6.5.17 is easily developed.

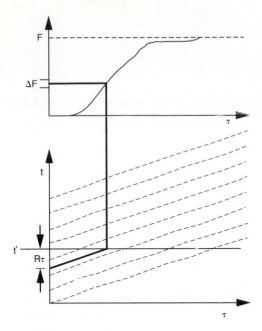

FIGURE 6.5.6 Computational Algorithm for Effluent Concentration

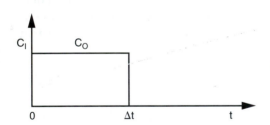

FIGURE 6.5.7 Influent Curve for Pulse Input

EXAMPLE   **6.5.1   Pulse Injection of a Tracer**
PROBLEM   For a steady flow system, consider the injection of a tracer as a pulse at a concentration c_o over a time period Δt, followed by tracer-free water. Calculate the effluent concentration.

The influent concentration curve is shown in Figure 6.5.7 and is represented mathematically using the **Heaviside unit function** as

$$c_I(t) = c_0\{H(t) - H(t - \Delta t)\} \tag{6.5.18}$$

where $H(t)$ is the Heaviside function with argument t, which has the value 0 for negative arguments $(t < 0)$ and the value 1 for positive arguments. Using Equation 6.5.18 in the convolution integral Equation 6.5.16, we find

$$c_E(t) = \int_0^1 c_o\{H(t - R\tau) - H(t - R\tau - \Delta t)\}dF(\tau) = c_o\left\{F\left(\frac{t}{R}\right) - F\left(\frac{t - \Delta t}{R}\right)\right\} \tag{6.5.19}$$

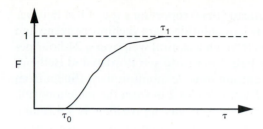

FIGURE 6.5.8 Fractional Breakthrough Curve for Example

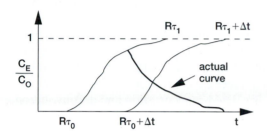

FIGURE 6.5.9 Development of Effluent Curve from the Fractional Breakthrough Curve

The meaning of this equation is that one can find the effluent concentration by using the fractional breakthrough curve $F(\tau)$ twice. For the first curve, the argument is $\tau = t/R$, so on the effluent curve, the corresponding time is $t = R\tau$. For the lagged curve, the argument is

$$\tau = \frac{t - \Delta t}{R}$$

so that the corresponding time on the effluent curve is $t = R\tau + \Delta t$.

To see how to use this idea, consider the fractional breakthrough curve shown in Figure 6.5.8. To find the effluent curve, we need to scale and lag the curve $F(\tau)$. The ratio c_E/c_o plays the role of F, and for the first curve, $t = R\tau$ plays the role of τ. Thus, the point with travel time τ_o on the breakthrough curve corresponds to the point with time $t = R\tau_o$ on the effluent curve, etc. The resulting procedure is shown in Figure 6.5.9. The first of the breakthrough curves shown in Figure 6.5.9 is the breakthrough of the tracer pulse in the effluent. The second breakthrough curve corresponds to the breakthrough of tracer-free fluid that displaces the pulse. Thus, the amount of tracer in the effluent at any time is found from the difference of the two curves.

This example also shows how one can estimate the fractional breakthrough curve from a pulse-input tracer test. If we inject a conservative tracer ($R = 1$) for a period Δt and measure the effluent curve, Equation 6.5.19 says that we can estimate the fractional breakthrough curve from

$$F(t) = \frac{c_E(t)}{c_o} + F(t - \Delta t) \tag{6.5.20}$$

For example, Wise and Charbeneau (1994) report on a tracer test that was performed at an industrial facility where chloride, para-dichlorobenzene and naphthaline were injected as tracers through a central well over a 24-hour period followed by injection of tracer-free water (also see Borden and Bedient, 1987). Conductivity and tracer concentrations were monitored in effluent from two adjacent recovery wells located a distance of 2 m from the injection well, with all three wells in a line. The injection rate was approximately 1.9 L/min. There was weak breakthrough of the chemicals in one of the recovery wells and strong breakthrough in the other (RU 29), showing that the flow pattern was nonideal and that use of analytic models would not be appropriate for analysis of the tracer data.

The tracer concentrations have to be normalized before they are used to estimate the breakthrough curve (residence time distribution curve). The ambient background concentrations of chloride and conductivity were 50 mg/L and 300 μMHO/L, while the injected tracer concentrations were 500 mg/L and 2240 μMHO/L, respectively. The normalized concentration and deconvolution equations are

$$f(t) = \frac{c_E(t) - c_a}{c_I - c_a} \; ; \; F(t) = f(t) + F(t - \Delta t) \qquad \textbf{(6.5.21)}$$

In Equation 6.5.21, $c_E(t)$ is the effluent concentration at time t, c_a is the ambient background concentration, and c_I is the injected tracer concentration. The normalized concentration curve for chloride and conductivity are shown in Figure 6.5.10, along with the calculated fractional breakthrough curve. The curves are shown as a function of volume produced from well RU 29, rather than as a function of time to allow for the time varying injection and production rates that occurred during the experiment. The rates decreased over the period of the experiment, although the ratio of injection to well recovery remained fairly uniform—suggesting that the unknown streamline pattern remained steady. This latter assumption is a requirement for the use of residence time distribution theory to analyze subsurface flow problems. Wise and Charbeneau (1994) use the calculated breakthrough curve along with the measured tracer influent and effluent concentrations and Equation 6.5.17 to estimate the sorption and first-order decay constants for the organic chemicals used in the experiment.

6.6   Standard Flow Patterns

In this section, we will present the residence time distribution functions for a couple of standard flow configurations. These functions are useful for analyzing the transport characteristics of the patterns in various applications.

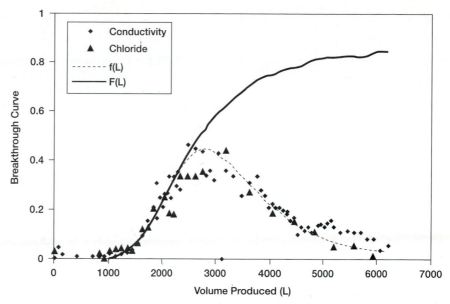

FIGURE 6.5.10   Estimation of fractional breakthrough curve (after Wise and Charbeneau, 1994)

6.6.1   Linear Variation in Detention Times

Of all the possible forms of fractional breakthrough curves, the linear curve is special (Huisman and Olsthoorn, 1983). It is possible to show that for any flow system with an influent concentration that has a periodic variation, the effluent concentration is constant with a value equal to the mean influent concentration if the duration of the breakthrough curve matches the period of the influent concentration variation. An example presented by Huisman and Olsthoorn (1983) is the Rhine River in The Netherlands, which is used for groundwater recharge. During years with normal precipitation, the Cl concentration varies from 80 mg/L during the early spring to 310 mg/L during the early fall. By developing a groundwater recovery system with a yearly variation in residence times, one can achieve an effluent with a constant chloride concentration of 190 mg/L. Moreover, with a linear breakthrough curve, the variability of the effluent concentration is the smallest possible if the influent concentration is variable or random. The following example shows how one might design a water reuse system to achieve a linear variation in detention times.

The fractional breakthrough curve for the system with a linear variation in detention times is shown in Figure 6.6.1. The stream tubes leading from the source to the sink have travel times varying from τ_0 to τ_1. The relationship between the travel time and fractional breakthrough for an arbitrary stream tube is

$$\tau = \tau_F = \tau_0 + (\tau_1 - \tau_0)F \leftrightarrow F = \frac{\tau - \tau_0}{\tau_1 - \tau_0} \qquad \textbf{(6.6.1)}$$

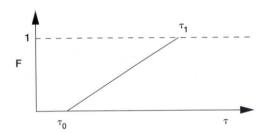

FIGURE 6.6.1 Fractional Breakthrough Curve for a Linear Variation in Travel Times

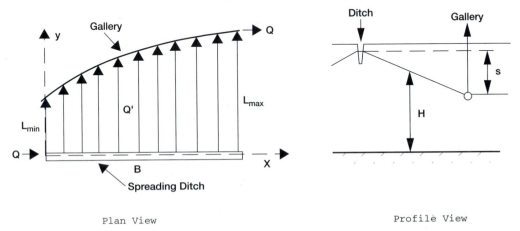

Plan View   Profile View

FIGURE 6.6.2 Flow System for a Linear Variation in Dentention Times

Huisman and Olsthoorn (1983) discuss how one might design a flow system to achieve a linear variation in travel time. For one of their systems, the basic design idea is to introduce water through a spreading ditch. From the ditch, the water will move laterally to be collected in a gallery (one on each side of the ditch). By varying the distance, y, between the spreading ditch and the gallery, one can achieve the desired variation in detention times. A plan view of one-half of the configuration is shown in Figure 6.6.2, where H is the average saturated thickness between the spreading ditch and the gallery and s is the drawdown between the same locations. Along any lateral flow path, Darcy's Law gives the velocity:

$$v = \frac{1}{n}K\frac{s}{y}$$

The travel time from the spreading ditch to the gallery is

$$\tau = \frac{y}{v} = \frac{ny^2}{Ks}$$

If τ_0 and τ_1 are chosen based on design considerations, then the distances L_{min} and L_{max} shown in Figure 6.6.2 are given by

$$L_{min} = \sqrt{\frac{Ks\tau_0}{n}} \; ; \; L_{max} = \sqrt{\frac{Ks\tau_1}{n}} \tag{6.6.2}$$

Huisman and Olsthoorn (1983) show that for a linear variation in detention times the relationship between the lateral distance, y, to the gallery from the spreading ditch as a function of the distance along the ditch, x, is given by

$$y^3 = \frac{3K^2Hs^2}{2nQ}(\tau_1 - \tau_0)x + \left(\frac{Ks\tau_0}{n}\right)^{3/2} \tag{6.6.3}$$

Equation 6.6.3 may be used to find the length of the spreading ditch:

$$B = \frac{2Q}{3H\sqrt{nKs}}\left(\frac{\tau_1^{3/2} - \tau_0^{3/2}}{\tau_1 - \tau_0}\right) \tag{6.6.4}$$

EXAMPLE PROBLEM

6.6.1 Convolution with a Linear Variation in Detention Times

Calculate the effluent concentration for a conservative chemical in a flow system with a linear variation in detention times.

Equation 6.5.16 provides the general convolution equation relating injection and effluent concentrations for a flow pattern. For a flow system with a linear variation in detention times, Equation 6.6.1 shows that

$$dF = \frac{d\tau}{\tau_1 - \tau_0}$$

The convolution equation becomes

$$c_E(t) = \frac{1}{\tau_1 - \tau_0}\int_{\tau_0}^{\tau_1} c_1(t - R\tau)d\tau \tag{6.6.5}$$

This equation shows directly that the effluent concentration is equal to the average of the influent concentration over the time interval $\Delta\tau = \tau_1 - \tau_0$. In particular, if $R = 1$ and the influent stream is periodic with period $\Delta\tau$, then the effluent concentration is the true average of the influent concentration. If the chemical species has a retardation factor $R=1$, then the required duration of the variation in detention times and the period of influent variation, T, are related through $\Delta\tau = T/R$ to obtain an effluent concentration which is an average of the influent concentration.

6.6.2 Injection-Production Well

Section 6.4.2 discussed the flow pattern and streamline travel time distribution for a constituent to move from an injection well to a production well in a source-sink pair. Figure 6.4.4 shows the flow pattern with an injection well located at the point $(x,y) = (-L,0)$ and a production well at $(L,0)$. The streamline $\Psi = Q/2$ corresponding to the point $F = 0$ on the breakthrough curve has a travel time τ_F equal to

$$\tau_0 = \frac{4\pi n b L^2}{3Q} \tag{6.6.6}$$

The streamlines $\Psi = Q/4$ and $\Psi = -Q/4$ correspond to the point $F = 0.5$ on the breakthrough curve and have a travel time equal to

$$\tau_{0.5} = 3\tau_0 = \frac{4\pi n b L^2}{Q} \tag{6.6.7}$$

The streamline $\Psi = 0$ has an infinite travel time.

In general, the travel time for each streamline and its point on the breakthrough curve are related through

$$\tau_F = \frac{4\pi n b L^2}{Q}\left(\frac{1 - \pi F \cot(\pi F)}{\sin^2(\pi F)}\right) \tag{6.6.8}$$

A pore volume cannot be defined for the source-sink pair, because it is infinite. One may normalize the travel time by dividing by the shortest travel time for a streamline (the travel time for streamline $\Psi = Q/2$). With this dimensionless travel time $T_F = \tau_F/\tau_0$, the fractional breakthrough curve is given by

$$T_F = 3\left(\frac{1 - \pi F \cot(\pi F)}{\sin^2(\pi F)}\right) \tag{6.6.9}$$

The fractional breakthrough curve is shown in Figure 6.6.3.

Equation 6.6.9 provides T_F as a function of F. For many applications, it is of interest to have the function $F(T_F)$, which is the inverse to the function $T_F(F)$. This relationship is presented in Table 6.6.1.

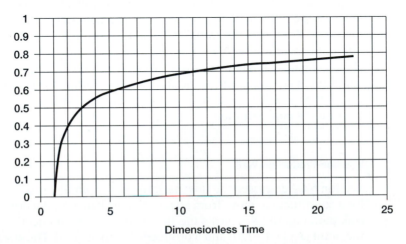

FIGURE 6.6.3 Fractional Breakthrough Curve for Source-Sink Pair

TABLE 6.6.1 Breakthrough Curve for a Source-Sink Pair

Dimensionless Time T_F	Fractional Breakthrough F	Dimensionless Time T_F	Fractional Breakthrough F
1.00	0.0000	3.50	0.5298
1.02	0.0722	4.00	0.5536
1.04	0.1001	4.50	0.5733
1.06	0.1211	5.00	0.5898
1.08	0.1395	6.00	0.6166
1.10	0.1548	7.00	0.6376
1.15	0.1868	8.00	0.6546
1.20	0.2130	9.00	0.6688
1.30	0.2544	10.00	0.6810
1.40	0.2873	12.00	0.7008
1.50	0.3141	14.00	0.7165
1.60	0.3372	16.00	0.7293
1.70	0.3573	18.00	0.7401
1.80	0.3750	20.00	0.7494
1.90	0.3909	25.00	0.7679
2.00	0.4052	30.00	0.7819
2.20	0.4301	40.00	0.8022
2.40	0.4512	50.00	0.8166
2.60	0.4696	60.00	0.8275
2.80	0.4857	80.00	0.8435
3.00	0.5000	100.00	0.8548

EXAMPLE PROBLEM

6.6.2 Analysis of Breakthrough Curve

An injection and production well are situated 40 ft apart in a confined aquifer of thickness 30 ft and porosity 0.30. Each well pumps water at a rate of 15,000 ft³/d (75 gpm). A tracer with a retardation factor of 1.5 is injected at a concentration of 100 mg/L for a period of 5 days, followed by injection of tracer-free water. What is the maximum concentration of the tracer in the production well effluent?

According to Equation 6.6.6, the shortest travel time for water through this system is given by $\tau_0 = 1.044$ days. The maximum concentration occurs just before the tracer-free water that is injected after 5 days breaks through at the well. Using Equation 6.5.13 with $t_I = 5$ days and $\tau = 1.044$ days, this event occurs at $t = 6.566$ days into the test. For the injection front ($t_I = 0$d), this value corresponds to dimensionless time

$$T_F = \frac{T_F}{\tau_0} = \frac{(t - t_I)/R}{\tau_0} = \frac{6.566 \text{ d}/1.5}{1.044 \text{ d}} = 4.2$$

and from Table 6.6.1, the corresponding fractional breakthrough is $F \cong 0.562$. The dimensionless time for the displacement front ($t_I = 5$d) is

$$T_F = \frac{(t - t_l)/R}{\tau_0} = \frac{(6.566 - 5.0)/1.5}{1.044} = 1.0$$

which corresponds to $F = 0$. Thus, using Equation 6.5.19, the maximum concentration is $c_E(6.57d) = 100 \text{ mg}/L \times 0.562 = 56.2 \text{ mg}/L$.

6.6.3   Repeated Five-Spot Pattern

The repeated **five-spot pattern** consists of a continuous distribution of injection and production wells, where both the injection and production wells are arranged in a continuous square pattern with a well of the opposite type located in the center. This arrangement is shown in Figure 6.6.4, where the spacing between neighboring injection wells and between neighboring production wells is L.

In order to develop mathematical equations for the potential and stream functions, an obvious first choice is to use the principle of superposition for the individual wells. Thus, for example, the potential function may be written

$$\Phi = \Phi_o + \sum_{M=-\infty}^{\infty} \sum_{N=-\infty}^{\infty} \frac{Q}{2\pi} \ln(r_{MN}) - \sum_{M=-\infty}^{\infty} \sum_{M=-\infty}^{\infty} \frac{Q}{2\pi} \ln(r'_{MN}) \quad \textbf{(6.6.10)}$$

where

$$r_{MN} = \sqrt{(x - ML)^2 + (y - NL)^2} \; ; \; r'_{MN} = \sqrt{(x - (M + 1/2)L)^2 + (y - (N + 1/2)L)^2}$$

Equation 6.6.10 may also be written

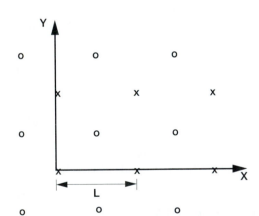

FIGURE 6.6.4   Repeated five-spot well pattern

x - production wells

o - injection wells

$$G(x,y) = \exp\left(\frac{2\pi(\Phi - \Phi_o)}{Q}\right) = \left(\frac{\displaystyle\prod_{M=-\infty}^{\infty}\prod_{N=-\infty}^{\infty} r_{MN}}{\displaystyle\prod_{M=-\infty}^{\infty}\prod_{N=-\infty}^{\infty} r'_{MN}}\right)$$ (6.6.11)

The function $G(x,y)$ defined by Equation 6.6.11 has the following properties. First, because Φ is an analytic function, so is $G(x,y)$, because it is an analytic function of an analytic function. Second, $G(x,y)$ has a simple zero at any location $x = \alpha L, y = \beta L$, where α and β are arbitrary integers. These locations of the simple zeros of the function correspond to the locations of the production wells in Figure 6.6.4. Third, the function $G(x,y)$ has a simple pole (function goes to infinity as $1/r$) at any location $x = (\alpha + \frac{1}{2})L, y = (\beta + \frac{1}{2})L$, where α and β are again arbitrary integers. These locations of the simple poles correspond to the locations of the injection wells in Figure 6.6.4. Finally, the function $G(x,y)$ is doubly periodic. This term means that $G(x + \alpha L, y) = G(x, y + \beta L) = G(x,y)$ for arbitrary integers α and β. The fact that this is true should be obvious from the figure.

Within the theory of analytic functions, it is known that the elliptic functions share the properties noted above for the function $G(x,y)$. In particular, the Jacobian elliptic functions have the required arrangement of zeros and poles (Milne-Thomson, 1950). Elliptic functions have been used for representing the hydraulics of five-spot patterns, line drives, and staggered line drives by a number of authors (Prats et al., 1955; Prats, 1956; Hauber, 1964; Morel-Seytoux, 1965, 1966, 1987). Our main interest in the five-spot pattern is its fractional breakthrough characteristics. For the repeated five-spot pattern, Prats et al. (1955) found that the breakthrough sweep efficiency, which is defined by

$$\Sigma = \frac{(\text{injection rate}) \times (\text{breakthrough time})}{\text{area affected by one injection well}}$$

is calculated to be

$$\Sigma = \frac{\pi^2}{4K^2(1/2)} = \frac{\pi^2}{4 \times 1.8541^2} = 0.7178$$ (6.6.12)

where $K(\ )$ is the complete elliptic integral of the first kind, and it is a tabulated function (see Milne-Thomson, 1950; Abramowitz and Stegun, 1964). Morel-Seytoux (1965) finds the travel time along an arbitrary streamline. His results may be used to derive the equation of the pore volume—fractional breakthrough curve as

$$PV = \frac{\pi}{2K^2(1/2)} K(\cos^2(\pi(1 - F)/2))$$ (6.6.13)

where one pore volume corresponds to the pore space volume within the region affected by one well, and $K(\;)$ is again the complete elliptic integral of the first kind. In the case of the five-spot pattern, the magnitude of one pore volume is given by

$$1PV = nbL^2$$

while the number of pore volumes injected at a given time t is

$$\#PV = \frac{Qt}{nbL^2} \qquad (6.6.14)$$

With Equations 6.6.13 and 6.6.14, the relationship between the fractional breakthrough from a five-spot pattern and travel time along a streamline is given by

$$\tau = \frac{\pi nbL^2}{2QK^2(1/2)} K(\cos^2(\pi(1 - F)/2)) \qquad (6.6.13)$$

Equation 6.6.15 is the main result of interest. The resulting pore-volume curve is shown in Figure 6.6.5. A brief tabulation is presented in Table 6.6.2.

6.7   Evaluating the Environmental Consequences of Groundwater Contamination

One of the goals of groundwater modeling is to be able to determine the environmental impacts of subsurface contamination and to convey clearly the en-

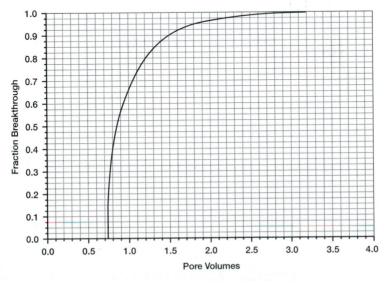

FIGURE 6.6.5   Fractional Breakthrough Curve for a Five-Spot Pattern

TABLE 6.6.2 Fractional Breakthrough Curve for a Five-Spot Pattern

Fractional Breakthrough	Pore Volumes	Fractional Breakthrough	Pore Volumes
0.00	0.7178	0.65	0.9681
0.05	0.7189	0.70	1.0256
0.10	0.7223	0.75	1.0973
0.15	0.7280	0.80	1.1900
0.20	0.7360	0.85	1.3131
0.25	0.7465	0.90	1.4899
0.30	0.7599	0.92	1.5934
0.35	0.7763	0.94	1.7172
0.40	0.7958	0.95	1.8014
0.45	0.8195	0.96	1.9041
0.50	0.8472	0.98	2.2341
0.55	0.8807	0.99	2.5544
0.60	0.9201	1.00*	3.1925

*This point is estimated from the condition that the area above the breakthrough curve is equal to unity (see Figure 6.5.3).

vironmental effects in an understandable fashion. The need of a basis for clear communication is essential when the environmental impact assessments are to be used by managers and public decision-makers that have little understanding of groundwater modeling fundamentals or practices.

In general, three basic factors are needed to assess the present or future impacts of groundwater contamination. These are the location of the contaminants, the arrival time of contaminants, and the quantity of contaminants. The location of the contaminants is important because it determines the risk potential. A contaminant isolated from man, both now and in the future, may represent little hazard, even when large quantities are present. On the other hand, even small amounts of contaminants arriving at critical locations over a short period may involve severe hazard. The arrival time of the contaminant is important because it determines whether corrective actions are required immediately or whether natural processes may be afforded the time to mitigate the hazard. Finally, the quantity of the contaminant is important because small amounts may be little consequence, while larger quantities may constitute serious hazards. These three factors may be interrelated through development of distributions showing the arrival time of contaminants at different locations and distributions providing the outflow quantities at these locations.

Nelson (1978) suggests that the environmental consequences of subsurface contamination problems can be evaluated completely and effectively by fulfilling the following five requirements:

1. determining each present or future outflow boundary of contaminated groundwater,
2. providing the location/arrival time distributions,
3. providing the location/outflow quantity distributions,

4. providing these distributions for each individual chemical or biological constituent of environmental importance, and

5. using the arrival distributions to determine the quantity and concentration of each contaminant that will interface with the environment as time passes.

Applications of these requirements are discussed in the following paragraphs using the results from Example 6.4.1. Nelson (1978) presents more detailed and complicated examples.

Example 6.4.1 applies potential theory to model the capture zone of a production well located in an aquifer with a regional gradient. Here we use these results to model the transport of a contaminant from a source treated as an injection well located in an aquifer with a regional gradient to a river which acts as the interface with the biosphere. The contaminant is considered to be conservative, not interacting with the porous matrix or undergoing biotic or abiotic transformations. Transport occurs as the contaminant is carried forward with the bulk fluid movement (advection). Other transport processes are neglected. We assume that all of the influent from the source is captured by a river located a distance L downgradient. Figure 6.7.1 shows the injection well, the river, and the injection fronts for different times. The following analysis is approximate in that the effects of an image well located a distance L on the other side of the river are not included. The potential function and stream function are given by

$$\Phi = \Phi_o - Ux - \frac{Q}{4\pi} \ln(x^2 + y^2) \tag{6.7.1}$$

$$\Psi = -Uy - \frac{Q}{2\pi} \tan^{-1}\left(\frac{y}{x}\right) = -Uy - \frac{Q}{2\pi}\theta \tag{6.7.2}$$

The equation for the streamline that forms the water divide may be found as follows. Far upstream this streamline coincides with the x-axis, for which $y = 0$ and $\theta = \pi$. Using these results in Equation 6.7.2 gives $\Psi = -0 - (Q/2\pi)$ $\pi = -Q/2$. Because this equation holds for the entire streamline, the equation for the maximum lateral extent of the contaminant downgradient from the stagnation point is

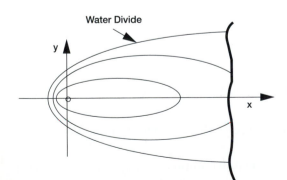

Water Divide

FIGURE 6.7.1   Injection fronts at different times for and injection well in a regional flow field

$$y_{max} = \frac{Q}{2U}\left(1 - \frac{\tan^{-1}(y_{max}/L)}{\pi}\right) \qquad (6.7.3)$$

where $\pm y_{max}$ is the maximum lateral extent of the contaminant at a location L downgradient from the source. Equation 6.7.3 must be solved iteratively.

The **location/outflow quantity distribution** gives the amount of outflow at various locations along the environmental interface. In general, the outflow quantity or flux will vary both with location and time. For this simple example, however, the outflow quantity is independent of time and is given along the length of the river by

$$U_x(L,y) = -\frac{\partial \Phi}{\partial x} = \frac{Q}{2\pi}\left(\frac{L}{L^2 + y^2}\right) + U \qquad (6.7.4)$$

The **location/arrival time distribution** gives the location at which the contaminant will reach the outflow boundary as a function of time. It provides the location of contaminant outflow as a function of time for the instantaneous pulse of traced fluid that departs from the contaminant source at time t_o. Under transient conditions, the location/arrival time distribution will be a function of the release time t_o. For a steady flow problem, it gives the time at which flow along a given streamline will reach the outflow boundary. For the present example, this time is given by Equation 6.4.2a:

$$t = \frac{nbL}{U} - \frac{nbQ}{2\pi U^2} ln\left(\cos\left(\frac{2\pi Uy}{Q}\right) + \frac{L}{y}\sin\left(\frac{2\pi Uy}{Q}\right)\right) \qquad (6.7.5)$$

where b is the average or effective saturated thickness of the aquifer. The first term in Equation 6.7.5 gives the advection travel time due to the regional gradient from the source to the river. The second term accounts for the shortened travel time caused by the injection well. Along the x-axis, Equation 6.7.5 reduces to

$$t = \frac{nbL}{U} - \frac{nbQ}{2\pi U^2} ln\left(1 + \frac{2\pi UL}{Q}\right) \qquad (6.7.6)$$

We designate the value of y which satisfies Equation 6.7.5 for a given time by $y^*(t)$. The function $y^*(t)$ gives the location/arrival time distribution for this example. For a continuous release of contaminant, at a given time t, the contaminant enters the river between the locations $-y^*(t)$ and $y^*(t)$. The ultimate extent of contamination is found by letting $t \to \infty$ in Equation 6.7.5, which again gives Equation 6.7.3.

Equations 6.7.4 and 6.7.5 provide the necessary information for calculation of the mass flux into the river as a function of time. The outflow quantity (mass per time, $\dot{m}$) is obtained by combining the location/outflow quantity distributions and the location/arrival time distributions. For the general case, this process will require numerical quadrature. For our example, however, this information is found by integrating the location/outflow quantity distribution from Equation 6.7.4 between the arrival time limits $y^*(t)$ found from Equation 6.7.5:

$$\dot{m} = c_o \int_{-y^*(t)}^{y^*(t)} U_x(y)\,dy$$

where c_o is the source concentration. With Equation 6.7.4, this equation gives

$$\dot{m}(t) = 2c_o U y^*(t) + \frac{Qc_o}{\pi} \tan^{-1}\!\left(\frac{y^*(t)}{L}\right) \qquad \text{(6.7.7)}$$

Equation 6.7.7 is a simple result that may be used to determine the environmental impacts of a release of contaminants through the recharge well. A specific application is shown in the following example.

EXAMPLE PROBLEM

6.7.1   Contaminant Influx from a Pulse Release

A recharge well receives a conservative contaminant in the effluent from a treatment plant. The contaminant is present at a concentration of 10 mg/L over duration of 14 days. Recharge from the well is captured by a river located 500 m downgradient through an aquifer with a porosity $n = 0.35$ and a saturated thickness of $b = 10$ m. The regional aquifer flux is $U = 5$ m^2/d, and the recharge well injection rate is $Q = 500$ m^3/d. Determine the contaminant influx to the river as a function of time. In particular, determine the maximum influx in kg/d.

The location/arrival time distribution for the injection well influent arriving at the river is shown in Figure 6.7.2. This curve was calculated using Equation 6.7.5, with the maximum lateral extent of contamination found from

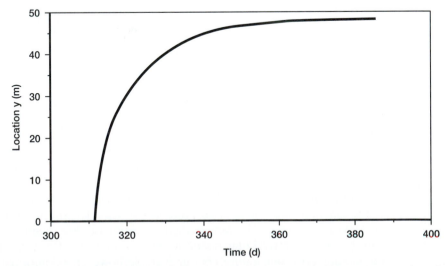

FIGURE 6.7.2   Location/arrival time distribution at the river for the contaminant from the injection well

Equation 6.7.3 to be $y_{max} = 48.5$ m. The contaminant first arrives at the river at a time of $t = 311$ days. Figure 6.7.2 shows that for locations further away from the center line from the injection well to the river, the travel time is longer. For locations approaching the groundwater divide, the travel time increases significantly. This statement is true because water which moves along these streamlines must approach near the stagnation point directly upgradient from the well. During this approach, the velocity decreases, causing a correspondingly longer travel time.

At a given time, the location/arrival time distribution from Figure 6.7.2 may be used to determine the region within which contaminated fluid arrives at the river, $-y^*(t) < y < y^*(t)$. The mass flux entering the river within this region is given by Equation 6.7.7. For this example, the contaminated water front is followed by the clean water front which has a location arrival time distribution identical in shape to that shown in Figure 6.7.2, except that it is shifted to the right by a time $\Delta t = 14$ days. The resulting mass flux curves for both fronts are shown in Figure 6.7.3. From a time of 311 days to 325 days, only contaminated water enters the river. At 325 days, one finds $y^* = 37$ m, and $\dot{m} = 3.81$ kg/d. This value is the maximum mass flux to the river. For times beyond 325 days, contaminated water continues to enter at further regions along the river, while clean water has displaced the contaminated water near the center line.

For example, at 330 days, the contaminated water enters the river between locations $-40.6 < y < -23.6$ m and $23.6 < y < 40.6$ m, while clean water enters over the region $-23.6 < y < 23.6$ m. The mass flux from Equation 6.7.7 with $y^* = 40.6$ is 4.19 kg/d, while that with $y^* = 23.6$ is 2.44 kg/d. Thus, the flux into the river is $\dot{m} = 4.19 - 2.44 = 1.75$ kg/d. Thus, the resulting influx is found

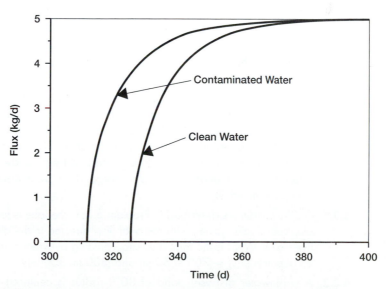

FIGURE 6.7.3 Contaminant influx to the river from the recharge well

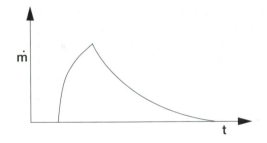

FIGURE 6.7.4   Resulting contaminant influx to river

by taking the difference between the contaminated water and clean water curves in Figure 6.7.3. The resulting distribution is shown in Figure 6.7.4.

Example 6.7.1 shows how groundwater modeling may be applied to evaluate the consequences of groundwater contamination. The picture is greatly simplified, however. In addition to the requirements suggested by Nelson, one also has to consider other processes that may influence or control the fate of subsurface contamination. Much effort goes into predicting what happens to contaminants while they are in the aquifer rather than estimating their arrival at environmental interfaces. Nevertheless, the issues raised by Nelson (1978) are important for consideration when making environmental assessments.

Problems

6.2.1. Briefly explain why Equation 6.2.3 with 6.2.2 is not an appropriate general form of Darcy's Law for application in an aquifer; e.g., what does it imply that is not physically possible?

6.2.2. Using Equation 6.2.18, calculate the travel time for radial flow from an injection well from radius $r_1 = 0.5$ ft to $r_2 = 40$ ft if the aquifer thickness is 25 ft, porosity is 0.35, and the injection rate is 10,000 ft^3/d. Verify your result using simple continuity arguments.

6.3.1. A lake that is roughly circular in shape with a radius of 600 m is situated in an alluvial valley and in direct communication with an unconfined aquifer, where the mean regional flow rate (aquifer flux) is 0.1 m^2/d. The lake is fed by a number of small creeks and water is lost by evaporation. The net recharge rate is estimated to be 300 m^3/d. Is the lake recharged by groundwater, or does it act solely as a source of groundwater recharge. (*Hint:* Does water enter the lake on the upstream side?)

6.3.2. For the conditions described in Problem 6.3.1, the lake is located at the origin of a coordinate system, with regional flow directed in the direction of the positive x-axis. What is the total discharge passing between monitoring wells located at points $(x,y) = (700$ m,50 m) and (800 m, 300 m)?

6.3.3. A wastewater disposal pond of 80 ft radius is centered at the origin of a coordinate system and directly recharges an unconfined aquifer at a rate of

15,000 ft^3/d. The aquifer saturated thickness at the pond is 60 ft. A production well located at (x = 300ft, y = 100 ft) has a discharge of 20,000 ft^3/d. The aquifer has a hydraulic conductivity of 12 ft/d and its regional flow occurs at a rate of 2 ft^2/d, directed along the x-axis. Write down the potential and stream functions for this problem and determine the saturated thickness at an observation well located at (x = 350 ft, y = −75ft).

6.3.4. Using *just* the potential and stream functions, estimate the direction (in both degrees and radians) and magnitude of the seepage velocity at point (x,y) in an unconfined aquifer. With a consistent set of units, the point (x,y) = (100,0), the hydraulic conductivity and porosity are 10 and 0.4, respectively, and the regional flow of magnitude U = 1 makes an angle $\alpha = \pi/4$ with the x-axis. A production well with radius 0.1 is located at the origin. The discharge rate of the well is 500π, and the saturated thickness at the well is 10. NO USE OF DIFFERENTIATION IS ALLOWED. (*Hint:* Evaluate Φ and Ψ at two neighboring points located along a line of constant Φ, and use the properties of these functions. It may be helpful to set the problem up on a spreadsheet.)

6.4.1. A well located at the origin of a coordinate system injects a chemical tracer into an aquifer with a regional gradient along the x-axis. The injection rate is Q = 10,000 ft^3/d, while the regional aquifer flux is U = 1 ft^2/d. If the porosity and average thickness of the aquifer are 0.35 and 50 ft, respectively. How long will it take for the tracer to reach an observation well located at the point (x,y) = (500 ft,200 ft) if mixing processes in the aquifer are negligible? How long would it take to reach this monitoring well in the absence of the regional gradient?

6.4.2. For the conditions specified in Problem 6.4.2, how long will it take for the tracer to reach a monitoring well located 600-ft downgradient from the injection well?

6.4.3. A well produces 550 m^3/d from an aquifer with a regional flow of U = 0.25 m^2/d. The aquifer porosity and thickness are 0.3 and 20 m. What is the minimum size of a rectangular area that will contain the 30-year wellhead protection area? (*Hint:* When trying to solve Equation 6.4.2a, choose a value of x and iterate to find the corresponding y-value by writing the equation as y_{i+1} = $x \sin(2\pi U y_i/Q)/(\exp((2\pi U x/Q) - (2\pi U^2 t/(nbQ)))\cos(2\pi U y_i/Q))$.)

6.4.4. A 200 m wide plume of contamination exists in a sandy aquifer, and a set of production wells is to be placed downgradient of the plume to prevent off-site migration. The aquifer's saturated thickness is 12 m, the hydraulic conductivity is 6 m/d, and the regional gradient is 0.004. In order to prevent excessive drawdown, the well production rate is limited to 30 m^3/d per well. How many wells are required to capture the plume?

6.4.5. An injection well is placed a distance of 100 m from a river that acts as a constant-head boundary. The injection rate is 200 m^2/d, and the aquifer's saturated thickness and porosity are 10 m and 0.35, respectively. Regional flow in the aquifer is negligible. What is the minimum travel time of a tracer between the injection well and the river?

6.4.6. Water flows laterally into a river from an unconfined aquifer at a rate of 0.6 m²/d from the direction perpendicular to the river. The hydraulic conductivity of the aquifer is 4 m/d, and the saturated thickness at the river is 12 m. If a production well is placed at a distance of 80 m from the river and pumps at a rate of 125 m³/d, will it be pulling any water from the river? What will be the saturated thickness at the well if its radius is 30 cm?

6.4.7. A well produces 30,000 ft³/d from an unconfined aquifer with a regional flow rate of $U = 2$ ft²/d directed toward a large surface water reservoir. The well is located 2,500 ft from the reservoir. Estimate the fraction of the well discharge that comes from the reservoir.

6.4.8. An aquifer passes beneath an agricultural field where it receives a nitrogen load resulting in an average nitrate concentration of 18 mg/L. The aquifer drains to a large river with an average nitrate concentration of 0.1 mg/L. The lateral flow to the river from the aquifer is 0.15 m²/d. A production well that will draw both river and aquifer water is to be placed near the river. The average well production rate is 350 m³/d. If a nitrate concentration of 5 mg/L or less is desired, how close to the river must the well be placed?

6.6.1. A water reuse system has been designed to achieve a nearly linear variation in travel times through the subsurface. The minimum travel is 90 days and the maximum time is 450 days. Due to failure of a pretreatment step, a conservative ($R = 1$) contaminant enters the subsurface system with a concentration 40 mg/L above normal during a period of 120 days. What is the maximum concentration above normal in the system effluent, and at what time does it occur?

6.6.2. A shallow aquifer is contaminated with *trichloroethene* (TCE) at residual saturation. You are to carry through a "very rough" calculation of groundwater remediation using pump-and-treat with recirculation of treated water through a continued five-spot pattern. The following information is given: the average saturated thickness is 2 m, the spacing between injection wells is 10 m, the aquifer porosity is 0.35, the residual TCE volumetric content is 0.02, and the injection/production rate from a single well is 15 m³/d. TCE has a density of 1460 g/L and solubility in water of 1.1 g/L. For your analyses, assume that the re-injected water enters with zero TCE concentration and that a sharp front exists between the contaminated part of the aquifer and the part from which TCE has been dissolved. The following sketch shows a control volume along an arbitrary stream tube of the flow pattern. The front separating the contaminated and swept areas of the aquifer is shown at time t and $t + \Delta t$, with the displaced area 'hatched'. It is assumed that water entering a contaminated area leaves with a concentration at the solubility limit. Show that a mass balance for the area gives $((n - \theta_{or})S_{TCE} + \theta_{or}r_{TCE})\Delta x = S_{TCE}q\Delta t = S_{TCE}nv\Delta t$, where θ_{or} is the residual TCE volumetric content, ρ_{TCE} is its density, and S_{TCE} is the TCE solubility. Develop an expression for the concentration front retardation factor, give its value for this condition, and estimate the volume of water and time required for reaching a TCE concentration of 25 mg/L in the production well effluent for a single five-spot.

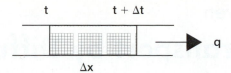

6.6.3. A groundwater management system provides a linear variation of detention times with a range of ½–year (e.g., $\Delta \tau = ½ - yr$). An influent stream is periodic with period T and a chemical of concern has a retardation factor of $R = 2$. If the influent concentration is

$$c_i(t) = c_a + c_b \sin\left(\frac{2\pi t}{T}\right)$$

calculate the time-variable effluent concentration when $T = 1 yr$.

6.7.1. An aquifer with an average saturated thickness of 45 ft has a porosity of 0.35 and a regional flux of 10 ft^2/d towards a surface water reservoir. The aquifer is recharged by a well that is located 5,000 ft from the reservoir at a rate of 4800 ft^3/d. A salinity pulse with a concentration of $1g/L = 1$ kg/m^3 is present in the recharge water for duration of 45 days. Determine i) the maximum lateral extent of the well recharge reaching the reservoir and ii) the time and maximum salinity influx to the reservoir.

Chapter Seven
Solute Transport by Diffusion

Diffusion is the process of mass transport associated with random molecular motions (Bird et al., 1960; Crank, 1975). These molecular motions result in the transport of mass from regions where the concentration is larger to regions where it is smaller. Because this transport mechanism is associated with molecular motions instead of bulk fluid movement, diffusion does not result in mass transport over large distances—especially for transport in liquids. Diffusive transport is important, however, in transport through low permeability soils, in volatilization of chemicals through soil air, in the fate of NAPLs trapped in fractures, and as a rate-limiting step in sorption of chemicals. The parameters describing diffusive transport are discussed in this chapter, and applications of diffusive transport are also presented.

7.1   Fick's Laws

Fick's laws describe mass transport by diffusion. **Fick's first law** states that the diffusion mass flux is proportional to the concentration gradient

$$\tilde{J}_{\text{diff}} = -D_m \text{grad}(c) \tag{7.1.1}$$

In Equation 7.1.1, D_m is the molecular diffusion coefficient. **Fick's second law** is the equation of conservation of mass for a fluid phase

$$\frac{\partial c}{\partial t} = -\nabla \cdot \tilde{J}_{\text{diff}} = D_m \nabla^2 c \tag{7.1.2}$$

Equations 7.1.1 and 7.1.2 have the same form as those for conduction of heat in solids, where the temperature replaces the concentration and Fourier's law of heat conduction is used instead of Fick's first law (Carslaw and Jaeger, 1959).

7.2   **Molecular Diffusion Coefficients**

The molecular diffusion coefficient is much larger for gases than for liquids. For the **gas phase**, kinetic theory and the ideal gas law yield the following expression for estimating the (self-) diffusion coefficient (Bird et al., 1960):

$$D_m(g) = \frac{2}{(3pd^2)} \sqrt{\frac{(\kappa T)^3}{\pi^3 m}}$$ **(7.2.1)**

In Equation 7.2.1, $D_m(g)$ is the gas phase molecular diffusion coefficient, p is the absolute pressure (1 *atmosphere* = $1.013(10^6)$ dyn/cm^2), d is the molecular diameter (assumed to be a rigid sphere), κ is Boltzmann's constant ($\kappa = 1.3805(10^{-16})$ erg/$^\circ K$), T is the absolute temperature, and m is the mass per molecule ($m = M/N$, where M is the molecular weight and N is Avogadro's number—6.023 (10^{23}) molecules per mole). Equation 7.2.1 shows that the molecular diffusion coefficient increases with increasing temperature but decreases with increasing molecular size and mass. For example, for O_2 under atmospheric pressure and 15 degrees C, $d = 2.96 (10^{-8})$ cm (from kinetic theory and gas viscosity measurements, Moore (1972)), and $m = 5.3 (10^{-23})$ g/molecule. With these values, Equation 7.2.1 gives $D_m(O_2) = 0.15$ cm^2/s. The measured self-diffusion coefficient for oxygen at 15 degrees C and one atmosphere pressure is 0.18 cm^2/s.

For liquids, the hydrodynamic theory leads to the Stokes-Einstein equation for predicting the diffusion coefficient

$$D_m(l) = \frac{\kappa T}{3\pi\mu d}$$ **(7.2.2)**

In Equation 7.2.2, $D_m(l)$ is the liquid phase molecular diffusion coefficient, μ is the dynamic viscosity of the liquid, and d is the molecular diameter of the diffusing substance. For the diffusion coefficient of O_2 in water at 20 degrees C, Equation 7.2.2 gives $1.45 (10^{-5})$ cm^2/s, as compared with a measure value of $1.80 (10^{-5})$ cm^2/s.

For a temperature of 25 degrees C, $D_m(g)$ varies between 0.15 and 0.25 cm^2/s (14 and 23 ft^2/d) for low molecular weight gases (e.g. O_2, CO_2). The diffusion coefficient decreases with increasing size of the diffusing molecule. The water phase molecular diffusion coefficient for gases that dissolve is about 10^{-4} smaller than the air phase molecular diffusion coefficient (Lyman et al, 1982). Johnson and Babb (1956) show that diffusion coefficients are on the same order of magnitude for a hydrocarbon phase solvent as for the aqueous phase. For pesticides of intermediate molecular weight (100–300 g/mol), Jury et al. (1980) recommend using an average value of about 0.05 cm^2/s for the air phase molecular diffusion coefficient and a value of about 5×10^{-6} cm^2/s for the water phase molecular diffusion coefficient. Cohen and Mercer (1993) tabulate the air and water diffusion coefficients for a number of *volatile organic chemicals* (VOCs). The air diffusion coefficients range from 0.10 cm^2/s for methylene chloride to

0.042 cm²/s for dibutyl phthalate. For the listed VOC's, the water phase diffusion coefficients were about four orders of magnitude smaller than the air phase coefficients.

7.3   Diffusion in Porous Media

Fick's laws also describe diffusive transport through porous media. The difficulty with describing diffusion in a porous medium comes from the phenomenological relationship for an effective diffusion coefficient that must be dependent on the pore structure and the saturation of the various phases that are present. **Fick's first law** for a saturated porous medium takes the form

$$\tilde{J}_{\text{diff}} = -nD_s \text{grad}(c) \tag{7.3.1}$$

where n is the (areal) porosity and D_s is the **effective diffusion coefficient in soil**. The diffusion coefficient in soil is smaller than the molecular diffusion coefficient of a fluid, D_m, because the diffusing substance must follow a tortuous path through the porous medium. The ratio of D_s to D_m is the **tortuosity** of the medium:

$$\frac{D_s}{D_m} = \tau \tag{7.3.2}$$

In developing a calculus for flow through porous media, at the continuum scale, one measures linear distances without regard to the presence of individual pores. Thus, the concentration gradient is modeled as the ratio , where L is the linear distance between two points along the direction of maximum concentration change. The actual concentration change, however, occurs over a larger distance L_e shown in Figure 7.3.1, and the pore scale gradient is given by the ratio . The tortuosity accounts for the decrease in ∇c by the factor L/L_e. In addition, the transport distance between two points is smaller on the continuum scale than the pore scale by this same factor, L/L_e. Thus, the tortuosity represents the average value of

$$\tau = \left\langle \frac{L}{L_e} \right\rangle^2_{\text{ave}} \tag{7.3.3}$$

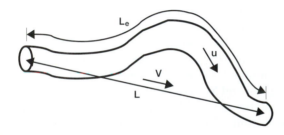

FIGURE 7.3.1   Tortuous Path through a Porous Medium

Also see the discussion in Section 2.1. Arguments related to the permeability of porous media suggest that τ varies from $1/3$ to $1/2$. Millington (1959) and Millington and Quirk (1961) present a theory suggesting that for a saturated porous solid

$$nD_s = n^{4/3}D_m$$

which implies that

$$\tau = n^{1/3} \tag{7.3.4}$$

For a porosity of 0.4, Equation 7.3.4 gives $\tau = 0.74$.

In petroleum literature, advantage is taken of the analogy between mass and electricity transfer. The analogy is complete if the fluid is an ionic solution. The **formation factor**, F, is defined as the ratio between the conductivity of the solution and that of the porous medium (Dullien, 1992). This ratio leads to

$$F = \frac{D_m}{nD_s} \tag{7.3.5}$$

"Archie's equation" (see Dullien, 1992) suggests that $F = n^{-m}$, where m is the cementation exponent with values between 1.3 and 2.5 for various types of rocks. For unconsolidated material, values of m lying near 1.5 are often observed. This situation leads to the estimate $\tau \cong \sqrt{n}$, which is comparable to the theory of Millington.

For unconsolidated alluvial materials, observations (Perkins and Johnston, 1963) suggest that $\tau \cong 2/3$, while for clays, $\tau \cong 1/10$ or smaller. Marsily (1986) notes a value of τ of 0.01 which was observed in highly compacted bentonite for gases, cesium, and strontium.

Substitution of Fick's first law from Equation 7.3.1 with Equation 7.3.2 in the continuity equation 5.4.7 gives

$$\frac{\partial m}{\partial t} + \nabla \cdot (-n\tau D_m \nabla c) = 0 \tag{7.3.6}$$

If n, τ, and ρ_b are constant, and if local equilibrium is assumed with a linear sorption model, then we have

$$R\frac{\partial c}{\partial t} = \tau D_m \nabla^2 c \tag{7.3.7}$$

Equation 7.3.7 may be written with the retarded soil diffusion coefficient $D'_s = D_s/R$

$$\frac{\partial c}{\partial t} = D'_s \nabla^2 c \tag{7.3.8}$$

which is **Fick's second law** written for diffusion in a porous medium. Note that we have assumed that the areal and volumetric porosity have the same value so that they cancel.

7.4   Diffusion in Multiphase Systems

The diffusive transport of gases such as O_2 and volatile organic constituents in the soil occurs partly in the gaseous phase and partly in the liquid phase. Diffusion through the gaseous phase is primarily responsible for the transport of O_2 and CO_2 for maintenance of soil respiration and for supply of oxygen for aerobic biodegradation of organic soil contaminants. Writing Equation 7.3.1 for the vertical direction gives

$$J_{da} = -nD_{sa}\frac{\partial c_a}{\partial z} \tag{7.4.1}$$

where J_{da} is the air-phase diffusion mass flux (mass diffusing in the air-phase across a unit area per unit time), D_{sa} is the **effective air-phase diffusion coefficient** in the soil, c_a is the air-phase concentration (mass per unit volume of air) of the diffusing substance, and $\partial c_a/\partial z$ is the vertical concentration gradient. The effective diffusion coefficient in the soil must be smaller than the bulk air-phase molecular diffusion coefficient, D_{ma}, owing to the limited fraction of the total volume occupied by continuous air-filled pores and also to the tortuous nature of these pores. One may write

$$nD_{sa} = \theta_a\tau_a D_{ma} \tag{7.4.2}$$

where θ_a is the volumetric air content and τ_a is the air-phase *tortuosity factor* which accounts for the decreased cross-sectional area and increased path length induced by a porous medium, which is a function of the volumetric air space available for transport (Nielsen, et al, 1972). Different workers have over the years found different relations between D_{sa} and D_{ma}, and Charbeneau and Daniel (1993) list various examples.

For a solute in a multiphase system under equilibrium conditions, the concentration gradients in all of the fluid phases are proportional, and diffusion will occur in each of the phases at a rate that is dependent on the effective diffusion coefficient for that phase. Use of the Millington formula for the air, water, and NAPL (OIL) phases gives

$$nD_{sa} = \frac{\theta_a^{10/3}}{n^2}D_{ma} \; ; \; nD_{sw} = \frac{\theta_w^{10/3}}{n^2}D_{mw} \; ; \; nD_{so} = \frac{\theta_o^{10/3}}{n^2}D_{mo} \tag{7.4.3}$$

where D_{ma} is the air-phase molecular diffusion coefficient, D_{mw} is the water-phase molecular diffusion coefficient, and D_{mo} is the NAPL-phase molecular diffusion coefficient. The total vertical diffusive flux including transport in the three fluid phases is calculated from

$$J_d = -D_{sT}\frac{\partial m}{\partial z} \tag{7.4.4}$$

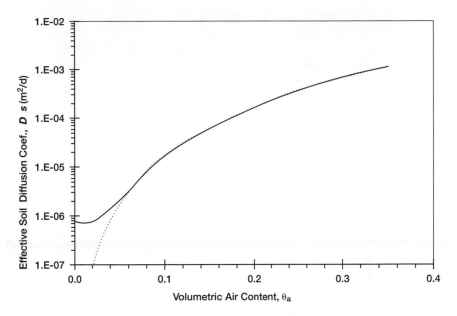

FIGURE 7.4.1 Effective Soil Diffusion Coefficient Ds for Benzene as a Fuction of Air Content for a Porosity $n = 0.40$ and a Fixed Volumetric NAPL Content of $0_o = 0.05$. Values are based on molecular diffusion coefficients $D_{ma} = 0.43$ m²/d and $D_{mw} = D_{mo} = 4.3 \times 10^{-5}$ m²/d. The dashed curve shows the effective soil diffusion coefficient based on air-phase transport only.

where D_{sT} is the **effective soil diffusion coefficient** and m is the bulk concentration. The effective soil diffusion coefficient is given by

$$D_{sT} = \frac{nD_{sa}}{B_a} + \frac{nD_{sw}}{B_w} + \frac{nD_{so}}{B_o} \qquad (7.4.5)$$

where B_a, B_w, B_o, are the bulk partitioning coefficients given by Equations 5.5.9 to 5.5.11. The effective soil diffusion coefficient is a function of the molecular diffusion coefficients for each of the fluid phases, the phase volumetric contents, and the solute partitioning coefficients.

Figure 7.4.1 shows the effective soil diffusion coefficient for benzene calculated from Equation 7.4.5 as a function of the volumetric air content. The calculations were made for a fixed NAPL content of 0.05 and a soil porosity of 0.40. The dashed curve in this figure shows the effective soil diffusion coefficient for air-phase transport. Comparison of the two curves suggests that the effective soil diffusion coefficient can be calculated from

$$D_{sT} \cong \frac{\theta_a^{10/3}}{n^2 B_a} D_{ma} \qquad (7.4.6)$$

The effective soil diffusion coefficient is over two orders of magnitude larger for a dry soil than for a wet soil, and for all but wet soils, only air-phase diffusion need be considered. Thus, under most conditions, Equation 7.4.6 may be used instead of Equation 7.4.5 for prediction of diffusion in the vadose zone, except near the capillary fringe or under conditions where the water table is located at a shallow depth.

For multiphase systems, Fick's second law is

$$\frac{\partial m}{\partial t} = D_{sT}\frac{\partial^2 m}{\partial z^2} \tag{7.4.7}$$

7.5   Applications of the Diffusion Equation

The diffusion equation (Fick's second law) is the classical partial differential equation of the parabolic type (along with the heat conduction equation). In this section, we consider two problems that are not only of interest in their own right, but they will also serve as a stepping stone for finding solutions to the advection-dispersion equation in the next chapter. The point source model will lead to a model for a spill (instantaneous release) of contaminants, while the constant source model will lead to a model for a plume of contaminants migrating through a porous medium.

7.5.1   Point Source in One-Dimension

Consider the release of mass M at a point x_o at time $t = 0$. The bulk concentration is then given by

$$m(x,0) = M\delta(x - x_0) \tag{7.5.1}$$

where $\delta(\ )$ is the **Dirac delta function** (Dirac measure). The property of this function which is important for us is that for any function $f(x)$, the following relationship holds

$$\int f(x)\delta(x - x_o)dx = f(x_o) \tag{7.5.2}$$

so that in particular,

$$\int \delta(x - x_o)dx = 1 \tag{7.5.3}$$

Use of Equations 7.5.1 and 7.5.3 correctly shows that for our initial condition,

$$\int_{-\infty}^{\infty} m(x,0)dx = M \tag{7.5.4}$$

The diffusion equation is written in terms of the aqueous concentration rather than the bulk concentration, so we need an initial condition on c rather than m. Equations 5.7.1 and 5.7.6 give

$$m = nRc \tag{7.5.5}$$

and the initial condition on c is

$$c(x,0) = \frac{M}{nR} \delta(x - x_0) \tag{7.5.6}$$

We want to solve see (Equation 7.3.8)

$$R\frac{\partial c}{\partial t} = D_s \frac{\partial^2 c}{\partial x^2} \tag{7.5.7}$$

subject to the initial condition of Equation 7.5.6. One way to solve this problem is with the Fourier transform defined by

$$\chi(k,t) = \frac{1}{\sqrt{2\pi}} \int_{-\infty}^{\infty} c(x,t)e^{iks} dx \tag{7.5.8}$$

Integration by parts shows that

$$\frac{1}{\sqrt{2\pi}} \int_{-\infty}^{\infty} \frac{\partial^2 c}{\partial x^2} e^{ikx} dx = \frac{-k^2}{\sqrt{2\pi}} \int_{-\infty}^{\infty} c(x,t)e^{ikx} dx = -k^2 \chi(k,t) \tag{7.5.9}$$

If we multiply the diffusion Equation 7.5.7 by

$$\frac{e^{ikx}}{\sqrt{2\pi}}$$

and integrate, we find

$$R\frac{d\chi}{dt} + k^2 D_s \chi = 0$$

while the initial condition becomes

$$\chi(k,0) = \frac{M}{nR\sqrt{2\pi}} e^{ikx_o}$$

The solution to this ODE is easily found to be

$$\chi(k,t) = \frac{M}{nR\sqrt{2\pi}} e^{ikx_0 - k^2 D_s t} \tag{7.5.10}$$

Equation 7.5.10 is the spectrum or characteristic function of the solution for the point source model. The inverse of the Fourier transform is defined by

$$c(x,t) = \frac{1}{\sqrt{2\pi}} \int_{-\infty}^{\infty} \chi(k,t)e^{ikx} dk \tag{7.5.11}$$

Substituting Equation 7.5.10 in 7.5.11 and evaluating the integral, we find

$$c(x,t) = \frac{M}{n\sqrt{4\pi RD_s t}} \exp\left(-\frac{R(x - x_o)^2}{4D_s t}\right) \tag{7.5.12}$$

One can easily verify that

$$\int_{-\infty}^{\infty} nRc \, dx = M \tag{7.5.13}$$

for all time.

　　There are a few points of interest with regard to this result. First, the units are consistent if we recognize that M is actually the mass release per unit area of a plane that is perpendicular to the x-axis. If the mass release is uniform on this plane specified by $(x = x_o, y, z)$, then the only non-zero concentration gradients are in the x-direction and a $1D$ diffusion model is appropriate. The second point of interest is that Equation 7.5.12 may be written

$$c(x,t) = \frac{M/(nR)}{\sqrt{4\pi D_s't}} \exp\left(-\frac{(x - x_o)^2}{4D_s't}\right) \tag{7.5.14}$$

Equation 7.5.14 uses the retarded soil diffusion coefficient and clearly shows that the mass release accounts for the limited pore space and for mass partitioning onto the soil solids. Finally, Equation 7.5.14 suggests a close link with the normal density function from probability theory, which may be written

$$\eta(x;\mu,\sigma^2) = \frac{1}{\sqrt{2\pi\sigma^2}} \exp\left(-\frac{(x - \mu)}{2\sigma^2}\right) \tag{7.5.15}$$

where the notation $\eta(x;\mu,\sigma^2)$ designates a variant x that is normally distributed with mean μ and variance σ^2. Comparison of Equations 7.5.14 and 7.5.15 suggests that the mass has mean location $x = x_o$, and that the variance of the mass distribution is equal to

$$\sigma^2 = 2D_s't \tag{7.5.16}$$

Einstein developed Equation 7.5.16 in 1905 during his development of the stochastic theory of Brownian motion. With diffusion mass transport, the spread of the mass distribution (proportional to the standard deviation, σ) increases as the square root of time, and that one-half times the rate of increase in the variance of the mass distribution is equal to the effective diffusion coefficient.

EXAMPLE PROBLEM　**7.5.1　Diffusion in 3D Space**

Develop the solution for $3D$ diffusion from an instantaneous point release of mass M using analogy with the normal density function, and verify that this solution satisfied the $3D$ continuity equation.

　　The multivariate normal density function is equal to the product of univariate normal density functions. Using the analogy between Equations 7.5.14 and 7.5.15, this information suggests that the solution for the problem of release of mass M at the point (x_o, y_o, z_o) at time $t = 0$ is given by

$$c(x,y,z,t) = \frac{M/(nR)}{(4\pi D_s't)^{3/2}} \exp\left(-\frac{(x - x_o)^2 + (y - y_o)^2 + (z - z_o)^2}{4D_s't}\right)$$

or

$$c(r,t) = \frac{M/(nR)}{(4\pi D_s't)^{3/2}} \exp\left(-\frac{r^2}{4D_s't}\right) \tag{7.5.17}$$

where r is the radial distance from the point of mass release. Because of symmetry, the $3D$ version of Equation 7.5.8 is

$$\frac{\partial c}{\partial t} = \frac{D_s'}{r^2}\frac{\partial}{\partial r}\left(r^2\frac{\partial c}{\partial r}\right)$$

(7.5.18)

Evaluating the partial derivatives and substituting into Equation 7.5.18 shows that Equation 7.5.17 satisfies the continuity equation. Furthermore, because

$$\int_0^\infty x^2 \exp\left(-\frac{x^2}{\alpha}\right)dx = \frac{\sqrt{\pi}}{4}\alpha^{3/2}$$

(7.5.19)

one can show that

$$\int_0^\infty 4\pi r^2 nRc(r,t)\,dr = M$$

(7.5.20)

for all time, which is analogous to Equation 7.5.13.

7.5.2 Diffusion from a Constant Concentration Source

Consider the problem where mass is introduced into a column by maintaining a constant concentration at $x = 0$ and having the mass diffuse into the domain $x > 0$. Figure 7.5.1 shows the concentration distribution at different times. We want to solve the $1D$ version of Equation 7.3.8

$$\frac{\partial c}{\partial t} = D_s'\frac{\partial^2 c}{\partial x^2}$$

(7.5.21)

with initial and boundary conditions given by $c(x,0) = c(\infty,t) = 0$; $c(0,t) = c_o$. The problem specified by Equation 7.5.21 and the given initial and boundary conditions is similar to many others in that the problem statement has no characteristic length or time scales. Physically, this means that length and time cannot appear in the solution as independent variables. Thus, we look for a self-similar solution that has distance and time collapsed into a single variable. The equation is second order in space and first order in time, so that a scaled variable proportional to $x/\sqrt{t}$ suggests itself. The conventional form comes from **Boltzmann's transformation**:

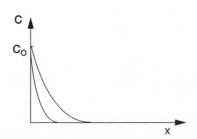

FIGURE 7.5.1   Diffusion from a Constant Concentration Source

$$\phi = \frac{x}{\sqrt{4D_s't}} \qquad (7.5.22)$$

Following the same steps leading to Equation 4.7.6, Equation 7.5.21 is transformed to

$$\frac{d^2c}{d\phi^2} + 2\phi \frac{dc}{d\phi} = 0 \qquad (7.5.23)$$

To integrate this ODE, let $c' = dc/d\phi$. Then Equation 7.5.15 may be written as $c'/c = -2\phi d\phi = -d\phi^2$. Direct integration gives $c' = Ae^{-\phi^2}$. Integrating again gives

$$c(\phi) = B + \int_0^\phi e^{-\zeta^2 d\zeta}$$

The choice of lower limit for the integral is arbitrary and only affects the value of the integration constant B. The initial and boundary conditions are transformed to $c(0) = c_o$ and $c(\infty) = 0$. The first of these shows that $B = c_o$. For the second, we need the result

$$\int_0^\infty e^{-\zeta^2} d\zeta = \sqrt{\pi}/2$$

to find that

$$A = -\frac{2}{\sqrt{\pi}} c_o$$

The solution to Equation 7.5.23 with stated boundary conditions is thus found to be

$$c(x,t) = c_0\left(1 - \frac{2}{\sqrt{\pi}} \int_0^\phi e^{-\zeta^2} d\zeta\right) \qquad (7.5.24)$$

where

$$\phi = \frac{x}{\sqrt{4D_s't}}$$

The function appearing in the solution shows up in many problems dealing with the diffusion equation and elsewhere in mathematical physics and is called the **complementary error function**, defined by

$$\text{erfc}(\phi) \equiv 1 - \frac{2}{\sqrt{\pi}}\int_0^\phi e^{-\zeta^2} d\zeta \equiv \frac{2}{\sqrt{\pi}} \int_\phi^\infty e^{-\zeta^2} d\zeta \qquad (7.5.25)$$

In terms of this function, Equation 7.5.24 may be written more concisely as

$$c(x,t) = c_o \operatorname{erfc}\left(\frac{x}{\sqrt{4D_s't}}\right) \tag{7.5.26}$$

The following example considers how Equation 7.5.26 may be used to estimate the chemical leachate flux into a soil liner.

EXAMPLE PROBLEM

Example 7.5.2   Flux into a Soil Liner

A soil liner is a layer of soil (usually compacted clay) that is placed at the base of a landfill or other facility to control releases of liquids, which in some cases could be contaminated. For this example, we consider a contaminated liquid with concentration c_o that is placed in a storage facility with a soil liner in place. It is assumed that the liner has sufficiently low permeability that advection transport may be neglected—at least as a first approximation. Furthermore, it is assumed that the liquid concentration within the facility remains constant at c_o. For this analysis, we treat the liner as if it were effectively of infinite thickness. Figure 7.5.2 shows a fluid ponded on a liner extending in the x-direction. Determine the diffusive flux into the liner and estimate the travel time for the contaminant to move through the liner system.

Equation 7.5.26 provides the solution for this problem as it is posed. We are interested in the total mass that has moved into the liner and the time of migration through the liner. The total mass which has moved into the liner may be found either by calculating the total mass present in the liner at a given time or from calculating the total flux into the liner across the liner surface at $x = 0$:

$$\frac{M(t)}{A} = \int_0^\infty nRc(x,t)dx = \int_0^t J(0,\tau)d\tau \tag{7.5.27}$$

The latter form is the easiest to evaluate. The diffusion flux is

$$J = -nD_s \frac{\partial c}{\partial x}$$

where the soil diffusion coefficient D_s is used for the flux calculation. The concentration gradient is calculated as

$$\frac{\partial c}{\partial x} = \frac{\partial}{\partial x}\left(c_o\left(1 - \frac{2}{\sqrt{\pi}}\int_0^\phi e^{-\zeta^2}d\zeta\right)\right) = -\frac{2c_o}{\sqrt{\pi}}e^{-\phi^2}\frac{\partial \phi}{\partial x} = -\frac{c_o}{\sqrt{\pi Dt}}\exp\left(-\frac{x^2}{4Dt}\right)$$

Thus, we find

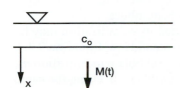

FIGURE 7.5.2   Chemical Leachate Ponded on a Liner

$$J(x = 0,t) = \frac{nD_s c_o}{\sqrt{\pi D_s' t}} = nc_o \sqrt{\frac{RD_s}{\pi t}} \qquad (7.5.28)$$

Using Equation 7.5.28 in Equation 7.5.27 and integrating, we find

$$\frac{M(t)}{A} = 2\, nc_o \sqrt{\frac{RD_s t}{\pi}} \qquad (7.5.29)$$

which shows that the total mass which diffuses into the liner increases as $\sqrt{t}$. These results should be compared with those from Section 4.7.1.

The second question concerns the speed at which the contaminant migrates through the liner. If we measure this speed by the speed of a particular concentration (isochore), say $c = 0.01 c_o$, then the problem is quite easy. Because c is a function only of ϕ, then the speed of a isochore must satisfy $\phi = \phi^*$, where ϕ^* is the argument for that particular isochore. In general, from Equation 7.5.22, we then have

$$x = \phi^* \sqrt{4D_s' t} \qquad (7.5.30)$$

In particular, for $c/c_o = 0.01$, then $\phi^* = 1.8$ (see Appendix G), and $x = 3.6\sqrt{D_s' t}$. For $D_s' = 10^{-4}\ \text{m}^2/\text{d}$, $x = 0.36$ m after 100 days, and $x = 1.54$ m after 5 years.

7.6   Volatilization Losses of Soil Contaminants

Volatilization is the mass transfer of chemical substances from soil to the atmosphere. Contaminants may leave the soil by vaporizing into the soil air and then exiting to the atmosphere by diffusion in soil air to the ground surface or by being carried in soil air as it is forced from the soil by vacuum extraction in wells. Because the mass must enter the atmosphere by traveling in soil air, it is often assumed that gaseous transport must dominate the mass transport process. Figure 7.4.2 suggests that this statement is true, except for when a soil is very wet.

The rate of volatilization is affected by many factors, such as soil properties, chemical properties, and environmental conditions. Its rate is ultimately limited by the chemical vapor concentration that is maintained at the soil surface and by the rate at which this vapor is carried away from the soil surface to the atmosphere. The mechanisms of volatilization are similar to those of evaporation of soil water. The factors that control the rate of volatilization have been studied mostly for pesticides, but there is little that distinguishes pesticides from other organics chemicals. One may assume that the observations based on pesticides are applicable to other organic chemicals (Spencer et al, 1973; Thomas, 1982; Spencer et al, 1982; Jury, 1986; Glotfelty and Schomburg, 1989).

The loss of chemicals from the soil profile due to volatilization may be calculated using the theory of diffusion and Fick's second law, Equation 7.4.7. It is assumed that the soil profile initially has a uniform bulk concentration m_o and that the surface concentration is maintained at $m(0,t) = 0$. Using the same procedures as in Section 7.5.2, one finds that the bulk concentration satisfies

$$m(z,t) = m_i \operatorname{erf}\left(\frac{z}{\sqrt{4D_s t}}\right) \qquad (7.6.1)$$

where erf( ) is the **error function** (see Appendix G). Equation 7.6.1 is essentially the same as Equation 7.5.26, except that the chemical of interest is leaving the medium rather than entering it. Again, one finds that the cumulative mass loss per unit ground surface area is given by

$$\frac{M_{vol}(t)}{A} = 2m_0 \sqrt{\frac{D_{sT} t}{\pi}} \qquad (7.6.2)$$

This model is essentially the model presented by Hamaker (1972), except that it accounts for partitioning and diffusive transport in all of the phases.

Moving Boundary Method

An alternative formulation for computing the volatilization loss of a chemical from soil, which is easier to generalize for some applications, considers a sharp moving boundary separating a contaminated region below from an uncontaminated region of diffusive transport above. Thibodeaux and Huang (1982) refer to the region below the boundary as the **wet zone** and that above the boundary as the **dry zone**. These zones are shown in Figure 7.6.1. The constituent concentration at the ground surface as assumed to be zero, and the volatile flux is proportional to the concentration difference $(m_o - 0)$ across the dry zone. As vaporization occurs at the plane separating the wet and dry zones, the location of the dry zone increases in depth. According to the moving boundary model, the volatilization flux is calculated by $J_{vol} = D_{sT} m_0/y$, where y is the depth to the interface separating the wet and dry zones. Application of the continuity principle leads to

$$m_i \frac{dy}{dt} = J_{vol}$$

The initial condition $y(0) = 0$ gives the result

$$y = \sqrt{2D_{sT} t} \qquad (7.6.3)$$

y = 0

DRY ZONE

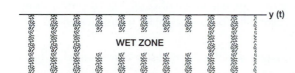

y (t)

WET ZONE

FIGURE 7.6.1 Configuration of wet and dry zone for the moving boundary model

Using this result, the volatilization flux and cumulative volatile loss are

$$J_{vol} = m_0 \sqrt{\frac{D_{sT}}{2t}} \tag{7.6.4}$$

$$\frac{M(t)}{A} = m_0 \sqrt{2D_{sT}t} \tag{7.6.5}$$

A direct comparison of the predictions given by the classical diffusion model of Equation 7.6.2 with the moving boundary model given by Equation 7.6.5 shows that the moving boundary model differs from the classical Fickian diffusion model by a factor of

$$\sqrt{\tfrac{2}{\pi}} \cong .08$$

This variation is much less than the accuracy with which most of the parameters can be measured in the field. The moving boundary model is also simpler.

Atmospheric Boundary Layer

Both the classical Fickian diffusion and the moving boundary models predict an infinite volatilization rate at time zero. The same thing occurs in evaporation models unless one considers that the atmosphere cannot accept an infinite moisture flux. Some sort of resistance must be considered for early times. Jury et al. (1983) suggests that the diffusion resistance through the air viscous sublayer at the soil surface be considered. Both the volatilization flux and the evaporation rate are limited by diffusion through a stagnant air layer of thickness δ above the soil surface, and Jury et al. (1983) suggest that the magnitude of δ be estimated from the evaporation rate. They developed an environmental screening model for pesticides that includes this effect, and they find that it can be important under circumstances where the pesticide is applied as an aqueous liquid, resulting in bulk upward fluid flow following pesticide application due to upward migration of the liquid phase. This condition forces the pesticide to accumulate at the ground surface, and the atmospheric sublayer resistance is necessary in order to obtain reasonable estimates of the environmental fate of the applied chemicals. For conditions where upward bulk-fluid flow is not an issue, however, Charbeneau and Daniel (1993) show that the effects of the atmospheric boundary layer are negligible, and this added complication does not need to be considered.

Problems

7.2.1. Oxygen gas has a molecular diffusion coefficient of $D_m(g) = 0.18$ cm^2/s. Using this value, estimate the molecular diffusion coefficient for a pesticide with a molecular weight of 244 g/mol.

7.2.2. A volatile chemical has a gas-phase molecular diffusion coefficient of $D_m(g) = 0.08$ cm^2/s when the soil air temperature is 20 degrees C. What is its diffusion coefficient if the soil air temperature is raised to 30° C?

7.2.3. A chemical has a liquid-phase molecular diffusion coefficient of $8 \, (10^{-6}) \, cm^2/s$ in water. What is its molecular diffusion coefficient in jet fuel that has a viscosity of 4 cp (centipoise)?

7.3.1. An indurated porous medium has a porosity of 0.15. For a chemical with a aqueous-phase molecular diffusion coefficient of $7(10^{-5}) \, m^2/d$, what is its effective diffusion coefficient (i.e., ratio of the mass flux to concentration gradient) within the water-filled medium?

7.3.2. Through compaction, the porosity of a sandy soil is reduced from 0.4 to 0.3. By what percent is its effective diffusion coefficient reduced?

7.4.1. What is the effective soil diffusion coefficient for diffusion of benzene in the vadose zone of a soil with porosity 0.4 if the soil organic matter fraction is 0.005 and the water content is 0.25?

7.4.2. For the conditions specified in problem 7.4.1, what is the percent change in effective soil diffusion coefficient if the water content is reduced to 0.15?

7.4.3. For diffusion of the gasoline additive MTBE through the vadose zone of a soil with porosity 0.35 and organic matter fraction $f_{oc} = 0.01$, at what volumetric water content does diffusion through the aqueous phase contribute 1 percent to the total diffusion flux? At what water content does it contribute 99 percent of the total diffusion mass flux?

7.5.1. For a point source in $1D$ diffusion, the curve of

$$\frac{nRc(x,t)}{M}$$

may be thought of as a probability distribution for finding an arbitrary solute molecule at a given location at a given time t if the molecule started at $(x = 0, t = 0)$, because the area under this curve is unity. Using this idea, determine the probability for a solute molecule to have moved a linear distance greater than 1 cm within a period of 1 day, and within a period of 1 week, if the molecule is nonsorbing $(R = 1)$ and if $D_s = 10^{-6} \, cm^2/s$. This problem is the classical problem of Brownian motion.

7.5.2. For the conditions of problem 7.5.1, what is the standard deviation of the solute distribution after 1 day? After 1 week?

7.5.3. A point source of mass M $(R = 1)$ is introduced into a soil with $D_s = 10^{-6} \, cm^2/s$. How long will it take before half the mass has diffused beyond a radial distance of 1 cm from the source location? Hint:

$$\int_0^{x_1} x^2 \exp\left(-\frac{x^2}{\alpha}\right) dx = \frac{\alpha}{4}\left(\sqrt{\pi} \, \mathrm{erf}\left(\frac{x_1}{\sqrt{\alpha}}\right) - 2x_1 \exp\left(-\frac{x_1^2}{\alpha}\right)\right)$$

7.5.4. For what value of ξ is the approximation

$$\mathrm{erfc}(\xi) \cong \frac{e^{-\xi^2}}{\xi\sqrt{\pi}}$$

valid to within 5 percent error? To within 1 percent error?

7.5.5. For what value of ξ is the approximation

$$\text{erfc}(\xi) \cong 1 - \frac{2\xi}{\sqrt{\pi}}$$

valid to within 5 percent error? To within 1 percent error?

7.5.6. A contaminated waste stream is retained within a storage pond with a liner consisting of a 1 m thick clay layer. The permeability of the liner and the gradient across it are small enough so that mass transport by advection is considered to be negligible. Assuming $D_s = 10^{-4}$ m^2/d, how long will it take before the 10 percent, 1 percent, 0.1 percent, and 10^{-6} percent concentrations reach across the layer for a nonsorbing chemical species?

7.5.7. A liquid waste is contained in a storage pond with a 20 cm thick compacted clay liner. The concentration of a particular constituent is 1,000 mg/L. This constituent has a liquid molecular diffusion coefficient of 10^{-5} cm^2/s, and it is nonsorbing. The porosity and tortuosity of the liner are estimated to be 0.4 and $^1/_{10}$, respectively. What is the constituent mass diffused into 1 m^2 area of the liner during the first month, and during the first year? How long will it take before a leachate concentration of 10 mg/L reaches the outside of the liner?

7.6.1. An aqueous mixture containing 1 g/L of methylene chloride is released to soil, contaminating the upper 25 cm depth (a total of 5 cm of the chemical mixture infiltrated). The methylene chloride vapor has a molecular diffusion coefficient of 0.11 cm^2/s. The porosity and water content are 0.40 and 0.20, respectively, and the soil organic fraction is 0.005. How long will it take methylene chloride to volatilize from the soil if the temperature is 25 degrees C?

7.6.2. An aqueous mixture containing benzene is released to a soil with conditions specified in problem 7.4.1. How long does it take for the "dry zone" to reach a depth of 10 cm? What is the mass flux at this time if the initial source concentration of benzene is 100 mg/L?

Chapter Eight
Advection-Dispersion Transport and Models

The advection-dispersion equation is the basic relationship that is used to describe mass transport in porous media. This chapter describes the parameters that control mass transport behavior and how they may be measured in the laboratory and field. Applications of the advection-dispersion equation to problems involving chemical spills, contaminant plumes, and mass transport through aquifers are described. Aspects of numerical modeling of solute transport also are presented. The influences of nonideal factors are briefly discussed, and transport through the vadose zone is described.

8.1   Introduction

Advection and diffusion respectively describe mass transport by bulk fluid movement caused by fluid energy gradients and mixing of chemical constituents caused by the presence of concentration gradients and random molecular motions. In this chapter, we will address transport processes associated with bulk fluid movement and mixing. Together, they describe the mass transport of solutes by advection and dispersion, which also includes diffusive transport.

As noted in Section 5.3, the basic transport mechanism associated with bulk movement is advection, the transport of a chemical species as it moves along with the fluid flow. Because of our fundamental inability to describe the advection transport on the pore scale, however, we are forced to describe this transport in an average sense. The transport associated with the average bulk fluid movement is what is called **advective transport**. The transport associated with deviations from the average is called **mechanical dispersion**. On top of these two processes, diffusion still operates—so that the advection-dispersion equation includes the processes of advection, mechanical dispersion, and diffusion. The latter two processes are often grouped as **hydrodynamic dispersion**.

For the present discussion, we consider the transport of a conservative species, or an ideal tracer. An ideal tracer is a substance that moves through a porous medium without interacting with the matrix or undergoing chemical or biotic transformations. Understanding the transport of a conservative substance is the first step toward understanding the fate and transport of hazardous and radioactive chemicals, which may be influenced by a wide range of physical/chemical processes.

From Section 5.7, the transport equation for a conservative species in a flow field without volumetric sources or sinks takes the following form:

$$\frac{\partial m}{\partial t} + \tilde{q} \cdot \text{grad}(c) = \text{div}(n\tilde{\tilde{D}}_{hd} \cdot \text{grad}(c)) \tag{8.1.1}$$

Equation 8.1.1 presents us with the first of the difficulties that we face in describing transport through a porous medium. For a conservative species, the bulk concentration in the first term is $m = n_v c$, where n_v is the **volumetric porosity** (which acts as a capacitance factor). In the second term, the Darcy velocity is related to the seepage velocity through the **effective or kinematic porosity**, which represents the areal cross-section per unit area through which water is free to circulate, $\tilde{q} = n_e\tilde{v}$. In the third term, the **effective areal porosity** is also used in the flux calculation. If we assume that the porous medium has uniform porosity values, then we have the following transport equation:

$$n_v\frac{\partial c}{\partial t} + n_e\tilde{v} \cdot \text{grad}(c) = n_e\text{div}(\tilde{\tilde{D}}_{hd} \cdot \text{grad}(c)) \tag{8.1.2}$$

From Equation 8.1.2, it is apparent that the porosity plays the role of a scaling factor for separate processes in the transport equation. We usually assume, as the first step in a subsurface fate and transport investigation, that

$$n_v = n_e \tag{8.1.3}$$

in which case

$$\frac{\partial c}{\partial t} + \tilde{v} \cdot \text{grad}(c) = \text{div}(\tilde{\tilde{D}}_{hd} \cdot \text{grad}(c)) \tag{8.1.4}$$

This equation is the form of the transport equation that is usually applied.

8.2   One-Dimensional Flow and Column Experiments

The simplest method for estimating the dispersion coefficient of a porous medium, and the method that has been used most often, is through **column experiments**. During these experiments, a chemical tracer moves through a column, generally upward to help in the displacement of entrapped air, and the effluent concentration is measured as a function of time. The problem is modeled mathematically, and the experimental data is fit against the mathematical solution to the problem to find the parameter values that provide the best fit between the theory and observed data.

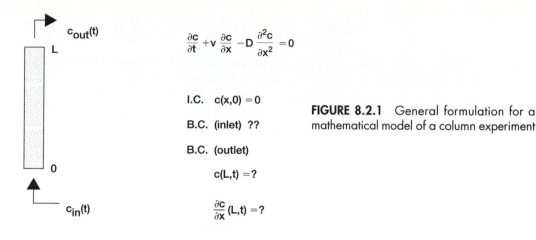

$$\frac{\partial c}{\partial t} + v\,\frac{\partial c}{\partial x} - D\,\frac{\partial^2 c}{\partial x^2} = 0$$

I.C. $c(x,0) = 0$

B.C. (inlet) ??

B.C. (outlet)

$c(L,t) = ?$

$\dfrac{\partial c}{\partial x}(L,t) = ?$

FIGURE 8.2.1 General formulation for a mathematical model of a column experiment

8.2.1 Background Theory

For column experiments with ideal tracers, the transport Equation 8.1.4 simplifies because the flow is one-dimensional. The corresponding form is

$$\frac{\partial c}{\partial t} + v\,\frac{\partial c}{\partial x} = D\,\frac{\partial^2 c}{\partial x^2} \tag{8.2.1}$$

where D is the combined mechanical dispersion and diffusion coefficient. These processes cannot be distinguished in a single, one-dimensional flow experiment. The general formulation for a mathematical model of column experiments is shown in Figure 8.2.1. Initially, the column is assumed to be clear of the tracer chemical. The choice of applicable boundary conditions is difficult. At the inlet end of the column, there are a number of choices. One may assume that the inlet concentration is constant, that a constant inlet flux is maintained, or that there are other alternatives.

Specification of boundary conditions at the outlet end is also difficult. One could assume that the concentration gradient is zero, which would imply that solute mass is carried from the column by advection—with no contribution from diffusion or dispersion. An alternative that is used later is to assume that the column is infinite in length and to specify a vanishing gradient at infinity. The effluent concentration is then given by the concentration predicted at the end of the column at location $x = L$, $c(L,t)$, and both advection and dispersion contribute to the mass flux from the column.

The simplest model is to assume that the inlet concentration is constant. This condition is called a **Dirichlet** or **type-one boundary condition**, and the inlet boundary condition takes the form

$$c(0,t) = c_o \tag{8.2.2}$$

where c_o is the concentration in the source reservoir. For this boundary condition, Ogata and Banks (1961) give the solution

$$\frac{c}{c_o} = \frac{1}{2}\left(\operatorname{erfc}\left(\frac{x - vt}{\sqrt{4Dt}}\right) + \exp\left(\frac{xv}{D}\right)\operatorname{erfc}\left(\frac{x + vt}{\sqrt{4Dt}}\right)\right) \qquad \textbf{(8.2.3)}$$

When xv/D is sufficiently large, the second term of Equation 8.2.3 may be neglected (e.g. with $xv/D > 500$, the error is less than 3 percent).

The second model assumes a constant flux from the source reservoir into the column. The source reservoir is well mixed so that the flux out of the reservoir is due only to advection. This process results in the boundary condition given as

$$qc_o = qc - nD\frac{\partial c}{\partial x}\bigg|_{x=0} \qquad \textbf{(8.2.4)}$$

Because Equation 8.2.4 includes both the concentration and its gradient, it is called a **Cauchy** or **type-three boundary condition**. The problem with this boundary condition has been considered by Lindstrom et al (1967), with the solution given as

$$\frac{c}{c_o} = \frac{1}{2}\left\{\operatorname{erfc}\left(\frac{x - vt}{\sqrt{4Dt}}\right) + \sqrt{\frac{4v^2t}{\pi D}}\exp\left(-\frac{(x - vt)^2}{4Dt}\right) - \left(1 + \frac{vx}{D} + \frac{v^2t}{D}\right)\exp\left(\frac{vx}{D}\right)\operatorname{erfc}\left(\frac{x + vt}{\sqrt{4Dt}}\right)\right\} \qquad \textbf{(8.2.5)}$$

The third model assumes that mass is introduced at the origin at a constant rate—but allows the mass to diffuse upstream as well as downstream. At the same time, mass is being advected downstream. For the present discussion, we will refer to this as a **type-two boundary condition**.[1] The boundary condition takes the form

$$\int_{-\infty}^{\infty} n\, c(x,t)dx = q\, c_o t \qquad \textbf{(8.2.6)}$$

This problem has been analyzed by Sauty (1980), who presents the solution as

$$\frac{c}{c_o} = \frac{1}{2}\left(\operatorname{erfc}\left(\frac{x - vt}{\sqrt{4Dt}}\right) - \exp\left(\frac{xv}{D}\right)\operatorname{erfc}\left(\frac{x + vt}{\sqrt{4Dt}}\right)\right) \qquad \textbf{(8.2.7)}$$

Equation 8.2.7 is the same as Equation 8.2.3, except for the sign of the second term on the right side.

The fourth and last model formulation which we consider is both the simplest and most useful, as we will see later. It may be developed directly as follows: Consider a column that is of infinite length in both directions. Initially, the concentration is $c = c_o$ for $x < 0$ and $c = 0$ for $x > 0$. A barrier at $x = 0$ keeps these solutions from mixing. This set-up is shown schematically in Figure 8.2.2. At time zero, the barrier is removed and a velocity v is established throughout

---

[1] A type-two or Neumann boundary condition refers to a specified flux or normal gradient of a potential.

the column. The transport equation is given by Equation 8.2.1 and we consider a frame of reference that moves with the velocity v. In this new primed reference system we have $x' = x - vt$ and $t' = t$, and in terms of this moving coordinate system, Equation 8.2.1 takes the form

$$\frac{\partial c}{\partial t'} = D\,\frac{\partial^2 c}{\partial x'^2} \tag{8.2.8}$$

Equation 8.2.8 says that from the point of view of our moving coordinate system, the problem appears to be one of diffusion, and by symmetry, the concentration at $x' = 0$ will remain at $c = c_o/2$. But this situation is precisely the problem solved in Section 7.5.2, and the solution is given by

$$c(x',t') = \frac{c_o}{2}\,\mathrm{erfc}\!\left(\frac{x'}{\sqrt{4Dt'}}\right)$$

Transforming back to the original (x,t) coordinate system, we obtain

$$\frac{c}{c_o} = \frac{1}{2}\,\mathrm{erfc}\!\left(\frac{x - vt}{\sqrt{4Dt}}\right) \tag{8.2.9}$$

For the present discussion, Equation 8.2.9 is referred to as the "simple model." One recognizes that Equation 8.2.9 is one component of each of the other solutions. The effective inlet condition is also shown in Figure 8.2.2. For early times, the concentration drops below c_o at the origin, violating Equation 8.2.2. After a short time, however, the entire concentration front has moved forward a sufficient distance for the concentration to return to c_o at $x = 0$. The violation is apparent only for early times and near the origin. The net effect, however, is that less mass enters the column for solution 8.2.9 than for Equation 8.2.3.

Equations 8.2.9, 8.2.3, 8.2.7, and 8.2.5 are all solutions to Equation 8.2.1 for different boundary conditions. The difference between the solutions is most easily seen by examining the predicted concentration at the origin ($x = 0$) as a

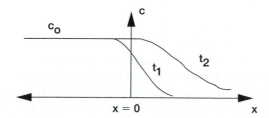

FIGURE 8.2.2 Origin of approximate model specified by Equation 8.2.9

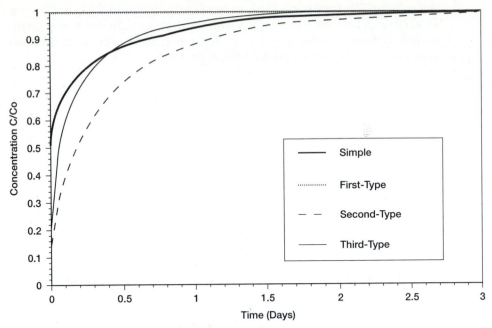

FIGURE 8.2.3 Inlet concentrations predicted by column experiment models for conditions with $v = 0.5$ m/d and $D = 0.05$ m²/d

function of time. These four solutions, called the simple, type-one, type-two, and type-three results, give (respectively)

$$c(0,t) = \frac{c_o}{2}\left\{1 + \operatorname{erf}\left(\frac{v^2 t}{4D}\right)\right\} \tag{8.2.10}$$

$$c(0,t) = c_o \tag{8.2.11}$$

$$c(0,t) = c_o \operatorname{erf}\left(\sqrt{\frac{v^2 t}{4D}}\right) \tag{8.2.12}$$

$$c(0,t) = c_o\left\{\operatorname{erf}\left(\sqrt{\frac{v^2 t}{4D}}\right) + \sqrt{\frac{v^2 t}{\pi D}}\exp\left(-\frac{v^2 t}{4D}\right) - \frac{v^2 t}{2D}\operatorname{erfc}\left(\sqrt{\frac{v^2 t}{4D}}\right)\right\} \tag{8.2.13}$$

Figure 8.2.3 shows the inlet concentrations predicted by Equations 8.2.10–8.2.13 for a case with $v = 0.5$ m/d, and $D = 0.05$ m²/d. For the simple model, the concentration starts at $c_o/2$ and increases to c_o. The type-one model has a concentration with starts and stays at c_o. For both the type-two and type-three models, the concentrations start at zero and eventually increase to c_o. Both models correspond to a constant flux, although the type-three model only allows the flux into the positive x-domain, and the concentration at the origin increases more rapidly. The comparison between the simple and type-three models is fairly close.

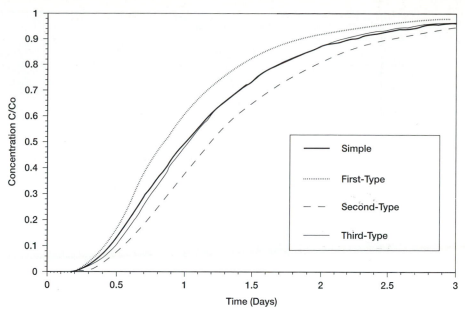

FIGURE 8.2.4 Effluent concentrations predicted by column experiment models for a case with L= 0.5 m, v= 0.5 m/d, and D = 0.05 m²/d

Figure 8.2.4 shows the predicted effluent concentrations from the four models for a column that is 0.5 m in length. The type-one model has the earliest breakthrough and highest concentrations for earlier times. This situation corresponds to its higher inlet concentrations, so more mass enters the column more rapidly. Predictions from the simple and type-three models are close, as might be expected from Figure 8.2.3—where they are shown to have similar inlet concentrations. The type-two model has the lowest concentrations, corresponding to its allowance of mass transport upstream of the inlet.

For comparison, Figure 8.2.5 shows the breakthrough curves for a column with L = 0.5 m, v = 0.5 m/d, and D = 0.005 m²/d. The predicted effluent concentrations from the four models are similar.

The only variables that appear in this problem are L, v, and D, and their dimensionless grouping is Lv/D. This grouping must characterize the breakthrough behavior predicted from the four models. Figures 8.2.4 and 8.2.5 suggest that for larger values of this dimensionless grouping, the difference between the models is less. Normal column experiments in the laboratory are characterized by a large magnitude of this group. For usual laboratory experiments, the four models give essentially the same result. Given this situation, one must certainly prefer to use the simpler model, and Equation 8.2.9 will be adopted as the basic model for analysis of one-dimensional column breakthrough. It is noted, however, that for larger-scale field problems that may be represented by one-dimensional models, the magnitude of the dimensionless group can be much less—and the choice of model becomes important.

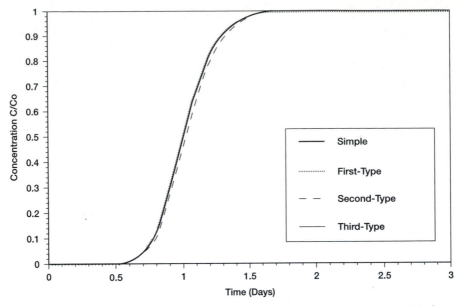

FIGURE 8.2.5 Effluent concentrations predicted by column experiment models for a case with L = 0.5 m, v = 0.5 m/d, and D = 0.005 m²/d

The previous discussion has shown that for large magnitudes of the dimensionless grouping Lv/D, the results predicted by all four models are nearly the same, and one might just as well use the simple Equation 8.2.9 for analysis. Parker and Van Genuchten (1984), who note the distinction between volume-average and flux-average concentrations, highlight the distinction between the various one-dimensional solutions. The volume-average concentrations are those obtained by averaging the pore water fluid concentration over an appropriate volume or REV (see Chapter 1). Flux-average concentrations are defined so that the flux past a given location is calculated from

$$J = q\,C_f$$

instead of from the usual advection-dispersion flux given by

$$J = q\,c - n\,D\frac{\partial c}{\partial x}$$

where c denotes the volume-average concentration. The flux-average concentration may be more appropriate for interpretation of laboratory and field data under conditions where the flow may by pass part of the medium (preferential flow), and such conditions are to be accounted for through the dispersion term of the transport equation. For example, if flow occurs through a series of fractures, or if breakthrough is measured from a soil column using a large pan lysimeter, interpretation of the resulting concentration might use the flux-average concentration. Parker and van Genuchten argue that Equation 8.2.3 is appropriate for interpretation of flux-averaged concentrations, while Equation 8.2.5 is appropriate for volume-averaged concentrations. They also discuss the application of these concepts to laboratory and field studies. For typical un-consolidated media where the dispersion term is small, the difference between the predicted concentrations is also small, and the distinction between flux- and volume-average concentrations is unnecessary.

Analysis of laboratory data is most easily carried out if the problem of transport through a packed column is cast in dimensionless form. If the column is of length L, then L provides a characteristic length scale, and L/v provides a characteristic time scale—the mean travel time through the column. One may multiply the transport equation by L/v to find

$$\frac{\partial c}{\partial T} + \frac{\partial c}{\partial X} = \frac{1}{P_L}\frac{\partial^2 c}{\partial X^2} \; ; \; X = \frac{x}{L} \; ; \; T = \frac{tv}{L} \; ; \; P_L = \frac{Lv}{D} \qquad \textbf{(8.2.14)}$$

The dimensionless number P_L is called the **Peclet number**. It is a measure of the relative importance of advection to dispersion (mixing) as transport processes:

$$P_L = \frac{\text{advection}}{\text{mixing}} \qquad \textbf{(8.2.16)}$$

For large Peclet numbers, the transport equation becomes

$$\frac{\partial c}{\partial T} + \frac{\partial c}{\partial X} = 0$$

which is a 'hyperbolic' *partial differential equation* (PDE) describing advection transport. For small Peclet numbers, the transport equation becomes

$$\frac{\partial C}{\partial T} = \frac{1}{P_L}\frac{\partial^2 C}{\partial X^2}$$

which is the diffusion equation, a 'parabolic' PDE describing diffusive transport.

The major characteristic that appears from this discussion is that the transport equation contains both hyperbolic and parabolic features (although fundamentally it remains a parabolic PDE). This concept is extremely important when we look toward numerical solutions for the transport equation. There are a large number of techniques that work well for parabolic PDEs and a fair number of techniques that work well for hyperbolic PDEs. It has been more difficult, however, to develop techniques that work well for PDEs that exhibit both types of behavior. This topic will be discussed later.

The **Peclet number** presented above was scaled upon the length of the column. Physically, a better scaling parameter might be the mean grain size, d, with Peclet number P_d. Another alternative is to scale length using the permeability, $\sqrt{k}$, with Peclet number $P_{\sqrt{k}}$. Other length characteristics suggest themselves upon occasion.

Equation 8.2.9 may be written in dimensionless form. In such a form, breakthrough occurs at $X = 1$, and for the breakthrough concentration, Equation 8.2.9 becomes

$$\frac{c(X = 1, T)}{c_o} = \frac{1}{2} \, \text{erfc}\left((1 - T)\sqrt{\frac{P_L}{4T}}\right) \tag{8.2.17}$$

Following Brigham et al. (1961) and Brigham (1974), we introduce the **scaled-time parameter**

$$\zeta = \frac{T - 1}{\sqrt{2T}} \tag{8.2.18}$$

and then Equation 8.2.17 may be written as follows:

$$\frac{c(X = 1, \zeta)}{c_o} = \frac{1}{2} \, \text{erfc}\left(-\zeta\sqrt{\frac{P_L}{2}}\right) \tag{8.2.19}$$

Using the relationship between the complementary error function and the **cumulative normal distribution**, that is $\text{erfc}(X) = 2N(-\sqrt{2}X)$, Equation 8.2.19 may also be written

$$\frac{c(X = 1, \zeta)}{c_o} = N(\zeta\sqrt{P_L}) \tag{8.2.20}$$

Use of the cumulative normal distribution is helpful for analysis of experimental data.

8.2.2   Experimental Results

The developments presented in Section 8.2.1 set the stage for looking at the results from column experiments. If D is a constant for a given medium and solute, then the breakthrough curve for experiments with differing velocities should plot differently. For example, according to Equation 8.2.16, larger values of P_L imply increased importance of advection versus mixing processes, and the resulting breakthrough curve will be narrower. In addition, the curves should be offset when plotted against time, as shown in Figure 8.2.6.

Laboratory studies have shown, however, that the amount of spreading appears to be independent of the velocity, but it is a function of the length of the column. Furthermore, when plotted against ζ from Equation 8.2.18, all of the breakthrough curves collapsed upon each other, as shown in Figure 8.2.7.

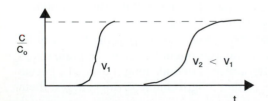

FIGURE 8.2.6   Expected breakthrough curves if D is a constant independent of v

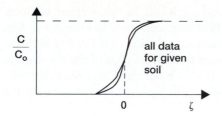

FIGURE 8.2.7 Observed breakthrough curves for "dispersion" transport

The conclusion to be reached from these observations is that the breakthrough must be independent of the flow rate. This idea means that the Peclet number remains constant even when the velocity changes. Thus,

$$\frac{D}{v} = \frac{L}{P_L} = a_L = \text{characteristic for soil}$$

Thus, the model suggested is

$$D = a_L v \tag{8.2.21}$$

where a_L is the **longitudinal dispersivity**. This model is the accepted model for longitudinal mechanical dispersion at the laboratory scale. Measured laboratory values of a_L range from 0.01 to 0.1 cm, which is on the order of the pore size. Furthermore, the spread or variance of the breakthrough curve is given by Einstein's relationship from his studies of Brownian motion (see Section 7.5.1 and 8.6) as follows:

$$\sigma^2 = 2Dt \tag{8.2.22}$$

Because the midpoint on the breakthrough curve arrives at $t = L/v$, we have

$$\sigma^2 = 2a_L L \tag{8.2.23}$$

showing that the spread is a function of the length of the column but not of the flow rate.

The general model that includes the effects of both diffusion and mechanical dispersion is

$$D = \tau D_m + a_L v \tag{8.2.24}$$

where D is now called the **hydrodynamic dispersion coefficient** (Bear, 1972) and τ and D_m are the tortuosity and molecular diffusion coefficient.

8.2.3   Analysis of Experimental Data

This section looks at methods for estimating D or a_L from experimental data. The best approach is to plot the entire data set and the theoretical breakthrough curve for a range of values of D and choose the value that gives the best overall fit. In some cases, we may look for simpler methods of analysis. One method is to use the slope of the breakthrough curve near $c/c_o = 0.5$. This approach will be outlined later when field methods are discussed. In this section, we look at a method which uses two points on the breakthrough curve to estimate a_L. Equations 8.2.17 and 8.2.20 provide approximate solutions for the breakthrough curve. The method of analysis is developed from properties of the standard normal distribution and Equation 8.2.20. If Z is a standard unit normal random variable, then increments of ΔZ are related to those of the scaled-time parameter ζ through

$$\Delta Z = \sqrt{P_L}\Delta\zeta$$

Using the definition of the Peclet number, this equation gives

$$D = vL\frac{(\Delta\zeta)^2}{(\Delta Z)^2} \qquad \text{(8.2.25)}$$

A simple form of this result comes when one takes the points one standard deviation above and below the mean, $c/c_o = 0.5$. This equation gives $Z = -1 \rightarrow N(Z) = 0.16$ and $Z = 1 \rightarrow N(Z) = 0.84$. On the breakthrough curve, these two points correspond to $c/c_o = 0.16$ and $c/c_o = 0.84$ with corresponding ζ values, $\zeta_{.16}$ and $\zeta_{.84}$. Thus,

$$D = \frac{vL}{4}(\zeta_{.84} - \zeta_{.16})^2 \qquad \text{(8.2.26)}$$

D is the hydrodynamic dispersion coefficient for one-dimensional flow and represents the sum of mechanical dispersion plus molecular diffusion (Equation 8.2.24), and thus

$$a_L \equiv \frac{D - \tau D_m}{v} \qquad \text{(8.2.27)}$$

EXAMPLE PROBLEM   **8.2.1   Estimation of Longitudinal Dispersivities (Modified from Pickens and Grisak, 1981)**

The following data is obtained from a laboratory column study of mixing in a porous medium using chlorides as a tracer: $L = 30$ cm, $v = 9.26(10^{-4})$ cm/s, and $t = 8.526$ hrs when $c/c_o = 0.16$ and $t = 9.499$ hrs when $c/c_o = 0.84$. Estimate the dispersivity for this data.

This data gives $T_{.16} = 0.9474$ and $\zeta_{.16} = -0.0382$ and $T_{.84} = 1.0555$ and $\zeta_{.84} = 0.0382$. Thus, $\zeta_{.84} - \zeta_{.16} = 0.0764$, and with Equation 8.2.26, this data gives $D = 4.1(10^{-5})$ cm²/s. For chlorides $\tau D_m \cong 1(10^{-5})$ cm²/s, so Equation 8.2.27 provides

$$a_L = \frac{4.1 \times 10^{-5} - 1 \times 10^{-5}}{9.26 \times 10^{-4}} = 0.033 \text{ cm}$$

This value represents a typical magnitude for a_L from other laboratory studies as well.

8.2.4   Diffusion versus Dispersion ?

It is of interest to consider the relative importance of diffusion and mechanical dispersion in mass transport through porous media. The data from Example 8.2.1 shows that mechanical dispersion and molecular diffusion are of the same magnitude for velocities

$$v = \frac{\tau D_m}{a_L} = \frac{10^{-5}}{0.033} = 3 \times 10^{-4} \text{ cm/s} = 26 \text{ cm/d}$$

Such a velocity is small but not unreasonable for laboratory studies of transport through alluvial materials, although it is quite large for flow through clay materials. This idea suggests that molecular diffusion processes dominate transport in clays, while mechanical dispersion becomes important for transport in coarser materials—at least, in the laboratory. If we carry the same arguments to the field with $a_L \cong 1$ m, then we find $v = 10^{-5}/100 = 10^{-7}$ cm/s = 3 cm/yr. This velocity is a small velocity, and it suggests why molecular diffusion is almost never considered for field simulations, except under conditions of flow in deep hydrogeologic basins where a velocity of 3 cm/yr might actually be large.

The results of numerous experiments by various authors for granular (unconsolidated) porous media have been collected and analyzed by Pfannkuch (1963; also see Bear, 1972) in terms of the Peclet number

$$P_d = \frac{vd}{D_m} \tag{8.2.28}$$

where d is the mean or effective grain diameter. For small $P_d < 0.4$, the transport process is pure molecular diffusion. For $0.4 < P_d < 6$, the effect of molecular diffusion is of the same order of magnitude as that of mechanical dispersion. For $6 < P_d < 250$ the main spreading is caused by mechanical dispersion combined with transversal molecular diffusion across the pores (rather than along the length of the pores). For $250 < P_d < 18,000$, the mixing process is dominated by mechanical dispersion. Within this region, we have

$$\frac{D}{D_m} \cong 1.8 P_d$$

or along with Equation 8.2.28, $D \cong 1.8$ dv, which suggests that $a_L \cong 1.8$ d. Fried and Combarnous (1971) suggest

$$a_L = (1.8 \pm 0.4)d \tag{8.2.29}$$

which gives the order of magnitude of dispersivity for such media. For still larger P_d, the mixing is still dominated by mechanical dispersion, although the flow lies beyond the range of Darcy's Law and inertial effects are important.

8.3   Radial Flow from a Well

We now turn from estimation of dispersivities in the laboratory to field methods of estimation. There are a number of types of field tests that can be carried out. These include natural gradient tracer tests, single well injection tests, single well injection-production tests, and multiple well tracer tests. During a natural gradient tracer test, a tracer is introduced into the aquifer, and its transport with the natural groundwater flow is monitored through measurements at down-gradient observation wells. Natural gradient tests generally require long time periods because of the slow groundwater flow associated with most aquifers. During a single well injection test, a tracer is injected into a well, and its breakthrough is monitored in an observation well located near the injection well. During a single well injection-production test, the tracer is first injected into the aquifer and is then produced from the same well with the effluent concentration being monitored. For a multiple well tracer test, the tracer is introduced at one well and is produced from one or a number of other wells. For all these tests, a mathematical model of the flow system is developed, and the parameters are adjusted so that the best fit is developed between the model predictions and observed concentrations. We first consider single well tracer tests.

For the single well injection test, the dispersivity is estimated from the width or variance of the displacement front as it moves past a monitoring well. The discussion follows de Jong (see Bear, 1972, p 637–38). The model is based on two assumptions. The first of these concerns the basic form of the solution, while the second concerns the mechanisms of spreading of a concentration front in radial flow.

We assume that the tracer distribution satisfies our approximate error function solution given by Equation 8.2.9 for linear flow:

$$\frac{c(x,t)}{c_o} = \frac{1}{2}\,\mathrm{erfc}\!\left(\frac{x - vt}{\sqrt{4Dt}}\right)$$

Equation 8.2.22, however, shows that the variance of the front is $\sigma^2 = 2Dt$, and the mean front displacement is vt. To make this model appropriate for radial flow, we *let R be the mean displacement of the front* and write

$$\frac{c(r,t)}{c_o} = \frac{1}{2}\,\mathrm{erfc}\!\left(\frac{r - R}{\sqrt{2\sigma^2}}\right) = N\!\left(\frac{R - r}{\sigma}\right) \tag{8.3.1}$$

where r is the radial distance from the well and $N(\ )$ is the **cumulative normal distribution** function. If the tracer spreads radially within a thickness b of the aquifer, then continuity gives $Q\,t = \pi\,\mathrm{n}\,R^2\,b$, or

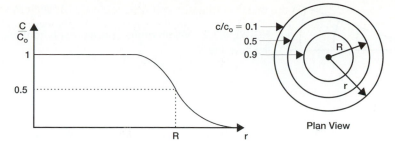

FIGURE 8.3.1 Approximate concentration distribution in radial flow from an injection well

$$R = \sqrt{\frac{Q t}{\pi n b}} \qquad (8.3.2)$$

Because erfc(0) = 1, the concentration at a radius $r = R$ is $c/c_o = 0.5$. The concentration distribution is sketched in Figure 8.3.1.

The second assumption is that mixing of the tracer is a linear sum of two effects: 1) longitudinal dispersion and 2) divergence of streamlines,

$$d\sigma = d\sigma_1 + d\sigma_2 \qquad (8.3.3)$$

where σ_1 refers to longitudinal dispersion and σ_2 to divergence of the streamlines. For longitudinal dispersion with $D = a_L v$, we have

$$\sigma_1 = \sqrt{2Dt} = \sqrt{2a_L R} \qquad (8.3.4)$$

where for radial flow, we have

$$R = \int_0^t v(r = R,t)dt$$

with $v(r = R,t)$ representing the velocity at the mean position of the tracer front at the given time. Differentiation of Equation 8.3.4 gives

$$\frac{d\sigma_1}{dR} = \sqrt{\frac{a_L}{2R}} = \frac{a_L}{\sigma_1} \qquad (8.3.5)$$

Now consider the effect of divergence of flow lines. We may take σ_2 as a measure of the width of the concentration front, as shown in Figure 8.3.2. The shaded regions in this figure show the volume of fluid at different times as it is displaced from the well. Considering only divergence of the stream tube, the volumes are the same, so

$$2\pi R\sigma_2 = \text{constant}$$

and

$$\frac{d\sigma_2}{dR} = -\frac{\sigma_2}{R} \qquad (8.3.6)$$

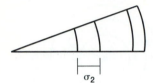

FIGURE 8.3.2 Width of the concentration front due to divergence of the flow lines

Combining Equations 8.3.3, 8.3.5, and 8.3.6, we find that

$$\frac{d\sigma^2}{dR} = 2\left(a_L - \frac{\sigma^2}{R}\right) \tag{8.3.7}$$

Equation 8.3.7 is a nonhomogeneous ODE for the variance of the front. The complementary and particular solutions are A/R^2 and $2a_L R/3$ respectively. With the initial condition $\sigma^2 = 0$ at $R = r_w$, where r_w is the radius of the well, this condition gives

$$\sigma^2 = \frac{2}{3} a_L \left(R - \frac{r_w^3}{R^2}\right) \tag{8.3.8}$$

For $R \gg r_w$, Equation 8.3.8 reduces to

$$\sigma^2 = \frac{2}{3} a_L R \tag{8.3.9}$$

Combining Equations 8.3.1, 8.3.2, and 8.3.9 gives the concentration distribution as

$$F(r,t) = \frac{c(r,t)}{c_o} = \frac{1}{2} \operatorname{erfc}\left(\frac{r - \sqrt{\frac{Qt}{\pi n b}}}{\sqrt{\frac{4}{3} a_L \sqrt{\frac{Qt}{\pi n b}}}}\right) \tag{8.3.10}$$

where $F(r,t)$ is the fractional breakthrough function which rises from 0 to 1.

Equation 8.3.10 is used to estimate the longitudinal dispersivity from the breakthrough curve measured at an observation well located at $r = r_{ow}$. One of the least-sensitive statistics of the curve is its slope at the mean displacement time $t = t_m$ (where $R = r_{ow}$). From Figure 8.3.3, the slope at $t = t_m$ is found from

$$\frac{dF(r_{ow}, t_m)}{dt} = \frac{\Delta F}{\Delta t} = \frac{1}{\Delta t}$$

At $r = r_{ow}$, F changes as R changes, and

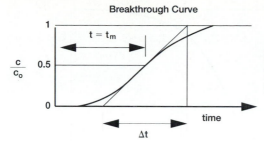

FIGURE 8.3.3 Estimation of the longitudinal dispersivity from the breakthrough curve at an observation well in radial flow

$$\frac{dF}{dt} = \frac{dF}{dR}\frac{dR}{dt}$$

At $t = t_m$, however, R also equals r_{ow}, so that dF/dR is evaluated at $r = R$.

$$\frac{dF}{dR} = \frac{1}{\sqrt{\frac{4\pi}{3}a_L R}} \; ;$$

$$\frac{dR}{dt} = \sqrt{\frac{Q}{4\pi n b t}} = \frac{R}{2t_m} = \frac{r_{ow}}{2 t_m} \rightarrow \frac{dF}{dt} = \frac{1}{\Delta t} = \frac{r_{ow}}{2t_m \sqrt{\frac{4\pi}{3}a_L r_{ow}}}$$

These calculations give

$$a_L = \frac{3\, r_{ow}}{16\,\pi}\left(\frac{\Delta t}{t_m}\right)^2 \tag{8.3.11}$$

Equation 8.3.11 is the desired result. One needs the distance from the injection well to the observation well, an estimate of the mean travel time to the well (from the point where $c/c_o = 0.5$), and an estimate of the slope at the midpoint (which is the same as $1/\Delta t$).

EXAMPLE
PROBLEM

8.3.1 Field Breakthrough Curves in Radial Flow Experiments

A field experiment was performed with an injection rate of 4 m³/d, and measurements were taken in a monitoring well located 3 m from the injection well. The injection interval was 4 m thick with a porosity of 0.3. When the entire interval was packed off, the estimated longitudinal dispersivity was 0.1 m. When an interval with the same hydraulic conductivity as the mean for the entire interval was isolated, the estimated dispersivity was 0.03 m. Compare these two breakthrough curves.

For these conditions, the mean breakthrough time occurs when $R = r_{ow}$, which is at $t = 8.48$ days. The breakthrough curves corresponding to the two dispersivities are shown in Figure 8.3.4. As would be expected, the two curves cross at a time of 8.48 days. Note the greater width for the curve, however, with

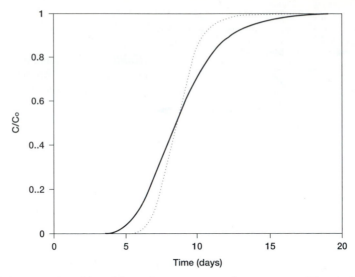

FIGURE 8.3.4 Predicted breakthrough curves at an observation well located 3 m from the injection well with dispersivities of 0.1 m (solid line) and 0.03 m (dashed line)

$a_L = 0.1$ m. This equation corresponds to the rapid movement through more permeable layers and longer time needed to displace the tracer from the low permeability layers that lie within the interval.

Equation 8.3.10 may be used to estimate the concentration at an observation well during an injection test. A more general development of the effects of longitudinal dispersion on groundwater transport is presented by Gelhar and Collins (1971), who apply methods of singular perturbation theory from boundary layer theory of fluid mechanics. Comparison with this more general solution, as well as with numerical models, shows that Equation 8.3.10 works quite well. Gelhar and Collins (1971) also present a model for the concentration at a well during an injection-production (push-pull) test. These tests are also useful for estimating the dispersivity in the field. The result presented by Gelhar and Collins (1971) for the well effluent concentration during the production phase of a test is:

$$\frac{c}{c_o} = \frac{1}{2}\,\mathrm{erfc}\left\{ \frac{\dfrac{V_p}{V_i} - 1}{\sqrt{\dfrac{16}{3}\left(\dfrac{a_L}{R_m}\right)\left[2 - \left|1 - \dfrac{V_p}{V_i}\right|^{1/2}\left(1 - \dfrac{V_p}{V_i}\right)\right]}} \right\} \qquad \textbf{(8.3.12)}$$

where

V_p = Cumulative volume of water produced at time t

V_i = Total volume of water injected during the injection phase

R_m = Average frontal position of the injected water at the end of the injection period, which is defined by

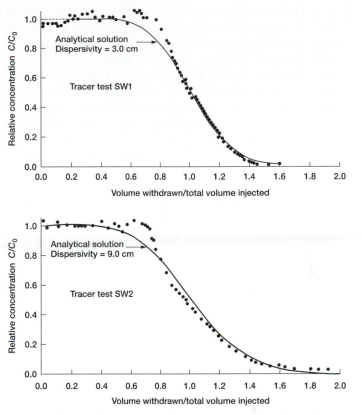

FIGURE 8.3.5 Single well breakthough data from an injection-production test (from Pickens and Grisak, 1981)

$$R_m = \sqrt{\frac{V_i}{\pi\, n\, b}}$$

To use Equation 8.3.12 to estimate the field dispersivity, the observed data is plotted with c/c_o versus V_p/V_i. One then plots Equation 8.3.12 for various values of a_L and chooses the value that provides the best fit between the observed data and the model.

For example, Figure 8.3.5 shows the results from two field tests reported by Pickens and Grisak (1981). The aquifer has a thickness of about 8.2 m, and its porosity is estimated to be 0.38. During the first test (SW1), a volume 95.6 m³ of a radioactive tracer (^{131}I) was injected over a period of 1.25 days at a rate of 0.886 L/s. The tracer was then produced at the same rate for a period of 2.0 days. For this first test, the average radial front position at the end of the injection phase was 3.13 m, and the best-fit dispersivity was found to be $a_L = 3$ cm. During the second test (SW2), a volume of 244 m³ was injected over a period of 3.93 days, resulting in an average radial front position at the end of the injection phase of 4.99 m. Analysis of the production data using Equation 8.3.12 gives a best-fit dispersivity of $a_L = 9$ cm.

During these same tracer tests, observation wells were located at radial distances of 1, 2, 3, and 4 m from the injection/production well and were equipped with multi-level samplers, through which one could record the breakthrough concentrations at different depths. Analysis of this breakthrough data shows that the mean travel time to the observation wells varies considerably with depth. These data can be used to estimate the layer hydraulic conductivity based on the assumption that the mean arrival time to a given radius is in an inverse proportion to the layer hydraulic conductivity and the square of its distance from the injection well. Analysis of breakthrough data for each layer at its multi-level sampler yields dispersivities that generally range from about 0.3 to 0.5 cm, with some trend towards increasing dispersivities with increasing radial distance. These dispersivity values are not much larger than those typically found from laboratory studies. On the other hand, the dispersivity values estimated from the injection-production test that samples all layers at once yields much larger dispersivity values. This disparity suggests that for field problems where one is interested in transport within the aquifer as a whole, dispersivity values must reflect the variable hydraulic properties of the different aquifer strata.

8.4   Transverse Dispersion

Experiments to investigate transverse dispersion are much more difficult to set up and carry out than column experiments for longitudinal dispersion. Some experimental results for granular material were collected and reported by List and Brooks (1967). They note that the results from these studies are not as clear and reproducible as results from laboratory studies of longitudinal dispersion.

One approach to analyze transverse dispersion is to set up a sand box with uni-directional flow and introduce a tracer at a port at one end and let steady-state conditions be achieved. This procedure is shown in Figure 8.4.1. Because the field is steady, and because transverse gradients greatly exceed longitudinal gradients, the transport equation may be approximated by

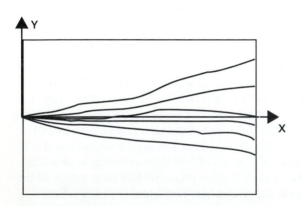

FIGURE 8.4.1   Laboratory set-up fro study of transverse mixing in a porous medium

$$v \frac{\partial c}{\partial x} = D \frac{\partial^2 c}{\partial y^2} \tag{8.4.1}$$

where D is now the transverse dispersion coefficient. The boundary condition is

$$c(0,y) = M' \delta(y) \tag{8.4.2}$$

where M' is the "mass release rate" at $x = 0$. Equations 8.4.1 and 8.4.2 again lead us to the problem considered in Section 7.5.1. Comparison with that section shows that the solution is

$$c(x,y) = \frac{M'}{\sqrt{4\pi \frac{D}{v} x}} \exp\left(-\frac{y^2}{4\frac{D}{v}x} \right) \tag{8.4.3}$$

See Section 8.8.2 for a dimensionally consistent formulation of this model.

Harleman and Rumer (1973) consider a similar problem with Equation 8.4.1 and boundary conditions given by $c(0,y) = c_o$ for $-\infty < y < 0$ and $c(0,y) = 0$ for $0 < y < \infty$. The solution for this problem is given by

$$\frac{c(x,y)}{c_o} = \frac{1}{2} \operatorname{erfc}\left(\frac{y}{\sqrt{4\frac{D}{v}x}} \right) \tag{8.4.4}$$

Both Equations 8.4.3 and 8.4.4 may be used in the laboratory to estimate the transverse dispersion coefficient.

Laboratory experiments show that both Equations 8.4.3 and 8.4.4 are satisfied under most conditions. The data presented by List and Brooks (1967) shows that for $P_d < 1$, molecular diffusion dominates, and the apparent magnitude of D is inversely proportional to P_d. Within the approximate range $1 < P_d < 100$, both molecular diffusion and mechanical dispersion are important. For $P_d > 100$ (or so), the data suggests that the concentration distribution is independent of the flow rate, which suggests in turn that

$$\frac{D}{v} = a_T = \text{constant} \tag{8.4.5}$$

where a_T is the **transverse dispersivity**. In the laboratory, transverse dispersivities are measured to be 5 to 100 times smaller than longitudinal dispersivities.

$$a_T \cong 0.04d \tag{8.4.6}$$

When compared with Equation 8.2.29, we see that for laboratory conditions with homogeneous media, transverse mixing is much less than longitudinal mixing—at least, for large magnitudes of P_d.

8.5   Mechanical Dispersion Tensor

At the same time that laboratory investigations were providing empirical information as to the nature of dispersion in porous media, a number of researchers were involved with developing an appropriate theory. Scheidegger (1954) presented a statistical random walk model that suggested that dispersion was isotropic. De Jong (1958) presented a random walk model which includes directional characteristics for the channels and the mean direction of flow. The resulting model suggested that dispersive mixing would be greater in the direction of flow than transverse to that direction. Bear (1961) introduced the hypothesis that the only significant components of the velocity are either parallel or normal to the mean flow direction. This concept leads to the following dispersion model,

$$D_{ij} = a_{ijkl} \frac{v_k v_1}{|v|} \tag{8.5.1}$$

where D_{ij} is a component of the second-order dispersion tensor, a_{ijkl} is a fourth-order dispersivity tensor with 81 components, and v_k and v_l are components of the velocity vector. Because of certain symmetries, the tensor a_{ijkl} actually has only 36 (or 21) non-zero components. Scheidegger (1961), who went on to show that for an isotropic medium there are only two characteristic constants, discussed these characteristics. Scheidegger's two characteristic constants are our longitudinal and transverse dispersivities introduced earlier.

We will develop Scheidegger's model for an isotropic medium using much simpler arguments. The basic premise is that for uniform flow fields the dispersion model must reduce to

$$D_L = a_L v \tag{8.5.2}$$

$$D_T = a_T v \tag{8.5.3}$$

for the longitudinal and transverse components, respectively.

To start with, we recognize that the formulation must be in terms of a tensor, because the underlying process is independent of the coordinate system, and such invariance with respect to transformation of coordinate system is the basis of tensor properties. In addition, one can argue that the tensor must be of even order (a zero-order tensor is a scalar, and we have already seen that a scalar is the appropriate description for diffusion which is isotropic and independent of direction). Finally, we restrict ourselves to media whose physical characteristics are homogeneous and isotropic. The general form of a second-order tensor that is proportional to the velocity is

$$D_{ij} = \alpha \, v \, \delta_{ij} + \beta \frac{v_i v_j}{v} + \gamma \, v_k \, \varepsilon_{ijk} \tag{8.5.4}$$

where α, β, and γ are parameters to be determined, δ_{ij} is the **Kronecker unit tensor** (or Kronecker delta function; $\delta_{ij} = 1$ if $i = j$ and 0 otherwise) and ε_{ijk} is the

alternating tensor ($\varepsilon_{ijk} = 1$ if i,j,k is an even permutation of x,y,z; $= -1$ for an odd permutation; and $= 0$ otherwise).

To determine the parameters α, β, and γ in Equation 8.5.4, we note that because the field is isotropic, the tensor must be symmetric so that

$$D_{ij} = D_{ji}$$

which implies that $\gamma = 0$ (because $\varepsilon_{ijk} = -\varepsilon_{ijk}$). Now consider a flow field with a uniform velocity in the x-direction, so that $v = v_x$ and $v_y = v_z = 0$. For this case, with Equation 8.5.3 and using Equation 8.5.4, we must have

$$D_{yy} = a_T v = \alpha v + \beta \frac{0 \times 0}{v}$$

so that $\alpha = a_T$. With Equation 8.5.2 and using Equation 8.5.4, we find that

$$D_{xx} = a_L v = a_T v + \beta \frac{v \times v}{v}$$

Thus, $\beta = a_L - a_T$, and we have the general result

$$D_{ij} = a_T v \delta_{ij} + (a_L - a_T)\frac{v_i v_j}{v} \qquad \textbf{(8.5.5)}$$

Equation 8.5.5 is the form of the **mechanical dispersion** tensor that has been used in most subsequent investigations, including most numerical models.

In order to better understand the form of Equation 8.5.5, consider the two-dimensional velocity field with the velocity making an angle θ with the x-axis, as shown in Figure 8.5.1. The components of the dispersion tensor are

$$D_{xx} = a_T v\,(1) + (a_L - a_T)\frac{v^2 \cos^2(\theta)}{v} = a_L v \cos^2(\theta) + a_T v(1 - \cos^2(\theta))$$

$$= a_L v \cos^2(\theta) + a_T v \sin^2(\theta) = a_L \frac{v_x^{\,2}}{v} + a_T \frac{v_y^{\,2}}{v}$$

Similarly,

$$D_{yy} = a_L v\,\sin^2(\theta) + a_T v \cos^2(\theta) = a_L \frac{v_y^{\,2}}{v} + a_T \frac{v_x^{\,2}}{v}$$

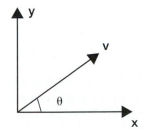

FIGURE 8.5.1   Velocity field with a general orientation with respect to the x-axis

and

$$D_{xy} = D_{yx} = (a_L - a_T)\frac{v_x v_y}{v} = (a_L - a_T)v\frac{\sin(2\theta)}{2}$$

In the transport equation itself, the dispersion term gives

$$\text{div}(\tilde{\tilde{D}} \cdot \text{grad}(c)) = \frac{\partial}{\partial x}\left(D_{xx}\frac{\partial c}{\partial x}\right) + \frac{\partial}{\partial x}\left(D_{xy}\frac{\partial c}{\partial y}\right) + \frac{\partial}{\partial x}\left(D_{xz}\frac{\partial c}{\partial z}\right) + \ldots + \frac{\partial}{\partial z}\left(D_{zz}\frac{\partial c}{\partial z}\right)$$

for a total of nine separate terms.

As noted in Section 8.1, the longitudinal dispersivity is usually considered to be 5 to 100 times larger than the transverse dispersivity. In the laboratory, a_L has been found to vary from 0.1 to 10 mm. In the field, the dispersivities are sometimes measured through single and multiple well tracer tests such as that described in Section 8.3. More often, however, what is usually done is that measured field concentrations are simulated with mathematical models, and the coefficients are adjusted to get an adequate match. The values found in this fashion are usually much larger than laboratory values. Recent literature has shown field values of a_L to vary from 1 to 100 m or larger. These values of a_L are larger than laboratory values by a factor of up to 10^5. Pickens and Grisak (1981) describe detailed laboratory and field experiments to measure the dispersivity. These field experiments show that the measured dispersivities depend on how the measurements were made. If a point sample is obtained for an individual layer during a tracer test, then the breakthrough curve is rather sharp, which corresponds to a small dispersivity on the order of centimeters. On the other hand, if measurements are made from a well which is screened over several intervals, then the well averages the flow characteristics of all the intervals and the resulting dispersivity is much larger, on the order of meters. In addition, these larger dispersivities are scale-dependent, with the apparent dispersivity increasing linearly with distance of transport. Such results are consistent with recent theories of flow in stratified and random media.

Figure 8.5.2 shows field data for longitudinal dispersivities as a function of scale (distance). This data, which was collected by Gelhar et al. (1992), shows clearly that the dispersivity increases with the scale of the observation. Gelhar et al. (1992) judged the reliability of the data as low, intermediate, and high, based on the type of test or observation and the method of analysis. The solid symbols are for porous media while the open symbols are for fractured media. There is no apparent distinction between these two types of media. One should note that almost at any scale, the observed dispersivities vary by over two orders of magnitude, showing that scale is only one variable in determining the effective dispersivity for a medium.

Field observations have lead to the development of estimation models for the dispersivity and its scale dependence. In development of the land disposal regulations for hazardous wastes, the EPA (Federal Register, 51(9), pg. 1652, 1986) suggests that

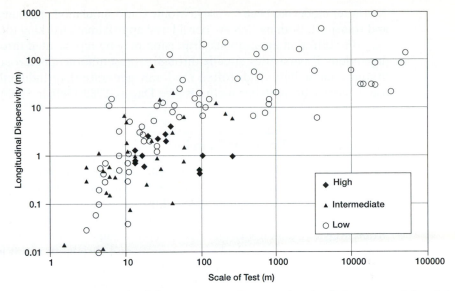

FIGURE 8.5.2   Longitudinal dispersivity as a function of scale of observation (after Gelhar et al., 1992)

$$a_L = 0.1 \, X_r \; ; \; a_T = 1/3 \, a_L \qquad\qquad (8.5.8)$$

where X_r is the distance to a receptor well. The vertical dispersivity, a_v, is considered to be random and uniformly distributed between 0.0125 and 0.1 of the longitudinal dispersivity.

   In the background documents for EPA's *Composite Landfill Model* (EPACML), an alternative representation is presented which is attributed to Gelhar. The longitudinal dispersivity is randomly distributed within three classes. For a receptor well located 500 ft from the point of release, the representation takes the form

Class	1	2	3	
a_L (m)	0.1–1	1–10	1–100	(8.5.9)
Probability	0.1	0.6	0.3	

where the dispersivity is uniformly distributed within each class. For distances other than 500 ft, the following equation is used:

$$a_L(X_r) = a_L(500 \text{ ft}) \sqrt{\frac{X_r}{500}} \qquad\qquad (8.5.10)$$

where X_r is the distance to the receptor well, in feet. The transverse and vertical dispersivity are assumed to have the following values:

$$a_T = \frac{1}{8} a_L \; ; \; a_v = \frac{1}{160} a_L \qquad\qquad (8.5.11)$$

This latter model has been for use in Monte Carlo simulation of groundwater fate and transport. Both models are considered approximate working tools.

For different applications, dispersion may be represented through scale-dependent dispersivities, directly through representation of heterogeneous hydraulic conductivity fields (often randomly generated), or indirectly through use of stochastic modeling methods (see Dagan, 1989; Gelhar, 1993).

8.6   Moments of the Transport Equation

The methods for parameter estimation that have been presented in the previous sections fit experimental data to a mathematical solution of the transport equation and find the parameters by adjusting them to give a best-fit between the data and the mathematical model. For analysis of groundwater transport problems where data from a number of observation wells are available, the field data often shows a great deal of uncertainty and variability, and it is useful to develop more robust methods for estimation of transport parameters. An approach that has found a wide range of application is to work with the statistical moments of the data. The statistical moments of a data set are easily calculated, and they are transport model-independent. If they can be related to specific transport parameters, then they provide powerful tools for parameter estimation. This section introduces the method of moments description of transport processes.

In developing the **method of moments**, we will consider two problems simultaneously. The lower part of Figure 8.6.1 shows the plan view of a point source at the origin and a finite source of width W located along the y-axis. Flow is in the x-direction with a Darcy velocity q_x, and the aquifer has a constant thickness b. The magnitude of the point source is M_o (mass), and the source is represented as an initial condition along the boundary using delta functions as $M_o \, \delta(y) \, \delta(t)$. The finite source has a magnitude $\dot{m}$ (mass/time) and concentration c^* spread uniformly over a length W of the y-axis and depth b (so that $J_x Wb = \dot{m}$, where J_x is constant over this part of the boundary and vanishes elsewhere). This relationship represents a vertical plane source model, as shown in the top of Figure 8.6.1. The mathematical moments for the bulk concentration and the aqueous concentration are defined by

$$M_{\alpha\beta}(t) = \int_\Omega x^\alpha y^\beta \, m(x,y,t) \, d\Omega \;\; ; \;\; C_{\alpha\beta}(t) = \int_\Omega x^\alpha y^\beta \, c(x,y,t) \, d\Omega \qquad \textbf{(8.6.1)}$$

In Equation 8.6.1, Ω is the domain ($x > 0$, $-\infty < y < \infty$, $0 < z < b$), and $d\Omega = b \, dx \, dy$. Equation 8.6.1 is the same type of moments that one considers in probability theory (first moment being the mean and second moment about the mean being the variance) and in mechanics (first and second moments of inertia). Equation 7.5.5 shows that the two moments are related through

$$M_{\alpha\beta}(t) = n \, R \, C_{\alpha\beta}(t) \qquad \textbf{(8.6.2)}$$

and thus, moments behave in this respect just as concentrations behave.

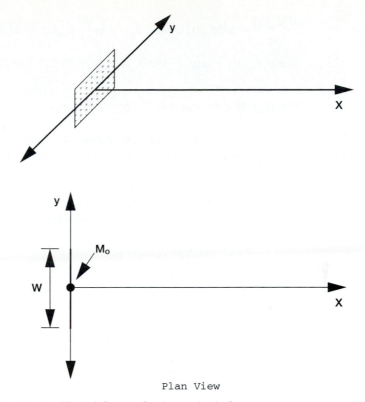

Plan View

FIGURE 8.6.1   Problem definition for the method of moments

The moment equations are developed directly from the transport Equation 5.3.7, assuming first-order decay for the source strength $(S^+ = -\lambda m)$. The transport equation is multiplied by $x^\alpha y^\beta$ and is integrated over the domain

$$\int_\Omega x^\alpha y^\beta \left(\frac{\partial m}{\partial t} + \nabla \cdot \tilde{J} + \lambda m \right) d\Omega = 0 \qquad (8.6.3)$$

Because the order of integration and differentiation with respect to time can be reversed, the first and last terms give

$$\frac{dM_{\alpha\beta}(t)}{dt} + \lambda \, M_{\alpha\beta}(t)$$

To deal with the middle term, we apply Green's First Identity that says

$$\int_\Omega x^\alpha y^\beta \, \nabla \cdot \tilde{J} \, d\Omega = \int_\Gamma x^\alpha y^\beta \tilde{J} \cdot \hat{n} \, d\Gamma - \int_\Omega (\alpha \, x^{\alpha-1} y^\beta J_x + \beta \, x^\alpha y^{\beta-1} J_y) d\Omega$$

where Γ represents the boundary of the domain and $\hat{n}$ is the outward unit normal. The dispersion terms are handled through integration by parts. Evaluating all the terms and collecting the pieces, we finally have

$$\frac{dM_{\alpha\beta}(t)}{dt} + \lambda \, M_{\alpha\beta}(t) = \dot{m}\delta_{\alpha0}\,\delta_{\beta0} + \dot{m}\frac{W^2}{12}\,\delta_{\alpha0}\delta_{\beta2} + M_o\delta(t)\delta_{\alpha0}\delta_{\beta0}$$

$$+ \alpha \, q_x \, C_{\alpha-1,\beta} + \alpha(\alpha-1)nD_{xx}\,C_{\alpha-2,\beta} + \beta(\beta-1)n\,D_{yy}\,C_{\alpha,\beta-2}$$

$$+ n\,D_{xx}b\,W\,c^*\,\delta_{\alpha1}\,\delta_{\beta0} + n\,D_{xx}\,b\frac{W^3}{12}\,c^*\,\delta_{\alpha1}\,\delta_{\beta2} \tag{8.6.4}$$

Equation 8.6.4 is the general result. The **Kronecker unit tensor**, $\delta_{\alpha\beta}$, serves to identify the moments to which each of the terms contribute. For any given problem, the right-hand-side of Equation 8.6.4 contains source characterization terms and lower order moments of the aqueous concentration distribution that are related to the bulk concentration moments through Equation 8.6.2.

To solve Equation 8.6.4 for a particular problem, it is useful to recall that the general solution to the equation

$$\frac{dM_{\alpha\beta}(t)}{dt} + \lambda \, M_{\alpha\beta}(t) = f(t) \tag{8.6.5}$$

is

$$M_{\alpha\beta}(t) = e^{-\lambda t}\left(\int^{t} f(\tau)e^{\lambda t}\,d\tau + A\right) \tag{8.6.6}$$

where A is the constant of integration which affects only the lower limit of the integral.

As an example, consider the **point source model**, where we want to calculate the lower order moments for an instantaneous point release of mass. For the zero-order moment, Equation 8.6.4 becomes

$$\frac{dM_{00}(t)}{dt} + \lambda \, M_{00}(t) = M_o\,\delta(t) \tag{8.6.7}$$

The delta function is interpreted to mean that the initial condition is $M_{00}(0) = M_o$, and Equation 8.6.7 may be integrated to give

$$M_{00}(t) = M_o\,e^{-\lambda t} \tag{8.6.8}$$

which says that the total mass decays exponentially. For the first moment, Equation 8.6.4 becomes

$$\frac{dM_{10}(t)}{dt} + \lambda \, M_{10}(t) = q_x\,C_{00} = \frac{q_x M_o e^{-\lambda t}}{n\,R} \tag{8.6.9}$$

where Equation 8.6.2 has been used. The integral of Equation 8.6.9 is

$$M_{10}(t) = \frac{q_x\,M_o\,t\,e^{-\lambda t}}{n\,R} \tag{8.6.10}$$

so that the mean location μ is given by

$$\mu_x = \frac{M_{10}}{M_{00}} = \frac{q_x t}{n R} = \frac{v_x t}{R} \qquad (8.6.11)$$

For the second moment about the y-axis, Equation 8.6.4 becomes

$$\frac{dM_{20}(t)}{dt} + \lambda M_{20}(t) = 2 q_x C_{10} + 2 n D_{xx} C_{00} = \frac{2M_o e^{-\lambda t}}{n R}\left(\frac{q_x^2 t}{n R} + n D_{xx}\right) \qquad (8.6.12)$$

The solution to this equation is:

$$M_{20}(t) = \frac{2 M_o e^{-\lambda t}}{n R}\left(\frac{q_x^2 t^2}{2 n R} + n D_{xx} t\right) \qquad (8.6.13)$$

The variance of the solution is:

$$\sigma_x^2 = \frac{M_{20}}{M_{00}} - \mu_x^2 = \frac{2 D_{xx} t}{R} \qquad (8.6.14)$$

In particular, for a conservative species, one obtains Einstein's result relating the variance of the concentration distribution to the dispersion coefficient:

$$D_{xx} = \frac{1}{2}\frac{d\sigma_x^2}{dt} \qquad (8.6.15)$$

Borden Landfill Natural Gradient Tracer Test

A natural gradient tracer test was performed at the Canadian Forces Base Borden over a two year period (Mackay et al. 1986; Freyberg 1986; and Roberts et al. 1986). The site hydraulic characteristics were described in Section 2.2.3. At the start of the experiment, 12 m³ of water containing a number of solutes was introduced to the aquifer through several injection wells. The migration of the chemical tracers was monitored by sampling of 275 multilevel wells, each having from 14 to 18 sampling ports. More than 15,000 samples were analyzed during the experiment. During each sampling interval, the aqueous concentration moments were calculated according to Equation 8.6.1. The resulting vertically averaged concentration distribution of Cl- is shown in Figure 8.6.2. This figure shows that mixing is much greater in the longitudinal as compared with the transverse direction.

Equation 8.6.1 may be used to estimate the spatial moments of the concentration distribution for the various constituents. Each sampling location is weighted by its effective area or volume when evaluating the integrals numerically. The resulting spatial concentration variances are shown in Figure 8.6.3. The data points shown in this figure are the average values for the chloride and bromide sampling, as reported by Freyberg (1986). Both species are considered to be conservative. It is clear that the longitudinal and traverse variances increase with time. The effective dispersion coefficient is related to the spatial variance through

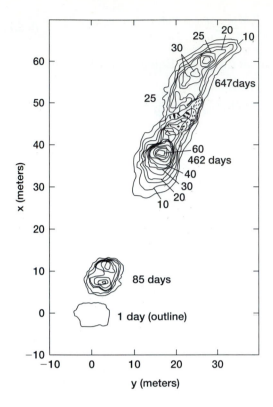

FIGURE 8.6.2 Vertically average concentration distribution of Cl; 1y at various times after injection (after Mackay et al., 1986)

$$D = \frac{1}{2} \frac{d\sigma^2}{dt}$$

Using the average slope from the data, Freyberg estimates that $a_L = D_L/v = 0.36$ m while $a_T = 0.039$ m, where the average seepage velocity is estimated to be $v = 0.091$ m/d.

Review of the data in Figure 8.6.3 suggests that the slope of the longitudinal variance versus time relationship might increase with time. This idea corresponds to the scale effect, where the effective longitudinal dispersivity increases with distance of migration. Recent research on transport through heterogeneous porous media suggests that the dispersivity is related to the local variability of the hydraulic conductivity and its scale of correlation (see Dagan, 1989, and references therein). Specifically, this research suggests that in the far-field where the solute has migrated over a distance of many hydraulic conductivity integral scales, the longitudinal dispersivity is estimated by

$$a_L = \sigma^2_{ln(K)}\lambda_{ln(K)} \tag{8.6.16}$$

where $\sigma^2_{ln(K)}$ is the variance of the logarithm of the hydraulic conductivity field and $\lambda_{ln(K)}$ is the integral scale of the field (as the name suggests, the integral scale is equal to the integral of the field's spatial correlation function—it serves

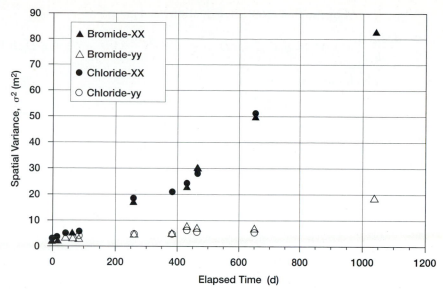

FIGURE 8.6.3 Calculated vertically averaged concentration variances from the Borden data (after Freyberg, 1986)

as a measure of the correlation distance). For a vertically averaged field, Dagan further suggests that the variance should be reduced by a factor of 0.74 due to spatial averaging. Thus, the data presented in Section 2.2.3 suggests that the far-field value of a_L could be as large as $a_L = 0.74 \times 0.285 \times 2.8 = 0.59$ m.

EXAMPLE PROBLEM

8.6.1 Estimation of the Length of a Decaying Plume

Estimate the length of a plume with a constant source rate that decays according to first-order kinetics.

For the constant source model with decay, Equation 8.6.4 gives

$$\frac{dM_{00}(t)}{dt} + \lambda M_{00} = \dot{m}$$

which integrates to

$$M_{00}(t) = \frac{\dot{m}}{\lambda}\left(1 - e^{-\lambda t}\right) \tag{8.6.17}$$

This equation shows that the maximum total mass that can be found in the aquifer is

$$M_{00}(\infty) = \frac{\dot{m}}{\lambda} \tag{8.6.18}$$

For the first moment, Equation 8.6.4 gives

$$\frac{dM_{10}(t)}{dt} + \lambda\, M_{10} = q_x\, C_{00} + n\, D_{xx}b\, W\, c^* = \frac{q_x\dot{m}}{\lambda\, n\, R}\left(1 - e^{-\lambda t}\right) + n\, D_{xx}b\, W\, c^* \quad \textbf{(8.6.19)}$$

which integrates to

$$M_{10}(t) = \frac{q_x\dot{m}}{\lambda\, n\, R}\left(\frac{1 - e^{-\lambda t}}{\lambda} - t\, e^{-\lambda t}\right) + \frac{n\, D_{xx}b\, W\, c^*}{\lambda}\left(1 - e^{-\lambda t}\right) \quad \textbf{(8.6.20)}$$

and the mean location of the plume satisfies

$$\mu_x = \frac{q_x}{n\, R}\left(\frac{1}{\lambda} - \frac{t}{e^{\lambda t} - 1}\right) + \frac{n\, D_{xx}b\, W\, c^*}{\dot{m}} \quad \textbf{(8.6.21)}$$

8.7   Analytical Models of Chemical Spills (Instantaneous Release)

In this section and the next, we will look at analytical solutions of the advection-dispersion equation. These solutions are useful in many applications. They may be used to assess the potential impacts of releases of contaminants to groundwater, either through spills or continual releases that result in the creation of contaminant plumes. The analytical solutions provide a means to estimate the potential exposure concentrations at various locations. Analytical models also provide tools for use in parameter estimation. The models incorporate parameters such as dispersivities and retardation factors, and through fitting of these models to experimental data, estimates of the parameters may be made. These applications have been discussed for one-dimensional flow systems in the earlier sections of the chapter. Analytical models also may be used as useful screening tools for establishing soil clean-up level. Critical exposure locations are identified and health-based exposure concentrations are established for these locations. Then, the models may be used "backwards" to calculate the source concentration that would result in this critical exposure. This procedure could establish the minimum clean-up standard. Analytical models also are commonly used as one component of a risk or radiological assessment model. The general framework for these models is shown in Figure 5.1.6. They include models for pollutant release to the environment, models for pollutant migration through various environmental pathways, and models to calculate exposure and dose. Analytical models are often used for calculation of the subsurface migration of chemicals. Finally, analytical models may be used to evaluate the performance of numerical simulation models. Because analytical models are exact, they provide a benchmark against which numerical models can be compared.

In this section, we will focus on a few models of chemical spills. Mathematically, spills are represented as initial value problems. The initial distribution of chemical concentrations is specified, and the models determine how the concentration changes through time and space. The limitations on analytical models are severe. Parameter fields must be assumed to be uniform. Most often, the flow field is one-dimensional, though dispersion transport can occur in three dimensions. The models differ principally in the number of dimensions considered and in the different initial distributions of chemicals.

We first consider the general solution for an initial value problem. This solution will apply for any source condition in any number of dimensions. We then consider a simple point-source spill model and how various forms of the model develop from one another. Finally, we consider how more complicated source conditions may be developed.

8.7.1   Fourier Analysis of Initial Value Problems in Solute Transport

In this section, we want to find the general solution to the problem of transport of a solute in a uniform flow field in the x-direction, with mixing in all three directions for an arbitrary initial distribution of contaminants. We want to find the general solution to the solute transport equation

$$\frac{\partial c}{\partial t} + v'\frac{\partial c}{\partial x} + \lambda c = D'_{xx}\frac{\partial^2 c}{\partial x^2} + D'_{yy}\frac{\partial^2 c}{\partial y^2} + D'_{zz}\frac{\partial^2 c}{\partial z^2} \qquad (8.7.1)$$

for arbitrary initial conditions. In Equation 8.7.1, $v' = v/R$ is the **retarded velocity**, $D'_{xx} = D_{xx}/R$ is the **retarded x-dispersion** coefficient, etc. The simplest approach to solving this equation is to use Fourier methods. The basic idea in **Fourier analysis** is to develop the solution in terms of waves of various frequencies and wave numbers. A general Fourier component takes the following form,

$$c \sim A\, e^{i(\tilde{k}\cdot\tilde{x} - \omega t)} \qquad (8.7.2)$$

where A is the wave amplitude, $i = \sqrt{-1}$, $\tilde{k}$ is the wave number vector (with components k_x, k_y, k_z), and ω is the wave frequency. The dispersion relation[2], $\omega = \omega(k)$, is the relation between the wave frequency and wave number, and follows directly from the process model being investigated. The relation is found by substituting the general Fourier component Equation 8.7.2 into the solute transport Equation 8.7.1 and evaluating the derivatives. After canceling the wave component and dividing through by the complex i, this equation gives

$$\omega = k_x v' + i\lambda + i(D'_{xx}k_x^2 + D'_{yy}k_y^2 + D'_{zz}k_z^2) \qquad (8.7.3)$$

Substituting the dispersion relation back into the general Fourier component to eliminate the wave frequency, ω, gives the following equation:

$$c \sim A e^{i[k_x(x - v't) + k_y y + k_z z]} \cdot e^{-[D'_{xx}k_x^2 + D'_{yy}k_y^2 + D'_{xx}k_{xx}^2 + \lambda]t} \qquad (8.7.4)$$

A few conclusions follow directly from Equation 8.7.4. First, the general component decays exponentially at a rate corresponding to λ, regardless of its other characteristics. Second, physical dispersion results in a "decay" with the shorter wavelengths (larger k's) dissipating first. This process results in a smoothing of the concentration distribution over time, because the "lumps" which are

---

[2]Dispersion relation is a term used in Fourier analysis and is not associated with the physical process of dispersion.

made up of small waves decay away rapidly leaving the smoother distribution which is made up from the longer waves (smaller k's). Third, advection does not result in a change in magnitude, but simply generates a translation in space (the magnitude of $|e^{i\theta}| = 1$).

The next question concerns how to use these results to solve a particular problem. First, consider a *one-dimensional* ($1D$) case. The general solution comes from superposition of waves with different wave numbers (wavelengths). If we consider the amplitude to be a function of the wavelength, $A(k)$, the general solution for the $1D$ case may be developed from

$$c(x,t) = \int_{-\infty}^{\infty} A(k)e^{ik(x-v't)-(D'k^2+\lambda)t}\,dk \qquad (8.7.5)$$

where the integral is over all wave numbers and the directional subscripts have been dropped. The initial conditions must satisfy

$$c(x,0) = \int_{-\infty}^{\infty} A(k)e^{ikx}\,dx \qquad (8.7.6)$$

which shows that $c(x,0)$ is the Fourier transform of $A(k)$. $A(k)$ is called the **spectrum** of the function $c(x,0)$. The Fourier inversion theorem states that the function $A(k)$ is given by

$$A(k) = \frac{1}{2\pi}\int_{-\infty}^{\infty} c(\xi,0)e^{-ik\xi}\,d\xi \qquad (8.7.7)$$

Substituting Equation 8.7.7 into the general solution of Equation 8.7.5 gives the formal solution as

$$c(x,t) = \int_{-\infty}^{\infty}\left(\frac{1}{2\pi}\int_{-\infty}^{\infty} c(\xi,0)e^{-ik\xi}\,d\xi\right)e^{ik(x-v't)-(D'k^2+\lambda)t}\,dk \qquad (8.7.8)$$

To transform Equation 8.7.8 to the desired form, we change the order of integration, complete the squares of the argument as a quadratic in k, and evaluate the resulting integral. This process ultimately gives

$$c(x,t) = \int_{-\infty}^{\infty} c(\xi,0)G(x,t|\xi,0)d\xi \cdot e^{-\lambda t} \qquad (8.7.9)$$

In Equation 8.7.9,

$$G(x,t|\xi,0) \equiv \frac{1}{\sqrt{4\pi D't}}\exp\left(-\frac{(x-\xi-v't)^2}{\sqrt{4D't}}\right) \qquad (8.7.10)$$

Equation 8.7.9 shows that the function from Equation 8.7.10 plays the part of an impulse response function for the partial differential Equation 8.7.1. It is called a **Green's function**, and it gives the contribution to the solution at point x and time t due to the initial unit mass at point $x = \xi$. Equation 8.7.9 also shows that first-order decay may be completely factored out of an initial value problem. The solution with decay is equal to the solution without decay times the exponential decay term.

Returning now to the three-dimensional case, note that the general Fourier component can be written as the product of terms involving exponentials with x, y, and z separately. Repeating the steps shown earlier, one can arrive at the product of Green's functions. These are:

$$G_x(x,t|\xi,0) \equiv \frac{\exp(-\frac{(x - \xi - v't)^2}{4D'_{xx}t})}{\sqrt{4\pi D'_{xx}t}}$$

$$G_y(y,t|\eta,0) \equiv \frac{\exp(-\frac{(y - \eta)^2}{4D'_{yy}t})}{\sqrt{4\pi D'_{yy}t}} \qquad \text{(8.7.11)}$$

$$G_z(z,t|\zeta,0) \equiv \frac{\exp(-\frac{(z - \zeta)^2}{4D'_{zz}t})}{\sqrt{4\pi D'_{zz}t}}$$

The general solution to the *3D* problem may be written

$$c(x,y,z,t) = \int\int\int c(\xi,\eta,\zeta,0)G_x\,G_y\,G_z\,d\xi\,d\eta\,d\zeta \cdot e^{-\lambda t} \qquad \text{(8.7.12)}$$

where G_x, G_y, and G_z are the individual Green's functions defined in Equation 8.7.11. More concisely, this relationship is written in vector notation with $G(x,y,z,t|\xi,\eta,\zeta,0) = G_x G_y G_z$ as

$$c(\tilde{x},t) = \int_{-\infty}^{\infty} c(\tilde{\xi},0)G(\tilde{x},t|\tilde{\xi},0)d\tilde{\xi} \cdot e^{-\lambda t} \qquad \text{(8.7.13)}$$

Equation 8.7.13 provides the general solution for an initial value problem with uniform flow in a three-dimensional domain.

8.7.2 Point Spill Model

Consider the release of a volume V_o centered at point $\tilde{x} = 0$ and containing a contaminant at a concentration c_o. If the time period of the release is short compared to the time period of interest for transport, and if the volume is small enough to not significantly influence the groundwater flow pattern near the release point, then the release may be considered as having occurred instantaneously at a single point $\tilde{x} = 0$. The mass released is $M = c_o V_o$. The initial conditions on c are found from

$$M = c_o V_o = \int\int\int m(\tilde{x},0)dV = (n + \rho_b K_d)\int\int\int c(\tilde{x},0)\,dV$$

so that

$$\int\int\int c\,dV = \frac{M}{n + \rho_b K_d} = \frac{c_o V_o}{n\,R} \qquad \text{(8.7.14)}$$

where R is the retardation factor. The notion of release at a point physically corresponds to that of a Dirac delta function. Thus, the initial condition may be written

$$c(\tilde{\xi},0) = \frac{V_o c_o}{n\,R}\,\delta(\tilde{\xi}) \qquad (8.7.15)$$

Substituting the initial condition of Equation 8.7.15 in the general solution of Equation 8.7.13 gives

$$c(\tilde{x},t) = \frac{V_o c_o}{nR}\,G(\tilde{x},t|\xi,0)e^{-\lambda t} \qquad (8.7.16)$$

or explicitly, in terms of the individual Green's functions,

$$c(\tilde{x},t) = \frac{V_o c_o \exp\!\left(-\dfrac{(x-v't)^2}{4D'_{xx}t} - \dfrac{y^2}{4D'_{yy}t} - \dfrac{z^2}{4D'_{zz}t}\right)}{nR\sqrt{64\pi^3 D'_{xx}D'_{yy}D'_{zz}t^3}} \cdot e^{-\lambda t} \qquad (8.7.17)$$

This solution is quite important. Notice that at any time, the maximum concentration is

$$c_{max}(t) = \frac{V_o c_o}{nR\sqrt{64\pi^3 D'_{xx}D'_{yy}D'_{zz}t^3}} \cdot e^{-\lambda t} \;\;;\;\; \tilde{x} = (x,y,z) = (v't,0,0) \qquad (8.7.18)$$

This equation shows that the maximum concentration decreases as

$$c_{max} \propto t^{-3/2}e^{-\lambda t} \qquad (8.7.19)$$

Furthermore, the concentration is constant at a value $c = c^*$ on the surface

$$\frac{(x-v't)^2}{D'_{xx}} + \frac{y^2}{D'_{yy}} + \frac{z^2}{D'_{zz}} = 4t\,ln\!\left(\frac{V_o c_o}{c^* nR\sqrt{64\pi^3 D'_{xx}D'_{yy}D'_{zz}t^3}}\right) - \lambda t \qquad (8.7.20)$$

Equation 8.7.20 shows that the surfaces of constant concentration are ellipsoids with axes increasing with time, and if first-order decay occurs, the axes eventually decrease with time. The concentrations spread, while the maximum concentration decreases.

8.7.3   Vertically Mixed Spill Model

We want to consider the same model as in Section 8.7.2, except that the release is mixed over the thickness of the aquifer; for example, if the release occurs through a fully penetrating and screened well. The transport equation remains the same, except that there are no vertical gradients and the term involving z does not appear. For the initial condition, we have

$$M = V_o c_o = \int\int m\,b\,dA$$

where b is the aquifer thickness and the integration is over the "plan view" area of the aquifer. Thus,

$$\iint c \, dA = \frac{V_o c_o}{n \, R \, b} \tag{8.7.21}$$

Again, with the notion of a point release, we have

$$c(\xi, \eta, t = 0) = \frac{V_o c_o}{n \, R \, b} \, \delta(\xi, \eta) \tag{8.7.22}$$

where the concentration is now a function of only x, y, and t, and $\delta(\xi, \eta)$ is the 2D Dirac delta function centered at the origin.

Repeating the arguments of Section 8.7.1, one can also show that the Green's function for two dimensions is simply the product of G_x and G_y from Equation 8.7.11. Thus, Equation 8.7.15 becomes

$$c(\tilde{x}, t) = \frac{V_o c_o \exp\left(-\dfrac{(x - v't)^2}{4 D'_{xx} t} - \dfrac{y^2}{4 D'_{yy} t}\right)}{4 \pi \, n \, b \, t \sqrt{D_{xx} D_{yy}}} \cdot e^{-\lambda t} \tag{8.7.23}$$

where R has canceled from the denominator (the retarded dispersion coefficients are no longer being used here, though they do appear within the exponential term). Equation 8.7.23 shows that the maximum concentration decreases as

$$c_{\max} \propto t^{-1} e^{-\lambda t} \tag{8.7.24}$$

which should be compared with Equation 8.7.19. Mixing in two dimensions does not result in as fast of a concentration reduction as in three dimensions.

8.7.4 Vertical Mixing Region

Again, consider the point release of a contaminant at the top of the aquifer. Near the source, the spreading is in three dimensions. The solution presented in Section 8.7.2, however, is not quite correct for this case, because it has half of the mass spreading above the water table. The correct solution near the source is twice that of Equation 8.7.17, and only that part of the solution below the water table is considered. The solution then has mass $Me^{-\lambda t}$ below the water table. The resulting equation is:

$$c(\tilde{x}, t) = \frac{V_o c_o \exp\left(-\dfrac{(x - v't)^2}{4 D'_{xx} t} - \dfrac{y^2}{4 D'_{yy} t} - \dfrac{z^2}{4 D'_{zz} t}\right)}{4 \, n \, R \sqrt{\pi^3 D'_{xx} D'_{yy} D'_{zz} t^3}} \cdot e^{-\lambda t} \tag{8.7.25}$$

Far from the source, the contaminant has spread over the thickness of the aquifer and is only able to continue to spread in the two lateral directions. This process is shown schematically in Figure 8.7.1, where the far-field concentration distribution is described by Equation 8.7.23. The question of fundamental interest concerns the horizontal distance to where the 3D solution ceases to be valid, because the base of the aquifer prevents further vertical migration and the distance to where the 2D solution becomes valid. Between these two distances, we must use a solution that recognizes that the water table and the bottom of

FIGURE 8.7.1   Schematic view of spreading of a point release model showing the near field and far-field distributions

the aquifer are boundaries with zero vertical concentration gradients (no vertical flux across these boundaries).

The easiest way to solve this problem is to take the 3D solution, which we know satisfies the transport equation and the initial conditions, and build up the solution that satisfies the boundary condition through superposition of image sources. In order to make the bottom of the aquifer a zero flux boundary, a source must be placed at $z = 2b$. Together, with a source at $z = 0$ and one at $z = 2b$, the level $z = b$ has a zero vertical gradient across it. This source does cause a gradient across the water table at $z = 0$, however. To cancel this gradient, we need a source at $z = -2b$. But now, this source will cause a gradient across $z = b$, so we need another source at $z = 4b \ldots$, etc. Thus, the solution requires an infinite number of sources released at $x = 0$, $y = 0$, and $z = \ldots, -4b, -2b, 0, 2b, 4b, \ldots$. This idea is shown in Figure 8.7.2. The general solution is due to the superposition of these sources, and as long as we only look at the solution for $0 \le z \le b$, this general solution must reduce to both the 3D and the 2D solutions within the appropriate ranges. Thus, the general solution is

$$c(x,y,z,t) = \frac{V_o c_o \exp\left(-\dfrac{(x - v't)^2}{4 D'_{xx} t} - \dfrac{y^2}{4 D'_{yy} t}\right)}{4 n R \sqrt{\pi^3 D'_{xx} D'_{yy} D'_{zz} t^3}} \cdot e^{-\lambda t} \sum_{j=-\infty}^{\infty} \exp\left(-\frac{(z + 2 j b)^2}{4 D'_{zz} t}\right) \quad (8.7.26)$$

Equation 8.7.26 is appropriate both for the near-field and far-field solutions. It is inconvenient to have to use the infinite system of images, however, and we are interested when we can get by with the simple 3D and 2D solutions in place of Equation 8.7.26.

In order to examine the range of validity of Equation 8.7.23 and when it may be used in place of Equation 8.7.26, we write Equation 8.7.26 in the form of Equation 8.7.23 multiplied by a correction term. Then, we can look at the correction term to identify the domain of vertical mixing. Write Equation 8.7.26 as

$$c(x,y,z,t) = \frac{V_o c_o \exp\left(-\dfrac{(x-v't)^2}{4 D'_{xx} t} - \dfrac{y^2}{4 D'_{yy} t}\right)}{4\pi n b t \sqrt{D_{xx} D_{yy}}} \cdot e^{-\lambda t} \left\{ \frac{b}{\sqrt{\pi D'_{zz} t}} \sum_{j=-\infty}^{\infty} \exp\left(-\frac{(z+2 j b)^2}{4 D'_{zz} t}\right) \right\} \quad (8.7.27)$$

The correction function is

$$F(\zeta,\phi) \equiv \frac{1}{\sqrt{\pi \phi}} \sum_{j=-\infty}^{\infty} \exp\left(-\frac{(\zeta + 2 j)^2}{4 \phi}\right) \quad (8.7.28)$$

where the dimensionless parameters

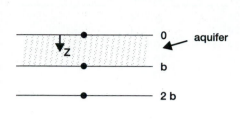

-2 b

0   ← aquifer

b

2 b

4 b

FIGURE 8.7.2 Development of the system of $3D$ sources to represent the finite thickness of the aquifer

$$\zeta = \frac{z}{b} \; ; \; \phi = \frac{D_{zz}'t}{b^2} \tag{8.7.29}$$

have been introduced. In terms of this function, the general solution may be written as follows,

$$c(x,y,z,t) = c_{2D}(x,y,t) \cdot F(\zeta,\phi) \tag{8.7.30}$$

where $c_{2D}(x,y,t)$ is the $2D$ vertically mixed solution of Equation 8.7.23. Thus, the question of vertical mixing concerns how rapidly the function

$$F(\zeta,\phi) \to 1$$

for a given $\zeta, 0 \le \zeta \le 1$. Consider first the top of the aquifer ($\zeta = 0$). Here,

$$F(0,\phi)\frac{1 + 2e^{-1/\phi} + 2e^{-4/\phi} + \dots}{\sqrt{\pi\phi}} \tag{8.7.31}$$

which corresponds to the terms $j = 0, j = 1$ and $-1, j = 2$ and $-2, \dots$. For small times, $\phi \to 0$ and $F(0,\phi) \to \infty$. This relationship implies that the $2D$ model Equation 8.7.23 greatly under-predicts the concentration for small times. For the bottom of the aquifer, we have

$$F(1,\phi) \cong \frac{2e^{1/4\phi} + 2e^{-9/4\phi} + 2e^{-25/4\phi} + \dots}{\sqrt{\pi\phi}} \tag{8.7.32}$$

which corresponds to the terms $j = 0$ and $-1, j = 1$ and $-2, j = 2$ and $-3, \dots$. Table 8.7.1 provides a brief listing of values of the mixing correction function, $F(\zeta,\phi)$.

Table 8.7.1 shows that the spill becomes vertically mixed, for practical purposes, after $\phi \cong 0.8$, or more conservatively, after

$$\phi = 1 \, ; \frac{D_{zz}'t}{b^2} \cong 1$$

So, for

TABLE 8.7.1 Vertical Mixing Function

ϕ	$F(0,\phi)$	$F(1,\phi)$
0.1	1.784	0.293
0.2	1.279	0.723
0.4	1.039	0.961
0.6	1.005	0.995
0.8	1.001	0.999
1.0	1.000	1.000

$$t > \frac{R\,b^2}{D_{zz}} \tag{8.7.33a}$$

the $2D$ solution can be used. By this time, the centroid of the spill will have moved a distance

$$x = \frac{v\,t}{R} > \frac{v\,b^2}{D_{zz}} \tag{8.7.33b}$$

In particular, if mechanical dispersion dominates the mixing process, the $2D$ solution is valid after

$$x > \frac{b^2}{a_v} \tag{8.7.33c}$$

where a_v is the vertical dispersivity.

Next, we evaluate the range of validity of the $3D$ model without correcting for boundaries. This model is the model provided by Equation 8.7.25. Proceeding as before, we write the general solution given by Equation 8.7.26 as the $3D$ model of Equation 8.7.25 multiplied by a correction function:

$$c(\tilde{x},t) = c(\tilde{x},t)_{\text{single}} \times F(z,t)$$

In this case, the correction function becomes

$$F(z,t) = F(\zeta,\phi) = 1 + \frac{\displaystyle\sum_{\substack{j=-\infty \\ j \neq 0}}^{\infty} \exp\left(-\frac{(\zeta - 2j)^2}{4\,\phi}\right)}{\exp\left(-\frac{\zeta^2}{4\,\phi}\right)}$$

where ζ and ϕ are defined by Equation 8.7.29. For the top of the aquifer, the correction function reduces to

$$F(0,\phi) = 1 + 2e^{-1/\phi} + 2e^{-4/\phi} + \ldots$$

while for the base of the aquifer it becomes

$$F(1,\phi) = 2\{1 + e^{-2/\phi} + e^{-6/\phi} + e^{-12/\phi} + \ldots\}$$

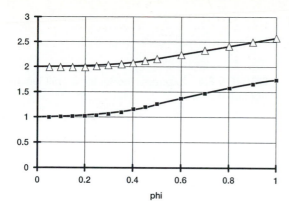

FIGURE 8.7.3 Correction function for single image approximation for spill model. The squares are for the function $(0,\phi)$ and the triangles are the function $(1,\phi)$

These functions are plotted in Figure 8.7.3. The correction for the base of the aquifer never does equal unity, because even for small times and for those close to the source, the general solution at the base of the aquifer is due to the sum of the source at the top of the aquifer and the image at $z = 2b$ ($j = -1$). Judging from the $F(0,\phi)$, it appears that the single source solution is acceptable for

$$\phi = \frac{D_{zz}\,t}{R\,b^2} < 0.2$$

$$x < \frac{0.2\,v\,b^2}{D_{zz}} \qquad\qquad (8.7.34)$$

This discussion shows that the general solution of Equation 8.7.26 need only be used for

$$\frac{0.2\,v\,b^2}{D_{zz}} < x < \frac{v\,b^2}{D_{zz}} \qquad\qquad (8.7.35)$$

For shorter and longer distances, one can use the single source $3D$ and $2D$ solutions, respectively.

8.7.5 A Simple Exposure Model

One point that is important to notice is that the vertically averaged horizontal flux for the $3D$ model is the same as that for the $2D$ model. Thus, in order to evaluate the flux past a given boundary, for example into a fully penetrating river, one can work with the $2D$ solution regardless of how close the spill is to the river. In order to evaluate the flux, we need to evaluate

$$\frac{\partial c(x, y, t)}{\partial x} = \frac{\partial}{\partial x}\left(\frac{V_o c_o\,\exp\left(-\dfrac{(x - v't)^2}{4\,D'_{xx}t} - \dfrac{y^2}{4\,D'_{yy}t}\right)}{4\,\pi\,n\,b\,t\sqrt{D_{xx}D_{yy}}} \cdot e^{-\lambda t}\right) = -\left(\frac{x - v't}{2\,D'_{xx}t}\right)c(x, y, t)$$

so that the flux in the x-direction is

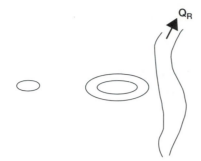

FIGURE 8.7.4   Exposure model for a spill leaching into a river

$$J_x(x, y, t) = q\,c - n\,D_{xx}\frac{\partial c}{\partial x} = \left(q + n\,D_{xx}\left(\frac{x - v't}{2\,D'_{xx}t}\right)\right)c = \left(q + \frac{n\,R\,x}{t}\right)\frac{c(x, y, t)}{2} \quad \textbf{(8.7.36)}$$

The flux past the surface at $x = L$, for example, is

$$\dot{m}(x = L, t) = \int_{-\infty}^{\infty} b\,J_x(L, y, t)\,dy = \frac{V_o c_o e^{-\lambda t}}{\sqrt{16\,\pi\,R\,D_{xx}t}}\left(v + \frac{RL}{t}\right)\exp\left(-\frac{(L - v't)^2}{4\,D'_{xx}t}\right) \quad \textbf{(8.7.37)}$$

If the surface $x = L$ represents a river with discharge Q_R, as shown in Figure 8.7.4, then the resulting concentration can be calculated from a simple mass balance, giving

$$c_R(t) = \frac{\dot{m}(L, t)}{Q_R} \quad \textbf{(8.7.38)}$$

Equation 8.7.38 gives the concentration in the river as a function of time through the mass flux into the river given by Equation 8.7.37.

8.7.6   **Rectangular Source Model**

In the previous subsections, we have considered an instantaneous release of mass at a single point as a model for a localized spill of contaminants. In this section, we look at a release which has a finite size. A simple realization of these conditions is a zone of previous contamination that has become well mixed, and we want to look at the potential exposure concentration at a downgradient location. The initial conditions consist of a mass

$$M = n\,R\,c_o\,L\,W\,b \quad \textbf{(8.7.39)}$$

which is spread over a rectangular region of length L in the direction of flow, width W transverse to this direction, and uniformly spread over the aquifer thickness, b. The uniform initial concentration within this region is c_o. This initial condition is shown in Figure 8.7.5.

The initial conditions may be written in mathematical form as

$$c(\xi, \eta) = c_o\{H(\xi + \tfrac{L}{2}) - H(\xi - \tfrac{L}{2})\}\{H(\eta + \tfrac{W}{2}) - H(\eta - \tfrac{W}{2})\} \quad \textbf{(8.7.40)}$$

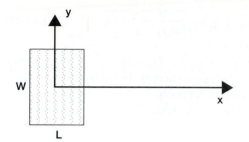

W

FIGURE 8.7.5 Initial condition for a rectangular source model

where $H(\ )$ is the **Heaviside unit function**. With the two-dimensional Green's functions from Equation 8.7.11, the general solution may be written

$$c(x, y, t) = \int\int c(\xi, \eta) G_x G_y \, d\xi d\eta \cdot e^{-\lambda} \qquad (8.7.41)$$

which is analogous to Equation 8.7.12. The role of the Heaviside functions is to set the limits of integration. Introducing the initial condition from Equation 8.7.40, we find

$$c(x, y, t) = c_o \int_{-L/2}^{L/2} G_x \, d\xi \int_{\approx W/2}^{W/2} G_y \, d\eta \cdot e^{-\lambda t} \qquad (8.7.42)$$

$$= \frac{c_o}{4} \left\{ \mathrm{erf}\left(\frac{x - v't + L/2}{\sqrt{4 D'_{xx} t}}\right) - \mathrm{erf}\left(\frac{x - v't - L/2}{\sqrt{4 D'_{xx} t}}\right) \right\} \cdot \left\{ \mathrm{erf}\left(\frac{y + W/2}{\sqrt{4 D'_{yy} t}}\right) - \mathrm{erf}\left(\frac{y - W/2}{\sqrt{4 D'_{yy} t}}\right) \right\} \cdot e^{-\lambda t}$$

Equation 8.7.42 gives the groundwater concentration for a single rectangular source that is uniformly mixed over the aquifer thickness. It represents a two-dimensional solution. Using superposition, solutions for other source areas that can be built up from simple rectangles can be found.

After a sufficient time has passed, one should not be able to recognize the initial source distribution, and Equation 8.7.42 should look like the 2D point spill model of Equation 8.7.23. Indeed, it is not difficult to show that Equation 8.7.42 approaches

$$c(x, y, t) = \frac{L\, Wc_o \exp\left(-\dfrac{(x - v't)^2}{4 D'_{xx} t} - \dfrac{y^2}{4 D'_{yy} t}\right)}{4 \pi t \sqrt{D'_{xx} D'_{yy}}} \cdot e^{-\lambda t} \qquad (8.7.43)$$

To see that this model is actually the point source model of Equation 8.7.23, note that

$$L\, W c_o = \frac{M}{n\, R\, b} \qquad (8.7.44)$$

where M is, again, the total mass present within the contaminated region at time zero.

One may use Taylor's series with the definition of the error function to show that Equation 8.7.43 is a good approximation to Equation 8.7.42, as long as

$$\left(-2 + \frac{(x - v't)^2}{D'_{xx}t}\right)\frac{L^2}{96 \, D'_{xx}t} \ll 1 \; ; \; \left(-2 + \frac{y^2}{D'_{yy}t}\right)\frac{W^2}{96D'_{yy}t} \ll 1 \qquad \text{(8.7.45)}$$

Note that these terms vanish where $(x - v't)^2 = 2 \, D_{xx}'t$ and $y^2 = 2 \, D_{yy}'t$, which corresponds to a distance of one standard deviation from the centroid. Where these terms vanish, one must also look at terms of higher order. At the centroid, where the concentration would be greatest, the first-order terms are

$$\frac{-L^2}{48 \, D'_{xx}t} \; ; \; \frac{-W^2}{48 \, D'_{yy}t}$$

Thus, the point source model is a useful approximation when these terms are both small.

8.8   Analytical Models of Contaminant Plumes (Continuous Release)

We now turn from spill models that represent an instantaneous release of a contaminant to models which represent conditions where the contaminant is released on a continuing basis, once the release has started. The contaminant distribution appears as a plume that initially grows in length and width and eventually reaches a steady-state distribution, if the source strength continues at a constant rate. First, we consider the general mathematical framework from which plume models may be developed. Then, we will look at the steady-state concentration distribution near a source that has been releasing mass at a constant rate for a long time period. Later, we will look at models that can simulate transient effects associated with ephemeral sources or the build-up to a steady-state condition.

8.8.1   Mathematical Considerations in the Development of Contaminant Plume Models

In the last section, we considered analytical models for contaminant spills. These were mathematically represented as initial value problems for the advection-dispersion equation. The simplest way of visualizing the development of a contaminant plume model is to think of the plume as being derived from a continuing series of spills. This idea is the concept of convolution for linear mathematical systems. For our present discussion, let $c_s{}^*(x,y,z,t)$ represent the mathematical model for a chemical spill of unit mass ($M = 1$), which is given as the solution of an initial value problem. Then, the solution for a plume that develops from a source of the same configuration but with a time-variable mass release rate $\dot{m}(t)$ is given by

$$c_p(x, y, z, t) = \int_0^t \dot{m}(\tau)\, c_s^*(x, y, z, t - \tau)\, d\tau \qquad (8.8.1)$$

In Equation 8.8.1, $c_p(x,y,z,t)$ gives the plume concentration distribution at time t. In particular, if the mass release rate is constant $\dot{m}$, then Equation 8.8.1 becomes

$$c_p(x, y, z, t) = \dot{m} \int_0^t c_s^*(x, y, z, \tau)\, d\tau \qquad (8.8.2)$$

Equations 8.8.1 and 8.8.2 may also be used to develop two-dimensional plume models, where for an aquifer, the variable z is eliminated.

Section 8.7.1 presents the general solution of an initial value problem that may be used to represent a contaminant spill. The solution is presented in terms of a Green's function integral. Let $\chi(\tilde{\xi})$ represent an indicator variable for the source region in Equation 8.7.13. This definition means that if the point $\tilde{\xi}$ is within the initial source zone of the spill, then $\chi(\tilde{\xi}) = 1$. Otherwise, $\chi(\tilde{\xi}) = 0$. Also, let $V(\Omega)$ represent the volume of the source region. Then, Equation 8.7.13 may be written

$$c_s^*(\tilde{x}, t) = \frac{1}{V(\Omega)} \int_\Omega \chi(\tilde{\xi})\, G(\tilde{x}, t | \tilde{\xi}, 0)\, d\tilde{\xi} \cdot e^{-\lambda t} \qquad (8.8.3)$$

Equation 8.8.3 assumes that the initial concentration is uniform (and of magnitude one) over the initial source domain Ω. Substituting Equation 8.8.3 into Equation 8.8.1 gives the following equation:

$$c_p(\tilde{x}, t) = \frac{1}{V(\Omega)} \int_0^t \dot{m}(\tau) \int_\Omega \chi(\tilde{\xi}) G(\tilde{x}, t - \tau | \tilde{\xi}, 0)\, d\tilde{\xi} \cdot e^{-\lambda(t-\tau)} d\tau \qquad (8.8.4)$$

Equation 8.8.4 represents the formal solution for a contaminant **plume model** that is derived from an initial value problem formulation of a spill model, where the source strength is variable through time.

Another way that plume models are developed is through solution of initial-boundary value problems for the advection-dispersion equation. For initial-boundary value problems, the solution domain is usually considered to be the half-space (or half-plane), and a boundary condition that specifies the characteristics of the plume is specified along the bounding surface. The initial condition is zero contaminant concentration within the solution domain at time zero. If a solution for the problem with a constant boundary condition can be found, then **Duhamel's theorem** may be used to find the contaminant plume with time-variable source strength (Carslaw and Jaeger, 1959, p. 31). According to this theorem, if $c = F(x, y, z, t)$ represents the concentration at point (x, y, z) at time t in an aquifer in which the initial concentration is zero, while its "boundary" concentration is the constant function $\phi(0, y, z)$, then the solution of the problem in which the initial concentration is zero, and the surface concentration is $B(t)$ $\phi(0, y, z)$ is given by

$$c_p(x, y, z, t) = \int_0^t B(t) \frac{\partial F(x, y, z, t - \tau)}{\partial t} \, d\tau = \int_0^t B(t - \tau) \frac{\partial F(x, y, z, \tau)}{\partial t} \, d\tau \quad \textbf{(8.8.5)}$$

A careful consideration of Equation 8.8.5 for $F(x, y, z, t)$ shows that there is not much conceptual difference between the result of Equation 8.8.5 and that of Equation 8.8.1 (or Equation 8.8.4).

8.8.2   A Simple Plume Model

The simplest model for a plume was already considered in Section 8.4 with regard to transverse mixing and is applicable only for steady-state conditions. Consider a constant mass release rate $\dot{m}$ occurring at the point $(x,y) = (0,0)$, as shown in Figure 8.8.1. This model considers only advection and transverse mixing based on the assumption that longitudinal gradients are small and longitudinal dispersion is negligible. During the time interval δt, the width of the slab of water containing contaminants that passes the $x = 0$ plane is

$$\frac{v}{R} \, \delta t$$

and the mass which enters this slab is $\dot{m} \, \delta t$. This concept is shown schematically in the lower part of Figure 8.8.1. Because longitudinal mixing is neglected in this model, each slab of water is independent, and mass balance dictates that

$$\dot{m} \, \delta t = \frac{v}{R} \, \delta t \, b \int m \, dy = \frac{v}{R} \, \delta t \, b \, n \, R \int c \, dy$$

This equation gives

$$\int c \, dy = \frac{\dot{m}}{qb}$$

The concentration along $x = 0$ takes the form of the Dirac delta function in y, so that the "initial condition" is

$$c(0, y) = \frac{\dot{m}}{q \, b} \, \delta(y) \quad \textbf{(8.8.6)}$$

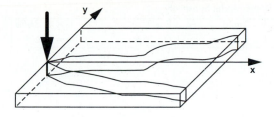

FIGURE 8.8.1 Configuration for the simple plume model

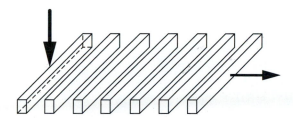

The spike at the origin in Figure 8.8.1 represents Equation 8.8.6. For steady flow, the steady-state transport equation is

$$v\frac{\partial c}{\partial x} - D_{yy}\frac{\partial^2 c}{\partial y^2} + \lambda R c = 0 \tag{8.8.7}$$

and the solution of Equation 8.8.7 with 8.8.6 is

$$c(x,y) = \frac{\dot{m}}{qb}\frac{1}{\sqrt{4\pi\dfrac{D_{yy}}{v}x}}\exp\left(-\frac{y^2}{4\dfrac{D_{yy}}{v}x}\right)\exp\left(-\frac{\lambda R x}{v}\right) \tag{8.8.8}$$

Equation 8.8.8 is essentially the same as Equation 8.4.3. The first term is the mass that enters each **slab** of water that passes beneath the source. The second and third terms account for transverse Gaussian dispersion within the slab as it is advected downgradient. The last term accounts for the decay of mass as the section moves downgradient.

EXAMPLE   **8.8.1**   **Effects of Parameter Variability**
PROBLEM   Examine the influence of the various parameters appearing in Equation 8.8.8 considering a release with basic parameters given in Table 8.8.1. The half-life gives a corresponding first-order loss rate constant $\lambda = 0.0019 d^{-1}$. Note that the

TABLE 8.8.1   Base Case Parameters for Plume Model Sensitivity

n = 0.3	b = 5m	v = 1m/d	R = 1	$T_{1/2}$ = 365d
a_L = 10m	a_T = 1m	$D_{xx} = a_L v$	$D_{w} = a_T v$	$\dot{m}$ = 5kg/d

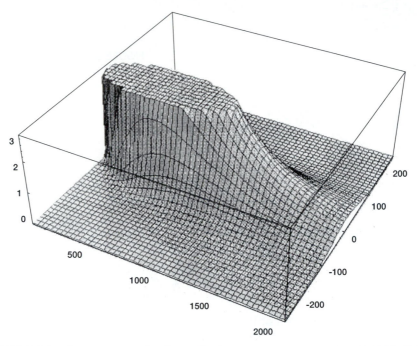

FIGURE 8.8.2   Concentration distribution for base case paramters given in Table 8.8.1 for the simple plume model

simple plume model does not include longitudinal dispersion. The longitudinal dispersivity is included in Table 8.8.1 for comparison with other models that will be discussed in the following sections.

One of the first features to notice is that the solution for the concentration distribution given by Equation 8.8.8 provides the distribution as a geometric surface, such as that shown in Figure 8.8.2. This figure shows a three-dimensional view of the concentration distribution, with the vertical axis corresponding to the concentration magnitude. The surface is cut off at an elevation corresponding to a concentration of 3 mg/L. The intersection of this horizontal cut with the concentration distribution surface gives the **concentration contour** corresponding to 3 mg/L, as a function of (x, y) coordinates.

Figure 8.8.3 shows a concentration contour map for the base case. The contours correspond to concentrations of 0.1, 1, 2, 3, and 5 mg/L, with the smaller contours representing the higher concentrations. As might be expected, the longitudinal dimensions of the contaminant plume greatly exceed the transverse dimensions. In addition, one may observe that the regions with higher concentrations are much more restricted in area extent.

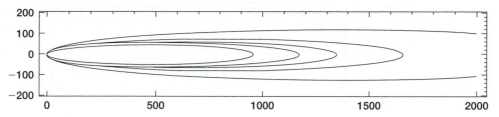

FIGURE 8.8.3 Simple plume model concentration contours for base case paramters. Concentrations = 0.1, 1, 2, 3 and 5 mg/L.

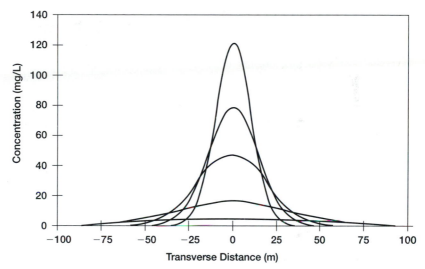

FIGURE 8.8.4 Simple plum model cross-section concentration distribution for base case parmeters at locations of 50, 100, 200, 500, 1000 and 2000 m along the ongitudinal acis of the plume.

A vertical slice perpendicular to the longitudinal axis of the concentration distribution gives a view of the cross-section concentration distribution for the plume. Such a view is shown in Figure 8.8.4 for cross-sections located at distances of 50, 100, 200, 500, 1,000 and 2,000 meters along the longitudinal axis of the plume. For each cross-section, the concentration distribution takes the form of a normal or Gaussian distribution. Furthermore, the standard deviation (σ) of the distribution increases as the square-root of the longitudinal distance:

$$\sigma_y = \sqrt{2\,a_T\,x} \qquad (8.8.9)$$

This relationship is exactly analogous to the standard deviation of the concentration distribution from a point release increasing as the square root of time. One feature that is clear from the cross-section view (that may not be clear from a concentration contour view such as that shown in Figure 8.8.3) is that the

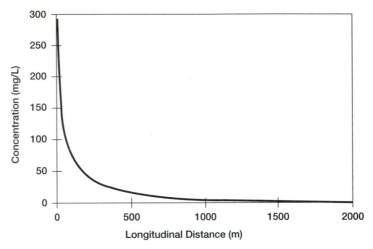

FIGURE 8.8.5   Centerline concentration for the simple plume model

concentration magnitudes may greatly exceed the maximum concentration contour shown on a map. Another feature that may not be so obvious is that the area under each of the cross-section curves decreases with increasing longitudinal distance. This happens because of constituent decay assumed for the base case set of parameters. If the constituent was conservative ($\lambda = 0$), then the area under each curve would be the same for the simple plume model.

Still another way to view the concentration distribution is to show the maximum concentration as a function of longitudinal distance along the plume centerline. Such a view is shown in Figure 8.8.5. This figure shows how rapidly the maximum concentration decreases with distance from the source for small distances, and yet how slowly it decreases with distance for larger distances.

One of the more interesting parameters to look at when evaluating the sensitivity of the simple plume model contaminant distribution is the magnitude of the seepage velocity. Figure 8.8.6 shows a set of centerline concentrations for three different seepage velocities. All of the other parameters correspond to the base conditions in Table 8.8.1. For the low velocity, $v = 0.1$ m/d, the maximum concentration is much greater near the source but rapidly decreases with longitudinal distance. For the high velocity, $v = 10$ m/d, the concentration is lowest near the source but retains those values for a much greater distance from the source. Two factors explain this behavior. The first is that the mass release rate is assumed to be constant. With the constant source term, the greater aquifer velocity means that more water moves beneath the source per unit time, that the source is diluted to a greater extent, and that the concentrations near the source are smaller. On the other hand, a smaller aquifer velocity means that the source term is added to a smaller quantity of water and the corresponding concentrations are larger. The second factor that explains the behavior observed in Figure 8.8.6 is that the model considers first-order losses with a one-year half-life.

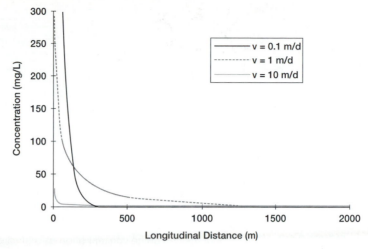

FIGURE 8.8.6   Sensitivity of the centerline concentration as a function of seepage velocity for the simple plume model

With the smaller velocity, the constituent is lost to decay before it can move very far. On the other hand, for the greater velocity, the contaminant can move significant distances before appreciable losses are experienced.

This example has important implications for representing different types of source terms. If the source is located above the aquifer, such as the release of leachate from a landfill, then the representation used to develop Figure 8.8.6 is appropriate with the source term magnitude independent of the aquifer velocity. For this condition, the magnitude of the aquifer velocity determines the extent to which the source term is diluted upon entering the aquifer. On the other hand, if the source is located within the aquifer, such as contaminated soil or a floating hydrocarbon lens, then the source term mass transfer will increase with increasing velocity. The ratio $\dot{m}/q$ will show much less variability as q or v changes, and the concentrations near the source will be similar.

Figure 8.8.7 shows the centerline concentration for three conditions with different loss-rate half-lives. As one might expect, the extent of the plume decreases with decreasing half-life. It is important to note that even a small loss-rate coefficient (large half-life) is important to include, instead of assuming that the first-order losses are negligible. In addition, when the constituent half-life is on the order of months, then the area extent of the resulting contaminant plume is very limited. Such loss rates are typical of BTEX compounds (benzene, toluene, ethyl-benzene, and xylene) from gasoline releases, and this idea helps explain why large plumes are seldom associated with spills of petroleum hydrocarbons.

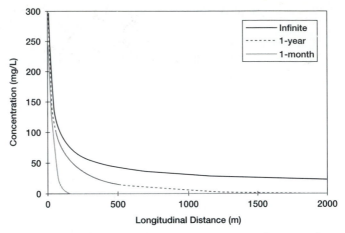

FIGURE 8.8.7 Sensitivity of the centerline concentration as a function of consituent half-life for the simple plume model

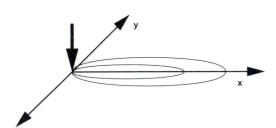

FIGURE 8.8.8 Configuration for the point-source plume model

8.8.3 Point-Source Plume Model

The second plume model to be considered is also a point-source model, but this formulation includes both longitudinal dispersion and the transient period of development of the plume. The configuration is shown in Figure 8.8.8, where the source is located at the origin. Bear (1972, 1979), Hunt (1978), and Wilson and Miller (1978) have discussed this model in the literature.

The model for a plume is developed from convolution of the point-source spill model presented in Section 8.7.3. Substitute Equation 8.7.23 with $V_o c_o = 1$ into Equation 8.8.2 and perform the following change in variables:

$$r^2 = \frac{x^2}{D'_{xx}} + \frac{y^2}{D'_{yy}} \; ; \; B^2 = \frac{1}{\lambda + \frac{v'^2}{4 D'_{xx}}} \; ; \; w = \frac{r^2}{4(t - \tau)} \quad \textbf{(8.8.10)}$$

The result may be written

$$c_p(x, y, t) = \frac{\dot{m} \exp\left(\frac{v x}{2 D_{xx}}\right)}{4 \pi n b \sqrt{D_{xx} D_{yy}}} \int_{r^2/4t}^{\infty} \frac{1}{w} \exp\left(-\left(w + \frac{r^2}{4 w B^2}\right)\right) dw \quad \textbf{(8.8.11)}$$

where $\dot{m}$ is the constant mass release rate (mass per unit time). If we look back at Section 3.5.1, we see that the integral in Equation 8.8.11 is the leaky aquifer well function of Hantush (compare with Equation 3.5.2). With this function, the solution for the plume model (Equation 8.8.11) may be written as follows:

$$c_p(x, y, t) = \frac{\dot{m}\,\exp\!\left(\dfrac{v\,x}{2\,D_{xx}}\right)}{4\,\pi\,n\,b\sqrt{D_{xx}D_{yy}}}\,W\!\left(\frac{r^2}{4\,t},\frac{r}{B}\right) \qquad (8.8.12)$$

Also, because as $u \to 0$ ($t \to \infty$), the well function becomes

$$W\!\left(u, \frac{r}{B}\right) \to 2K_0\!\left(\frac{r}{B}\right)$$

where $K_0(\ )$ is the modified Bessel function of the second kind of order zero, the steady-state plume solution is

$$c_p(x, y, t \to \infty) = \frac{\dot{m}\,\exp\!\left(\dfrac{v\,x}{2\,D_{xx}}\right)}{2\,\pi\,n\,b\sqrt{D_{xx}D_{yy}}}\,K_0\!\left(\frac{r}{B}\right) \qquad (8.8.13)$$

The following approximations are good for *large X*:

$$W(u, X) \cong \sqrt{\frac{\pi}{2\,X}}\,e^{-x}\,\mathrm{erfc}\!\left(\sqrt{u} - \frac{X}{2\sqrt{u}}\right)\ ;\ K_0(X) \cong \sqrt{\frac{\pi}{2X}}\,e^{-x} \qquad (8.8.14)$$

The approximation given by Equation 8.8.14 are good to within 10 percent error for X greater than 1, and 1 percent error for X greater than 10. The second approximation follows directly from the first, and the first is found through asymptotic expansion of the Hantush well function. Using this approximation, the steady-state (maximum) concentration along the x-axis is

$$c_p(x, y = 0, t \to \infty) = \frac{\dot{m}\,\exp\!\left(\dfrac{v\,x}{2\,D_{xx}}\left(1 - \sqrt{1 + \dfrac{4\,\lambda\,R\,D_{xx}}{v^2}}\right)\right)}{n\,b\sqrt{4\,\pi\,D_{yy}\,v\,x}\sqrt{1 + \dfrac{4\,\lambda\,R\,D_{xx}}{v^2}}} \qquad (8.8.15)$$

Note that if $\lambda = 0$, Equations 8.8.8 and 8.8.15 become identical for the solution along the *x*-axis.

The module presented in Appendix D for the Hantush well function may be used to evaluate the point-source plume model solution given in this section.

8.8.4 Gaussian-Source Plume Model

The point-source plume model considered in the previous section is useful for predicting the time development of a plume from a continuous source, although it has the disadvantage of predicting infinite concentrations at the source. The concentrations at the source must be infinite in order to introduce a finite mass flux to the aquifer through a single point. If one is interested in predicting concentrations near the source as well as in the far field, then a source of finite

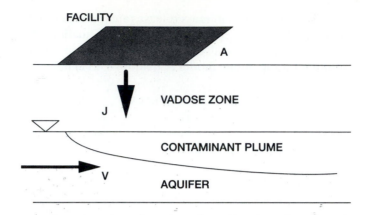

FACILITY

A

VADOSE ZONE

J

CONTAMINANT PLUME

V

AQUIFER

FIGURE 8.8.9   Basic setup of the Gaussian-source plume model

size must be considered. As a third plume model, we consider the Gaussian-source plume model, which provides a useful representation for this purpose. The model presented in this section is similar to EPACML, the composite land-fill model developed for the EPA. The basis for EPACML is presented by Huyakorn et al. (1982).

In the Gaussian-source plume model, the leachate from a surface facility is assumed to migrate through the unsaturated zone and mix with groundwater flowing beneath the facility. This process is shown schematically in Figure 8.8.9. The groundwater model is set up with a Gaussian source placed at the down-gradient end of the facility as a boundary condition. Questions of interest concern the depth of penetration of the leachate into the aquifer, and the coupling of the facility release with the aquifer source so that mass balance is achieved.

As shown in Figure 8.8.10, the facility has length L and width W with respect to the mean flow direction, and the total penetration depth of leachate at the down-gradient end of the facility is H. Penetration is caused both by the vertical advection of water as it moves beneath the facility—and by vertical dispersion. The vertical component of the groundwater velocity is assumed to vary linearly from a value i_f/n at the water table to zero at the base of the aquifer, where i_f is the infiltration through the facility. As groundwater moves laterally beneath the facility, the water is displaced downward within this velocity field, and this process gives the advection component to the penetration depth. The vertical dispersion component of the penetration depth is assumed to be one standard deviation of the vertical mixing component of a solute as it is advected beneath the facility. Together, these give

$$H = H_{adv} + H_{dis} = b\left(1 - \exp\left(-\frac{i_f L}{q_x b}\right)\right) + \sqrt{2\, a_v L} \qquad \textbf{(8.8.16)}$$

Equation 8.8.16 gives the depth of penetration of the contaminant into the aquifer beneath the facility. If the value of H calculated with Equation 8.8.16 exceeds b ($H > b$), then in the plume calculations, we take $H = b$.

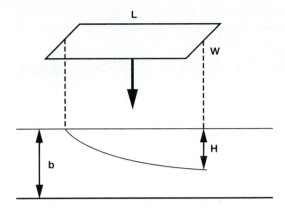

FIGURE 8.8.10 Development of mixing zone beneath the facility

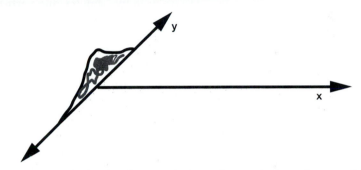

FIGURE 8.8.11 Gaussian distribution which is taken as the boundary condition at the downstream extent of the area beneath the facility

In the Gaussian-source plume model, the source is specified by a boundary condition along the $x = 0$ axis, which takes the shape of a Gaussian distribution and is specified by

$$c(x = 0, y, 0 < z < H, t) = c_o \exp\left(-\frac{y^2}{2\,\sigma^2}\right) \qquad (8.8.17)$$

where c_o is the maximum concentration and the standard deviation, σ, is a measure of the width of the source. This relationship is shown in Figure 8.8.11.

The boundary condition is coupled with the facility by requiring that the mass flux from the facility equal the advection and dispersion flux from the boundary. Considering just the advection flux, with reference to Figure 8.8.9 we have

$$\dot{m} = J\,A = i_f\,c_w A = q_x H \int_{-\infty}^{\infty} c_o \exp\left(-\frac{y^2}{2\,\sigma^2}\right)\,dy = \sqrt{2\pi}\ q_x H c_o\,\sigma \quad (8.8.18)$$

When both advection and dispersion are considered, Sudicky (background documents for EPACML) has shown that the models are coupled through

$$\dot{m} = \int_{-\infty}^{\infty} \left(q_x c - n D_{xx} \frac{\partial c}{\partial x} \right)\Bigg|_{x=0} b\, dy = \sqrt{\frac{\pi}{2}}\, q_x H\, c_o\, \sigma \left(1 + \sqrt{1 + \frac{4\lambda^* D_{xx} R}{v^2}} \right) \quad \textbf{(8.8.19)}$$

With Equation 8.8.19, we see that the peak concentration beneath the facility is related to the mass rate of flow through

$$c_o = \frac{\dot{m}}{\sqrt{\dfrac{\pi}{2}}\, q_x H\, \sigma \left(1 + \sqrt{1 + \dfrac{4\lambda^* D_{xx} R}{v^2}} \right)} \quad \textbf{(8.8.20)}$$

A similar relation may be written from Equation 8.8.18.

Transport is assumed to occur in two dimensions; vertical dispersion is assumed to be negligible beyond the source zone where it contributes to the vertical penetration of the chemical into the aquifer. In addition, we now want to have the possibility of adding on the effects of dilution from infiltration of surface recharge into the plume, at least in an approximate manner. The alternative formulation considers that infiltration forces the plume to move deeper into the aquifer and the plume thickness remains constant at H. This latter formulation is more conservative. For our analyses, we generally assume that if $H < b$, then recharge forces the plume deeper into the aquifer and H remains constant, while if $H = b$, then we assume that recharge serves to dilute the plume and acts as an equivalent decay term. In this latter case, the transport equation is

$$\frac{\partial c}{\partial t} + v' \frac{\partial c}{\partial x} + \left(\lambda + \frac{i_r}{n R H} \right) c = D'_{xx} \frac{\partial^2 c}{\partial x^2} + D'_{yy} \frac{\partial^2 c}{\partial y^2} \quad \textbf{(8.8.21)}$$

where i_r is the diffuse recharge rate outside of the facility. The effects of first-order decay and plume dilution may be combined into a single effective decay coefficient,

$$\lambda^* = \lambda + \frac{i_r}{n R H}$$

Equation 8.8.21 assumes that the flow is steady and that the velocity remains uniform in the x-direction. In addition to Equation 8.8.17, Equation 8.8.21 is subject to the following initial and boundary conditions: $c(x,y,t = 0) = 0$, $c(\infty,y,t) = c(x, -\infty,t) = c(x,\infty,t) = 0$.

The solution for the **steady-state problem** may be developed using Fourier transforms. Application of asymptotic analysis to the result gives (Smith and Charbeneau, 1990)

$$c(x,y,t \rightarrow \infty) = \cfrac{c_o \exp\left(\cfrac{vx}{2 D_{xx}}\left(1 - \sqrt{1 + \cfrac{4\lambda^* R D_{xx}}{v^2}}\right) - \cfrac{y^2}{2\sigma^2\left(1 + \cfrac{2 x D_{yy}}{\sigma^2 v \sqrt{1 + \cfrac{4\lambda^* R D_{xx}}{v^2}}}\right)}\right)}{\sqrt{1 + \cfrac{2 x D_{yy}}{\sigma^2 v \sqrt{1 + \cfrac{4\lambda^* R D_{xx}}{v^2}}}}} \qquad (8.8.22)$$

EXAMPLE   **8.8.2**   **Influence of Source Size**

PROBLEM   Determine the sensitivity of the plume centerline concentration to the size of the source for the conditions of Example 8.8.1. The base parameters are given in Table 8.8.1.

    The new variable that appears in the Gaussian-source model is the size of the source, designated by the standard deviation of the plume at the source. The parameter σ should be related to the width of the source, W, in a direct way. In the EPACML model, it is assumed that $W = 6\sigma$. This equation is equivalent to assuming that 99.7 percent of the mass resides beneath the facility at its downgradient extent. It is expected that mass that enters the aquifer on the upgradient end of the facility, however, will begin to move laterally, and the condition imposed by EPACML appears to be stringent. Here, it is assumed that $W = 4\sigma$, which is equivalent to assuming that 95.6 percent of the mass within the plume resides directly beneath the facility at its downgradient end.

    Figure 8.8.12 shows the centerline plume concentrations for the same base conditions as Example 8.8.1, but with σ values of 0.1 m, 1m, and 10 m. Also shown in this figure is the solution from the simple plume model, which is nearly coincident with the solution for $\sigma = 0.1$ m. From this figure, it appears that one cannot distinguish between sources of size 0.4 m and 4 m (4σ) at distances beyond 10 m under the assumed conditions. A source of size 40 m, however, appears as a point source only beyond distances of a few hundred meters.

EXAMPLE   **8.8.3**   **Comparison of Plume Models**

PROBLEM   Compare the solutions obtained from the simple plume model, the point-source plume model, and the Gaussian-source plume model for steady-state conditions using the base parameters presented in Table 8.8.1.

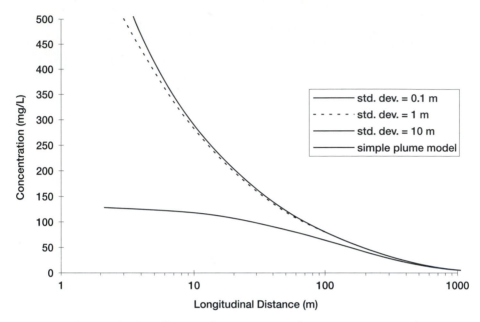

FIGURE 8.8.12 Influence of source size on plume concentrations. The upper curve corresponds to σ = 0.1 m, while the lower curves correspond to σ=1m and 10 m, respectively.

A comparison of solutions from the simple plume model and the Gaussian plume model has already been presented in Figure 8.8.12. It was noted that the solutions are nearly coincident for σ = 0.1 m. Perhaps the most significant feature to note is that the Gaussian model uses a longitudinal dispersivity of 10 m, while longitudinal dispersion is neglected in the simple plume model. The similarity in solutions highlights the minor role played by longitudinal dispersion in the contaminant plume distribution under steady-state conditions. Certainly, however, longitudinal dispersion is important in describing the concentration distribution near the leading edge of the plume under transient conditions as the plume is developing.

Figure 8. 8.13 shows the concentration contours for the simple plume model, the point-source plume model, and the Gaussian-source plume model using the base parameters given in Table 8.8.1. The most significant feature is the similarity between the contours, suggesting equivalence in solutions. Again, this display highlights the minor importance of longitudinal dispersion in determining steady-state plume concentration distributions. It may be concluded that as long as the source is small and one is only interested in maximum concentrations from a continual source, then the simple plume model is adequate. If transient effects are important and the source is still small, then the point-source plume model might be preferred because the calculations are simple. If the size of the source is important, then the Gaussian-source model can be used. Its formulation for transient conditions is discussed next.

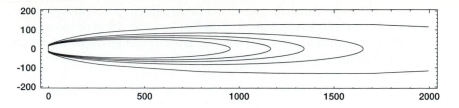

a) Simple Plume Model

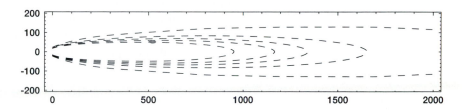

b) Point-Source Plume Model

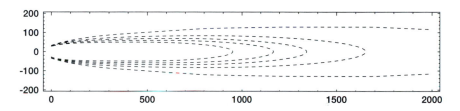

c) Gaussian-Source Plume Model

FIGURE 8.8.13 Steady-state plume model concentration contours a)simple plume model, b) point-source plume model, and c) Gaussian-source plume model using base paramters give in Table 8.8.1.

To find the solution to the **transient problem**, we first simplify notation by introducing the following dimensionless variables:

$$X = \frac{v'x}{D'_{xx}} \; ; \; Y = \frac{y}{\sigma} \; ; \; T = \frac{v'^2 t}{D'_{xx}} \; ; \; \Lambda = \frac{\lambda^* D'_{xx}}{v'^2} \; ; \; D = \frac{D'_{xx} D'_{yy}}{\sigma^2 v'^2} \; ; \; C = \frac{c}{c_o} \quad (8.8.23)$$

In terms of these variables, Equations 8.8.21 and 8.8.17 become

$$\frac{\partial C}{\partial T} + \frac{\partial C}{\partial X} - \frac{\partial^2 C}{\partial X^2} - D \frac{\partial^2 C}{\partial Y^2} + \Lambda C = 0 \quad (8.8.24)$$

$$C(0, Y, T) = \exp\left(-\frac{Y^2}{2}\right) \quad (8.8.25)$$

The other boundary conditions remain the same.

The mathematical statement of the transient problem is given in Equations 8.8.24 and 8.8.25. Application of the Laplace transform reduces the equation to the steady state form, whose solution is found using Fourier transform methods. Application of the inverse transforms then gives

$$C(X,Y,T) = X \int_0^T \frac{\exp\left(-\dfrac{X^2}{4\,\omega} - \dfrac{Y^2}{2 + 4\,D\,\omega} + \dfrac{X}{2} - \dfrac{1 + 4\Lambda}{4}\omega\right)}{\sqrt{4\,\pi\,\omega^3(1 + 2\,D\,\omega)}}\,d\omega \quad \textbf{(8.8.26)}$$

This equation is the general solution for constant boundary conditions. The solution for a time-variable boundary condition may be found using Duhamel's theorem (8.8.5). Because the transient solution of Equation 8.8.26 is an integral with T appearing only in the upper limit, the partial derivative with respect to T is simply the integrand. Recognizing this fact, Duhamel's theorem gives

$$C(X,Y,T) = \frac{X \exp\left(\dfrac{X}{2}\right)}{\sqrt{4\,\pi}} \int_0^T \frac{B(T - \omega)\exp\left(-\dfrac{X^2}{4\omega} - \dfrac{Y^2}{2 + 4\omega} - \dfrac{1 + 4\Lambda}{4}\omega\right)}{\sqrt{\omega^3(1 + 2D\,\omega)}}\,d\omega \quad \textbf{(8.8.27)}$$

Equation 8.8.27 forms the basis of the *Transient Source Gaussian Plume* (TS-GPLUME) model. Equation 8.8.25 gives the **surface** concentration, and Equation 8.8.26 corresponds to the case $B(T) = 1$. If a time variable mass flux to the aquifer is known, $\dot{m}(T)$, then Equation 8.8.20 may be used to relate this value to the time variable source concentration, and one uses $B(T - \omega) = c_o(T - \omega)$ in Equation 8.8.27. The integral may be evaluated using Romberg integration or other numerical quadrature techniques. The transient-source Gaussian plume model is one component of the *Hydrocarbon Spill Screening Model* (HSSM), which is discussed in Chapter 9, and an example application is presented in that chapter.

8.8.5   Horizontal Plane Source (HPS) Plume Model

The last plume model we will look at considers a horizontal plane source located at the top of the aquifer, as shown in Figure 8.8.14. This model is a transient model that includes steady advection along the x-axis, mixing in all three directions, decay, and a time-variable source rate. Through superposition, multiple sources can be considered so that the actual source shape may be "L-shaped" or any other combination that can be made from rectangles.

The source condition is specified through an infiltration rate and concentration over the region of the rectangle. Thus, if the rectangle has length L and width W, then the mass release rate is

$$\dot{m} = i_f\, L\, W\, c_0 \quad \textbf{(8.8.28)}$$

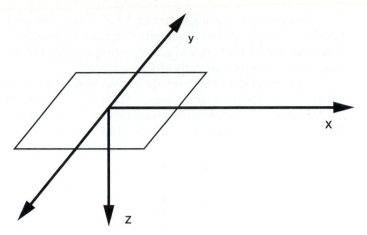

FIGURE 8.8.14   Horizontal plane source model

where c_o is the concentration in the waste leachate and i_f is the infiltration rate from an overlying facility.

The solution for the horizontal plane source may be developed from the solutions we have already obtained. A horizontal plane "spill" of **unit mass** is given by a combination of the solution for a rectangular source in the horizontal plane located at the aquifer surface (Section 8.7.6) and a point source in the vertical (Section 8.7.4):

$$c_s(x,y,z,t) = \frac{2e^{-\lambda t}}{4\,n\,R\,L\,Wb\,\sqrt{4\,\pi\,D_{zz}{'}t}}\left(\text{erf}\left(\frac{x - v't + L/2}{\sqrt{4 D_{xx}{'}t}}\right) - \text{erf}\left(\frac{x - v't - L/2}{\sqrt{4 D_{xx}{'}t}}\right)\right)$$

$$\times \left(\text{erf}\left(\frac{y + W/2}{\sqrt{4 D_{yy}{'}t}}\right) - \text{erf}\left(\frac{y - W/2}{\sqrt{4 D_{yy}{'}t}}\right)\right) \times \sum_{j=-\infty}^{\infty} \exp\left(-\frac{(z + 2j\,b)^2}{4 D'_{zz}t}\right) \qquad \textbf{(8.8.29)}$$

Galya (1987) suggests that Equation 8.8.29 is most useful for short times where only a few terms in the summation need to be used. For longer times, however, Carslaw and Jaeger (1959, p. 361) show that another equivalent form is given by

$$c_s(x,y,z,t) = \frac{e^{-\lambda t}}{4\,n\,R\,L\,Wb\,\sqrt{4\pi D'_{zz}t}}\left(\text{erf}\left(\frac{x - v't + L/2}{\sqrt{4 D_{xx}{'}t}}\right) - \text{erf}\left(\frac{x - v't - L/2}{\sqrt{4 D_{xx}{'}t}}\right)\right)$$

$$\times \left(\text{erf}\left(\frac{y + W/2}{\sqrt{4 D_{yy}{'}t}}\right) - \text{erf}\left(\frac{y - W/2}{\sqrt{4 D_{yy}{'}t}}\right)\right) \times \left(1 + 2\sum_{j=1}^{\infty} \exp\left(-\frac{j^2\,\pi^2\,D'_{zz}t}{b^2}\right)\cos\left(\frac{j\pi z}{b}\right)\right) \qquad \textbf{(8.8.30)}$$

With either Equations 8.8.29 or 8.8.30, the solution for the plume with a mass release rate $\dot{m}(t)$ is found by convolution as

$$c_p(x, y, z, t) = \int_0^t \dot{m}(\tau)\,c_s(x, y, z, t - \tau)\,d\tau \qquad \textbf{(8.8.31)}$$

Equation 8.8.31 with Equation 8.8.30 is the solution presented by Galya (1987). If one recognizes the limits discussed in Section 8.7.2, however, then Equations 8.8.31 with 8.8.29 and appropriate approximating expressions provides a more computationally efficient model.

Johnson et al. (1998) provide a series of nomographs that can be used to estimate *dilution attenuation factors* (DAFs), which are the ratio of leachate concentration in contact with contaminated soil to the average receptor concentration. These nomographs are based on the HPS model for conditions when the contaminated soil is located within the vadose zone above the water table.

EXAMPLE PROBLEM   **8.8.4   Exposure Concentrations from Contaminated Soil**

As an example application of the HPS model, consider the situation shown schematically in Figure 8.8.15. A layer of soil contaminated with residual petroleum hydrocarbon is located within a region at the ground surface. Benzene is leached from the soil to an underlying aquifer and is transported to a downgradient well. Evaluate the effects of well screen interval on the maximum exposure concentration.

Models for leachate generation are described in Chapters 5 and 9. Using a simple model for multiphase partitioning, mass transfer, and advection transport, such as presented in Section 5.4, the mass flux to the aquifer is calculated as shown in Figure 8.8.16. Time zero corresponds to the leachate first arriving at the water table. Degradation in the vadose zone is not considered. The initial benzene mass flux to the water table is about 146 g/yr and decreases to about 70 g/yr after 60 years. The curve shown in Figure 8.8.16 provides the function $\dot{m}(t)$.

The results from Figure 8.8.16 are used in Equation 8.8.31 to calculate the concentration at any point (x, y, z) and time t. To calculate the concentration for an observation well that is screened from the top of the aquifer to a depth z_w, an averaging must be applied:

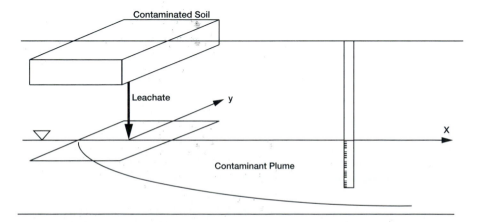

FIGURE 8.8.15   Groundwater Exposures from Contaminated Soil

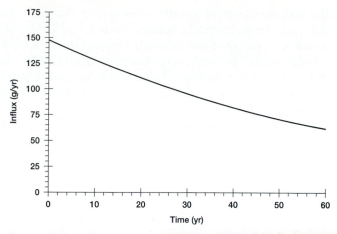

FIGURE 8.8.16 Benzene Influx to the Saturated Zone from Contaminated Soil Using a Simple Leaching Model

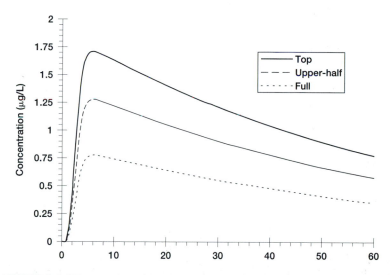

FIGURE 8.8.17 Benzene Concentration in a Downgradient Observation Well

$$c_w\left(x, y, t; z_w\right) = \frac{1}{z_w} \int_0^{z_w} c_p\left(x, y, z, t\right) dz \qquad \textbf{(8.8.32)}$$

For this example, the following parameters are assumed: $n = 0.3$, $b = 10$ m, $v = 100$ m/yr, $a_L = 10$ m, $a_T = 1$ m, $a_v = 0.1$ m, $R = 4$, and $T_{1/2} = 1$ yr. The source area is assumed to be 10 m by 10 m. The results for a well located at ($x = 100$ m, $y = 0$ m) are shown in Figure 8.8.17 for three conditions. First, the concentration at the top of the aquifer ($z = 0$) is calculated. The maximum concentration at the top of the aquifer is 1.7 µg/L at a time of 6 years. The benzene concentration exceeds 1 µg/L for a period of 41 years. The second curve in Figure 8.8.17 shows the concentration for a well screened over the upper half of the aquifer. For this case, the maximum concentration is 1.3 µg/L, also at a time of 6 years. For a well screened over this interval, the concentration exceeds 1 µg/L for a period of 21 years. The third curve shown in Figure 8.8.17 is for a well screened over the entire aquifer thickness. This result is the same result that one would obtain from a two-dimensional model with the same source term. For this case, the maximum concentration is 0.8 µg/L at 6 years.

One may use results such as that shown in Figure 8.8.17 to calculate the well concentration for other screened intervals. Hemstreet (1996) shows that the concentration for a well screened between the depth z_{w1} and z_{w2} is given by

$$c_w(x, y, t; z_{w1}, z_{w2}) = \frac{1}{z_{w2} - z_{w1}}(z_{w2} \times c_w(x, y, t; z_{w2}) - z_{w1} \times c_w(x, y, t; z_{w1})) \quad \textbf{(8.8.33)}$$

For example, if the well was screened over the lower half of the aquifer, then Equation 8.8.33 may be used to show that the maximum concentration would be

$$c_w(100,0,t_{max};5,10) = \frac{1}{5}(10 \times 0.8 - 5 \times 13) = 0.3 \mu g/L$$

This example shows that for certain types of problems, the dimensionality of the model is important, as is the ability to deal with time variable source terms.

8.9   Numerical Simulation of Solute Transport

Use of numerical models for simulating groundwater flow is described in Section 2.5. Simulation of solute transport presents other types of numerical difficulties beyond those associated with site characterization, parameter selection, and model calibration. Some of these difficulties are described in this section, and possible modeling approaches are presented.

8.9.1   Numerical Simulation

In order to consider some of the problems that arise in numerical simulation of solute transport, we first look at the $1D$ transport equation that takes the following form:

$$\frac{\partial c}{\partial t} + v \frac{\partial c}{\partial x} - D \frac{\partial^2 c}{\partial x^2} = 0 \quad \textbf{(8.9.1)}$$

Let the concentration at node point location $x = i\Delta x$ and time $t = m\Delta t$ be designated by $c^m{}_i$. A straightforward application of finite differencing gives

$$\frac{c_i^{m+1} - c_i^m}{\Delta t} + \alpha v \frac{c_{i+1}^{m+1} - c_{i-1}^{m+1}}{2\,\Delta x} + (1 - \alpha)\, v \frac{c_{i+1}^m - c_{i-1}^m}{2\,\Delta x}$$

$$- \alpha\, D \frac{c_{i+1}^{m+1} - 2\,c_i^{m+1} + c_{i-1}^{m+1}}{\Delta x^2} - (1 - \alpha) D \frac{c_{i+1}^m - 2c_i^m + c_{i-1}^m}{\Delta x^2} = 0 \quad \textbf{(8.9.2)}$$

In Equation 8.9.2, α is a weighting factor with values between 0 and 1 that is used to determine the character of the solution technique.

For $\alpha = 0$, this equation is an **explicit** formulation which is $O(\Delta t, \Delta x^2)$. The notation $O(\Delta t, \Delta x^2)$ is read "of order delta-t, delta-x-squared", and it specifies the rate at which the truncation error of the finite difference expression vanishes as Δt and Δx decrease in magnitude. The formulation is numerically stable so long as

$$\frac{2\,D\Delta t}{\Delta x^2} + \frac{v\,\Delta t}{\Delta x} \leq 1 \qquad\qquad \textbf{(8.9.3)}$$

For most practical problems, this equation leads to a short time step.

For $\alpha = 1$, this equation is an **implicit** scheme and is stable for all time steps. It is also $O(\Delta t, \Delta x^2)$. For $\alpha = \frac{1}{2}$, this equation is called a **Crank-Nicolson** formulation. It is $O(\Delta t^2, \Delta x^2)$, and it is always stable.

The problem with all of these schemes is they exhibit **overshoot** (concentrations greater than the maximum possible) and **undershoot** (concentrations less than zero) near concentration fronts, as shown in Figure 8.9.1. The negative concentrations are especially a problem when chemical reactions are added to the model.

The overshoot and undershoot are primarily associated with the space-centered form of the advection term approximation. A downstream one-sided approximation is found to be inherently unstable. An upstream (upwind, from its development for meteorological simulations) approximation, however, overcomes the problems of overshoot and undershoot. With this formulation, one uses

$$\frac{\partial c}{\partial x} \cong \frac{c_i^m - c_{i-1}^m}{\Delta x}$$

for an explicit formulation and

$$\frac{\partial c}{\partial x} \cong \frac{c_i^{m+1} - c_{i-1}^{m+1}}{\Delta x}$$

for an implicit formulation. While the upwind form overcomes overshoot and undershoot problems, they do create **numerical dispersion**. An analysis shows

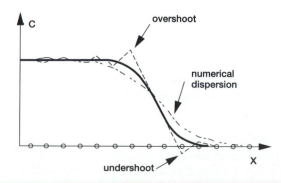

FIGURE 8.9.1 Computational problems with numerical simulation of the transport equation

that the upwind difference formulations have truncation errors for the explicit model and for the implicit model, respectively, of

$$\left(\frac{v\Delta x}{2} - \frac{v^2\,\Delta t}{2} \right) \frac{\partial^2 c}{\partial x^2} \; ; \; \left(\frac{v\,\Delta x}{2} + \frac{v^2\,\Delta t}{2} \right) \frac{\partial^2 c}{\partial c^2}$$

These truncation errors add to the apparent dispersion seen in the simulation, and thus the name. For a $1D$ simulation one may overcome the problem of numerical dispersion by adjusting the dispersion coefficient to

$$D^* = D - \frac{v\,\Delta x}{2} + \frac{v^2\,\Delta t}{2} \; ; \; D^* = D - \frac{v\,\Delta x}{2} - \frac{v^2\,\Delta t}{2} \qquad \textbf{(8.9.4)}$$

for the explicit model and implicit model, respectively (see Marsily, 1986, p. 390). This approach is cumbersome in practice unless v, Δx, and Δt are constants in space and time. Alternatively, one chooses a time step and mesh size so that $v = \Delta x/\Delta t$. This special case corresponds to the situation where all of the water is displaced from each finite difference cell during each time step.

From this discussion, it appears that one should use the upwind form with appropriate corrections for simple $1D$ simulations. For non-simple and multi-dimensional problems, however, the upwind formulation is not attractive.

The underlying problem comes from the mathematical form of the advection-dispersion equation. As noted previously, it exhibits both parabolic and hyperbolic characteristics. The space-centered form of the difference equations works well for parabolic partial differential equations. This form corresponds to the dispersion and source/sink parts of the problem, as viewed from an Eulerian system. For the advection part of the problem formulation, the *method of characteristics* (MOC) is most appropriate. The MOC form corresponds to the Lagrangian formulation of the problem. An approach towards a computational algorithm is to use a hybrid of the two formulations.

8.9.2   Hybrid Eulerian-Lagrangian Model Formulations

In this section, we will look at mixed **Eulerian-Lagrangian models** for simulation of solute transport. Numerical simulation for the flow equation was described in Section 2.5, based on Equations 2.3.32 or 2.3.33. The general mass transport equation (Equation 5.3.7) that includes sorption, volumetric sources, and first-order decay may be written

$$\frac{\partial m}{\partial t} + \nabla \cdot (\tilde{q}\,c - n\tilde{\tilde{D}} \cdot \nabla c) + \lambda\,m - W'c' = 0 \qquad \textbf{(8.9.5)}$$

where W' is the volumetric source strength (volume/time/volume) and c' is the source concentration. We first want to place Equation 8.9.5 in a form that is useful for numerical approximation.

Development of the Mass Transport Equation

If we assume that the distribution coefficient and solids density are constant and recall that $\rho_b = (1 - n)\rho_s$, then the first term in Equation 8.9.5 may be written

$$\frac{\partial m}{\partial t} = n\,R\frac{\partial c}{\partial t} + c(1 - \rho_s K_d)\frac{\partial n}{\partial t}$$

Furthermore, if the advection term in Equation 8.9.5 is expanded, then we have

$$n\,R\frac{\partial c}{\partial t} + c\,(1 - \rho_s K_d)\frac{\partial n}{\partial t} + \tilde{q}\cdot\nabla c + c\nabla\cdot\tilde{q} - \nabla\cdot(n\tilde{\tilde{D}}\cdot\nabla c) + \lambda\,n\,R\,c - W'c' = 0$$

Equation 2.3.32 may be substituted to eliminate the divergence of the Darcy velocity field, which gives after dividing through by the porosity

$$R\frac{\partial c}{\partial t} + \tilde{v}\cdot\nabla c = \frac{1}{n}\nabla\cdot(\tilde{\tilde{D}}\cdot\nabla c) - \lambda\,R\,c + \frac{W'}{n}(c' - c) + c\left(S_s\frac{\partial h}{\partial t} - (1 - \rho_s\,K_d)\frac{\partial n}{\partial t}\right) = F(c) \quad \textbf{(8.9.6)}$$

This formulation is the Eulerian formulation that gives the concentration at any point as a function of time. If W' is positive (a volumetric source) and the source concentration is different from the local concentration, then the source can cause a change in the local concentration. On the other hand, if W' is negative, then the withdrawn water may have the local concentration and $c' = c$, meaning that the term vanishes. This situation is not always the case. For example, if the volume sink is associated with evaporation of water, then $c' = 0$ and the term gives a first-order source (rather than decay). The last term on the right of Equation 8.9.6 is small and is usually neglected, especially considering that during most of the time period of solute transport, the flow conditions are considered steady. The function $F(c)$ in Equation 8.9.6 includes the dispersion, first-order decay and source/sink terms.

Development of the Lagrangian Equations

The Lagrangian formulation looks at the problem from the point of view of particles which move along with the flow at a "characteristic velocity" and gives the concentration change for these moving particles. The method of characteristics provides a numerical method for solution of the transport problem in the Lagrangian formulation.

The characteristic form of the transport Equation 8.9.6 is

$$\frac{dt}{R} = \frac{dx}{v_x} = \frac{dy}{v_y} = \frac{dz}{v_z} = \frac{dc}{F(c)} \quad \textbf{(8.9.7)}$$

This equation says that for a fluid particle that moves with velocity

$$\frac{d\tilde{x}}{dt} = \frac{\tilde{v}}{R} = \tilde{v}' \quad \textbf{(8.9.8)}$$

the concentration change for the particle is given by

$$\frac{dc}{dt} = \frac{F(c)}{R} \quad \textbf{(8.9.9)}$$

In Equation 8.9.8, $\tilde{v}'$ is the retarded seepage velocity. Equations 8.9.8 and 8.9.9 apply for moving fluid particles, and it is Equation 8.9.9 that presents conceptual problems for solution. Combined with Equation 8.9.6, it may be written

$$R\frac{dc}{dt} = \frac{1}{n}\nabla \cdot (\tilde{D} \cdot \nabla c) - \lambda R c + \frac{W'}{n}(c' - c) \tag{8.9.10}$$

Here, the left side is written for moving particles (Lagrangian), while the right side involves spatial gradients at specific points (Eulerian). Equation 8.9.10 shows that application of the method of characteristics leads to a mixed Eulerian-Lagrangian formulation. Various solution algorithms are discussed in the following paragraphs.

Forward-Moving Particle Algorithms

One approach to the solute transport problem is to work with both particles and cells. Initially, a uniform distribution of particles in cells is established, as shown in Figure 8.9.2a—where there are four particles per cell for a $2D$ domain. Each particle in a cell is initially assigned the cell's initial concentration. The particles are then moved according to the local velocity field to simulate transport due to advection. Figure 8.9.2b shows this situation for two advection time steps. For each particle during each time step, the advection transport equation Equation 8.9.7 gives

$$\tilde{X}_{t+\Delta t} = \tilde{X}_t + \tilde{V}'_{t+\Delta t}\Delta t \tag{8.9.10}$$

where $\tilde{X}_{t+\Delta t}$ is the location of the particle at the end of the step, $\tilde{X}_t$ is the location of the particle at the beginning of the step, and $\tilde{V}'_{t+\Delta t}$ is the retarded seepage velocity of the particle as it moves from $\tilde{X}_t$ to $\tilde{X}_{t+\Delta t}$ (the particles Lagrangian velocity evaluated at time $t + \Delta t$). The particle velocity is determined by interpolation of the velocities calculated from the flow model. During this advection, step the particle concentration remains $c(\tilde{X}_t)$, which is the particle concentration at its initial location and beginning of the time step.

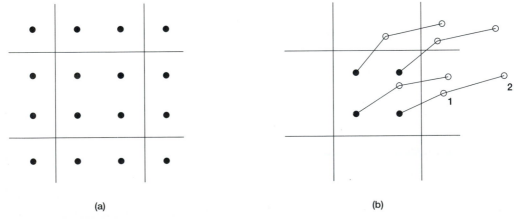

(a)   (b)

FIGURE 8.9.2   Initial distribution of four particles per cell

At the end of the advection step, the concentration for each cell is calculated by averaging the particle concentrations for all particles that are located within the cell. This new cell concentration is designated $c^I_{i,j,k}$. Once the cell concentrations are estimated, the concentration change for each cell due to dispersion and decay is then found from Equation 8.9.8 written

$$\Delta c^{II}_{i,j,k} = \frac{F(c^I_{i,j,k})}{R}\Delta t \tag{8.9.11}$$

In evaluating $F(c^I_{i,j,k})$ a central difference approximation is used for the dispersion terms and the entire term is treated explicitly. For a particle within cell (i,j,k), its final concentration at time $t + \Delta t$ is then found from

$$c(\tilde{X}_{t+\Delta t}) = c(\tilde{X}_t) + \Delta c^{II}_{i,j,k} \tag{8.9.12}$$

These values are used for the next time step.

The algorithm presented by Equations 8.9.10 through 8.9.12 was introduced by Garder et al. (1964) and was applied in the widely used MOC code developed by the United States Geological Survey (Konikow and Bredehoeft, 1978). Updates to the USGS MOC code are described by Goode and Konikow (1989) and by Konikow et al. (1994). More recently, Konikow et al. (1996) have presented a 3D version of MOC (MOC3D) that is directly integrated with the MODFLOW model introduced in Section 2.5.

The MOC algorithm lets the particles move forward along the streamlines or streak lines of the flow field for the advection term. To solve the remaining dispersion and source/sink part, an arithmetic average of the particle concentrations within the cell is used. A potential difficulty with this approach is that when the resulting distribution of particles is not uniform, one would really like to use volume averages of the particle concentrations where less weight is assigned to particles that are clumped together. The resulting simulation model may have significant mass balance errors. These mass balance errors, however, can be partially controlled by re-establishing the particle grid if the particle spacing becomes too large.

Backward-Interpolation Algorithms

Russell and Wheeler (1983) presented an alternative to the forward particle-tracking algorithm. Rather than track a large number of particles forward in time, which requires keeping track of their position and concentration, the alternative approach places a particle at a node point at time $t + \Delta t$, and the particle is tracked backward to find its position at the old time level t. The particle concentration at time t is then interpolated from concentrations at neighboring nodal points. This concentration is used as $c^I_{i,j,k}$, and Equation 8.9.11 is used to find the node point concentration change for the time step.

This modified MOC (MMOC) algorithm uses fewer particles than the MOC method, and because the particle location at each new time step is known, there is no need to store the particle identities. The resulting MMOC algorithm is faster than the MOC algorithm, although because of the interpolation involved in finding the particle concentration at the beginning of the time step, the method does introduce numerical dispersion.

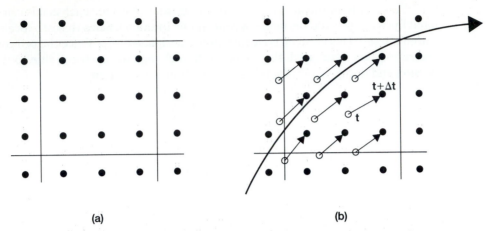

(a)   (b)

FIGURE 8.9.3   Backward-tracking of particles in the MMOC algorithm

Introducing a larger number of particles per cell and applying the MMOC algorithm can minimize numerical dispersion. For example, Figure 8.9.3 shows a cell with 9 particles whose positions are known at time $t + \Delta t$. Each particle is tracked backward to find its location at time t, and the particle concentration is interpolated from neighboring particles rather than neighboring cells. The average of the particle concentrations in each cell provides $c^I_{i,j,k}$, and Equations 8.9.11 and 8.9.12 complete the algorithm. Because the particles are uniformly spaced at time $t + \Delta t$, use of an arithmetic average for the cell concentration should not introduce mass balance errors.

Hybrid MOC Algorithms

Both the MOC and MMOC algorithms have advantages for certain classes of problems. The MMOC algorithm is generally more computationally efficient, while the MOC algorithm is more accurate in tracking concentration fronts. Neuman (1981, 1983) has suggested combining these methods to take advantages of both attributes. Using an automatic adaptive scheme, particles are introduced and tracked forward near concentration fronts. Away from these fronts where the concentration field is more uniform, a backward tracking algorithm is implemented with one particle per cell. Criteria are developed to select between the MOC and MMOC algorithms over different parts of the simulation domain.

Zheng (1990) used this hybrid method in developing his mass transport *three-dimensional algorithm* (MT3D) for the EPA. This model was developed to use the flow field simulated by MODFLOW. The MT3D model uses an automatic adaptive procedure where particles are introduced and move with the flow field in regions where concentration gradients are large, and away from such fronts, nodal point concentrations are directly tracked backward in time. The concentration of the particle which ends up at the center of the cell at the end of the time step is determined by finding the location of the particle at the beginning of the time step looking backwards along the flow path, and interpolating to find its concentration.

Application of MOC3D

Numerical solute transport models may be used to simulate performance of groundwater remediation systems as part of system design. Once a model has been calibrated to field conditions, various alternative designs including well locations and pumping rates may be evaluated, and the performance sensitivity to uncertain variables can be estimated. As an example, Figure 8.9.4a shows the numerical model representation for a landfill located near a river. The 2D-model domain uses 38 rows and 19 columns, with each cell 75 by 75 meters. A constant head of 23 m is specified along row 1, and the average head along the river is 17 m. The aquifer's saturated thickness varies between that of the constant head boundary and the river, with the uniform base of the aquifer selected as the elevation datum. The porosity is 0.35 and a uniform effective hydraulic conductivity of 15 m/d is assumed. These variables give an average seepage velocity of about 36 m/yr for the aquifer. The longitudinal and transverse dispersivities are 10 m and 1 m, respectively.

Three cells represent the landfill, and the chemical release rate is 0.4 kg/d from each cell (1.2 kg/d total). The landfill source is modeled as an injection well with a discharge of 1 m^3/d and concentration 400 mg/L for each cell. The release occurs over a period of approximately 30 years (10,000 days). The MODFLOW/MOC3D models were used for the simulation. The resulting concentration distribution from this period of mass release is shown in Figure 8.9.4b, where the maximum concentration is 16.1 mg/L near the source area. The contaminant plume extends nearly 1.5 km from the source.

Following the period of release, the stress period of recovery is simulated using 3 wells pumping at individual rates of 500 m^3/d. A 10-year restoration period is simulated. During this time period, the simulation model predicts that more than 10,900 kg of the chemical constituent will be removed, and approximately 1,950 kg will remain within the aquifer. The maximum concentration remaining in the aquifer is 2.6 mg/L. Because little mass is predicted to leave the aquifer through the river, the simulation mass balance error is 850 kg of chemical, which is -7 percent of the mass released. The mass balance error at the end of the release period is -0.16 percent.

8.10   Nonideal Flow in Saturated Porous Media

The advection-dispersion transport models described in this chapter are based on a number of assumptions. First, it is assumed that effective homogeneous parameter values can be identified that characterized the porous medium's hydraulic conductivity and porosity. Although numerical models allow representation of large-scale variability in hydraulic conductivity, effective values must still be found for the conductivity of each cell. Smaller-scale heterogeneity cannot be directly included, although it may influence mass transport.

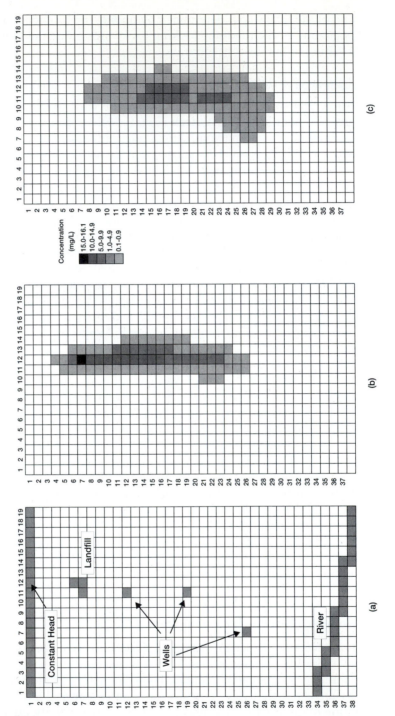

FIGURE 8.9.4 Simulation of landfill leachate migration and contminant recovery using the MOC3D model

The second basic assumption is that interphase mass transfer during sorption is sufficiently fast compared with the hydraulic residence time so that mass-transfer interaction can be considered instantaneous. This idea leads to the local equilibrium assumption wherein the sorbed concentration is always related to the aqueous concentration through the sorption isotherm, $c_s = F(c)$. Furthermore, it has been assumed that the sorption isotherm is linear. That is, $F(c) = K_d c$, and the resulting retardation factor is constant.

Finally, it has been assumed that first-order rate models may describe reactions such as decay. This fact is certainly the case for radioactive decay. It is not always true for biodegradation reactions, however, as was discussed in Section 5.5.

Solute transport problems that are adequately described by these assumptions may be considered to be ideal. It is of interest to briefly describe what impact nonideal conditions have on mass transport, where some or all of these basic assumptions are not satisfied. Aspects of nonideal transport are discussed in the following sections.

8.10.1   Spatially Variable Hydraulic Conductivity

Spatially variable hydraulic conductivity fields are discussed in Chapter 2 and again at numerous points in this chapter. The influence of layered heterogeneity may be described by using small dispersivities (centimeters) for each layer if the individual layers are modeled. If only fully screened monitoring wells are used, however, the influence of the layers shows up with an early breakthrough, excessive tailing, and an apparently large dispersivity. For non-structured, spatially-variable hydraulic conductivity fields, their influence may be included through use of a larger dispersivity whose magnitude increases with distance of migration. For large distances, the magnitude of the dispersivity depends on the statistical description of the hydraulic conductivity field (see Section 8.6 for an example). Anderson (1997) provides a description of geological settings and the difficulties they pose in site characterization.

The presence of regions with much smaller hydraulic conductivity, such as clay lenses, within the flow domain presents a special problem for groundwater remediation efforts. Low hydraulic conductivity means that diffusion rather than advection dominates mass transport. Given the long time frames that contaminants remain in contact with these regions following a release, however, a significant mass may diffuse into the low-permeability units. Then, during remediation efforts, groundwater readily flows past and around the low permeability regions, but the mass transfer from the low permeability region is slow and diffusion controlled. Excessive tailing of contaminants occurs—and long remediation times are required to completely remove pollutants.

8.10.2   Rate-Limited Sorption

There is ample field evidence that the local equilibrium assumption does not always apply. For example, in the Borden tracer experiments reported by Roberts et al. (1986), organic tracers that were introduced along with Cl⁻ and Br⁻ were found to decelerate when compared to the migration of the halides. This slowing-down was attributed to **rate-limited sorption**. Eventually, after a period of a month or so, the rate of migration of the sorbed species was smaller than that of the halides by an amount equal to the equilibrium retardation factor for the species. For shorter times, however, sorption was limited by mass transfer kinetics, presumably associated with intraparticle diffusion to the sorption sites.

Characteristically, rate-limited sorption results in early breakthrough and tailing of concentration fronts and deceleration of contaminant spills and plumes. Rate-limited desorption results in tailing of contaminants during remediation efforts. Representation of rate-limited sorption in groundwater models requires that one model both the sorbed and aqueous concentrations. For example, a simple, one-dimensional nonequilibrium model corresponding to a linear isotherm would take the form

$$n\frac{\partial c}{\partial t} + \rho_b\frac{\partial c_s}{\partial t} + q\frac{\partial c}{\partial x} - nD\frac{\partial^2 c}{\partial x^2} = 0 \; ; \; \frac{\partial c_s}{\partial t} = k_s(K_d c - c_s) \qquad \textbf{(8.10.1)}$$

where k_s is the mass transfer coefficient (units of time^{-1}). The larger the magnitude of k_s, the more rapidly local equilibrium conditions is approached. The model formulation of Equation 8.10.1 is similar to dual-porosity models, such as those used for flow in fractured media and for flow through macropores. Van Genuchten and Wierenga (1976, 1977) discuss such models.

EXAMPLE PROBLEM **8.10.1   Sorption Model**

A soil with a porosity $n = 0.4$ and 1 percent organic carbon is used in a batch experiment to study sorption kinetics using Equation 8.10.1. A solution containing 100 mg/L benzene ($K_{oc} = 83$ ml/g) is mixed with the soil. Estimate the time to equilibrium with $k_s = 1, 0.1, 0.01$ and 0.001 d^{-1}.

The equilibrium distribution coefficient is $K_d = f_{oc}K_{oc} = 0.83$ ml/g. For a batch reactor, there is no transport, and x is not a variable. The first part of Equation 8.10.1 states that

$$n\frac{dc}{dt} + \rho_b\frac{dc_s}{dt} = \frac{dm}{dt} = 0$$

so that the bulk concentration remains constant, say $m = m_o$. Thus, at any time,

$$c = \frac{m_o - \rho_b c_s}{n}$$

Substituting this equation into the second part of Equation 8.10.1 gives

$$\frac{dc_s}{dt} + k_s\left(1 + \frac{\rho_b K_d}{n}\right)c_s = \frac{k_s K_d m_o}{n}$$

With $c_s(0) = 0$, the solution is

$$c_s = \frac{K_d m_o}{n + \rho_b K_d} (1 - e^{-k_s Rt})$$

If we measure the time to equilibrium by the effective half-life where

$$c_s = \frac{c_s(\max)}{2} = \frac{K_d m_o}{2(n + \rho_b K_d)}$$

then this last equation gives the half-life $t_{1/2}$.

$$t_{1/2} = \frac{\ln(2)}{k_s R} \cong \frac{0.693\, n}{k_s(n + \rho_b K_d)}$$

The solutions for this problem are $t_{1/2}$ = 0.16, 1.6, 16 and 160 days for k_s = 1, 0.1, 0.01 and 0.001 d^{-1}, respectively. The important point to note is that the combination $k_s R$ may be interpreted as the kinetic parameter in this model.

8.10.3   Nonlinear Isotherms

With local equilibrium and linear sorption isotherms, the effects of sorption are to retard the movement of the sorbing solutes as compared with the movement of groundwater. Sorption isotherms for nonpolar organic solutes are usually linear so long as the concentration remains below about one-half its aqueous saturation value. At higher concentrations, however, and for sorption of polar organics, metals, and cations, **nonlinear isotherms** arise. The effects of nonlinear isotherms on transport may be included through use of retardation functions, except that now the magnitude of the retardation factor depends on the aqueous concentration.

There are many sorption models that result in nonlinear sorption isotherms. One of the simplest of these is the **Langmuir isotherm**, which may be developed from kinetic arguments (Adamson, 1978). This isotherm takes the form

$$c_s = \frac{\Gamma c}{K_L + c} \tag{8.10.2}$$

where Γ is the sorption capacity (mass sorbed per mass of soil) and K_L is the Langmuir constant which may be interpreted as the ratio of the rate coefficients for desorption and sorption, and is related to the binding energy of the sorbed constituent. When $c = K_L$, Equation 8.10.2 shows that $c_s = \Gamma/2$.

Another function that is often used to fit sorption data is the **Freundlich isotherm**. This isotherm takes the form

$$c_s = K_f c^{1/N} \tag{8.10.3}$$

where K_f and N are constants with $1/N \leq 1$. While it is possible to derive the Freundlich isotherm theoretically, it is most often taken as an empirical equation and the parameters are not given a physical interpretation. The special case with $N = 1$ reduces to the linear isotherm. For general reviews describing sorption isotherms, see Travis and Etner (1981) and Kinniburgh (1986).

The simplest way to see the effects of a nonlinear isotherm on solute transport is to consider advection-sorption models. The use of advection models in the nonlinear theory of chromatography dates back to the work of DeVault (1943). He noted that the transport equation

$$n\frac{\partial c}{\partial t} + \rho_b \frac{\partial c_s}{\partial t} + q\frac{\partial c}{\partial s} = 0 \qquad (8.10.4)$$

may be solved using the method of characteristics where s is the distance along the streamline. The sorption isotherm gives c_s as a function of the solute concentration, $c_s = c_s(c)$. With the "chain rule", Equation 8.10.4 becomes

$$\left(1 + \frac{\rho_b}{n}\frac{dc_s}{dc}\right)\frac{\partial c}{\partial t} + v\frac{\partial c}{\partial s} = 0 \qquad (8.10.5)$$

Equation 8.10.5 is a linear or **quasi-linear hyperbolic partial differential equation**, depending on whether c_s is a linear or nonlinear function of c. For a linear isotherm the solution is $c(s,t) = c_o(s - v't)$, where $c_o(s)$ is the initial condition along the streamline and v' is the retarded velocity ($v' = v/R$, where R is the retardation coefficient). This solution describes translation of the initial concentration distribution along the streamline without change in shape. If the isotherm is nonlinear, then things change dramatically. A nonlinear isotherm means that

$$\frac{dc_s}{dc} = f(c)$$

(i.e., the slope of the isotherm depends on the concentration). The **method of characteristic** formulation that is equivalent to Equation 8.10.5 is

$$\frac{dt}{1 + \dfrac{\rho_b}{n}\dfrac{dc_s}{dc}} = \frac{ds}{v} = \frac{dc}{0} \qquad (8.10.6)$$

which gives the solution

$$c = \text{constant along } \frac{ds}{dt} = \frac{v}{1 + \dfrac{\rho_b}{n}\dfrac{dc_s}{dc}} \qquad (8.10.7)$$

With Equation 6.5.10, this relationship may be written

$$c = \text{constant along } \frac{dt}{d\tau} = 1 + \frac{\rho_b}{n}\frac{dc_s}{dc} \qquad (8.10.8)$$

Equations 8.10.7 and 8.10.8 are the **method of characteristic** solutions, and the right side of Equation 8.10.8 gives the retardation function for the characteristic (see Equation 6.5.12). The path of a characteristic in the s-t plane or τ-t plane is called the **base characteristic**. Equation 8.10.8 states that the concentration is constant along a base characteristic. Furthermore, because c is constant, so is

dc_s/dc, and the retardation of the characteristic $dt/d\tau$ is also constant. This fact means that the base characteristic is a straight line in the τ-t plane. The rate of movement of the concentration c is equal to the retarded velocity, v', where

$$v' = \frac{v}{1 + \dfrac{\rho_b}{n}\dfrac{dc_s}{dc}} \tag{8.10.9}$$

If c_s is a nonlinear function of c, then v' is also a function of c, $v' = v'(c)$. This fact means that different concentrations move at different speeds, and base characteristics with different concentrations have different slopes in the τ-t plane. In turn, this means that the concentration distribution in space changes shape from one time period to another. That is, spatial concentration distributions or **waves** may either sharpen or spread as they move, solely due to the effects of nonlinear sorption isotherms.

For realistic isotherms,

$$\frac{dc_s}{dc} = f(c)$$

is either constant or a decreasing function of c; that is,

$$\frac{df}{dc} = \frac{d^2c_s}{dc^2} \le 0$$

and the curvature of the isotherm is negative ($f(c)$ is a concave function). For these isotherms, Equation 8.10.9 shows that higher concentrations move at faster speeds than lower concentrations. Thus, if the solute is being displaced from a region, the concentration front will tend to spread with higher concentrations moving at faster speeds. This spreading is in addition to that caused by dispersion. On the other hand, an injection front of such a chemical species will tend to sharpen due to sorption alone. For an injection front, this idea would suggest that the higher concentrations overtake the lower concentrations and their base characteristics cross. An appropriate physical analogy is that of a wave breaking on a beach, where the wave height corresponds to the solute concentration.

Breaking waves on a beach are a physical reality. Chemical waves do not break, however: one cannot measure two concentrations at the same location. Physically, dispersion prevents the development of sharp concentration fronts. Nonlinear sorption effects tend to sharpen a concentration front as it moves through the subsurface, and dispersion increases as the concentration gradients increase. The result is that a stable concentration profile develops which is advected through the subsurface at a characteristic speed that is a function of the sorbed and aqueous concentrations upstream and downstream of the front. It may be shown (see Whitham, 1974; Smoller, 1983) that the speed of a self-sharpening concentration front is given by

$$\frac{ds}{dt} = \frac{v}{1 + \dfrac{\rho_b}{n}\dfrac{\Delta c_s}{\Delta c}}$$

(8.10.10)

where Δc_s and Δc are the concentration changes across the front. For a further discussion of advection-sorption models, see Charbeneau (1981, 1982, 1988).

There are other factors that influence the transport and fate of chemicals in the subsurface environment. Brusseau (1994) reviews many of these. Analyses of factors that cause "nonideal" transport remain important areas of research. In this chapter, we have presented those important mechanisms that provide a basic framework for interpreting and understanding field observations of groundwater pollutant transport.

8.11   Subsurface Mass Transport through the Vadose Zone

Chemical and radiological contaminants are often released near the ground surface. Assessment of the potential migration of these pollutants is important for management of cleanup requirements and estimation of potential risks to exposed populations and ecosystems. In order for such releases to impact groundwater resources, the chemicals must migrate through the vadose zone. In this section, we consider significant issues associated with vadose zone mass transport.

Figure 8.11.1 shows a region of contaminated soil and the factors and processes that affect the fate and transport of chemicals. As rainfall percolates into the soil it will mobilize dissolved contaminants. Infiltration of rainfall carries contamination into the soil, and the net infiltration transports pollutants through the vadose zone to the water table where they become groundwater

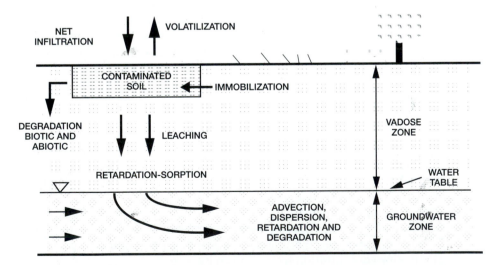

FIGURE 8.11.1   Processes which affect the movement and fate of contaminants in the subsurface environment

TABLE 8.11.1 Important Fate and Transport Processes in the Unsaturated Zone

---

Processes that affect Losses:

• Degradation—biological, chemical, photochemical
• Volatilization

Processes that affect Retardation:

• Immobilization
• Sorption
• Ion exchange

Processes that affect Transport

• Advection
• Diffusion and dispersion
• Residual saturation
• Preferential flow

---

contaminants. Within groundwater, these chemicals may be transported later-ally for distances of thousands of meters. The presence of soil air complicates not only water flow but also flow of immiscible fluids such as hydrocarbons that may vaporize. **Losses** through adsorption of the contamination on the soil, volatilization to the atmosphere, degradation by microorganisms, or through other physical, chemical, or biological processes may prevent the contamina-tion from reaching the water table.

Table 8.11.1 lists the factors that affect transport and potential contami-nation from releases of chemicals near the ground surface. These factors deter-mine the ability of the soil to adsorb and degrade wastes (the soils' **assimilative capacity**) and whether chemicals are likely to accumulate within the soil profile or leach through the profile and contaminate groundwater. Understanding these factors helps in the identification of proper waste disposal sites and in deter-mining suitable remediation methods for contaminated sites. In addition, these factors determine what happens to chemicals after old or abandoned waste sites are closed, and the type and quantity of chemical emissions to the atmosphere that may occur at sites that have recently been contaminated by volatile chem-icals. Finally, these factors determine the appropriate mathematical models to predict transport and fate of chemicals in the unsaturated zone (Loehr and Char-beneau, 1988). For protection of public health and the environment, particular-ly groundwater, it is desirable to enhance losses and retardation of contamination in the soil. This procedure may be done through proper design and operation of a waste management site, or through development of structural and engineer-ing controls at a site under closure or remediation.

Observations of Solute Transport through Soils

One of the first problem areas where chemical movement in soil water was in-vestigated concerned salinity control in soils under irrigation. Irrigation water carries dissolved salt that builds up in the soil as evapotranspiration consumes soil water. If not controlled, salt can accumulate in the soil to the point where plants can no longer grow and the land becomes barren. To control soil salinization, additional irrigation water, called a **leaching requirement**, is added to flush or leach the salt through the crop root zone to the intermediate vadose zone and groundwater where the salt will remain in solution. Early estimates (United States Salinity Laboratory, 1954) of leaching requirements were based on the quantity of water leaving the root zone and the mass balance of salt within a given depth of soil. Use of simple mass balance and plug flow assumptions neglects how solute movement is related to water movement. Solutes are displaced differently in different soils and for different rates of water movement (Biggar and Nielsen, 1960, 1962; Miller et al. 1965).

Miller et al. (1965) present results from a field study of solute displacement behavior under different rates of water movement. They used three water application methods to displace chemically pure KCl applied on the soil surface (Panoche clay loam) at a loading of about 1 kg/m^2. Based on a solubility of 264 kg/m^3 (Seidell, 1958), the KCl would dissolve with a 0.38 cm (0.15 inch) application of water. The three water application methods used by Miller et al. (1965) included continuous surface ponding, intermittent ponding with 6 inches (15 cm) of water, and intermittent ponding with 2 inches (5 cm) of water.

Water application through **continuous surface ponding** required much more water than the other application methods to displace the peak chloride concentration to a given depth. Under continuous ponding conditions, most of the water movement occurs in the larger pores, and the chloride contained within the smaller pores is bypassed. **Intermittent application** of water allows more time for the salt concentration to equilibrate across the range in pore sizes at a given depth, so the constituent moves more uniformly through the soil and the resulting displacement is much more efficient (less bypassing of salt contained in the smaller pores). This effect is shown in Figure 8.11.2. When drainage occurs,

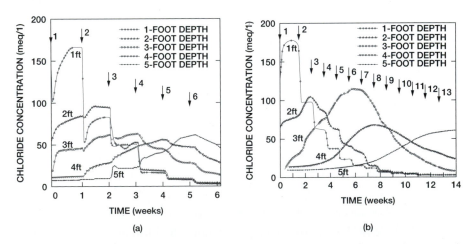

FIGURE 8.11.2 Chloride concentration vs. Time for Panoche clay loam with (a) 6 inch water applications and (b) 2 inch water applications. Arrows represent the times for each subsequent application of ponded water at the soil surface. (After Miller et al., 1965)

the flow first comes primarily from the larger pore sequences that contain the recently ponded water and the extracted solution may not represent the average soil solution.

Experiments such as that of Miller et al. (1965) show that solute displacement behavior depends on the water application method. Attempts to rapidly flush constituents through the soil profile using large application rates result in rapid water movement through the larger pores, bypassing the chemical mass contained in the smaller pores. The resulting displacement is not efficient, because larger quantities of water are required to displace the constituent from the profile. In addition, the constituent is spread over a greater depth. Small application rates, which are more typical of infiltration from natural rainfall events, result in a more uniform displacement of the chemical through the soil profile. The smaller infiltration rate results in lower water content and requires smaller quantities of water to displace the constituent from the profile.

The results of many other investigations have shown that the long-term displacement of solutes through field soils may be modeled using average water flow rates and average water contents, instead of modeling the transient transport of solutes through a heterogeneous soil profile in response to individual rainfall or infiltration events. Simplified models can provide appropriate accuracy for many practical applications.

Solute Transport Models for a Conservative Species

For many applications, it is necessary to estimate the travel time for solutes from the ground surface to the water table and the rate at which leachate leaves the unsaturated zone to become groundwater contamination. This estimate may be based on **advection** or simple displacement of constituent by water, without considering dispersion within the soil column. Three pieces of information are required: the net groundwater recharge rate, the average soil water content, and the depth to the water table. Section 4.9 discusses these factors and provides estimates of typical seepage velocities, assuming that uniform Darcy flow is occurring throughout the soil profile.

Most analyses of transport in the unsaturated zone assume one-dimensional vertical flow. The basic flow equation is

$$\frac{\partial \theta}{\partial t} + \frac{\partial q}{\partial z} = 0 \qquad \textbf{(8.11.1)}$$

where θ is the volumetric water content, t is time, z is depth measured positive downward, and q is the volumetric flow of water given by Darcy's Law. Water uptake by plant roots is neglected in Equation 8.11.1.

For a conservative solute, the one-dimensional transport equation takes the following form

$$\frac{\partial(\theta c)}{\partial t} + \frac{\partial}{\partial z}\left(q\, c - \theta\, D\, \frac{\partial c}{\partial z}\right) = 0 \qquad \textbf{(8.11.2)}$$

where c is the solute concentration and D is the dispersion coefficient. Combining Equations 8.11.1 and 8.11.2, one obtains

$$\theta\frac{\partial c}{\partial t} + q\frac{\partial c}{\partial z} = \frac{\partial}{\partial z}\left(\theta D\frac{\partial c}{\partial z}\right) \qquad (8.11.3)$$

which is the governing equation for both steady and transient flow. Use of an average uniform water content, which observations have shown to be an appropriate simplification for many applications, allows Equation 8.11.3 to be written in its simplified form

$$\frac{\partial c}{\partial t} + v\frac{\partial c}{\partial z} = D\frac{\partial^2 c}{\partial z^2} \qquad (8.11.4)$$

where $v = q/\theta$ as in Equation 5.2.4.

Dispersion Coefficient

The value of D reflects both molecular diffusion and mechanical dispersion resulting from the fact that local fluid velocities within individual pores deviate from the average fluid flux both in magnitude and direction. Nielsen et al. (1986) note that in practice, D is used as an empirical parameter that includes all of the solute spreading mechanisms that are not directly included in Equation 8.11.3. These mechanisms include, for aggregated soils, the nonequilibrium concentrations between the inter- and intraaggregate pore space, the presence of macropores, and for non-conservative solutes, the effects of nonlinear sorption or decay. As such, D plays the role of a "fudge factor" incorporating those processes not directly considered in the simple advection-dispersion model formulation.

Nevertheless, the conceptual model is that D accounts for the two additive phenomena of molecular diffusion and mechanical dispersion, and in a one-dimensional system the dispersion coefficient D is written as

$$D = \tau D_m + \alpha|v|^m \qquad (8.11.5)$$

where v is the pore water velocity, τ is a **tortuosity factor** that depends on the water content (Kemper and van Schaik, 1966) but not on the pore water velocity v, D_m is the **molecular diffusion coefficient**, and α and m are empirical constants. The tortuosity factor was discussed in Section 7.3. For homogeneous, saturated soil, the exponent m is approximately unity (Saffman, 1959), and the parameter α is then the **dispersivity**. For unsaturated soils, the value of α ranges from about 0.5 cm or less for laboratory-scale experiments involving disturbed soils, to about 10 cm or more for field-scale experiments (Biggar and Nielsen, 1976; Van de Pol et al., 1977; Jury and Sposito, 1985).

Solving for Concentration Profiles

Solving the differential Equation 8.11.4 for the concentration as a function of depth and time, $c(z,t)$, requires an **initial condition**, which is usually assumed to be $c(z,0) = 0$ (an initially clean soil), and two boundary conditions, one at the top of the soil column and one at the bottom. At the bottom, it is usual to assume that there is no contaminant, $c(\infty,t) = 0$. The solution will then depend on the boundary condition at the soil surface. Many results are available, including those for a constant surface concentration value $c(0,t) = c_o$ (which is called a **Type 1 boundary condition**, whose solution is presented by Ogata and Banks (1961)) and a for constant surface flux

$$qc_o = qc - \theta D \frac{\partial c}{\partial z}\bigg|_{z=0}$$

(which is a **Type 3 boundary condition**, where solution is given by Lindstrom et al. (1967)). These are the same solutions that were described in Section 8.2 with regard to laboratory column studies. A simple approximate solution for both of these boundary conditions is given by

$$\frac{c}{c_o} = \frac{1}{2} \operatorname{erfc}\left(\frac{z - vt}{\sqrt{4Dt}}\right) \qquad (8.11.6)$$

Observations and Modeling of Transport in Spatially Variable Soils

It is well known that soils vary significantly from point to point in their textural composition and hydraulic properties. This fact means that most of the parameters characterizing solute transport in the unsaturated zone vary both laterally and vertically. The spatial variability in soil water properties within various texture classes has been investigated on a national basis (for the United States) by Rawls and Brakensiek (1982, 1989) and by Carsel and Parrish (1988). These authors present equations for estimating soil water retention and hydraulic properties from data on soil texture, bulk density, organic matter and clay contents, and they provide information on the variability of these parameters within soil texture classes. The results of these investigations are useful in application of generic models for solute transport in the unsaturated zone, but for many problems one is interested in the variability of transport over a field at a particular site.

Statistical Variation of Soil Water Properties

The type of variability in transport behavior that is observed in the field is shown in the study of Van de Pol et al. (1977), who measured the movement of chloride and tritium at several locations within a field plot. They compared the measured solute movement rates with the rate at which water was applied to the plot. Van de Pol et al. (1977) used an 8 m by 8 m plot and a uniform steady infiltration rate of 2 cm per day. The soil profile consisted of layers of clay, silty clay, silty loam, and medium to fine sand. The site was instrumented with 24 sampling locations to a depth of 1.5 m. For each sampling location, the concentration curves were fit to Equation 8.2.5, and the best-fit values of v and D were determined. Frequency histograms for v and D were analyzed, and it was found that for both chloride and tritium they were lognormally distributed.

If a variable X is lognormally distributed, then $Y = ln(X)$ is normally distributed with a probability density function given by

$$f(Y) = \frac{\exp\left(-\dfrac{(Y - m_y)^2}{2s_y^{\,2}}\right)}{X\sqrt{2\pi s_y^{\,2}}} \tag{8.11.7}$$

where m_y and and s_y are the mean and standard deviation of the natural logarithm of the variate ($s_y^{\,2}$ is the variance). For the X-variate, its mean and variance are given by

$$m_x = \exp\left(m_y + \frac{s_y^{\,2}}{2}\right) \tag{8.11.8}$$

$$s_x^{\,2} = \exp(2m_y + s_y^{\,2})(\exp(s_y^{\,2}) - 1) \tag{8.11.9}$$

The data from Van de Pol et al. (1977) for chloride showed that the velocity v had log-mean and log-standard deviation values of $m_{ln(v)} = 1.203$ and $s_{ln(v)} = 0.564$, while the log-mean and log-standard deviation values of D were found to be $m_{ln(D)} = 2.963$ and $s_{ln(D)} = 1.132$, respectively. The velocity distribution predicted from these parameters is shown in Figure 8.11.3. This distribution has a mean velocity of $m_v = 3.90$ cm/d. Review of Figure 8.11.3 shows that while in most of the field the seepage velocity is less than this mean value, there is a substantial part of the field where the pore water velocity is more than three times the mean value. The mean chloride dispersion coefficient is from Equation 8.11.8, $m_D = 36.74$ cm²/d. For tritium, the values of v and D are log-normally distributed with mean values of 3.78 cm/d and 36.65 cm²/d, respectively. These data show pore water velocities varying over an order of magnitude at a single field site. It may be concluded that calculation of solute flux from $\bar{q}c$, where $\bar{q}$

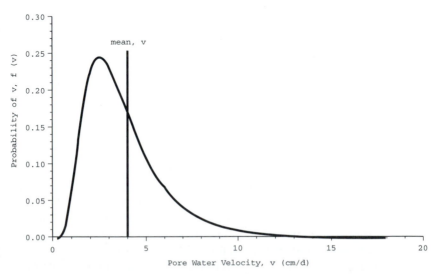

FIGURE 8.11.3 Log-normal velocity distribution from Van De Pol, et al. (1977) with a mean $mv = 3.90$ cm/d and standard deviation $sv = 2.39$ cm/d.

is the mean infiltration rate and c is solute concentration, can lead to substantial errors in estimation of the amount of solute leaching past a given soil depth. This idea suggests that spatial variability should be explicitly included in field estimates of solute transport.

Summary of Field Studies

Jury et al. (1987) summarizes results from a number of field studies of the spatial variability of saturated hydraulic conductivity. They find that while the mean value varies over nearly two orders of magnitude (from 4 to 316 cm/d), the variance of $ln(K_s)$ lies within the range of 0.5 to 1.6. Jury et al. (1987) also determine the spatial variability of infiltration rate and show that it has less variability than the saturated hydraulic conductivity. The saturated hydraulic conductivity and the infiltration rate are lognormally distributed. Values of the porosity are normally distributed. Nielsen et al. (1973) reports a mean porosity of 0.454 and a standard deviation of 0.048. There is somewhat less information available on correlation scales in the unsaturated zone. Russo and Bresler (1981) report that for the Harma field in Israel, the horizontal integral scale for hydraulic conductivity decreases from 34 m at the soil surface to 14 m at a depth of 0.9 m. The porosity integral scale decreases from 76 m to 28 m over this same range. A detailed statistical investigation (Jury et al., 1987), however, of the three-dimensional spatial structure of the Harma field and the Panoche field studied by Nielsen et al. (1973) shows that both fields were best described by three-dimensional spherical covariance's with linear three-dimensional drift functions. The saturated hydraulic conductivity values had comparable integral scales (14.5 m for the Harma field and 8.1 m for the Panoche field), but the Harma field had a much larger drift component and residual stochastic component (nugget variance) of 40 percent, compared with 13 percent for the Panoche field. Vertical correlation scales are generally much smaller than horizontal ones.

Preferential Flow

Preferential flow refers to the rapid movement of solutes through fractures, root holes, and other heterogeneities, at rates much greater than expected from consideration of the porous medium as a whole. Preferential flow is much more important in vadose zone transport than in transport within saturated media. This characteristic is due, primarily, to the greater prevalence of preferential flow channels in the shallow subsurface. In addition, the hydraulic gradient driving flow is much greater in the vadose zone (the gravitational gradient is unity, because water moves vertically downward) than in the zone of saturation (the gradient in groundwater is more often 0.01 or lower, and the water moves nearly horizontally).

Dyes and other chemical tracers have been used to study contaminant movement through the vadose zone (Steenhuis et al., 1990; Ghodrati and Jury, 1990; McCord et al., 1991). Kung (1990) notes that funneling due to heterogeneities may result in flow occurring through less than 1 percent of the soil.

Unstable flow and fingering occurs in layered soils and may cause lateral spreading of contaminants. Factors such as these make flow and transport difficult to predict and measure; point samples can easily miss narrow fingers of solute. Similar problems occur with transport through unsaturated fractured rock (Kung, 1990).

Unfortunately, there are no proven predictive models for vadose zone transport with preferential flow. Because most solutes that are transported below the root zone will eventually be transported to the water table, however, because the loss rates by adsorption and volatilization decrease rapidly with depth, approximate models are adequate for many applications.

Monte Carlo Methods

Calculation of solute transport in the presence of spatial variability in processes and parameters can be performed using stochastic modeling methods. The simplest approach to stochastic modeling is through application of Monte Carlo methods. The approach is to pick values of v and D from the estimated frequency distributions, and then use analytical models such as Equation 8.11.6 to predict the breakthrough curve at the depth of interest for the given parameters (random choice of v and D). The procedure is repeated many times and the resulting breakthrough curves are averaged. As an alternative to considering v and D as independent variables, one may use models correlating their values. For example, from tracer experiments on 20 subplots of a 150-ha field to a depth of 6 feet, Biggar and Nielsen (1976) have found the approximate relation

$$D = 0.6 + 2.93 \, v^{1.11} \tag{8.11.10}$$

where D is in cm^2/d and v in cm/d. With an equation such as 8.11.10, one could use a single estimated distribution of v to carry through the analysis. Carsel and Parrish (1988) discuss the correlation between soil water retention and hydraulic parameters that are important for application of Monte Carlo methods.

Amoozegar-Fard et al. (1982) used the Monte Carlo approach with lognormal distributions of v and D measured by Biggar and Nielsen (1976). Their simulations showed that the average rate at which salt is leached from the field hinges on the variability of the pore water velocity and not on that of the apparent diffusion coefficient, especially for greater depths. The relative concentration remaining near the soil surface for a pulse input was also found to be dramatically different when a random pore water velocity was used rather than an average deterministic value, and the resulting concentration distribution compared favorably with the field averaged measurements of Biggar and Nielsen (1976).

Stochastic-Advection Transport Models

There are a number of alternatives to Monte Carlo analysis for estimation of transport through spatially variable soils. Probabilistic models have been developed by a number of researchers. For example, Dagan and Bresler (1979) developed a model assuming vertical flow and that rainfall and soil properties varied laterally but not vertically. Their results show that at the field scale, the breakthrough curve resulting from a spatially variable random hydraulic conductivity field is similar to the breakthrough curve predicted by Equation 8.11.6 with a large dispersivity. For a lognormal hydraulic conductivity field with $\sigma_y^2 = 1.96$ ($y = lnK$), the apparent dispersivity for breakthrough at a depth of 5 meters is 0.32 meters.

Transport with a spatially variable velocity will result in the spreading of a concentration front, similar to the spreading associated with hydrodynamic dispersion. There is an important difference between these two models, however, that appears when one considers the breakthrough curves at two different depths. According to advection-dispersion theory, the width of the breakthrough curve satisfies the Einstein relation

$$\sigma \approx \sqrt{2\,D\,t} = \sqrt{2\,a_L\,L}$$

so that the width increases as the square-root of either distance or time. The result for a spatially variable velocity, or stochastic-advection field is different. The spreading of a front is caused by the velocity field directly, and the width of a front increases with either distance or time to the first power. In comparison with the advection-dispersion model, this equation gives a dispersivity that increases linearly with distance

$$a_L = \alpha L$$

This equation is the **stochastic-advection transport model**. Simmons (1982) and Jury, et al. (1991) have characterized this behavior through the transformation properties of the residence time density function (see Equation 6.5.2) as follows:

$$E_z(I) = \frac{L}{z}E_L\left(\frac{IL}{z}\right) \tag{8.11.11}$$

where $E_L(I)$ is the residence time density function for a soil depth L and I is the cumulative infiltration.

Transfer Function Models

A different modeling approach, based on Danckwerts' (1953; also see Levenspiel, 1972) work on residence time distribution theory, was introduced for modeling subsurface transport by Jury (1982) under the name of a transfer function (a term which comes from applications of systems theory in electrical engineering). The basic ideas have been presented in Section 6.5. Jury begins with the point of view that the many causes of spatial variability of water and solute transport renders measurement of the hydraulic and retention parameters of a field soil all but impossible. As a consequence, the deterministic approach to modeling chemical transport is abandoned in favor a general transfer function model. This model provides a method for transforming an input function (solutes added to the soil surface) into an output function (solutes moving through the soil) for a system consisting of a field soil. Residence time distribution theory is usually described by way of the residence time density function $E(\tau)$, which is the probability density function for the age of a fluid element when it leaves a flow system if the element has zero age when it enters the system. Thus, the probability that a fluid element will reside within a flow system for a time period between τ and $\tau + d\tau$ is $E(\tau)d\tau$. The residence time distribution function $F(\tau)$ is related to the residence time density function through $dF(\tau) = E(\tau)d\tau$. $F(\tau)$ is the cumulative distribution function for residence times and rises from zero at the shortest residence time to (possibly) 1 at the longest. For a step tracer input, $F(\tau)$ is often called the breakthrough curve and may be measured for a flow system as the normalized effluent response to a step change in the influent composition for an ideal tracer. Jury (1982) notes that in applications it is often more useful to develop the distribution curves in terms of the net amount of water entering the soil surface (infiltration minus evaporation). If I represents the net amount of water entering the soil surface, then the general result from residence time distribution theory provides the concentration being leached from a depth L in terms of the inflow concentration at the soil surface through

$$C_L(I) = \int_0^\infty C_{in}(I - I')E(I')dI' = \int_0^1 C_{in}(I - I')dF(I') \qquad \textbf{(8.11.12)}$$

This equation is basically the same result as Equation 6.5.16. Jury (1982) shows how Equation 8.11.12 must be generalized to account for spatially variable recharge, while Jury, et al (1986, also see Sposito and Jury, 1988) generalize the model to account for solute transformations during transport which may vary both in space and time. Rainwater, et al (1987) provides an example dealing with chemically reactive solutes that is also applicable to the vadose zone. Applications of the transfer function model are described by Jury et al. (1982), Jury and Sposito (1985), and White et al. (1986), while Sposito et al. (1986) highlight the relationship between the transfer function model and the advection-dispersion equation. The major strength of the residence time distribution theory is that all physically based models may be cast within its form. In this regard, Equation 8.11.12 is just the convolution theorem for the classical solution of the advection-dispersion equation. The major difficulty in application of this theory is estimation of the residence time distribution function, or its related lifetime probability density function. Both theoretical arguments and field measurements suggest that in many applications, this density function may be lognormal in nature. In this case the appropriate field calibration problem consists of estimation of the parameters of the corresponding lognormal distribution. Charbeneau (1988) shows that if the residence time distribution function can be estimated for a flow problem, then the resulting theory provides a computationally efficient method for prediction of effluent concentrations from the system under very general conditions.

Problems

8.2.1. A laboratory tracer experiment is conducted using a cylindrical column (10 cm in diameter and 30 cm long) packed with uniform sand. The column has porosity 0.35 and the steady flow rate is 1 L/hr under a hydraulic gradient of 0.1. The $c/c_o = 0.5$ point on the breakthrough curve arrives 0.8 hours after the tracer initially entered the column. Likewise, the 0.25 and 0.75 points arrived at 0.7 and 0.9 hours, respectively. Estimate the dispersivity of the conservative tracer in the sand, assuming that $\tau D_m \cong 10^5$ cm^2/s. At what time would you estimate the $c/c_o = 0.95$ point would appear on the breakthrough curve?

8.2.2. A laboratory column that is 6 cm in diameter and 40 cm in length is packed with fine sand resulting in a porosity of 0.35. A tracer is passed through the column at a rate of 1 L/hr. The following effluent data is recorded:

time (hr)	c/c_o
0.35	0.075
0.37	0.215
0.385	0.37
0.396	0.5
0.41	0.65
0.43	0.83
0.44	0.89
0.46	0.96

What is the longitudinal dispersivity (cm) for the soil in the column?

8.2.3. A 50 cm long laboratory column is packed with fine sand with mean grain-size diameter of 0.5 mm, and a seepage velocity of .01 cm/s is established. A tracer is introduced into the column at a concentration of 100 mg/L as a pulse of 10 minutes duration. Predict the maximum effluent concentration during this experiment. (*Hint*: If the solution for a step change in influent concentration is $c = c_o f(x,t)$, then the corresponding solution for a tracer influent pulse of duration Δt is given by superposition as $c = c_o[f(x,t) - f(x,t - \Delta t)]$, which may be though of as superposition of the solution for the leading edge of the tracer plus the solution for the leading edge of the clean water which displaces the tracer.)

8.2.4. If the experiment of problem 8.2.2 is repeated with a pulse of concentration c_o introduced for a 6 minute period followed by tracer-free water, what is the maximum value of c/c_o to be expected in the column effluent?

8.2.5. A contaminant is migrating through an aquifer composed of medium-grained sand (mean grain size of 1 mm). The average hydraulic gradient is 0.01 and a representative value of the hydraulic conductivity is 10^{-3} cm/s. Is the movement of an ideal tracer influenced primarily by advection and mechanical dispersion, or by molecular diffusion? State your assumptions clearly and explain your answer.

8.2.6. For an exposure and risk study it is desired to estimate the maximum concentration which would result from a failure of a waste containment facility. The facility is large enough and the exposure points are close enough so that a 1D model is thought to be appropriate. The following characteristics are assumed: the initial source strength is 100 mg/L, the hydraulic gradient is 0.005, the hydraulic conductivity is 20 m/d, and $n = 0.25$. Both the source and the solute during transport decay with a half-life of 100 days. The retardation factor is 2.0 and the longitudinal dispersivity is 0.05 times the distance to the exposure point. Determine the maximum exposure concentrations at distances of 20 and 50 meters. Repeat the analysis for the case with a constant source strength of $c_o = 100$ mg/L with the solute still decaying during transport. (*Hint*: for the conditions stated, Equation 8.2.9 multiplied by $e^{-\lambda t}$ may be used for the first part while the last part requires that you solve the transport equation $v\, dc/dx - D\, d^2c/dx^2 + \lambda Rc = 0$ with the boundary conditions $c(0) = c_o$ and $c(\infty) = 0$.)

8.3.1. During a characterization study of a shallow aquifer, a single well tracer test was performed to estimate the mixing characteristics of the aquifer. The saturated thickness of the aquifer is about 18 ft and the porosity is estimated to be 0.35. Chloride was injected at a concentration that averaged 80 mg/L and the concentration was measured in an observation well 28 ft from the injection well. The background chloride concentration was 20 mg/L. The injection rate averaged 2 ft^3/min and the aquifer is assumed to be homogeneous. From the following data, estimate the longitudinal dispersivity of the aquifer.

Time (hrs)	Conc (mg/L)
56.6	20
69.5	22
95.3	28
111.6	39
123.7	47
131.7	52
137.2	55
146	60

8.3.2. For the conditions described in Example 8.3.1, what is the maximum monitoring well concentration if the tracer is injected at 100 mg/L for 4 days and then displaced by clean water at the same injection rate, assuming a longitudinal dispersivity of 0.1 m?

8.3.3. A conservative tracer is injected at a concentration of 100 mg/L over a period of 3 days at a rate of 8 m^3/d, and followed by clean water. The breakthrough is measured in a monitoring well that is located 6 m from the pumping well. The injection interval is 2 m thick, and the porosity is 0.35. The following data is measured. What is the effective dispersivity of this interval?

Time (d)	Conc (mg/L)
4.78	1.5
6.02	6.6
6.58	11.1
7.18	16.3
8.37	28.0
9.57	35.2
10.31	36.8
11.40	35.0
12.58	30.0
14.30	20.9
15.55	15.1
19.70	3.8

8.5.1. Groundwater flows in a confined aquifer at an angle of 25 degrees with respect to the x-axis coordinate system. The seepage velocity is 1 m/d and the longitudinal and transverse dispersivities are 2 m and 0.2 m, respectively. What are the magnitudes of the 2D components of the mechanical dispersion tensor?

8.5.2. What are the mean or average principal components of the mechanical dispersion tensor for a receptor distance of 1,000 ft if the seepage velocity is 1 ft/d and the EPACML model of Equations 8.5.9 to 8.5.11 is assumed?

8.6.1. An organic chemical is released from a disposal site at a rate of 3 kg/d. If this chemical has a half-life in an underlying aquifer of 180 days, estimate the maximum build-up (total mass) of the chemical within the aquifer. What is the effective residence time for the chemical in the aquifer?

8.6.2. For a point source release of mass M_o, use the method of moments to calculate the mean μ_y and variance σ_y^2 for lateral displacements for the case without decay ($\lambda = 0$).

8.6.3. A conservative tracer is released at a rate of 1 kg/d from a source of width W = 10m. The Darcy velocity, porosity, and lateral (transverse) dispersivity are 1 m/d, 0.25, and 0.4 m, respectively. Use the method of moments to calculate the effective width of the resulting tracer plume as a function of time, where the width is estimated as $4\sigma_y$. At what time does the effective width of the plume double compared with its initial value?

8.6.4. How does the width of a plume vary along its length under steady-state conditions if longitudinal dispersion and decay can be neglected? You may assume a point source at the origin so that the boundary condition may be written $c(0,0) = M\delta(x)\delta(y)$. (Hint: define the jth moment by $\{C_j = \int y^j c(x,y)dy\}$ and use this in the equation $\{v\partial c/\partial x - D_{yy}\partial^2 c/\partial y^2 = 0\}$ to find how the standard deviation varies with x.)

8.7.1. One cubic meter of a conservative tracer is released near the middle of a thick aquifer. The initial concentration is 100 mg/L. The seepage velocity and porosity are 1 m/d and 0.3, while the longitudinal, lateral transverse and vertical dispersivities are 1 m, 0.2 m and 0.05 m, respectively. How long will it take for the maximum concentration to decrease below 10 mg/L?

8.7.2. A truck hauling liquid waste has overturned on the highway. The waste is not recovered at the time of the spill, and we are interested in the concentration reaching a small stream located at a distance of 150 meters from the highway and running parallel with it. The stream carries a discharge of 5 cfs. The waste moves through a sandy aquifer that is 9 meters thick and has a porosity of 0.35. The longitudinal dispersivity is 3.0 m while the vertical and horizontal transverse dispersivities are 0.10 and 0.6 m, respectively. The groundwater seepage velocity is 0.8 m/d and is directed transverse to the highway and stream. 4,000 kilograms of waste were spilled. For each of the following cases, state the maximum concentration of waste reaching the stream, and plot the stream concentration that will be observed by downstream users of surface water from the stream. a) The waste is conservative ($R = 1.0$ and $\lambda = 0.0$). b) The waste is radioactive with a half-life of 120 days ($R = 1.0$). c) The waste adsorbs onto the soil with $R = 3.5(\lambda = 0.0)$.

8.7.3. A spill of 5,000 kg of liquid waste occurs over an unconfined aquifer with a shallow water table. The saturated thickness of the aquifer is 8 m, its porosity and hydraulic conductivity are estimated to be 0.35 and 5 m/d, and the gradient towards a small creek located 200 m from the site of the spill is 0.02. The creek has a discharge of 2 cfs. The waste has a distribution coefficient of $K_d = 0.1 L/kg$ and a conservative half-life is estimated to be 500 days. Estimate the peak concentration at the top, middle, and base of the aquifer at the location of the creek. If the creek captures all of the aquifer flow, what is the maximum concentration that would occur in the creek? Use the average dispersivities from Equation 8.5.9 (see the solution to problem 8.5.2).

8.8.1. Leachate is escaping from a waste-storage facility and being transported in a shallow unconfined aquifer. The following characteristics of the site are estimated from available data: $K = 12$ ft/d; $n = 0.35$; $I = 0.008$; $b = 25$ ft; $a_L = 40$ ft; $a_T = 6$ ft. The dispersivity values are estimated based on a travel distance of 500 ft. If a conservative species is released at a rate of 15 kg/day, what is the maximum concentration at distances of 100 ft, 500 ft, 1,000 ft, and 5,000 ft from the source? How long will it take for the concentration to reach 95 percent of this maximum value at each of these distances? Use the point source model for this problem.

8.8.2. Leachate is released from a waste-storage facility at a rate of 10 kg/d into a shallow unconfined aquifer over a period of 4 years. The saturated thickness of the aquifer is 6 m, its porosity and hydraulic conductivity are estimated to be 0.30 and 6 m/d, and a uniform hydraulic gradient 0.015 exists at the site. A regulatory compliance boundary is located at a distance of 300 m from the facility. Assuming that the waste is conservative, estimate the maximum exposure concentration that will occur at the compliance boundary and the time of its occurrence after the beginning of the release. Use the average dispersivities from Equation 8.5.9 (see the solution to Problem 8.5.2).

8.8.3. An industrial facility has a 2 acre on-site landfill for disposal of solid waste. The landfill is roughly square in plan view. A 120 ft thick sandy aquifer lies beneath the facility and the upper part of the aquifer has a permeability of 35 ft/d and a porosity of 0.3. The average hydraulic gradient is 0.01. A regulatory compliance well is located on the site boundary, 1,000 ft from the landfill. What is the maximum landfill leachate concentration for a conservative species if the permissible well concentration is 10 mg/L? What is the maximum landfill leachate concentration for a species with a retardation factor of 2 and a hydrolysis half-life of 600 days if the permissible well concentration is 0.1 mg/L? Assume that the infiltration rate through the landfill is 6 inches per year, and that the longitudinal, transverse, and vertical dispersivities are 90 ft, 11 ft, and 0.6 ft, respectively. Neglect transport in the vadose zone.

8.8.4. For the conditions of problem 8.8.3, plot and compare the steady-state concentration distributions at $x = 1,000$ ft (as a function of y) for the simple plume model, the point-source model, and the Gaussian-source model for conditions with landfill leachate concentration (assumed conservative) 1,500 mg/L. What is the approximate width of the plume at this location?

8.11.1. A conservative tracer is applied to the ground surface and leached through the vadose zone by rainfall. The net infiltration rate (Darcy velocity) is estimated to be 25 cm/yr and the average volumetric water content is 0.30 over the upper 3 meters and 0.20 over the remaining 2 meters above the water table. If the dispersivity is estimated to be 10 cm, estimate the time between when the concentration ratios of $c/c_o = 0.1$ and $c/c_o = 0.9$ reach the water table (where c_o is the initial tracer concentration).

8.11.2. During a tracer experiment in the vadose zone it is found that the velocity is log-normally distributed with a mean $m_v = 3$ cm/d and a standard deviation $s_v = 3.6$ cm/d. Over what fraction of the field would you expect the velocity to exceed 10 cm/d?

Chapter Nine

Multiphase Flow and Free Product Recovery

Nonaqueous phase liquids (NAPLs) are known to be present at numerous industrial and waste disposal sites, and they are suspected to exist at many more. Due to the numerous variables influencing NAPL transport and fate in the subsurface, they are likely to go undetected, and yet they are likely to be a significant limiting factor in site remediation. This statement is especially true for *dense NAPLs* (DNAPLs), such as chlorinated solvents, for which unfortunately there are few proven remediation technologies available (Huling and Weaver, 1991).

For environmental engineers and hydrogeologists, it is important to understand the many processes that control the migration and entrapment of NAPLs in the subsurface, although in general, it is less important that they develop an understanding of methods for modeling NAPL flow in multiphase systems. An understanding of the transport mechanisms provides insight to the potential NAPL distribution at a site. For most cases at subsurface remediation sites, however, by the time the contaminant is detected and remediation efforts start, the NAPL is present only at residual or low saturations, and NAPL migration is not a significant issue. An exception occurs at some petroleum refineries and locations with leaking underground storage tanks where significant amounts of light NAPLs (LNAPLs or OILs) may be present floating on the water table.

In this chapter, we will focus on the processes and mechanisms that control the transport and fate of NAPLs in the subsurface. The principles of NAPL flow are presented, and capillary trapping and the formation of residual saturations are discussed. The distribution of LNAPLs and DNAPLs are described, including DNAPLs trapped in fractured clays and rock. A model for NAPL infiltration is presented, and screening models for NAPL transport and fate are discussed. Finally, soil vapor extraction and free-product recovery systems for subsurface remediation are discussed—and a procedure for design of free-product recovery systems is presented.

9.1 Principles of Multiphase Flow

NAPL is a term used to describe the physical and chemical differences between a hydrocarbon liquid and water that results in a distinct physical interface that will separate the two phases in a mixture. The term **OIL (organic immiscible liquid**) designates the same condition. This latter term has some advantages, because organic liquid chemicals such as alcohols are NAPLs, although they are miscible with water and are not the subject of concern in this chapter. The interface, which is visually observable, divides the bulk phases of the two liquids, but individual compounds may solubilize from the NAPL into the groundwater.

NAPLs are generally divided into two classes based on their densities. We will use the term **DNAPL** to designate nonaqueous phase liquids that are denser than water and OIL to designate organic immiscible liquids, typically petroleum hydrocarbon liquids, which are less dense than water. The term NAPL will be retained for general discussion. OILs will pool and spread as a floating free product layer upon the water table if they are released to the subsurface in sufficient quantities. Typical examples of OILs are gasoline, fuel oils, and most petroleum hydrocarbon mixtures. Those nonaqueous phase liquids with densities greater than water (DNAPLs) can pass across the water table if they are present in sufficient quantities, and they may be found at great depths within the saturated zone of a groundwater aquifer. DNAPLs are typically chlorinated solvents such as *trichloroethene* (TCE) and *tetrachloroethene* (PCE), wood preserving process wastes (creosote and pentachlorophenol), coal tars, and pesticides.

The features that make transport and fate characteristics of NAPLs unique from those of miscible constituents are directly or indirectly associated with the presence of the interface separating the two phases. These features are also present in the unsaturated zone where the immiscible fluids are air and water. The characteristics of multiphase flow have been studied for some time by petroleum engineers and by soil scientists. Only recently have environmental engineers been drawn to this subject as they address problems related to subsurface contamination and remediation.

In this chapter, both two- and three-phase fluid systems will be considered. The two-phase system will generally consist of water and NAPL phases. The three-phase system is a medium filled with a water-NAPL-air mixture. We will first review some of the concepts from the theory of capillarity, which were briefly introduced in Chapter 4.

9.1.1 Capillarity

For a multiphase system, part of the pore space is filled with water, part with NAPL, and the rest with air (if it is present). In the study of multiphase fluid systems in porous media, it is more common to work with phase saturation values rather than volumetric fluid contents, and this convention is adopted in this chapter. The sum of the fluid saturation values is equal to unity:

$$S_w + S_o + S_a = 1 \tag{9.1.1}$$

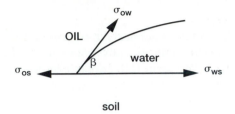

FIGURE 9.1.1 Contact angle used to classify wetting characteristics.

where S_w is the water saturation, S_o is the NAPL saturation, and S_a is the air saturation. As noted in Chapter 4, the new processes that occur in a medium saturated by more than one immiscible fluid are associated with the presence of the interface separating one fluid phase from another. The fluid that shows the greatest preference towards contact with the solid matrix is called the **wetting fluid**. For most soils, this substance is water. The fluid that shows the least preference for the solid phase is called the **nonwetting fluid**. In a three-phase system, water is usually the wetting fluid, air the nonwetting fluid, and NAPL has intermediate wettability.

In a multiphase system, the **contact angle**, β, shown in Figure 9.1.1, is of great interest because it determines the wetting properties of the porous matrix. In Figure 9.1.1, there are three phases present along with three interfaces, and each has its own interfacial energy. The interfacial energy and contact angle are related through Young's equation (see Adamson, 1978):

$$\cos(\beta) = \frac{\sigma_{os} - \sigma_{ws}}{\sigma_{ow}} \qquad \textbf{(9.1.2)}$$

where σ_{ij} is the interfacial tension between phases i and j. If the contact angle between the solid and the interface for a phase is less than 90 degrees, then the phase is the wetting phase. The phase with contact angle greater than 90 degrees is the nonwetting phase.

For a curved interface separating two fluid phases i and j, there is an associated pressure difference across that interface. This pressure difference is given by

$$p_c = \frac{2\sigma_{ij}}{r_c} \qquad \textbf{(9.1.3)}$$

where r_c is the average radii of curvature. Equation 9.1.3 is called Laplace's equation (Adamson, 1978). The pressure difference across the interface separating the wetting and nonwetting phases is called the **capillary pressure**, and it is defined by

$$p_c = p_{nw} - p_w \qquad \textbf{(9.1.4)}$$

where p_{nw} is the nonwetting phase pressure and p_w is the wetting phase pressure. The notation p_w is also used to denote the water pressure, but because water is the wetting phase, there is no confusion.

Leverett (1941) has investigated how the capillary pressure varies with elevation under conditions of **vertical equilibrium**. The term vertical equilibrium is introduced to designate those conditions where the vertical pressure distribution satisfies the hydrostatic pressure equation for the water and NAPL phases. Vertical equilibrium holds under fluid static conditions, and it also is a useful approximation when fluid flow is primarily horizontal. Leverett chose an elevation datum where $p_c = 0$ and showed that

$$p_c = \Delta\rho g z \qquad\qquad (9.1.5)$$

where $\Delta\rho$ is the density difference between the heavier and lighter fluid. Equation 9.1.5, which is used in Section 9.4, shows that the capillary pressure depends upon the elevation in a fluid system under conditions of vertical equilibrium.

9.1.2   Forces on Fluids in Multiphase Systems

Fluids in porous media will move only if they are subjected to impressed forces. Thus, a first step in understanding the migration of fluids in a multiphase system is to consider the **forces** that act on the various phases and how they relate to each other.

As noted in Chapter 2, forces acting on fluids in porous media consist of forces due to pressure gradients and to gravity. The force per unit mass for the water and NAPL phase may be written

$$\tilde{F}_w = -g\hat{k} - \frac{1}{\rho_w}\nabla p_w \qquad\qquad (9.1.6)$$

$$\tilde{F}_o = -g\hat{k} - \frac{1}{\rho_o}\nabla p_o \qquad\qquad (9.1.7)$$

where $\hat{k}$ is the upward unit vector. Using the definition of the capillary pressure from Equation 9.1.4, these may be combined to give

$$\tilde{F}_o = -g\hat{k} - \frac{\nabla p_c}{\rho_o} + \frac{\rho_w}{\rho_o}\left(\tilde{F}_w + g\hat{k}\right) \qquad\qquad (9.1.8)$$

Equation 9.1.8, which was obtained by Hubbert (1954), is an important result.

To see the effects of **buoyancy**, consider the case where capillary gradients are small and the water phase is under conditions of hydrostatic equilibrium ($\tilde{F}_w = 0$). In this case, Equation 9.1.8 reduces to

$$\tilde{F}_o = -g\hat{k} + \frac{\rho_w}{\rho_o}g\hat{k} \qquad\qquad (9.1.9)$$

which shows that the force on the NAPL phase is in the vertical direction. For a DNAPL ($\rho_o > \rho_w$), the force is downward because the second term is smaller than the first term. The opposite is true for OIL, which will tend to rise.

Another case of interest concerns **capillary forces** by themselves. For this case, Equation 9.1.8 becomes

$$\tilde{F}_o = -\frac{\nabla p_c}{\rho_o}$$

Significant capillary forces are developed near regions with significant changes in soil texture and near regions showing significant saturation changes, such as near wetting fronts. The regions with significant changes in soil texture are important for understanding **capillary trapping** of NAPLs. To see the effects of variations in soil texture, consider a soil region showing a strong gradation from fine- to coarse-grained texture. In the Laplace Equation (9.1.3), we assume that the mean radius of curvature is proportional to the **mean grain size**, d. Then Equation 9.1.3 becomes

$$p_c = \frac{C\sigma}{d}$$

where C is a constant of proportionality. With this form of the Laplace equation, the force on the NAPL is

$$\tilde{F}_o = \frac{C\sigma}{\rho_o d^2}\nabla d \qquad\qquad \textbf{(9.1.10)}$$

Equation 9.1.10 shows that the force is in the direction of increasing grain size, that is, towards the regions of coarser texture. This implies that for a heterogeneous media the hydrocarbon phase will tend to accumulate within the coarse-texture regions, including sand lenses, cavities created by decayed roots, and in fractures.

A third example of application of Equation 9.1.8 concerns **hydraulic control** of DNAPL migration through pumping of groundwater. It is assumed that capillary pressure gradients are negligible. For the water phase, the hydraulic gradient is defined by

$$\tilde{I}_w = -\nabla h_w = -\hat{k} - \frac{\nabla p_w}{\rho_w g}$$

Comparison with Equation 9.1.6 shows that

$$\tilde{F}_w = g\tilde{I}_w$$

Under the assumed conditions, Equation 9.1.8 takes the form

$$\tilde{F}_o = \left(\frac{\rho_w}{\rho_o} - 1\right)g\hat{k} + \frac{\rho_w}{\rho_o}g\tilde{I}_w \qquad\qquad \textbf{(9.1.11)}$$

Equation 9.1.11 shows that the migration of the NAPL is controlled by the influence of buoyancy (the first term on the right) and by the hydraulic gradient associated with water movement. For example, Equation 9.1.11 suggests that the downward migration of DNAPLs may be controlled through pumping of water from up-dip locations, thus causing an upward hydraulic gradient to balance the effects of buoyancy. Such a situation is shown in Figure 9.1.2. A DNAPL lens is located at the base of an aquifer on an inclined confining bed.

If the DNAPL cannot penetrate the confining bed (see Equation 9.1.10), then it will migrate down-dip along the bed due to buoyancy forces. If water production wells were located upgradient, however, such as at locations A and B in the plan view, then they might be able to establish sufficient hydraulic gradient in the water to balance the buoyancy forces.

EXAMPLE
PROBLEM

Example 9.1.1   Hydraulic Control of Downgradient DNAPL Migration

DNAPL migrates across a confining bed that has a downgradient slope of angle α, such as shown in Figure 9.1.2. The unit vector $\hat{l}$ points in the downgradient direction. Determine the identical pumping rates from wells A and B in order to stop downgradient migration.

Under conditions where the hydraulic gradient just stops the potential for downward migration, the force on the DNAPL acting in the downgradient direction must vanish: $\tilde{F}_o \cdot \hat{l} = 0$. With Equation 9.1.11, this situation gives

$$\left(\frac{\rho_w}{\rho_o} - 1\right)g\hat{k} \cdot \hat{l} + \frac{\rho_w}{\rho_o}g\tilde{I}_w \cdot \hat{l} = 0$$

To the first order, $\hat{k} \cdot \hat{l} = -\sin(\alpha) \cong -\alpha$, and $\tilde{I}_w \cdot \hat{l} = -I_h$, the horizontal component of the hydraulic gradient in the uphill direction (if the hydraulic gradient is applied to stop the DNAPL migration, its direction must be in the uphill direction, and thus the minus sign). This relationship gives

$$I_h = \left(\frac{\rho_o}{\rho_w} - 1\right)\alpha \tag{9.1.12}$$

For steady flow to a well, the Thiem equation (Chapter 3) gives

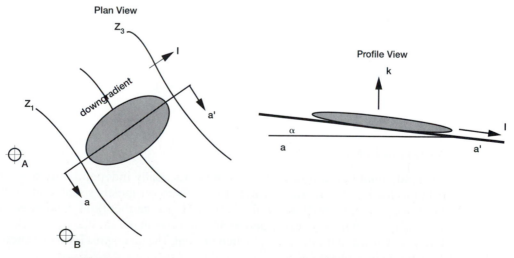

FIGURE 9.1.2   DNAPL lens on inclined confining bed

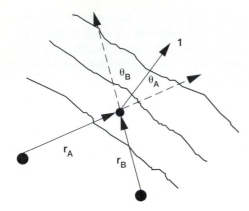

FIGURE 9.1.3 Hydraulic control of DNAPL migration

$$h = \frac{Q}{2\pi T} ln\left(\frac{r}{R}\right) + h_R$$

where R is the radius of influence and h_R is the head at radius R. The hydraulic gradient created by the well is

$$\tilde{I}_w = -\nabla h = -\frac{Q}{2\pi Tr}\hat{r}$$

where $\hat{r}$ is a unit vector in the radial direction from the well. The component of the gradient in the upstream direction is found by taking the scalar product with $\hat{l}$. As shown in Figure 9.1.3, this situation gives

$$\tilde{I}_w \cdot \hat{l} = -\frac{Q}{2\pi Tr}\hat{r} \cdot \hat{l} = -\frac{Q}{2\pi Tr}\cos(\theta)$$

If the flow rate from each well is the same, the necessary well production rate is given by

$$Q \geq \frac{2\pi T\alpha\left(\dfrac{\rho_o}{\rho_w} - 1\right)}{\dfrac{\cos(\theta_A)}{r_A} + \dfrac{\cos(\theta_B)}{r_B}} \tag{9.1.13}$$

9.1.3   Darcy's Law for Multiphase Flow

In steady multiphase flow, the fluids flow practically independently of each other as the interfacial boundaries between them are mostly situated in capillaries where there is no flow (Dullien, 1992). Hence, neither fluid influences the flow behavior of the other one, and each fluid flows in the capillary network allotted to it, just as if it were the only fluid present. The analogous form of Darcy's Law for a given phase i is

$$\tilde{q}_i = -\frac{k_{ei}}{\mu_i}(\nabla p_i + \rho_i g \hat{k}) \qquad (9.1.14)$$

which is the same as Darcy's Law for a saturated medium, except that in place of the intrinsic permeability k, the **effective permeability** k_{ei} for phase i appears. The effective permeability may be written as a function of the capillary pressure or the phase saturations. The extension of Darcy's Law given by Equation 9.1.14 appears to have been first suggested by Muskat and co-workers (Muskat and Meres, 1936; Wyckoff and Botset, 1936; Muskat el al., 1937).

It is customary to express the effective permeability as a fraction of the medium's intrinsic permeability, k. This statement defines the **relative permeability** for the ith phase:

$$k_{ri} = \frac{k_{ei}}{k} \qquad (9.1.15)$$

The extension of Darcy's Law may be written

$$\tilde{q}_i = -\frac{k k_{ri}}{\mu_i}(\nabla p_i + \rho_i g \hat{k}) \qquad (9.1.16)$$

Further, if the density of phase i is constant, then one may define a 'head' for this phase by

$$h_i = \frac{p_i}{\rho_i g} + z \qquad (9.1.17)$$

and Equation 9.1.16 becomes

$$\tilde{q}_i = -K_{is}k_{ri}\nabla h_i \qquad (9.1.18)$$

where the saturated hydraulic conductivity for phase i is defined by

$$K_{is} = \frac{\rho_i g k}{\mu_i} \qquad (9.1.19)$$

Equation 9.1.18 is the familiar form of Darcy's Law used in groundwater hydraulics, except for the presence of the relative permeability. The physical content of the equation is much different and more difficult, however, because of capillary pressure hysteresis effects on h_i, and k_{ri} being a nonlinear function of the saturations of the phases present.

The relative permeability varies from 1 under saturated conditions to 0 at residual or irreducible saturations. The $k_{ri}(S)$ relation does not exhibit significant hysteresis. With the soil characteristic curve, one may also write the relative permeability as a function of the capillary pressure head, Ψ. The $k_{ri}(\Psi)$ function does show considerable hysteresis, however.

The considerations of Section 4.4 with regard to measurement of soil properties hold true for multiphase systems in general. Estimation of the relative permeability function using the pore size distribution is conceptually based on

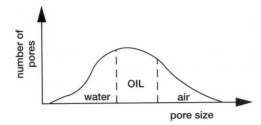

FIGURE 9.1.4   Phase content as a function of pore size for a three-phase fluid system

Figure 9.1.4 (compare with Figure 4.1.6). This figure shows that the smaller pores are filled with the wetting fluid (water) while the larger pores are filled with the nonwetting fluid (air). If an effective permeability can be assigned to pores of a given size, through Poiseuille's law for example, then one may integrate across the range of pores that are filled with a given fluid to assign a medium permeability to that fluid. The Poiseuille formula shows that the fluid velocity in a capillary tube is proportional to its radius squared, while the Laplace equation shows that the pore radius is inversely proportional to the capillary head. Concepts and models such as these allow one to derive theories for predicting the relative permeability function from a measured or estimated capillary pressure curve. As noted in Chapter 4, the most widely used models are those of Burdine (1953) and Mualem (1976).

The **Brooks-Corey-Burdine model** for the NAPL phase relative permeability in a three-phase water-NAPL-air system combines the Poiseuille equation relating the permeability of a capillary to the pore size with the Brooks and Corey capillary pressure model (see Section 4.4.3). The NAPL relative permeability is estimated by integrating dS/p_c^2 from S_w to $S_t = S_w + S_o$ (the **total liquid saturation**), as suggested by Figure 9.1.4, resulting in

$$k_{ro}(S_o, S_w) = \left(\frac{S_o - S_{or}}{1 - S_{or}}\right)^2 \left(\left(\frac{S_t - S_{wr}}{1 - S_{wr}}\right)^{\varepsilon - 2} - \left(\frac{S_w - S_{wr}}{1 - S_{wr}}\right)^{\varepsilon - 2}\right) \quad \textbf{(9.1.20)}$$

The leading term in Equation 9.1.20 accounts for the NAPL phase tortuosity and $\varepsilon = 3 + 2/\lambda$. The corresponding water and air phase relative permeability are given by Equations 4.4.8 and 4.4.9, where $\theta = \theta_t$ in Equation 4.4.9 for the reduced total liquid saturation.

The NAPL relative permeability from Equation 9.1.20 is shown in Figure 9.1.5 for a sandy loam soil with $\varepsilon = 6.0$, $n = 0.4$, $S_{wr} = 0.10$, and $S_{or} = 0$. The relative permeability curves are shown for four different water saturations, $S_w = 0.40, 0.30, 0.20$, and 0.10. A number of important features are shown on this figure. First, the maximum possible NAPL saturation decreases with increasing water saturation, as it should. More importantly, at a given NAPL saturation, the NAPL relative permeability increases with water saturation. The reason for this is that water occupies the smallest pores, and with more water present, the NAPL is forced into larger pores. The NAPL is present in the larger pores for larger water saturation, and the larger pores have a larger permeability associated with them. A third feature shown in Figure 9.1.5 is that even

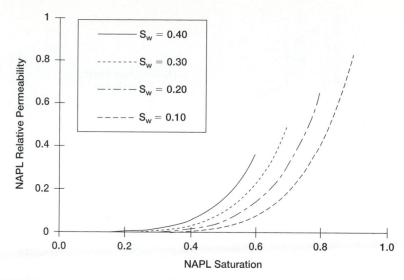

FIGURE 9.1.5   Hydrocarbon relative permeability for varying water saturations

with water at irreducible saturation, the NAPL relative permeability cannot equal unity—due to the increased NAPL-phase tortuosity associated with the water that is present. A fourth feature is that each of the relative permeability curves for a constant water content takes a form very similar to a power law. This model takes the form

$$k_{ro} = a(S_0 - S_{or})^b \tag{9.1.21}$$

where a and b are constants.

Another view of the same model given by Equation 9.1.20 is shown in Figure 9.1.6. This figure shows contours of constant NAPL relative permeability as a function of water and NAPL saturation. Each point on the diagram corresponds to a particular saturation distribution in the three-phase system. The 45 degree line connecting saturation's $(S_w, S_o) = (1,0)$ and $(0,1)$ corresponds to zero air saturation $(S_a = 0)$. The two parallel dashed lines correspond to air saturations $S_a = 0.2$ and 0.4. The vertical dashed line at $S_w = 0.1$ corresponds to the irreducible water saturation. The curves for NAPL relative permeabilities $k_{ro} = 0.4, 0.2, 0.1, 0.05,$ *and* 0.01 are shown. The figure shows that for the soil with the given properties, $k_{ro} = 0.1$ can exist for NAPL saturations ranging from 0.34 to 0.64, depending on the saturations of the other phases. A value $k_{ro} = 0.4$ can exist only within the narrow range $s_o = 0.64$ to 0.8. For a hydrocarbon saturation $S_o = 0.2$, the NAPL relative permeability is $k_{ro} = 0.01$ or less. This observation suggests that even though the assumed residual NAPL saturation is zero $(S_{or} = 0)$, once the NAPL has been introduced to the soil, the soil will retain appreciable saturations (say $S_o = 0.2$ to 0.3) for long time periods, because the NAPL drainage will be slow under normal conditions. The drainage is slow

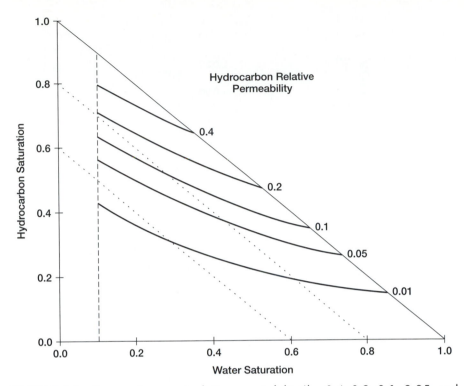

FIGURE 9.1.6 Curves of constant relative permeability (k_{ro}=0.4, 0.2, 0.1, 0.05, and 0.01) as a function of hydrobarbon and water saturation.

because the NAPL relative permeability becomes small as the NAPL drains from the soil. Thus, from the standpoint of transport and fate of NAPL liquids from spills, the residual NAPL saturation is probably not a significant parameter in determining the amount of NAPL that is held by the soil due to slow drainage. This retained NAPL could be called the NAPL **retention capacity** of the soil, in an analogous fashion to the soil's specific retention (or field capacity) for water. It should be clear, however, that the soil retention capacity is not a fixed and measurable quantity. Rather, it depends on the time for drainage.

The **effective hydraulic conductivity** for the NAPL phase is given by

$$K_o = \frac{k k_{ro} \rho_o g}{\mu_o} = K_{os} k_{ro} \qquad (9.1.22)$$

where k is the intrinsic permeability of the medium and K_{os} is the saturated hydraulic conductivity of the medium toward NAPL. The saturated NAPL conductivity is related to the saturated water conductivity through

$$K_{os} = \frac{\rho_o \mu_w}{\rho_w \mu_o} K_{ws}$$ **(9.1.23)**

where the saturated water hydraulic conductivity, K_{ws}, may be estimated from borehole or pumping tests, or from soil texture.

9.1.4 Continuity Equations for Multiphase Flow

The equations that govern multiphase flow in porous media are coupled, non-linear *partial differential equations* (PDEs). There is a mass conservation equation for each fluid and for each dissolved constituent in the system. There are also auxiliary relations that incorporate various physical phenomena into the PDEs. By nature, these latter relations are empirical, because the phenomena are highly complex and thus are not suited to exact mathematical expression. Taken together, these equations form a mathematical model of the flow and associated solute transport (e.g., Peaceman, 1977).

  If the liquids (water and NAPL) are assumed to be incompressible, then the continuity or phase conservation equation for liquid phase i is

$$n \frac{\partial S_i}{\partial t} + \nabla \cdot \tilde{q}_i = 0$$ **(9.1.24)**

Combining Equations 9.1.16 and 9.1.24 leads to the **conservation equation** for phase i:

$$n \frac{\partial S_i}{\partial t} = \nabla \cdot \left(\frac{k_{ri} k}{\mu_i} \left(\nabla p_i + \rho_i g \hat{k} \right) \right)$$ **(9.1.25)**

Equation 9.1.25 is a general form for the conservation of a fluid phase. Both the relative permeability and the pressure are functions of the fluid saturations. The resulting PDE can be very nonlinear. In a multiphase flow system, the fluids fill the available pore space. This idea is stated by the requirement of Equation 9.1.1. The combination of Equations 9.1.25 and 9.1.1 with appropriate capillary pressure functions relating the pressures between the phases, and with appropriate relative permeability functions provides a mathematical model for flow in a multiphase system. Even approximate solutions of this model set of equations require application of advanced numerical simulation techniques.

9.2 Capillary Trapping, Residual Saturation and Mass Transfer

A release of NAPL to the subsurface is often not detected until its presence is found at a potential exposure point such as a downgradient monitoring well or a water supply well. By the time the NAPL is detected, the released NAPL has spread to an extent determined by the magnitude of the release and the texture properties of the porous medium. Constituents from the released NAPL are usually found as miscible contaminants in groundwater. A separate NAPL phase is often not detected. The major issue in analysis of environmental exposure is

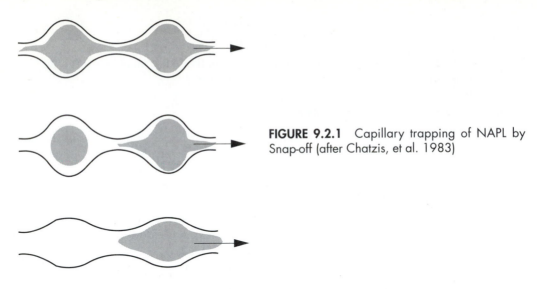

FIGURE 9.2.1 Capillary trapping of NAPL by Snap-off (after Chatzis, et al. 1983)

not with NAPL flow, but rather with the distribution of NAPL after a period of migration, and with the mass transfer of NAPL constituents to groundwater that flows past the contaminated region. In this section, we review the mechanisms associated with trapping of NAPL within a porous medium and the processes which control its dissolution release to groundwater.

9.2.1   Capillary Trapping and Distribution of NAPL

Two mechanisms are associated with **capillary trapping** of NAPL in porous media: snap-off and bypassing (Chatzis et al., 1983; Wilson et al., 1990). Figure 9.2.1 is a schematic view of the snap-off mechanism. The upper part of this figure shows a sequence of pore bodies and pore throats containing NAPL that is being displaced from the pore sequence by water (wetting fluid). The middle part of the figure shows what will happen if the aspect ratio between the size of pore bodies and throats is large. Capillary instability will cause the NAPL stream to break at the pore throat, leaving part of the NAPL trapped within the pore body. To describe this process, the displaced NAPL is said to experience **snap-off**, and the blob of NAPL trapped within a single pore body is called a **singlet**. Media with high pore body/throat aspect ratios will tend to trap NAPL as singlets.

The lower part of Figure 9.2.1 shows a pore sequence with a small aspect ratio. The distortion of the NAPL stream as it flows through the pore throat is not large enough to trigger capillary instabilities that cause snap-off, and the NAPL is displaced uniformly from the pore sequence. Thus, porous media with small pore body/throat aspect ratios will experience small residual saturation due to the snap-off process.

The wetting fluid moving around and cutting off a region of the medium from which NAPL is still draining causes the second mechanism of capillary trapping of NAPL during displacement from a porous medium. This **bypass**

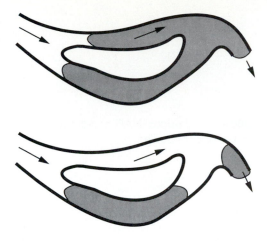

FIGURE 9.2.2 NAPL capillary trapping by wetting phase by-passing (after Chatzis, et al. 1983)

mechanism is represented schematically in Figure 9.2.2, which shows the NAPL being displaced by water through two pathways. If the displacement through one of the pathways is faster than through the other, and if the downstream junction allows breaking of the NAPL stream, then the NAPL contained within the slower drainage path will become trapped as the wetting fluid bypasses the region. The bypass mechanism may result in NAPL trapping in a single pore body, or in a group of connected pores that drain slowly. Chatzis et al. (1983) note that in general, the total volume of NAPL remaining in a porous medium because of bypassing will be larger than that left by snap-off.

The quantity of NAPL that remains trapped in a porous medium after displacement by a wetting fluid is called the **residual saturation**. The amount and distribution of residual NAPL are difficult to predict. Experiments suggest that, in terms of frequency of occurrence, most of the NAPL is trapped as singlets, although **doublets** (two pore bodies connected by residual NAPL through the pore throat) are also frequently found (Chatzis et al., 1983; Wilson et al., 1990; Powers et al., 1992). In addition, although they are less common, residual NAPL is trapped in a distribution containing a number of pore-bodies, such as that shown in Figure 9.2.3. Such a NAPL distribution is called a **ganglion**. A single NAPL ganglion contains much more volume than either a singlet or doublet, though they are found much less frequently. Collectively, singlets, doublets and ganglia are often simply called NAPL **blobs**.

Gravitational and viscous forces limit the sizes of residual ganglia. For example, in Figure 9.2.3, assume there is a DNAPL ganglia trapped in a static water column below the water table. The pressure difference through the water column is given by $\rho_w gL$. The corresponding pressure difference through the NAPL column is $\rho_o gL$. With the effective capillary radii shown in Figure 9.2.3, the condition for equilibrium is written

$$(\rho_o - \rho_w)gL = 2\sigma_{ow}\left(\frac{1}{r_1} - \frac{1}{r_2}\right) \tag{9.2.1}$$

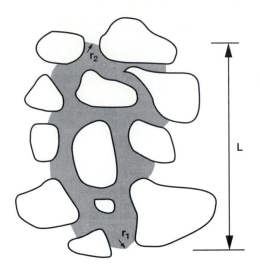

FIGURE 9.2.3 Residual NAPL ganglia containing a number of pore bodies

where the Laplace Equation 9.1.3 has been used for the capillary pressure. In the limiting case, with $r_2 \tau r_1$, the maximum ganglion length is

$$L_{max} = \frac{2\sigma_{ow}}{(\rho_o - \rho_w)gr_1} \tag{9.2.2}$$

From experiments with mixtures of glass beads, Chatzis et al. (1983) found residual nonwetting phase saturations ranging from about 14 to 16 percent. Over a range of bead sizes, the corresponding relative permeability to the wetting phase at residual nonwetting phase saturation was fairly constant with values ranging from 0.60 to 0.65. Heterogeneous packing that contains regions with clusters of small pores in a matrix of large pores, and different packing with clusters of large pores in a matrix of small pores were also investigated by Chatzis et al. (1983). For the packing with the cluster of small pores, the small pores occupied 19 percent of the volume, and they remained saturated with water. The overall residual nonwetting phase saturation was 11.6 percent, which suggests that the matrix had a residual content similar to that for the uniform bead packing. For the packing with the cluster of large pores, the large pores retained high nonwetting phase saturation. The large pores occupied 28.5 percent of the volume, and the measured residual nonwetting saturation was 35.7 percent. These results show that for heterogeneous media, fine-grained zones will remain saturated with water while coarse-grained regions will imbibe and hold the nonwetting phase during drainage. Effective nonwetting phase residual saturations can be expected to vary from 10 percent to 40 percent or more, depending on the resulting bead packing texture.

Wilson et al. (1990) measured the residual nonwetting phase saturations in three soils of aeolian, beach deposit, and fluvial deposit origin. For the three soils, they found average residual saturations of 27 percent, 18 percent, and 16 percent, and they concluded that prediction of residual saturation levels in a given soil is uncertain. From a review of the literature, Mercer and Cohen (1990) report residual nonwetting phase saturations ranging from 0.15 to 0.50 for the saturated zone.

Residual NAPL in the Vadose Zone

The distribution of residual NAPL in the vadose zone is significantly different than in the zone of saturation because of the presence of the third phase, air. For mineral soils, water is the wetting phase, air is the nonwetting phase, and NAPL is of intermediate wettability. This concept means that the soil grains and contact points will be covered with water (see Section 4.1). In this three-phase system, the NAPL may be able to spread as a film between the water and air phases. The spreading coefficient, Σ, (Adamson, 1978) provides a measure of the tendency of the NAPL to spread across an air-water interface, where the spreading coefficient is defined by (see Figure 9.2.4)

$$\Sigma = \sigma_{aw} - (\sigma_{ow} + \sigma_{ao})$$

(9.2.3)

NAPLs with $\Sigma > 0$ will spread across the air-water interface and may move distances much greater than would be possible through direct Darcy flow. NAPLs with $\Sigma < 0$, such as halogenated organic solvents, will **pool** or **bead** on the interface and will become trapped much as NAPL singlets in the saturated zone.

Wilson et al. (1990) note that in the vadose zone, the residual saturation of a spreading organic liquid consists of films, pendular rings, wedges surrounding aqueous pendular rings, and filled pore throats. Films of a spreading organic liquid maintain continuity of the NAPL phase, similar to the continuity of water at irreducible saturation. Residual saturations tend to be much lower in the vadose zone. For an aeolian soil, the measured residual saturation in the vadose zone was 9 percent, compared with 27 percent for the zone of saturation. Mercer and Cohen (1990) report residual NAPL saturations ranging from 10 percent to 20 percent in the vadose zone.

9.2.2 Displacement of Residual NAPL Saturation

Residual NAPL is trapped in porous media through capillary forces. **Residual NAPL** may be mobilized through either viscous or gravitational forces or a combination of the two. For example, if the Darcy velocity of the wetting phase is increased sufficiently, then the pressure gradient will squeeze the residual blobs

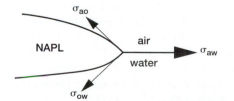

FIGURE 9.2.4 Capillary balance for the spreading coefficient

through pore throats. The leading edge will displace water (drainage) while the trailing edge of the blob will displace NAPL (imbibition). The dynamic pressure difference required to support mobilization is proportional to the difference in drainage and imbibition capillary pressures (Wilson et al., 1990).

A critical element in NAPL mobilization is the length of the blob in the direction of displacement. Longer blobs or ganglia are easier to mobilize because a greater pressure difference can be established across them. This idea is clearly seen with reference to Equation 9.2.2. If the surface tension is lowered (through the use of surfactants, for example), then the maximum length of a stable ganglion against gravity will decrease proportionally, and part of the residual NAPL will be mobilized.

Experimental data on mobilization of residual NAPL by viscous forces are shown in Figure 9.2.5 (after Morrow et al., 1988). Data are shown for sandstone correlation, bead correlation (bead sizes of 70, 97, 115, and 162 μm), and sintered bead data. The **capillary number** is the ratio of viscous to capillary forces and is defined by

$$N_c = \frac{q\mu}{\sigma} \qquad (9.2.4)$$

where q is the aqueous Darcy velocity, μ is the aqueous dynamic viscosity, and σ is the NAPL-water interfacial tension. The reduced residual saturation is given by

$$\frac{S_{or}}{S_{or}^*} \qquad (9.2.5)$$

where S_{or}^* is the residual saturation at low capillary numbers (low Darcy velocities) and S_{or} is the residual saturation at higher capillary numbers. For sandstone $S_{or}^* = 0.35$. Figure 9.2.5 suggests that for a packed bead bed, capillary numbers in excess of 0.01 are required before the residual NAPL is completely displace. To obtain some perspective on the required gradient, assume that the bed has $K_{ws} = 0.01$ cm/s, $k_{rw} = 0.6$ at residual NAPL saturation, $\mu = 0.01$ g/cm.sec, and $\sigma = 20$ dynes/cm. This relationship gives $q = N_c\sigma/\mu = 20$ cm/s, and from Equation 9.1.18, $I = q/(K_{sw}k_{rw}) = 3,300$. While such a gradient may be achieved in a laboratory column, it is certainly beyond the reasonable range for field conditions. Thus, it is suggested that residual saturations would be difficult to displace using hydraulic methods. This is consistent with the conclusions reached by Wilson and Conrad (1984).

9.2.3   Dissolution of Residual NAPL

For practical purposes, residual NAPL is immobilized. Except through enhanced physical/chemical means, it cannot be made to continue migration through a porous medium as a separate phase. This statement does not mean, however, that residual NAPL is of no further concern as a contaminant. All NAPL chemical constituents have some solubility in water, and the residual NAPL constituents will dissolve in groundwater flowing near the region of contamination. Constituents are transported as miscible contaminants in the aqueous phase. The following example shows why this idea is important.

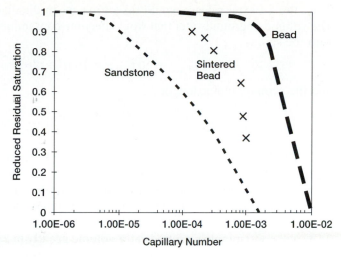

FIGURE 9.2.5 Reduced residual saturation vs. capillary number in various media (after Morrow et al., 1988)

EXAMPLE
PROBLEM

Example 9.2.1   TCE Release

As an example, consider a release of 1-barrel (42 gallons) TCE to groundwater. Because the density of TCE is 1.468 g/cm³, this value corresponds to a release of about 230 kg. From Table 5.4.1, the solubility of TCE in water is 1.1 g/L. It is assumed that the Darcy velocity is 10 ft/yr, the porosity is 0.35, and that the retention NAPL saturation is 0.25. This statement implies that the volume of soil contaminated with NAPL is

$$V_{\text{NAPL}} = \frac{42 \text{ gal}}{7.48 \text{ gal/ft}^3 \times 0.35 \times 0.25} = 64 \text{ ft}^3$$

It is further assumed that the NAPL distribution takes the shape of a cube. Thus, the cube length dimension is $L = V_{\text{NAPL}}^{1/3} = 4$ ft. The cross-section for groundwater flow through the contaminated region is $A = 4^2 = 16$ ft². If the water flowing through the contaminated region reaches its TCE solubility limit, then the mass flux from the source region is

$$\dot{m} = qAc = 10 \text{ ft/yr} \times 16 \text{ ft}^2 \times 1.1 \text{ g/L} \times 28.3 \text{ L/ft}^3 = 5000 \text{ gm/yr}$$

Assuming that there are no TCE losses through volatilization, biodegradation, or other processes, the miscible TCE release to groundwater will occur over a time period

$$T_{\text{release}} = \frac{230 \text{ kg}}{5 \text{ kg/yr}} = 46 \text{ years}$$

Thus, the 1-barrel release would take 46 years to dissolve into groundwater. The volume of groundwater that would become contaminated by direct contact with the spill is

$$V_{\text{water}} = qAT_{\text{release}} = 10 \text{ ft/yr} \times 16 \text{ ft}^2 \times 46 \text{ yr} = 7360 \text{ ft}^3$$

and the primary plume length is

$$L_{\text{plume}} = \frac{qT_{\text{release}}}{n} = \frac{10 \text{ ft/yr} \times 46 \text{ yr}}{0.35} = 1300 \text{ ft}^3$$

This cursory analysis suggests that a released volume of 1 barrel $= 5.61 \text{ ft}^3$ of TCE will directly contaminate a volume $7{,}360 \text{ ft}^3$ of groundwater. This deduction could have been made on solubility considerations alone. The calculation greatly underestimates the true contaminated volume of groundwater, however. Because of dispersion and diffusion, a release of 5000 g/yr of dissolved TCE will contaminate a groundwater volume greatly in excess of the estimate given previously. Furthermore, it is this latter unknown volume which must be considered in discussions of groundwater remediation efforts.

This example suggests that a small release of NAPL can result in large-scale contamination of groundwater. A question of fundamental interest concerns the mechanism and rate of mass transfer between residual NAPL and groundwater. Miller et al. (1990) note that most laboratory and field data suggest that the **local equilibrium assumption** is appropriate, according to which the aqueous concentration in contact with residual NAPL is equal to the thermodynamic equilibrium concentration. To evaluate this assumption, controlled laboratory dissolution mass transfer experiments have been carried out. A small section of a laboratory column is contaminated with residual NAPL, and clean water is transported through the section and sampled at the downgradient end of the column. The variables of interest include the magnitude of the mass transfer coefficient as a function of the groundwater velocity and the amount of residual NAPL present.

In order to interpret experimental results, a theoretical model must be formulated. The simplest model assumes that the mass transfer flux is proportion to the product of the equilibrium concentration deficit as a driving force and as a mass transfer coefficient:

$$J_{\text{diss}} = k_d(C_e - c) \tag{9.2.6}$$

where J_{diss} is the dissolution mass transfer ($M\,L^{-2}\,T^{-1}$) between the NAPL and aqueous phase, k_d is the dissolution mass transfer coefficient (L/T), and $[C_e - c]$ is the water phase concentration deficit ($M\,L^{-3}$), where C_e is the effective equilibrium water concentration of the constituent. Further consideration suggests that the mass transfer coefficient should depend on the interfacial area between the NAPL and water phases, and the groundwater velocity. A model for the local aqueous concentration change may be written

$$n(1 - S_{or})\frac{dc}{dt} = k_d a_{ow}(C_e - c) \tag{9.2.7}$$

where a_{ow} is the NAPL-water interfacial area per unit bulk volume (L^{-1}). Because of the unknown distribution of residual NAPL and the complex structure of the porous media, however, one cannot estimate either k_d or a_{ow} individually. The parameters may be lumped together as a mass transfer rate coefficient, κ. The rate coefficient has units T^{-1}.

In order to analyze experimental data, steady-state conditions are achieved, and the one-dimensional transport equation for the water phase is written

$$v\frac{dc}{dx} - D\frac{d^2c}{dx^2} - \kappa(C_e - c) = 0 \tag{9.2.8}$$

With the boundary conditions

$$c(0) = 0 \text{ and } \left.\frac{dc}{dx}\right|_{x=\infty} = 0$$

the solution is

$$\frac{c(x)}{C_e} = 1 - \exp\left\{\left(\frac{vx}{2D}\right)\left(1 - \sqrt{1 + \frac{4D\kappa}{v^2}}\right)\right\} \tag{9.2.9}$$

With Equation 9.2.9, measurements of the aqueous concentration at the end of the NAPL contaminated region may be used to estimate κ. Miller et al. (1990) used this approach. Their results suggest that there is no significant relation between κ and the mean particle size, but that κ does depend on the seepage velocity. Their data suggest the model

$$Sh = 12Re^{0.75}\theta_{or}^{0.6}Sc^{0.5} \tag{9.2.10}$$

In Equation 9.2.10, Sh is the Sherwood number, Re is the Reynolds number, and Sc is the Schmidt number, with

$$Sh = \frac{\kappa d_m^2}{D} \; ; \; Re = \frac{v\rho_w d_m}{\mu_w} \; ; \; Sc = \frac{\mu_w}{\rho_w D} \tag{9.2.11}$$

where d_m is the mean grain diameter. In Equation 9.2.10, the residual volumetric NAPL content is introduced as a measure of the interfacial area, a_{ow}, and Miller et al. (1990) note that there is significant uncertainty in the estimation of the power 0.6. The $^1/_2$ power dependency on the Schmidt number is based on analogy with other mass transfer problems (Levich, 1962), and is not based on direct experimental evidence.

Powers et al. (1992) reported a similar study. Their analysis was based on the mass transport equation neglecting dispersion

$$q\frac{dc}{dx} + \kappa(C_e - c) = 0 \tag{9.2.12}$$

where q is the Darcy velocity. With the boundary condition $c(0) = 0$ and a column of length L, Equation 9.2.12 yields

$$\kappa = -\left(\frac{q}{L}\right)ln\left(1 - \frac{c}{C_e}\right) \qquad (9.2.13)$$

While their experimental results showed poor correlation when θ_{or} was used as a variable, they did find strong correlation with the grain size distribution. The best correlation is given by

$$Sh = 57.7Re_q^{0.61}d_m^{0.64}U^{0.41} \qquad (9.2.14)$$

where Re_q is the Reynolds number based on the Darcy velocity rather than the seepage velocity, d_m is the median grain diameter (cm), and U is the soil uniformity index (d_{60}/d_{10}), where i percent of the particles are smaller than d_i.

EXAMPLE PROBLEM

Example 9.2.2   Comparison of Dissolution Mass Transfer Coefficients
To compare the models given by Equations 9.2.10 and 9.2.14, consider a packed bed with $d_m = 0.5$ mm, $n = 0.38$, $q = 1$ m/d, and residual toluene at $S_{or} = 0.15$. For toluene, $D_m = 0.8(10^{-5})$ cm²/s, so

$$Sc = \frac{\mu_w}{\rho_w D} = \frac{0.01 \text{ g/cm.s}}{1 \text{ g/cm}^3 \times 0.8(10^{-5})\text{ cm}^2\text{/s}} = 1250$$

The seepage velocity is

$$v = \frac{q}{n(1 - S_{or})} = 3.1 \text{ m/d} = 3.6(10^{-3}) \text{ cm/s}$$

This situation gives

$$Re = \frac{v\rho_w d_m}{\mu_w} = \frac{3.6(10^{-3})\text{cm/s} \times 1 \text{ g/cm} \times 0.05 \text{ cm}}{0.01 \text{ g/cm.s}} = 0.018$$

and, from Equation 9.2.10,

$$Sh = 12(0.018)^{0.75}(0.38 \times 0.15)^{0.6}(1250)^{0.5} = 3.74$$

Similarly, Equation 9.2.14 gives

$$Sh = 57.7(5.8 \times 10^{-3})^{0.61}(0.05)^{0.64}(1)^{0.41} = 0.37$$

which is one order-of-magnitude smaller. Using the smaller of these,

$$\kappa = \frac{D_m \times Sh}{d_m^2} = \frac{0.8(10^{-5})\text{cm}^2\text{/s} \times 0.37}{(0.05 \text{ cm})^2} = 1.2(10^{-3})s^{-1} = 100d^{-1}$$

Equation 9.2.13 may be used to find the length of contaminated region required before $c = 0.99C_e$. For this example, Equation 9.2.13 gives

$$L = -\left(\frac{q}{\kappa}\right)ln\left(1 - \frac{c}{C_e}\right) = -\frac{1 \text{ m/d}}{100 \text{ d}^{-1}} ln\left(1 - \frac{0.99C_e}{C_e}\right) = 0.046 \text{ m}$$

and use of the results from Equation 9.2.10 would give $L = 0.46$ cm. For both of the models considered in this example, it is clear that mass transfer is rapid compared with the speed of groundwater flow and that the local equilibrium assumption provides a useful and valid model for partitioning of residual NAPL.

DNAPL Screening Criteria

The previous discussion suggests that groundwater that comes into contact with DNAPL will rapidly dissolve chemical constituents to reach concentrations that correspond to the effective solubility limits. This idea does not mean, however, that one should expect to find groundwater monitoring sample concentrations near solubility limits if DNAPL is present at a site. Because of capillary phenomenon and heterogeneities, DNAPL tends to become isolated in only a small fraction of the porous media, and much of the water that is captured in a groundwater sample may not come into contact with DNAPL. As a result, soluble phase components of DNAPL are rarely found in excess of 10 percent of the solubility even when organic liquids are known or suspected to be present at a site (EPA, 1992).

It is of interest to develop **screening criteria** that may be used to suggest the presence of DNAPL at a site. Such criteria must recognize that because of the capabilities of DNAPL to segregate itself, it is unlikely that groundwater monitoring wells will be placed immediately adjacent to DNAPL bodies. Aqueous concentrations in excess of 1 percent of the effective solubility are suggestive of the presence of DNAPL, while concentrations that are less than this amount do no preclude the presence of DNAPL at a site (Cohen and Mercer, 1993; Pankow and Cherry, 1996).

9.3 NAPL Behavior in Fractured Media

The role of capillary forces in controlling the behavior of NAPLs is especially significant in fractured porous media. For fractured clays and fractured sedimentary bedrock such as sandstone, siltstone, shale, or carbonate rocks, the fractures represent a small fraction of the void space but they provide most of the hydraulic conductivity (see Section 2.2.4). The fracture aperture is generally larger than the size of the pores within the matrix, and thus the NAPL will first enter the media through the fractures because of their lower entry pressure. Once in the fracture network, if sufficient capillary pressure is not achieved, the NAPL will be confined to the fracture network, which could allow for rapid movement through the fracture system.

Consider the situation shown in Figure 9.3.1, where a pool of DNAPL resides within a sand geologic unit that is confined below by fractured clay unit. A sufficient DNAPL pool depth H_d must develop before the DNAPL is able to enter the fracture with aperture e, which represents the largest pore opening in the clay unit. If we assume that locally this irregular fracture opening takes the

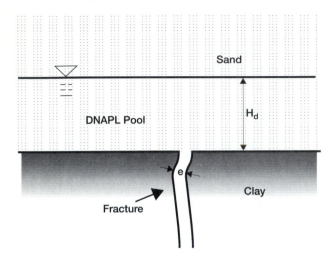

FIGURE 9.3.1   DNAPL entry into a fracture

shape of two parallel plates, then a force balance yields the following pool depth for initial entry into the fracture (Kueper and McWhorter, 1991)

$$H_d = \frac{2\sigma \cos(\beta)}{\Delta \rho g e} \qquad (9.3.1)$$

where σ is the interfacial tension between the DNAPL and water, β is the contact angle measure through water, $\Delta \rho$ is the density difference between the DNAPL and water, and e is the fracture aperture. It is usually assumed that water is perfectly wetting compared with DNAPL and $cos(\beta) = 1$. Kueper and McWhorter (1991) note that if the DNAPL pool spreads laterally across the clay to reach the fracture, then the DNAPL pressure must exceed the entry pressure of the porous medium, and Equation 9.3.1 is modified as

$$H_d = \frac{2\sigma}{\Delta \rho g e} - \Psi_{bow} \qquad (9.3.2)$$

where Ψ_{bow} is the entry head for DNAPL to displace water and enter the sand, as discussed in Section 9.4 (see Equation 9.4.5).

Once the DNAPL has entered the fractured medium it may move rapidly through the fracture system, at least until the source (DNAPL pool) is depleted. The DNAPL is able to move downward, even under conditions where the fracture aperture decreases with depth due of increasing earth stresses, because its effective pool depth also increases, as shown in Figure 9.3.2. On the left of the figure is a schematic view of DNAPL within a fracture network that is fed by the DNAPL pool located above. On the right is the hydrostatic pressure distribution for water and DNAPL. The DNAPL pressure increases at a greater rate with depth, as compared with water, because of its larger density. The effective DNAPL pool depth for entry into a fracture at point A is directly related to the elevation difference between DNAPL pool and point A. Under flowing

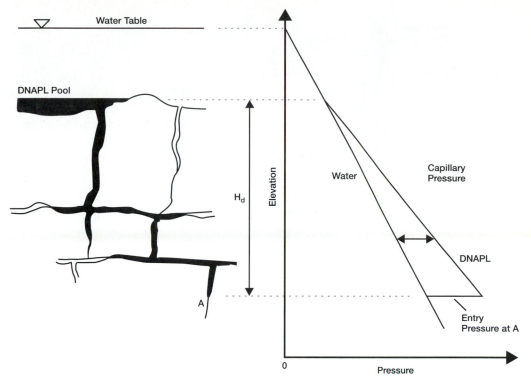

FIGURE 9.3.2 Pressure distribution for entry of DNAPL into a fracture network (after Kueper and McWhorter, 1991)

conditions, the pressure gradients would be smaller than those shown because of energy losses during Darcy flow.

Following source depletion, the DNAPL movement will eventually cease with the DNAPL either trapped within fractures or isolated within fractures at residual saturations. Because of the small fracture volume and the configuration of the fracture system, the DNAPL has a large surface area to volume ratio. Chemical constituents from the DNAPL will dissolve into the water retained in the fracture or water adjacent to the fracture in the porous matrix, with the aqueous concentration reaching the effective constituent solubility. Molecular diffusion will then cause the dissolved chemical to migrate laterally, away from the fracture into the porous matrix (Parker et al., 1994). This situation is shown schematically in Figure 9.3.3. The fracture contains residual DNAPL. The plots on either side of the fracture show the dissolved constituent concentration distributions as a function of lateral distance from the fracture at different times. The maximum concentration is located within and immediately adjacent to the fracture, and is equal to the effective solubility of the constituent. As time proceeds, there is an increasing mass transfer from the fracture into the porous matrix.

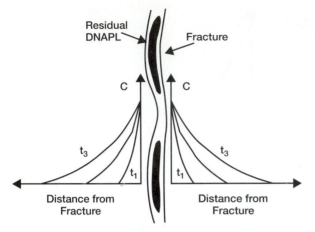

FIGURE 9.3.3 Lateral diffusion into the porous matrix of dissolved constituents from DNAPL trapped in a fracture (after Parker et al., 1994)

If we assume that the constituent concentration remains constant at its effective solubility limit, as long as residual NAPL is present within the fracture, then the problem of lateral diffusion of mass from the fracture into the porous matrix is the same as that considered in Section 7.5.2. The cumulative mass (per unit area of fracture face) that has moved from the fracture into the matrix is given by Equation 7.5.21, which may be written as follows:

$$\frac{M(t)}{A} = 4nc_0\sqrt{\frac{RD_s t}{\pi}} \qquad \textbf{(9.3.3)}$$

In Equation 9.3.3, n is the matrix porosity, c_o is the effective aqueous solubility of the chemical, R is the constituent retardation factor within the porous matrix, and D_s is the effective soil diffusion coefficient. The increase by a factor of two compared with Equation 7.5.21 accounts for diffusion occurring in both lateral directions. With fracture aperture e, constituent mass fraction f_i in the DNAPL, and DNAPL density ρ_o, the maximum initial constituent mass (per unit area of fracture face) contained in the fracture is $e f_i \rho_o$. With Equation 9.3.3, the DNAPL **disappearance time** is

$$t_D = \frac{\pi}{RD_s}\left(\frac{ef_i\rho_o}{4nc_o}\right)^2 \qquad \textbf{(9.3.4)}$$

Equation 9.3.4 assumes that the entire fracture is initially filled with DNAPL, and thus the model provides an upper bound on the disappearance time for a fracture of given aperture. The mass fraction f_i influences the effective solubility that is based on Raoult's law (Section 5.5), except that c_o is based on the mole fraction of the constituent solubility. Calculations show that the NAPL phase will disappear from the fracture network within a period of months to a few years (see Parker et al., 1994). Thus, after a sufficient time following a NAPL release to a fractured system, the NAPL will no longer be present, and much of the chemical is dissolved within the porous matrix—making remediation efforts more difficult. The same considerations are also true for an LNAPL within a fracture system, except that the fracture face in contact with the LNAPL may be more limited because of the tendency of LNAPL to accumulate near the water table.

<table>
<tr><td>EXAMPLE
PROBLEM</td><td>

Example 9.3.1   Dissolution Time for DNAPL from a Fracture

A chemical release has resulted in trapping of liquid-TCE within a fracture network below the water table. Estimate the time period for disappearance of the DNAPL if the average fracture aperture is 20 μm (0.002 cm).

From Table 5.4.1, the density and water solubility of TCE are 1470 mg/cm^3 and 1.1 mg/cm^3, and TCE has a molecular diffusion coefficient of 10^{-5} cm^2/s. For the porous matrix, it is assumed that the porosity and tortuosity both have the value 0.1 and that the retardation factor is 3.0. With these values, Equation 9.3.4 gives

$$t_D = \frac{\pi}{3.0 \times 0.1 \times 10^{-5} \text{ cm}^2/\text{s}} \left(\frac{0.002 \text{ cm} \times 1470 \text{ mg/cm}^3}{4 \times 0.1 \times 1.1 \text{ mg/cm}^3} \right)^2 \times \frac{1}{86400} \text{ d/s} = 54 \text{ d}$$

Thus, within a period of about 2 months, the TCE-liquid phase will dissolve from fractures of 20 μm size. For wider fractures of aperture 100 μm, the corresponding disappearance time is 3.7 years.

</td></tr>
</table>

9.4   Monitoring of Free-Product Petroleum Hydrocarbons

Groundwater contamination from spills and subsurface leakage of petroleum hydrocarbons and other lighter-than-water NAPLs (LNAPLs or OILs) is a widespread problem at service stations, industrial facilities, and petroleum refineries. When OIL reaches the water table as a free product, it depresses the water table under its own weight and eventually spreads laterally because of potential energy gradients within its phase. Free-product hydrocarbon may be recovered using trenches, skimmer wells, and pumping wells. One of the difficulties in assessing free-product recovery is determining the amount of OIL that is present and how much can be recovered. The free-product thickness in monitoring wells is a principle means for estimating the amount of OIL present.

Because of capillary forces, the OIL thickness in a monitoring well is not the same as that within the porous medium. Under equilibrium conditions, however, there is a direct relationship between monitoring well thickness and OIL distribution within the formation. Equilibrium conditions means that the vertical pressure distributions of the liquids (water and OIL) are hydrostatic. At least three factors determine the vertical distribution of free product under conditions of vertical equilibrium. These are the OIL density relative to that of water, the total amount of hydrocarbon present, and capillary pressure forces acting within the porous matrix.

The OIL density or specific gravity may be expressed on the basis of the following empirical equation (Perry and Chilton, 1973):

$$^\circ\text{API} = \frac{141.5}{\rho_r} - 135.5 \qquad (9.4.1)$$

where ρ_r is the OIL-to-water density ratio ($\rho_r = \rho_o/\rho_w$). The degree API is a scale measurement adopted by the *American Petroleum Institute* (API). The scale ranges from 0.0 (equivalent to a specific gravity of 1.076) to 100.0 (equivalent to a specific gravity 0.6112). Hydrocarbon fractions commonly found beneath refineries and other petroleum manufacturing sites range from °API 30 to 40, representing light paraffin oil to kerosene. Corresponding OIL viscosities are approximately 8 cp (centipoise) for °API 30, 4 cp for °API 35, and 2 cp for °API 40. These fractions present challenges to remediation because they are light enough to float on the water table, but they are not volatile enough to be easily remediated by the *in situ* vapor extraction processes (Charbeneau and Chiang, 1995).

This section presents the relationship between the monitoring well OIL thickness and the distribution of free-product in the formation. This relationship determines the volume of OIL present and is necessary for determining the mobility of the OIL during pumping, and the relationship between changes in free-product thickness and recovery volumes during free-product remediation using pumping wells.

9.4.1   Background and Theory

Under conditions of vertical equilibrium, the vertical pressure gradient within both the water and hydrocarbon phase satisfies the hydrostatic pressure equation, and distribution of free product may be determined from the water retention curve (soil moisture characteristic curve) of the soil using the method outlined by Schiegg (1985). This method directly relates the free-product distribution within the porous matrix to the product thickness one would find in a monitoring well in free communication with the aquifer fluids. This method is shown schematically in Figure 9.4.1. Within the formation, the fluids form a three-phase system. In the monitoring well, however, one finds three distinct fluid layers: a water layer below, a OIL layer floating on the water, and air above. z_{ao} is the elevation of the air and OIL interface in the monitoring well, z_{ow} is the elevation of the OIL and water interface, z_{aw} is the water table elevation that would exist in the absence of OIL, and b_o is the monitoring well OIL-layer thickness. The observation well OIL thickness is independent of capillary forces in the formation and corresponds to the same OIL and water hydraulic head under conditions of vertical equilibrium. For this reason, the monitoring well thickness provides a useful analog for estimation of free-product volumes present within the aquifer and for computation of free-product recovery by pumped wells.

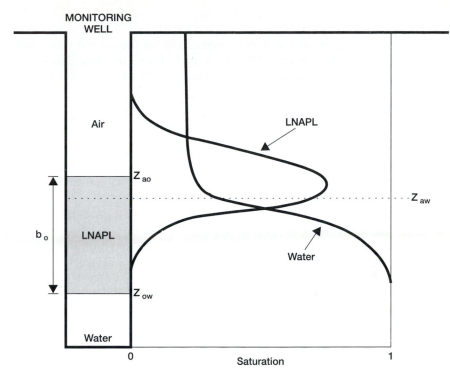

FIGURE 9.4.1 Liquid distribution within the formation and in a monitoring well for floating free-product hydrocarbon

Three assumptions provide the basis for calculating the vertical equilibrium liquid distribution. First, the air and water soil characteristic curve provides sufficient information describing the pore size distribution for the soil. Second, the characteristic curve for a different fluid system (i.e., OIL and water) may be estimated from the air and water curve using scaling parameters that depend only on properties of the fluids involved. Third, in a three-phase system, the capillary pressure between the OIL and water phase is a function of the water (wetting fluid) saturation, while the air and OIL capillary pressure is a function of the total liquid saturation (OIL plus water). This third assumption was suggested by Leverett (1941) and has been supported by experiment. With these assumptions, one can predict the vertical distribution of fluids in a three-phase system—and from this distribution, one can estimate the amount of hydrocarbon present for a given observation well thickness. The results from this theory as they apply to the Brooks and Corey power-law parametric model are outlined next.

Vertical Distribution of Fluid Saturation

In free-product recovery, it is assumed that OIL and water are incompressible and that within the vadose zone, air remains at atmospheric pressure. Under these conditions, the hydrostatic pressure equation gives

$$\Psi_{aw} = z - z_{aw} \tag{9.4.2}$$

where Ψ_{aw} is the capillary pressure head for the air and water fluid system, z is the elevation, and z_{aw} is the elevation of the water table. The Brooks and Corey (1964) power-law soil characteristic model from Equation 4.4.2 provides

$$\Theta_w = 1 \text{ for } z - z_{aw} \leq \Psi_{baw} \; ; \; \Theta_w = \left(\frac{\Psi_{baw}}{z - z_{aw}}\right)^{\lambda} \text{ for } z - z_{aw} > \Psi_{baw} \tag{9.4.3}$$

where Θ_w is the reduced water saturation and Ψ_{baw} is the air and water bubbling pressure head. Equation 9.4.3 gives the water saturation as a function of elevation above the water table under conditions of vertical equilibrium. According to this power law model, the pore size distribution index (λ) and the bubbling pressure head (a measure of the largest pore size) serve to determine the structure of the porous media.

To apply the Brooks and Corey model to a multiphase system including free-product OIL, it is assumed that changes in soil structure (e.g., swelling) are negligible and that differences in behavior from one fluid pair to another can be attributed only to differences in fluid properties. For multiple fluid systems, the subscripts w, o, and a are used to designate the water, OIL, and air phases. In addition, for fluid pairs, the subscript order is nonwetting fluid first and wetting fluid second. In general, for the ij-system ($ij = ao$ or ow), the bubbling pressure is related to that for the aw-system through

$$p_{bij} = p_{baw} \frac{\sigma_{ij}}{\sigma_{aw}} = \rho_w g \Psi_{baw} \frac{\sigma_{ij}}{\sigma_{aw}} \tag{9.4.4}$$

which follows because the maximum pore size remains constant and the bubbling pressure depends only on interfacial tension (σ). With the hydrostatic pressure equation $p_{cij} = \Delta\rho_{ij}g(z - z_{ij})$ (where z_{ij} is the elevation where the capillary pressure vanishes), Equation 9.4.4 gives

$$\Theta_j = \left(\frac{\rho_w \Psi_{baw} \sigma_{ij}}{\Delta\rho_{ij}\sigma_{aw}(z - z_{ij})}\right)^{\lambda} = \left(\frac{\Psi_{bij}}{z - z_{ij}}\right)^{\lambda} \tag{9.4.5}$$

for $(z - z_{ij}) \geq \Psi_{bij}$, where j is the wetting phase and

$$\Psi_{bij} = \frac{\rho_w \Psi_{baw} \sigma_{ij}}{\Delta\rho_{ij}\sigma_{aw}} \tag{9.4.6}$$

Similar scaling relationships were introduced by Leverett (1941) and later used by van Dam (1967), Schiegg (1985), Parker et al. (1987), Cary et al. (1989), and Demond and Roberts (1991). For the air and OIL system, one may use $\Delta\rho_{ao} = \rho_o$, because the density of air is small.

The **Leverett assumption** is that the water saturation in a three phase system depends on the OIL and water capillary pressure, while the total liquid saturation, $S_t = S_w + S_o$, is a function of the air and OIL capillary pressure. With the power-law soil characteristic model, this relationship may be written

$$\Theta_w(p_{cow}) = \left(\frac{\Psi_{bow}}{z - z_{ow}}\right)^\lambda \; ; \; \Theta_t(p_{cao}) = \left(\frac{\Psi_{bao}}{z - z_{ao}}\right)^\lambda \tag{9.4.7}$$

where z_{ow} and z_{ao} are the elevations at which the corresponding capillary pressures would vanish, and Θ_t is the total reduced liquid saturation.

Different OIL residual saturation values may exist above and below the free-product region near the water table. The appropriate scaling functions for the reduced saturation are as follows:

$$\Theta_w(p_{cow}) = \frac{S_w - S_{wr}}{1 - S_{wr} - S_{ors}} \; ; \; \Theta_t(p_{cao}) = \frac{S_w + S_o - S_{wr} - S_{orv}}{1 - S_{wr} - S_{orv}} \tag{9.4.8}$$

where S_{wr} is the irreducible water saturation and S_{ors} and S_{orv} are the residual OIL saturations in the saturated and vadose zones, respectively. One may use the water and OIL retention (or **field capacity**) values instead of the irreducible and residual saturations in the scaling relationships.

Equations 9.4.7 and 9.4.8 determine the fluid distribution near the water table. The capillary pressure datums z_{ow} and z_{ao} are the levels at which one would find the fluid interfaces in monitoring wells or in a gravel aquifer where formation capillary forces are absent. From Figure 9.4.1, the monitoring well OIL layer thickness is $b_o = (z_{ao}-z_{aw})-(z_{aw}-z_{ow}) = z_{ao}-z_{ow}$. Calculations from hydrostatics show that the various elevations are related through

$$z_{aw} - z_{ow} = \frac{\rho_o}{\rho_w} b_o \; ; \; z_{ao} - z_{aw} = \left(\frac{\rho_w - \rho_o}{\rho_w}\right) b_o \tag{9.4.9}$$

For example, if $b_o = 5$ ft and $\rho_o = 0.8$ g/cm^3, then from Equation 9.4.9, one finds that $z_{ao} - z_{aw} = 1$ ft while $z_{aw} - z_{ow} = 4$ ft. In addition, because the fluid head is given by

$$h = \frac{p}{\gamma} + z$$

one has the following equation for the difference in fluid heads,

$$h_o - h_w = z_{ao} - z_{aw} \tag{9.4.10}$$

or $h_o - h_w = 1$ ft (because $h_w = z_{aw}$). More generally, for conditions of vertical equilibrium, the head in the water phase is given by

$$h_w = \frac{p_{ow}}{\rho_w g} + z_{ow} = \left(1 - \frac{\rho_o}{\rho_w}\right) z_{ow} + \frac{\rho_o}{\rho_w} z_{ao} \tag{9.4.11}$$

where p_{ow} is the fluid pressure at the OIL and water interface in the monitoring well. The thickness of OIL observed in the well, b_o, is given by

$$b_o = z_{ao} - z_{ow} = \left(\frac{\rho_w}{\rho_w - \rho_o}\right) h_o \tag{9.4.12}$$

Van Dam (1967) presents similar relationships to Equations 9.4.9 through 9.4.12.

OIL and Water Distribution

Equations 9.4.7 and 9.4.8 provide the fluid saturation distribution within the free-product zone for a given monitoring well OIL-layer thickness b_o. If there is residual hydrocarbon beneath the free OIL layer down to an elevation z_{ors} and above the layer to an elevation z_{orv}, then the water distribution is given by

i) $\quad z < z_{ors}$, $\hspace{5cm} S_w = 1$

ii) $\quad z_{ors} < z < z_{ow} + \Psi_{bow}$, $\hspace{3.2cm} S_w = 1 - S_{ors}$ $\hspace{2cm}$ (9.4.13)

iii) $\quad z > z_{ow} + \Psi_{bow}$, $\hspace{1.3cm} S_w(z) = S_{wr} + (1 - S_{wr} - S_{ors})\left(\dfrac{\Psi_{bow}}{z - z_{ow}}\right)^{\lambda}$

Similarly, the OIL saturation distribution is given by

i) $\quad z < z_{ors}$, $\hspace{5cm} S_o = 0$

ii) $\quad z_{ors} < z < z_{ow} + \Psi_{bow}$, $\hspace{3.2cm} S_o = S_{ors}$ $\hspace{2cm}$ (9.4.14)

iii) $\quad z_{ow} + \Psi_{bow} < z < z_{ao} + \Psi_{bao}$, $\hspace{2cm} S(z) = 1 - S_w(z)$

iv) $\quad z_{ao} + \Psi_{bao} < z < z_r$, $\hspace{0.8cm} S_o(z) = (S_{wr} + S_{orv} - S_w(z)) + (1 - S_{wr} - S_{orv})\left(\dfrac{\Psi_{ao}}{z - z_{ao}}\right)^{\lambda}$

v) $\quad z_r < z < z_{orv}$ $\hspace{4.5cm} S_o = S_{orv}$

vi) $\quad z > z_{orv}$ $\hspace{4.8cm} S_o = 0$

The elevation z_r is the elevation where the water saturation curve and the total liquid saturation curve intersect. The OIL saturation decreases to its vadose zone residual value, S_{orv}, at this elevation.

$$z_r = \frac{\dfrac{\Delta\rho_{ao}}{\rho_w}\left(1 + \dfrac{\sigma_{ao}}{\sigma_{ow}}\left(\dfrac{1 - S_{wr} - S_{orv}}{1 - S_{wr} - S_{ors}}\right)^{1/\lambda}\right)}{1 - \dfrac{\Delta\rho_{ow}\sigma_{ao}}{\rho_o\sigma_{ow}}\left(\dfrac{1 - S_{wr} - S_{orv}}{1 - S_{wr} - S_{ors}}\right)^{1/\lambda}}\, b_o \hspace{2cm} (9.4.15)$$

The OIL is considered to be free product between the elevations $(z_{ow} + \Psi_{bow}) < z < z_r$.

Free-Product OIL Volume

The amount (volume within a unit area of the OIL lens) of hydrocarbon present in the free-product region (not including the hydrocarbon possibly present at residual saturation above or below the free product) is found by integrating the OIL saturation over the free-product region:

$$D_o = \int_{z_{ow} + \Psi_{\beta ow}}^{z_r} S_o(z)\,dz \hspace{2cm} (9.4.16)$$

This usage of the OIL free-product thickness, D_o, corresponds to that of Schwille (1967) who uses it for the ratio between the amount of hydrocarbon spreading laterally on the groundwater surface and the area occupied by it. When multiplied by the lens area, it represents the volume of mobile OIL and will be used to compute free-product recovery. Other authors refer to the hydrocarbon layer thickness as that which may be visually observed in a laboratory apparatus. This item is not the same as D_o.

With Equations 9.4.13 through 9.4.15, one may evaluate the integral given by Equation 9.4.16 resulting in

$$D_o = \xi + \eta(b_o)b_o \qquad (9.4.17)$$

where

$$\xi = \left[\frac{[\lambda(1 - S_{wr}) - S_{ors}]\Psi_{\beta ow} - [\lambda(1 - S_{wr} - S_{orv})]\Psi_{bao}}{1 - \lambda} \right] n \qquad (9.417a)$$

$$\eta(b_o) = \left[(1 - S_{wr}) + \frac{\chi}{1 - \chi} S_{orv} - \frac{(1 - S_{wr} - S_{ors})}{1 - \lambda} \left(\frac{(1 - \chi)\Psi_{bow}}{b_o} \right)^\lambda \right] n \qquad (9.4.17b)$$

$$\chi = \frac{\sigma_{ao}}{\sigma_{ow}} \left(\frac{\rho_w - \rho_o}{\rho_o} \right) \left(\frac{1 - S_{wr} - S_{orv}}{1 - S_{wr} - S_{ors}} \right)^{1/\lambda} \qquad (9.4.17c)$$

Charbeneau and Chiang (1995) presented these results. They differ from the earlier results of Farr et al. (1990) and Lenhard and Parker (1990) in that they allow different residual OIL saturation values for the saturated and vadose zone. In Equation 9.4.17, $\eta(b_o)$ has only a weak dependence on b_o, especially at moderate to large OIL-layer thicknesses. This statement implies that the relationship between D_o and b_o is nearly linear.

From this model, it appears that the total liquid saturation curve will intersect and cross the water saturation curve if $z_{ow} + \Psi_{bow} > z_{ao} + \Psi_{bao}$. Parker and Lenhard (1989) noted this result. The model predicts that the limiting condition for free OIL within the formation is $z_{ow} + \Psi_{bow} = z_{ao} + \Psi_{bao}$, or

$$\Delta\Psi = z_{ao} - z_{ow} = \Psi_{bow} - \Psi_{bao} \qquad (9.4.18)$$

Equation 9.4.18 may be interpreted to mean that a minimum thickness of OIL, $\Delta\Psi$, must be added to a monitoring well before the OIL has sufficient head to move outward into the formation. Stated another way, under vertical equilibrium conditions, the free product will preferentially accumulate within monitoring wells because the size of the well's **void space** is large and capillary pressures are negligible. Using the scaling relationship from Equation 9.4.6 and the fact that $\rho_a \cong 0$, Equation 9.4.18 may be written as follows:

$$\Delta\Psi = \left[\frac{\dfrac{\sigma_{ow}}{\sigma_{aw}}}{1 - \dfrac{\rho_o}{\rho_w}} - \frac{\dfrac{\sigma_{ao}}{\sigma_{aw}}}{\dfrac{\rho_o}{\rho_w}} \right] \Psi_{baw} \qquad (9.4.19)$$

Equation 9.4.19 is similar to the suggestion of Schiegg (1984) that an OIL thickness of $b_o = 2b_c$ must be developed within an observation well before the hydrocarbon has sufficient head to penetrate into the aquifer. According to Schiegg, b_c is the mean capillary rise of the soil saturation curve for water and air during drainage prior to the introduction of any OIL. Calculations for gasoline using Equation 9.4.19 give $\Delta\Psi = 1.56\Psi_{baw}$; calculations for 35° API fuel give $\Delta\Psi = 2.11\Psi_{baw}$; and calculations for recycled oil give $\Delta\Psi = 3.85\Psi_{baw}$. Schiegg's estimate of $\Delta\Psi = 2\Psi_{baw}$ appears reasonable.

When Equation 9.4.18 is satisfied, and if $S_{orv} = S_{ors}$, then $D_o = 0$—because $z_{ow} + \Psi_{bow} = z_{ao} + \Psi_{bao} = z_r$, and there is no free-product thickness. If $S_{orv} < S_{ors}$, then the model predicts a small value of D_o. For example, with 35° API fuel in a sand and $S_{ors} = 0.1$, $S_{orv} = 0.05$, calculations show that $b_o = 14.6cm$, $z_{ow} + \Psi_{bow} = z_{ao} + \Psi_{bao} = 5.31$ cm, and $z_r = 5.7$ cm. The resulting value of D_o is 0.029 cm.

9.4.2   Results for Different Soil Textures

The models described earlier for predicting the vertical distribution of OIL saturation are useful for showing the relationship between OIL thickness in a monitoring well and the amount (free-product volume, D_o) of OIL within the formation for different soil textures. For example, Figure 9.4.2 shows the formation OIL (35° API fuel) saturation for a sand and loam soil. The solid pair of horizontal lines shows the elevation of the air and OIL and OIL and water interfaces within a monitoring well for the sand soil. The horizontal pair of dashed lines shows the corresponding elevations for the loam soil. The formation saturation profiles contain the same quantities of OIL ($D_o = 0.3$ m = 1 ft) for both the sand and the loam distributions. A monitoring well in the sand soil,

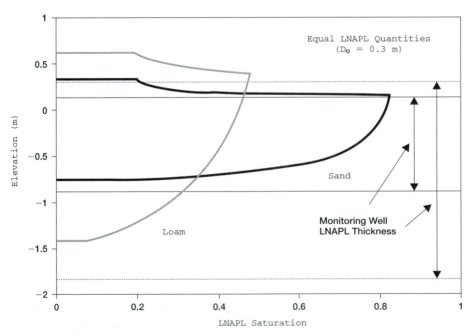

FIGURE 9.4.2   Vertical saturation distribution of LNAPL (35° API Fule) in a sand and loam soil where both profiles have the same LNAPL quantity

however, would have an OIL-layer thickness $b_o = 1.03$ m if it were in vertical equilibrium with formation fluids, while a monitoring well in the loam soil would have an OIL-layer thickness $b_o = 2.15$ m. It is clear that the relationship between monitoring well OIL-layer thickness and quantity of OIL present within the formation is sensitive to soil texture.

   Equation 9.4.17 may be used to find the formation free-product thickness (D_o) for different monitoring well thickness (b_o) values in different textured soils (Charbeneau et al., 1999). The soil and parameter values are similar to those listed in Table 4.4.1. The soils are listed for all classes shown on the texture triangle in Figure 1.1.2. The monitoring well and formation thickness for 35° API fuel are shown in Figure 9.4.3. For all cases, it is assumed that $S_{orv} = S_{ors} = 0.05$. The wide range in pore size for fine texture soils (i.e., silty clay loam, clay, and silty clay) results in natural smearing of the OIL over the formation, resulting in a large monitoring well thickness for a small formation thickness.

   Figure 9.4.3 shows that the relationship between monitoring well thickness and formation product thickness is approximately linear, except for small D_o values. This statement is true for all soil textures and for all OIL densities. Data such as that shown in Figure 9.4.3 were fit to the linear model (Charbeneau et al., 1999):

$$D_o = \beta(b_o - \alpha) \tag{9.4.20}$$

Equation 9.4.20 has a different form than Equation 9.4.17. In Equation 9.4.20, β is the slope of the linear relationship between formation free-product

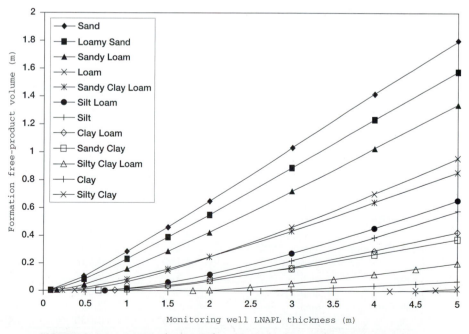

FIGURE 9.4.3   LNAPL Thickness for 35° API Fule in Different Soil Textures

volume and monitoring well OIL-layer thickness. The parameter β represents a capacitance factor for the OIL layer, analogous to the specific yield for water table fluctuations in an unconfined aquifer. The parameter α is the monitoring well OIL-layer thickness axis intercept and represents an extrapolated estimate of the OIL entry head. Because water is assumed to be the wetting fluid, a positive OIL head must be established before it can displace water from the pore space and enter the formation. In the Brooks and Corey model, this head is the bubbling pressure head or entry head. An analogous interpretation of the parameter α is that it represents the thickness of OIL that must be added to a monitoring well before sufficient head is developed to force the OIL out from the well and into the formation. Because α is an extrapolated limit, it is always larger than the actual OIL entry head for the soil (i.e., $\alpha > \Delta\Psi$).

Table 9.4.1 presents the best-fit parameters from Equation 9.4.17 to the mathematical model presented by Equation 9.4.20, from Charbeneau et al. (1999). The linear portion of the curve was fit, which corresponds to larger monitoring well thickness values (generally 3 to 5 meters). Results are presented for all soil textures and for specific gravity values of 0.70, 0.775 and 0.85. For all

TABLE 9.4.1   Best Linear Fit for OIL Thickness Using Equation 9.4.19

Soil Texture	Model Parameters	OIL Specific Gravity=0.70	OIL Specific Gravity=0.775	OIL Specific Gravity=0.85
Sand	$\alpha(m)$	0.10	0.20	0.30
	Slope, β	0.397	0.391	0.384
Loamy Sand	$\alpha(m)$	0.175	0.25	0.40
	Slope, β	0.363	0.352	0.344
Sandy Loam	$\alpha(m)$	0.325	0.44	0.65
	Slope, β	0.340	0.324	0.310
Loam	$\alpha(m)$	0.65	0.85	1.10
	Slope, β	0.303	0.278	0.247
Sandy Clay Loam	$\alpha(m)$	0.55	0.69	0.90
	Slope, β	0.252	0.232	0.211
Silt Loam	$\alpha(m)$	1.00	1.25	1.60
	Slope, β	0.273	0.237	0.195
Silt	$\alpha(m)$	1.12	1.45	1.80
	Slope, β	0.273	0.234	0.183
Clay Loam	$\alpha(m)$	1.07	1.35	1.75
	Slope, β	0.195	0.166	.134
Sandy Clay	$\alpha(m)$	1.07	1.35	1.75
	Slope, β	0.159	.134	0.110
Silty Clay Loam	$\alpha(m)$	1.47	1.85	2.50
	Slope, β	0.150	0.116	0.083
Clay	$\alpha(m)$	1.52	2.02	2.90
	Slope, β	0.071	0.052	0.036
Silty Clay	$\alpha(m)$	1.90	2.65	4.20
	Slope, β	0.056	0.038	0.024

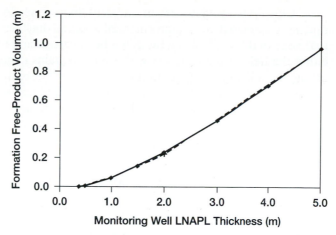

FIGURE 9.4.4 Piecewise Llnear Fit for Loam Soil

cases, it is assumed that $S_{orv} = S_{ors} = 0.05$. This table provides guidance to parameter selection for the free-product recovery models presented in Section 9.8. The parameters are not valid for thin OIL layers, however, especially for fine texture soils. To improve the fit of Equation 9.4.20 to Equation 9.4.17, the linear model can be applied piecewise over different ranges of b_o. For example, Figure 9.4.4 shows the curve and piecewise fit for loam soil. Equation 9.4.20 is used in three regions. For $0.385 < b_o < 1m$, $\alpha = 0.385$ m and $\beta = 0.101$; for $1 < b_o < 2$ m, $\alpha = 0.631$ m and $\beta = 0.168$; and for $2 < b_o < 5$ m, $\alpha = 1.055$ m and $\beta = 0.243$. Such applications of Equation 9.4.20 will be used in Section 9.8 to predict free-product recovery over a range of OIL-layer thickness, and a spreadsheet to facilitate these calculations is described in Appendix H. The spreadsheet *LNAPL Distribution.xls* allows one develop a piecewise fit to Equation 9.4.20 using either the Brooks and Corey model or the van Genuchten model, as described in Section 4.4, for the soil characteristic curve. The van Genuchten model has a zero air entry pressure head, so $b_o = 0$ when $D_o = 0$.

9.4.3   Discussion

The modeling approaches outlined in the preceding subsections assume that the fluids are always in vertical equilibrium, and they calculate the lateral potential gradients from the changes in elevation and free-product thickness that would be seen in a monitoring well. It is noted, however, that formation fluids are not always under conditions of vertical equilibrium. Indeed, Kemblowski and Chiang (1990) have shown that under dynamic conditions, the monitoring well thickness of free-product may show little relation to the formation distribution of fluids—but rather is a strong function of whether the water table is rising or falling.

EPA (1995) reviews seven different approaches for correlating product thickness measured in a monitoring well to actual thickness in the soil. They note that none of the available methods has been particularly reliable when tested in either the field or the laboratory. This issue remains a major obstacle in establishing the reliability of OIL monitoring. Nevertheless, the approach presented in this section is the only quantitative method that lends itself to development of predictive approaches for free-product recovery and system longevity.

9.5   NAPL Infiltration in the Vadose Zone

If a NAPL is released at or near the ground surface in sufficient quantity, it will move downward through the vadose zone under the force of gravity. Capillary pressure gradients will cause some lateral spreading and a diffuse NAPL saturation front. They are also primarily responsible for trapping of residual NAPL. The infiltrating NAPL will pool and spread across low permeability units. In addition, OIL will pool and spread as a lens when it reaches the water table. A DNAPL will also form a pool when it reaches the water table, until it builds up sufficient head to displace water from the pores and enter the zone of saturation—where it will continue to migrate downward to the base of the aquifer. DNAPL flow through the saturated zone will occur through fingers rather than as diffuse Darcy flow (see Figure 5.1.3). During multiphase transport, soil heterogeneities will influence the distribution of flow, and constituents contained in the NAPL will dissolve in water and volatilize into the soil air.

The development of multi-dimensional, multi-phase, multi-component mathematical models to describe the subsurface transport of NAPLs remains an interesting and challenging area for research. Advanced numerical methods are required, and special consideration must be given to the nonlinear constitutive relations that are used to parameterize flow and transport problems. Unfortunately, there are very few sites with sufficient reliable data to calibrate and test these models, and for this reason as well as their computational expense, such models have remained an area of research rather than a practical tool for field applications.

Nevertheless, for many applications, there is interest in approximate models that capture and describe significant features involved in NAPL infiltration and transport. The following restrictive assumptions are generally made. Richards' (1931) approach to soil moisture flow is followed wherein only a mass conservation equation for the water phase is used. For this approach to be a valid approximation, the flow of the air must not impede the flow of the water. One accounts for the presence of the air phase through use of an appropriate relative permeability function. When this approach is applied for the NAPL, it is assumed that the saturation of water is uniform and remains so, and the continuity equation for the water phase is eliminated. From these assumptions, Mull (1971, 1978), Raisbeck and Mohtadi (1974), Dracos (1978), Reible et al. (1990), and El-Kadi (1992) developed models for NAPL flow assuming that the

NAPL fills a fixed portion of the available pore space in a homogeneous medium. The models developed by Raisbeck and Mohtadi (1974) and Dracos (1978) share the limitation that they cannot simulate unsteady NAPL drainage in the soil after the release ends. The models of Mull (1971, 1978) and Reible et al. (1990) simulate drainage with arbitrary assumptions concerning the profile shape. Mull's (1971, 1978) model uses a series of rectangular profiles, while Reible et al. (1990) assume a zone of residual NAPL saturation behind a NAPL body moving within the profile. El-Kadi (1992) extends the approach of Mull to multiple dimensions.

During the redistribution (drainage) period, the NAPL saturation will be nonuniform. Furthermore, the models mentioned above have not included transport of soluble constituents of the NAPL phase. Charbeneau et al. (1989) and Weaver et al. (1994) present an approximate model to describe NAPL transport through the vadose zone using kinematic wave theory. The kinematic model, called *Kinematic Oily Pollutant Transport* (KOPT), is a gravity drainage model that neglects the effects of pressure gradients. Capillary forces may be included in the model through parameter selection for the residual NAPL saturation. For this model, the water content in the vadose zone is assumed to be constant. The average water saturation is found from the mean infiltration rate, $q_{\text{mean}} = W$, using the kinematic Equation 4.3.10 for the water phase and the Brooks-Corey-Burdine permeability model.

$$S_w = S_{wr} + (1 - S_{wr})\left(\frac{W}{K_{sw}}\right)^{1/\varepsilon} \qquad \textbf{(9.5.1)}$$

The water saturation is assumed to be constant for a given lithologic unit.

9.5.1   Kinematic Model for NAPL Infiltration: Theory

The one-dimensional continuity equation for NAPL flow is

$$n\frac{\partial S_o}{\partial t} + \frac{\partial q_o}{\partial z} = 0 \qquad \textbf{(9.5.2)}$$

When gravity forces dominate the mobility of the NAPL, as compared with capillary pressure gradients, the transport may be described by a **kinematic model** (unit gradient model) with

$$q_o = K_o(S_w, S_o) \qquad \textbf{(9.5.3)}$$

Equation 9.5.3 is the same as Equation 9.1.18 with a unit gradient and with z positive in the downward direction. Application of the **chain rule** gives

$$\frac{\partial S_o}{\partial t} + \frac{\partial K_o}{\partial S_w}\frac{\partial S_w}{\partial z} + \frac{\partial K_o}{\partial S_o}\frac{\partial S_o}{\partial z} = 0$$

which shows that the transport of the NAPL and water phases are coupled. We have assumed average water saturation corresponding to the average infiltration rate for water, however, and S_w is constant. This situation means that K_o is a function only of S_o, and the continuity equation reduces to the following equation:

$$n \frac{\partial S_o}{\partial t} + \frac{dK_o}{dS_o} \frac{\partial S_o}{\partial z} = 0 \qquad \text{(9.5.4)}$$

This equation is a first-order quasi-linear partial differential equation for the NAPL saturation. The factor $\dfrac{dK_o}{dS_o}$ is evaluated from the relative permeability function, with S_w serving as a parameter.

The NAPL continuity equation is most readily solved through application of the **method of characteristics** (MOC), as in Section 4.6.2. For Equation 9.5.4, the MOC equations are

$$\frac{dt}{1} = \frac{dz}{\dfrac{1}{n}\dfrac{dK_o}{dS_o}} = \frac{dS_o}{0} \qquad \text{(9.5.5)}$$

Instead of having a partial differential equation, we now have a set of *ordinary differential equations* (ODEs) to solve. While the solution $S_o(z,t)$ to Equation 9.5.4, however, is valid for all z and t for which the solution is continuous, Equation 9.5.5 is valid only for a particular curve which is called a **characteristic**. The first two equations give the **base characteristic**, which is the projection of the characteristic onto the z-t plane. Along the characteristic, the change in S_o is given by the first and third of Equation 9.5.5.

$$\frac{dz}{dt} = \frac{1}{n}\frac{dK_o}{dS_o} \;\; ; \;\; \frac{dS_o}{dt} = 0 \rightarrow S_o = \text{constant} \qquad \text{(9.5.6)}$$

Equation 9.5.6 is equivalent to the general solution $S_o(z,t)$ and may be used to construct it. Because K_o is only a function of S_o (S_w serves as a parameter), the slope of the base characteristics is constant and the characteristics are straight lines in the z-t plane.

The base characteristic Equation 9.5.6 shows that the speed of a given characteristic (dz/dt) is proportional to the tangent of the NAPL permeability curve $dK_o/d\theta_o$ at the corresponding NAPL saturation. A glance at the relative permeability curves (see Figure 9.1.3) shows that the tangent (slope) increases with NAPL saturation so that larger NAPL saturations have a greater speed and move faster through the soil profile. This situation corresponds to a drainage wave. For the leading edge of the infiltrating NAPL, the characteristic solution does not apply (larger NAPL contents cannot pass by smaller ones at the NAPL front, although their tendency is to overtake them). The problem is that the characteristic equations assume that the solution is continuous. If we had retained the capillary pressure terms in the flow equation, the solution would remain continuous, with the gravity component tending to sharpen the front and the capillary pressure gradients tending to spread it. A NAPL front distribution would develop and migrate downward through the profile. We capture this distribution within the kinematic model by allowing for the presence of discontinuous solutions—the so-called shock waves of gas dynamics (or hydraulic jumps from open channel flow). For an example showing a diffuse front and its approximating sharp front, see Figure 9.5.1.

Following the same arguments presented in Section 4.6.2, the continuity equation for displacement of the fictitious sharp NAPL front is

$$\frac{dz_f}{dt} = \frac{K_{o1} - K_{o2}}{n(S_{o1} - S_{o2})} \tag{9.5.7}$$

It is of interest to note that as the jump in NAPL saturation becomes infinitely small, then the NAPL-front equation becomes the same as the characteristic equation for the continuous solution.

Figure 9.5.2 shows the relationship between the sharp-front and characteristic models. The speed of the characteristic behind the front at NAPL saturation S_{o1} is given by the slope of the permeability curve at S_{o1}, and similarly for the speed of the characteristics ahead of the front at NAPL saturation S_{o2}. The slope of the chord connecting the curve at S_{o1} and S_{o2} gives the speed of the

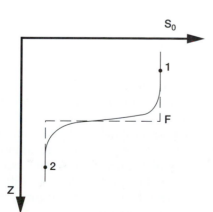

FIGURE 9.5.1   NAPL Front Moving Downward

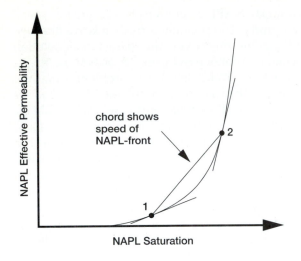

FIGURE 9.5.2 The chord determine NAPL front speed

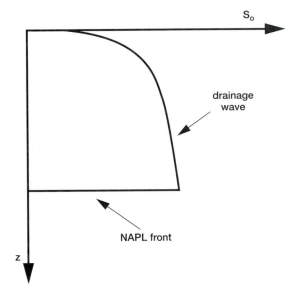

FIGURE 9.5.3 NAPL Infiltration Wave Representation in the KOPT Model

front. This arrangement shows that the speed of the front is intermediate between that of the characteristics behind and ahead of it. The characteristics behind the front move into it, while those ahead of the front are overtaken by the front itself. This **upstream control** is characteristic of shock-type phenomena.

For the kinematic NAPL infiltration model, there is only one equation for each front, while there is an infinite family of characteristic equations, one for each characteristic. The resulting NAPL saturation profile for a case with no NAPL initially within the soil column is shown schematically in Figure 9.5.3. Weaver (1988) describes a numerical implementation of the model.

9.5.2   **Kinematic Model for NAPL Infiltration: Implementation**

The method of characteristics provides an algorithm for construction the approximate kinematic solution to the problem of NAPL infiltration following a surface release. In general, the solution must be evaluated numerically because of the complex nonlinear form of the NAPL relative permeability model. If the NAPL relative permeability function may be represented through a power-law formulation, however, then the MOC formulation leads to results that can be presented analytically in a closed-form solution, similar to those presented in Section 4.6.2 for a rainfall event. This approach is followed in this section where we present a model for NAPL infiltration where the NAPL is ponded on the soil surface at a depth H_i for a duration t_c, and then the hydrocarbon is left to infiltrate.

The water saturation is estimated from the water infiltration rate $q_w = W$, as specified by Equation 9.5.1. Following the discussion in Chapter 4, the minimum air saturation (entrapped air) is estimated from the condition that the relative permeability to water would be one-half. This relationship gives

$$S_{ar} = (1 - S_w)(1 - (\tfrac{1}{2})^{1/\varepsilon})$$ **(9.5.8)**

The maximum NAPL saturation, S_{om}, is then calculated from

$$S_{om} = 1 - S_w - S_{ar}$$ **(9.5.9)**

With the water saturation known and constant, use of the Brooks-Corey-Burdine equation gives the maximum hydrocarbon relative permeability from Equation 9.1.20:

$$k_{rom} = \left(\frac{S_{om} - S_{or}}{1 - S_{or}}\right)^2\left(\left(\frac{S_{om} + S_w - S_{wr}}{1 - S_{wr}}\right)^{\varepsilon-2} - \left(\frac{S_w - S_{wr}}{1 - S_{wr}}\right)^{\varepsilon-2}\right)$$ **(9.5.10)**

With this maximum value known, the relative permeability function is fit to the following power-law form:

$$k_{ro} = k_{rom}\left(\frac{S_o - S_{or}}{S_{om} - S_{or}}\right)^{\zeta}$$ **(9.5.11)**

From Equation 9.5.11, it is apparent that when $S_o = S_{or}$, $k_{ro} = 0$, and when $S_o = S_{om}$, $k_{ro} = k_{rom}$.

As an example, consider the data presented by Weaver et al. (1993). These data give $n = 0.41$, $S_w = S_{wr} = 0.0588$, $\lambda = 4.84$, $\epsilon = 3 + 2/\lambda = 3.41$, $S_{ar} = 0.173$ (Equation 9.5.8), $S_{om} = 0.768$ (Equation 9.5.9), $S_{or} = 0.05$, and $k_{rom} = 0.429$ (Equation 9.5.10). To estimate the power ζ in Equation 9.5.11, the Brooks-Corey-Burdine equation was used to calculate a series of relative permeabilities. These are shown as the diamonds in Figure 9.5.4. Equation 9.5.11 was then fit to the data series, and a value of ζ was chosen which give the least squared difference.

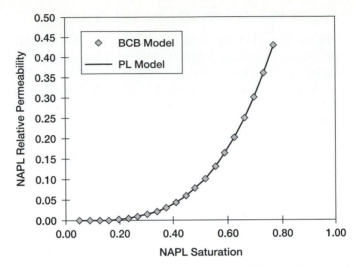

FIGURE 9.5.4 Comparison of Brooks-Corey-Burdine (BCB) and power-law (PL) relative permeability models

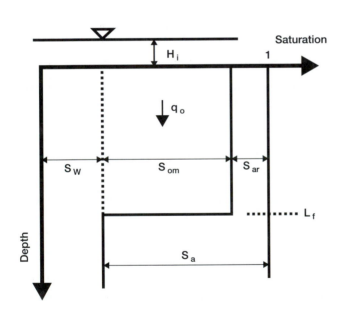

FIGURE 9.5.5 Schematic view for Green and Ampt NAPL infiltration model

The result is $\zeta = 3.3$. From Figure 9.5.4, it is evident that the power-law model is able to represent the Brooks-Corey-Burdine relative permeability model under the stated conditions. Additional experience suggests that this situation is true in general where one of the phase saturations is constant.

The model for NAPL infiltration combines the Green and Ampt infiltration model with the kinematic model for NAPL redistribution (see Chapter 4). A schematic view of the infiltration process is shown in Figure 9.5.5. NAPL is ponded on the ground surface at an initial depth H_i. The soil profile has uniform water saturation S_w, and during NAPL infiltration, soil air is trapped at saturation S_{ar}. The initial air saturation is S_a, and the maximum NAPL saturation is S_{om}. The NAPL flux q_o depends on the ponded depth H_i and on the total depth of infiltration, L_f, as well as soil and fluid properties.

With regard to the infiltration process and the NAPL profile, there are four distinct phases. During the first phase, NAPL is ponded at a constant depth H_i. The duration t_c of this phase may be zero. For the second period, the NAPL remains ponded, but its depth decreases as the NAPL infiltrates and the period ends at time t_{pond} when the ponded NAPL disappears. During this second period, the ponding depth is decreasing with $H(t) = H_i - (L_f - L_{fc})n\, S_{om}$ at a given time $t_c < t < t_{pond}$, where L_{fc} is the depth of NAPL penetration at the end of the period of constant ponding. During the third period, there is a region of constant NAPL saturation called the plateau region within the soil profile. The NAPL saturation plateau disappears at time t_p, and during the fourth period the NAPL profile undergoes drainage with a continually decreasing NAPL front saturation. The NAPL profile for these four periods is shown in Figure 9.5.6.

The Green and Ampt (1911) infiltration model for NAPL is the same as Equation 4.5.18:

$$n\, S_{om} \frac{dL_f}{dt} = q_o = K_{os}\, k_{rom}\left(\frac{H + \Psi_{fo} + L_f}{L_f}\right) \tag{9.5.12}$$

Ψ_{fo} is the NAPL capillary suction head for infiltration of the intermediate-wetting NAPL into unsaturated porous medium. Repeating the arguments leading to Equation 4.5.17, one may show that

$$\Psi_{fo} = \Psi_{om} + \int_{\Psi_{om}}^{\Psi(S_o=0,S_w)} \frac{k_{ro}}{k_{rom}}\, d\Psi \tag{9.5.13}$$

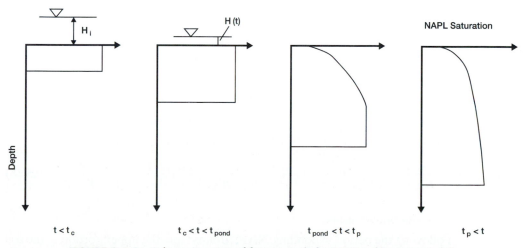

FIGURE 9.5.6 Schematic view of four periods for NAPL infiltration model

where $\Psi_{om} = \Psi(S_{om}, S_w)$. The NAPL relative permeability is given by Equation 9.5.11, and from Equations 9.4.6 and 9.4.7, the air-NAPL capillary pressure is

$$\Psi = \Psi_{bao}\left(\frac{S_w + S_o - S_{wr} - S_{or}}{1 - S_{wr} - S_{or}}\right)^{-1/\lambda}$$

Changing variables, Equation 9.5.13 is written

$$\Psi_{fo} = \Psi_{om} + \frac{\Psi_{bao}}{\lambda(1 - S_{wr} - S_{or})}\int_0^{S_{om}}\left(\frac{S_o - S_{or}}{S_{om} - S_{or}}\right)^{\zeta}\left(\frac{S_w + S_o - S_{wr} - S_{or}}{1 - S_{wr} - S_{or}}\right)^{-\left(\frac{1+\lambda}{\lambda}\right)}dS_o \quad \textbf{(9.5.14)}$$

Equation 9.5.14 is the general result, where the integrand equals 0 for $S_o < S_{or}$. Because $S_o = 0$ ahead of the NAPL front, we take $S_{or} = 0$. Furthermore, if $S_w \cong S_{wr}$, then Equation 9.5.14 may be integrated to give

$$\Psi_{fo} = \frac{\zeta\lambda\Psi_{bao}}{(\zeta\lambda - 1)}\left(\frac{S_{om}}{1 - S_{wr}}\right)^{-1/\lambda} \quad \textbf{(9.5.15)}$$

Using the data from Weaver et al. (1994) with $\Psi_{baw} = 24.8$ cm, $\rho_o = 0.79$ g/cm^3, and $\sigma_o = 25$ dynes/cm (along with Equation 9.4.5), Equation 9.5.15 gives $\Psi_{fo} = 13.4$ cm, which compares well with the value of 12.5 cm estimated by Weaver et al. (1994).

Period of Constant Ponding Depth

During the time period $0 < t < t_c$ the NAPL ponding depth is assumed constant at $H = H_i$. Using this information in Equation 9.5.12 gives

$$\frac{nS_{om}}{K_{os}k_{rom}}\left[L_f - (H_i + \Psi_{fo})ln\left(1 + \frac{L_f}{H_i + \Psi_{fo}}\right)\right] = t \quad \textbf{(9.5.16)}$$

Setting $t = t_c$ in Equation 9.5.16, one may (iteratively) solve for the NAPL front depth at the end of the period of constant ponding, L_{fc}.

Period of Decreasing Ponding Depth

During the period when the ponded NAPL is infiltrating ($t_c < t < t_{pond}$), the ponded head decreases according to the mass balance relationship $H(t) = H_i - (L_f - L_{fc})n\,S_{om}$. Using this relationship in Equation 9.5.12 with initial point (L_{fc}, t_c) gives

$$\frac{nS_{om}}{K_{os}k_{rom}}\left\{\frac{L_f - L_{fc}}{1 - nS_{om}} + \frac{H_i + nS_{om}L_{fc} + \Psi_{fo}}{(1 - nS_{om})^2}ln\left(\frac{H_i + L_{fc} + \Psi_{fo}}{H_i + (1 - nS_{om})L_f + nS_{om}L_{fc} + \Psi_{fo}}\right)\right\} = t - t_c \quad \textbf{(9.5.17)}$$

The time to when ponded NAPL disappears is found from

$$t_{pond} = t_c + \frac{nS_{om}}{K_{os}k_{rom}}\left\{\frac{H_i}{nS_{om}(1 - nS_{om})} + \frac{nS_{om}L_{fpond} + \Psi_{fo}}{(1 - nS_{om})^2}ln\left(\frac{H_i + L_{fc} + \Psi_{fo}}{L_{fpond} + \Psi_{fo}}\right)\right\} \quad \textbf{(9.5.18)}$$

Kinematic Wave Solution

Following the period of ponding, NAPL drains from the upper portion of the soil profile. Redistribution of NAPL is described by a kinematic wave following Equation 9.5.6. For the NAPL redistribution profile, the kinematic wave solution is

$$S_o(z,t) = S_{or} + (1 - S_w - S_{ar} - S_{or}) \left[\frac{n(1 - S_w - S_{ar} - S_{or})}{K_{os} k_{rom} \zeta} \left(\frac{z}{t - t_{pond}} \right) \right]^{\frac{1}{\zeta - 1}} \qquad (9.5.19)$$

$$q_o(z,t) = K_{os} k_{rom} \left[\frac{n(1 - S_w - S_{ar} - S_{or})}{K_{os} k_{rom} \zeta} \left(\frac{z}{t - t_{pond}} \right) \right]^{\frac{\zeta}{\zeta - 1}} \qquad (9.5.20)$$

Equations 9.5.19 and 9.5.20 provide the NAPL saturation and volume flux throughout the drainage part of the NAPL infiltration profile.

Displacement of the NAPL Front

The NAPL front moves at different speeds depending on the change in saturation across the front, according to Equation 9.5.7. While the region of constant NAPL saturation is present on the upstream side, the front moves at a constant speed. The time and depth to when the plateau disappears is given by

$$t_p = t_{pond} + \frac{n S_{om} L_{fpond}}{K_{os} k_{rom}} \left[\frac{S_{om} - S_{or}}{\zeta S_{om} - (S_{om} - S_{or})} \right] = t_{pond} + \frac{n(S_{om} - S_{or}) L_{fp}}{\zeta K_{os} k_{rom}} \qquad (9.5.21)$$

$$L_{fp} = \frac{S_{om} \zeta L_{fpond}}{(\zeta - 1) S_{om} + S_{or}} \qquad (9.5.22)$$

In Equations 9.5.21 and 9.5.22, t_p is the time that the NAPL saturation plateau region disappears, and L_{fp} is the NAPL front depth at time t_p.

Once the NAPL saturation plateau has disappeared, the change in NAPL saturation across the front gets smaller, and the speed of the front decreases. Using Equations 9.5.19 and 9.5.20 in Equation 9.5.7 gives the following for the displacement of the NAPL front following t_p:

$$t = t_{pond} = \frac{(L_f)^{\zeta}}{\left(L_p \left(\frac{L_p}{t_p - t_{pond}} \right)^{\frac{1}{\zeta - 1}} + \frac{(L_p - L_f) n S_{or} (K_{os} k_{rom})^{\frac{1}{\zeta - 1}}}{\zeta - 1} \left(\frac{\zeta}{n(1 - S_w - S_{ar} - S_{or})} \right)^{\frac{\zeta}{\zeta - 1}} \right)^{\zeta - 1}} \qquad (9.5.23)$$

Equation 9.5.23 gives the NAPL-front depth as a function of time. In particular, to find the time required for the NAPL to reach a given depth, then the depth L_f may be substituted in Equation 9.5.23 to find the time, t.

Figure 9.5.7 shows the results from the model presented in Equations 9.5.16 through 9.5.23 for the column experiment described by Weaver et al. (1994). Equation 9.5.18 predicts a time period of just over 6 minutes for the 6.5 cm of Soltrol ($\mu_o = 4.76 cp$), which is the same as what was observed. The region shown with $S_o = S_{om}$ has a constant NAPL saturation of 0.768. This constant S_o region (plateau) is bounded by a characteristic with the same saturation value. A total of six characteristics are shown, and each designates a constant NAPL saturation during NAPL redistribution, as described by Equation 9.5.19. The NAPL front reaches a depth of 60 cm after 100 minutes, which compares well with the observed value.

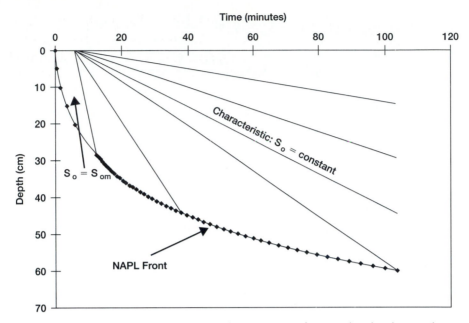

FIGURE 9.5.7 Base characteristic plan showing NAPL front and redistribution during column experiment

The spreadsheet *NAPL Infiltration.xls* has been developed to implement the NAPL infiltration calculations. The model supports estimates of k_{rom} and ζ from Equation 9.5.11. In addition, the water saturation, period of NAPL ponding, and ponding depth are required.

9.5.3 A Kinematic Model of Constituent Transport

The NAPL flow model presented in the last section considers only gravity forces, and results in a kinematic model. While a full multiphase advection-dispersion-equation based model could be used to describe the constituent transport, for many applications where data is limited, simple model formulations are of value. In this section, a transport model for the constituent is presented that also uses a kinematic model formulation. The transport equation for the constituent is

$$\frac{\partial m}{\partial t} + \frac{\partial J}{\partial z} = 0 \qquad (9.5.24)$$

where J is the mass flux of the constituent. With local equilibrium partitioning, this equation may be written in terms of the concentration in any phase. For generality, because the NAPL phase may become isolated within the unsaturated zone at residual saturation—with the constituent being transported to the water table with the infiltrating aqueous phase, we choose the concentration in the water phase as the primary variable. The appropriate relationships are

$$m = (n(S_w + S_o K_o) + \rho_b K_d)c_w \equiv B_w(S_w, S_o)c_w \; ; \; J = q_w c_w + q_o c_o = (q_w + q_o K_o)c_w \quad \text{(9.5.25)}$$

where B_w is the bulk water partition coefficient (K_o is the NAPL-water partition coefficient, not the NAPL hydraulic conductivity) which is a function of both the water and NAPL saturations, and the NAPL saturation varies both with depth and time. Likewise, the volumetric NAPL flux, q_o, is a function of depth and time. For consistency with the assumption that the flux of air is neglected, partitioning into the air phase is not used in the mass balance equation for the constituent. Substituting Equation 9.5.25 for the bulk concentration and mass flux into Equation 9.5.24 and expanding the derivatives gives

$$B_w \frac{\partial c_w}{\partial t} + (q_w + q_o K_o) \frac{\partial c_w}{\partial z} + c_w K_o \left(n \frac{\partial S_o}{\partial t} + \frac{\partial q_o}{\partial z} \right) = 0$$

With the continuity equation for the NAPL, however, this equation becomes

$$\frac{\partial c_w}{\partial t} + \frac{q_w + q_o K_o}{B_w} \frac{\partial c_w}{\partial z} = 0 \quad \text{(9.5.26)}$$

The MOC solution of Equation 9.5.26 is

$$\frac{dt}{1} = \frac{B_w dz}{q_w + q_o K_o} = \frac{dc_w}{0} \rightarrow c_w = \text{constant along} \; \frac{dz}{dt} = \frac{q_w + q_o K_o}{B_w} \quad \text{(9.5.27)}$$

Equation 9.5.27 is solved numerically after the solution $S_o(z,t)$ is obtained from the flow equation for the NAPL phase. Only a finite number of characteristics need to be computed, with intermediate values determined by interpolation between characteristics. Because the equations are linear, there is no need to distinguish between sharp fronts and continuous concentration distributions. Discontinuities also propagate along characteristics for linear hyperbolic equations. Weaver (1988) describes the model implementation.

In order to apply the solution for oily pollutants within the vadose zone, the models also require boundary conditions. If a surface flux of OIL is specified with concentration $c_{o(surf.)}$, then

$$c_{o(surf.)} q_o = c_{w(initial)} (q_w + q_o K_o)$$

which may be used to find the initial water phase concentration at the ground surface. If a depth D_o of OIL with concentration $c_{o(surf.)}$ is incorporated within the soil profile over a depth L_o, then the initial water content satisfies

$$c_{o(surf.)} D_o = c_{w(initial)} B_w(S_w, S_o) L_o$$

These boundary conditions are sufficient for most applications.

The model discussed in this section will provide the NAPL content within the vadose zone as a function of depth and time, as well as the constituent concentration in all phases. In addition, the flux of NAPL and constituent to the water table is determined. The model is approximate in that it is fully kinematic. A complete formulation would require retention of the capillary pressure gradient terms in the NAPL flow equation, and the dispersion terms for the

NAPL and water phases for the constituent transport equation. The result would be a pair of nonlinear partial differential equations of the parabolic type. Unless the constituent can affect the NAPL phase properties, however, the models are not coupled and can be solved consecutively, first for the NAPL distribution and flux and then for the constituent. Nevertheless, the required numerical techniques are much more difficult, computationally expensive, and require significant user training.

9.6   Screening Models for Fate/Transport of Organic Chemicals in Soil and Groundwater

Subsurface fate and transport models may be classified as either generic models or site specific models. Generic models are based on a simplified interpretation of the hydrogeology, including generally the assumption of uniform flow in a specified direction and homogeneous conditions for other parameters. These assumptions allow the use of analytic solutions to the transport problem. Analytic solutions have the advantage of simplicity and ease of computation. Site specific models are flexible enough to deal with the individual complexity of a given hydrogeologic setting and can include almost any level of detail in the simulation of important fate and transport processes. This flexibility and ability to address great levels of complexity comes at a price. Site specific models require numerical methods and may come at a great computational expense.

Screening models are generic models in the sense that they cannot be adapted to deal with many site-specific conditions. They may be used to evaluate the behavior of large numbers of chemicals in the environment. Screening models are tailored for specific applications by incorporating mathematical models of many important processes that affect the fate and transport of chemicals under the particular scenario for which the model was developed.

9.6.1   Screening Model for Soil Leaching

As an example, Short (1988) developed a screening model assessing the movement of constituents from **oily wastes** during land treatment operations. This *Regulatory and Investigative Treatment Zone* (RITZ) model utilizes a **moving boundary formulation**. A feature of the RITZ model is that it considers an OIL phase that is present at low saturation (**residual saturation**, so that it is immobile as a phase), and constituents from the oily phase may be lost due to volatilization and biodegradation. The immiscible oily phase may also degrade over time.

The oily waste is initially incorporated over a depth L_o, as shown in Figure 9.6.1(a). The computations in the RITZ model are concerned with the fate and transport of constituents contained within the oily waste; for example, benzene, toluene, or xylene in an oily sludge. The model can simulate the behavior of only one constituent, or **pollutant**, of the oily waste at a time. The following

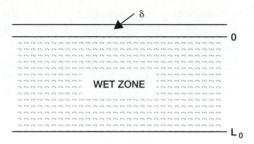

(a) Initial region of contamination to a depth l₀ with atmospheric viscous sublayer of thickness δ.

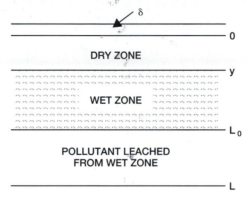

(b) Dryzone, wet zone, and leachate zone in a contaminated soil profile.

FIGURE 9.6.1   Soil Profile Modeled int he Regulatory and Investigative Treatment Zone (RITZ) Model

assumptions are made in the RITZ model : 1) the oily phase is immobile, 2) dispersion or diffusion is negligible, except as it contributes to volatilization, 3) equilibrium conditions are assumed for partitioning of the pollutant between the phases, 4) an effective degradation constant is used which accounts for degradation within the separate phases, and 5) the net infiltration of rainfall occurs at a constant rate. As time proceeds, a **dry zone** develops near the ground surface, marking the region influenced by volatilization. The base of the dry zone, shown at a depth y in Figure 9.6.1(b), proceeds downward at a rate controlled by the volatile flux from the profile and by the leaching rate of the pollutant. The pollutant is also leached from the wet zone and migrates downward into the region of the soil profile which does not contain the oily phase. The depth of the base of the zone of leachate is $z = L$. The concentration of the leachate from the wet zone changes with time, because both the pollutant concentration and the OIL saturation vary due to first-order degradation losses.

The model presented by Short (1988) includes the effects of diffusion resistance to volatilization across a viscous sublayer of thickness δ, as shown in Figure 9.6.1. This process has an effect only for short time periods and does not impact the overall model calculations. As noted in Section 7.6, this effect is significant only for conditions where the seepage velocity is upward and the chemical tends to accumulate at the ground surface. For the following discussion, the effects of the viscous sublayer are neglected.

The decay or degradation relations are assumed to be first-order for both the pollutant and the OIL phase, with first order rate constants of λ_p and λ_o, respectively. The bulk concentration and OIL saturation are given, respectively, by

$$m(t) = m(0)e^{-\lambda_p t} \qquad (9.6.1)$$

$$S_o(t) = S_o(0)e^{-\lambda_o t} \qquad (9.6.2)$$

As the OIL degrades, the capacity for the OIL phase to retain constituent mass decreases, resulting in mass transfer of the pollutant into the other phases as determined by the multiphase partitioning relationships. These partitioning relations are the same as in Section 5.5, except that the bulk water partition coefficient defined in Equation 5.5.8 is now time dependent.

$$B_w(t) = n(S_w + S_a K_H + S_o(0)e^{-\lambda_o t}K_o) + \rho_b K_d \qquad (9.6.3)$$

The bulk water partition coefficient is a function of time within the wet and dry zones, but it is a constant in the leachate zone below L_o, because the OIL phase is absent at that level. The aqueous pollutant concentration within the wet zone, which is the same as the concentration entering the leachate zone, is given by the following equation:

$$c_w(t) = \frac{m(0)e^{-\lambda_p t}}{B_w(t)} \qquad (9.6.4)$$

Depending on the relative magnitudes of the decay constants for the pollutant and the OIL phases, λ_p and λ_o, respectively, the aqueous concentration can either increase or decrease soon after the release occurs, although it must ultimately decrease exponentially.

The main focus of the wet zone model is to track the depth of the moving boundary y as a function of time. Mass balance for this interface gives

$$\frac{dy}{dt} = v_s + \frac{D_{sT}}{y} \qquad (9.6.5)$$

where the first term on the right-hand-side is the effective or **retarded velocity** of the constituent

$$v_s = \frac{q_w}{B_w(t)} \qquad (9.6.6)$$

(for a medium without an OIL phase, the factor $R = B_w/(nS_w)$ is the **retardation factor**), and the second term accounts for volatilization and upward diffusion across the dry zone. The effect of changing $S_o(t)$ on D_{sT} is neglected.

Equation 9.6.5 is the balance equation for the interface between the wet and dry zones. The variables may be separated and the equation may be integrated to give

$$y - \frac{D^*}{q} \ln\left(1 + \frac{q_w y}{D^*}\right) = \frac{q_w}{B_w(\infty)}\left(t + \frac{1}{\lambda_o} \ln\left(\frac{B_w(t)}{B_w(0)}\right)\right) \qquad \text{(9.6.7)}$$

where $D^* = n(D_{sw} + D_{sa}K_H + D_{so}K_o)$ and $B_w(\infty) = n(S_w + S_a K_H) + \rho_b K_d$. Equation 9.6.7 gives y as an implicit function of time. The maximum time for leachate generation from the initial region of oily waste contamination, t_m, may be found by substituting $y = L_o$ in Equation 9.6.7 and solving for $t = t_m$. With $y(t)$ known from Equation 9.6.7, the cumulative volatilization loss (per unit area) may be obtained through integration of

$$\frac{M_{vol}(t)}{A} = m(0)D_{sT}\int_0^t \frac{e^{-\lambda_p \tau}}{y(\tau)}\, d\tau \qquad \text{(9.6.8)}$$

Likewise, the cumulative leachate from the wet zone is calculated from

$$\frac{M_{leach}(t)}{A} = m(0)q_w \int_0^t \frac{e^{-\lambda_p \tau}}{B_w(\tau)}\, d\tau \qquad \text{(9.6.9)}$$

Finally, the amount of degradation from the wet zone may be calculated from

$$\frac{M_{deg}(t)}{A} = m(0)\lambda_p \int_0^t (L_o - y(\tau))e^{-\lambda_p \tau} d\tau \qquad \text{(9.6.10)}$$

Equations 9.6.8 to 9.6.10 provide the mass balance for the initial region of contamination by oily waste released into the soil, and they are valid as long as $y \le L_o$.

For land treatment systems and for other releases of contaminants, major questions concern when and at what concentration any leachate from the initial region of contamination will leave the **treatment zone**—in land treatment applications, the treatment zone consists of the upper 5 ft (1.5 m) of the soil profile—or enter an underlying aquifer at a depth L_{aq}. Within the leachate zone below the wet zone, the bulk water partition coefficient corresponds to that without an OIL phase present B_{wlz}. The leading edge of the contaminated pulse will reach depth L_{aq} after a time

$$t_{aq} = \frac{B_{wlz}(L_{aq} - L_o)}{q_w} \qquad \text{(9.6.11)}$$

For a given constituent, if the leading edge of the dry zone front catches the leachate front at a depth y before the time specified by Equation 9.6.11, then this constituent will not be transported to the aquifer. All of the constituent will be lost to the volatilization and degradation processes.

The travel time from the region of initial contamination to the water table for any **parcel** of pollutant is also equal to t_{aq} (this situation is due to the assumed linearity of the sorption isotherm). Thus, the concentration reaching the aquifer is equal to the concentration leaving the wet zone at a time t_{aq} earlier, multiplied by the factor $e^{-\lambda_p t_{aq}}$ to account for the degradation losses during transport from the wet zone to the water table. For the aqueous phase concentration reaching the aquifer at time t, this information gives

$$c_{waq}(t) = \frac{m(0)e^{-\lambda_p t}}{B_w(t - t_{aq})} \qquad (9.6.12)$$

The time for the pollutant to be completely removed from the profile is the time at which the dry zone interface reaches the water table. The continuity equation for the region beneath the zone of incorporation is still given by Equation 9.6.5, but now the **initial condition** is $y(t_m) = L_o$ and $B_w = B_{wlz}$. This equation may be integrated to find

$$y - L_o + \frac{D_{sT}}{v_s} \ln\left(\frac{v_s L_o + D_{sT}}{v_s y + D_{sT}}\right) = v_s(t - t_m) \qquad (9.6.13)$$

The maximum time for leachate from the vadose zone is found by setting $y = L_{aq}$ and solving for t in Equation 9.6.13.

RITZ Model Example

As an example of the application of the RITZ model, consider the leaching of benzene from a gasoline spill where the gasoline has contaminated the upper meter of the soil profile and has a saturation of $S_o = 0.125$. The partitioning of benzene from the gasoline is presented in Example 5.4.2. It is assumed that the average infiltration rate is 25 cm/yr, that the depth to the water table is 5 m, and that degradation of both benzene and gasoline is negligible. As shown in Figure 9.6.2, benzene first reaches the water table at 25 years after the release. Benzene is leached from the wet zone at 58 yrs, and it is removed from the vadose zone at 82 yrs. With a benzene concentration of 24 mg/L in water, the flux to the aquifer is 6 g/m^2/yr over the time period from 25 to 82 yrs. The volatilization loss is calculated to be 15 percent of the total benzene (60 g/m^2). Figure 9.6.2 also shows the depth of the dry zone for a case with negligible volatilization. For the non-volatile case, the constituent is not leached from the vadose zone until 93 years.

9.6.2   Hydrocarbon Spill Screening Model

Subsurface releases of OILs can lead to chemical exposures through many environmental pathways. Exposures to vapors through atmospheric pathways may be significant near the region of the release. Groundwater pathways may lead to surface water pathways that may lead to many other pathways. In this section, we consider only groundwater exposure pathways through direct exposure from groundwater wells. The purpose of this discussion is to examine some of the processes that must be considered in an analysis of NAPL release exposure assessment.

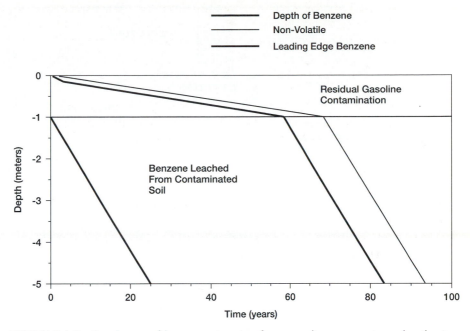

FIGURE 9.6.2 Simulation of benzen migration from gasoline contaminated soil using the RITZ model. Upper 1 m of the soil profile is contaminated by a spill with gasoline immobilized by capillary forces at a NAPL saturation $S_o = 0.125$. The recharge rate is 25 cm/yr.

When a hydrocarbon liquid that is less dense than water reaches the water table after a spill or release from a leaking tank or pipeline, it will pond in an NAPL lens which grows in thickness and spreads. After the source is cut off, the lens will spread until it reaches a thin layer within the capillary fringe. Much of the hydrocarbon will remain isolated both above and below the water table at residual saturation. The constituents from the OIL release can dissolve into groundwater that flows beneath the lens, thereby contaminating downgradient drinking water through miscible phase transport. The exposure assessment must quantify the magnitude of the release, the NAPL transport through the vadose zone, the capture of chemical constituents within the vadose zone, the NAPL transport along the water table, the residual saturations within the vadose zone and above and below the lens, the size of the NAPL lens, the dissolution mass transfer from the lens to groundwater, and the groundwater transport to the exposure location.

The *Hydrocarbon Spill Screening Model* (HSSM) addresses many of these features (Weaver et al. 1994). The basic configuration of HSSM is shown in Figure 9.6.3. NAPL infiltration and transport through the vadose zone are represented using the KOPT model described in Section 9.5. If a sufficiently large volume of NAPL is released, the OIL reaches the water table. If sufficient OIL

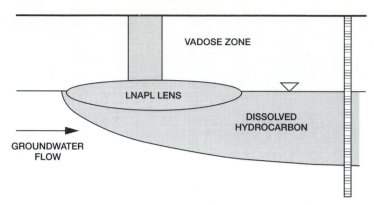

FIGURE 9.6.3 Hydrocarbon Spill Screening Model (HSSM)

head develops, the water table is displaced downward, lateral spreading begins, and the OIL lens part of the HSSM model is triggered. This component, called OILENS, assumes that spreading is radial and that the thickness of the lens is determined by buoyancy (Ghyben-Herzberg relations, see Section 3.10). The shape of the lens is given by the Dupuit assumptions, where the flow is assumed horizontal and the lateral gradient is independent of depth. The representation for mass transfer from the OIL lens to groundwater assumes that water in contact with the lens receives the NAPL-component's effective solubility concentration. Mass transfer is then based on vertical diffusion/dispersion as groundwater moves beneath the lens and on direct mass transfer as infiltrating water moves through the lens. The time-variable mass flux from the lens to groundwater is calculated and used by the *transient source Gaussian plume* (TSGPLUME) model component of HSSM. This component is described in Section 8.8.3. The TSGPLUME model calculates the time-variable exposure concentrations at downgradient locations within the aquifer. The theoretical background for the HSSM model is described by Charbeneau et al. (1995). Some of the model features and limitations are discussed in the following paragraphs.

The purpose of HSSM is use as a screening tool. For example, the model can be used to estimate the effects of OIL loading, partition coefficients, groundwater flow velocities, etc., on pollutant transport. Because approximations are used for developing the model, the model results must be viewed as approximations. If simulation of complex heterogeneous sites is needed, or other approximations made in the model are unacceptable, then a more inclusive model should be used instead of, or in addition to, HSSM.

For the OILENS component of HSSM, the assumptions of vertical equilibrium, radial flow, and a steady-state hydrocarbon distribution lead to a simplified representation of the lens. At any given time, the free-product distribution is specified by three variables: the effective lens OIL saturation, S_o, the lens head beneath the source, h_{os}, and the radius of the lens, R_t. The lens OIL saturation is specified as a constant input parameter and must be estimated from the conditions of the release. The remaining two variables, h_{os} and R_t, vary with time and must be calculated as part of the model. Their calculation is based on continuity principles. From the Dupuit equation, the OIL-layer head at any radius $r > R_s$ is given by

$$h_o(r) = h_{os}\sqrt{\frac{ln(R_t/r)}{ln(R_t/R_s)}} \qquad (9.6.14)$$

In Equation 9.6.14, R_s is the source radius and R_t is the radius of the OIL lens. As described in Charbeneau et al. (1995), application of continuity leads to formulation of two ordinary differential equations that are solved to yield h_{os} and R_t as a function of time. The model formulation also calculates the amount of NAPL and chemical constituent that remain trapped at residual saturation above and below the lens as the lens thickness decreases after the source is depleted.

Within the HSSM model, the OIL is treated as a two-component mixture. The OIL itself is assumed to be soluble in water and sorbing. Due to the effects of recharge water and contact with the groundwater, the OIL may be dissolved and transferred to groundwater. The OILs transport properties (density, viscosity, capillary pressure, relative permeability), however, are assumed to be unchanging. The second component is a chemical constituent that can partition between the OIL phase, water phase and the soil. This constituent of the OIL is considered the primary contaminant of interest. The mass flux of the second constituent into the aquifer comes from recharge water being contaminated by contact with the lens and from dissolution occurring as groundwater flows under the lens. The concentration of the chemical in the aquifer is limited by its effective water solubility.

For the dissolution mass transfer, it is assumed that the concentration of the contaminant at the base of the lens is equal to the contaminants effective solubility limit, as determined by Roult's law. As the migrating groundwater approaches the lens, it has no contaminant within it, and as the groundwater moves beneath the lens, the contaminant diffuses into the groundwater at a rate determined by continuity and vertical dispersion. This model is essentially the model presented by Hunt et al. (1988). The combined groundwater source term from infiltration and dissolution takes the following form:

$$\dot{m}_{source} = \left(q_{wi}\pi R_t^2 + 4qR_tI_d\sqrt{\frac{2R_ta_v}{\pi}}\right)c_{wo} \qquad (9.6.15)$$

In Equation 9.6.15, q_{wi} is the water infiltration rate vertically through the lens, q is the aquifer Darcy velocity, a_v is the vertical dispersivity, c_{wo} is the aqueous concentration in equilibrium with the NAPL constituent, and

$$I_d = \int_0^1 \sqrt[4]{1 - w^2}\, dw \cong 0.87402$$

It is apparent that the aquifer source term is dependent on the size of the lens, the infiltration rate and groundwater velocity, the constituent concentration within the lens, and the partitioning characteristics of the constituent between the OIL and water.

The required model input includes parameters specifying the type, extent and magnitude of the OIL release, the residual OIL contents for the unsaturated and saturated zones, the residual water content of the OIL lens, the transport properties of the water and OIL (density, viscosity, surface tension), the aquifer and soil water retention characteristics (vertical and horizontal hydraulic conductivity, porosity, irreducible water content, pore size distribution index, and air entry head), the dissolved constituent characteristics (initial concentration within the OIL, aqueous solubility, and the soil-water and OIL-water partition coefficients), and the aquifer transport characteristics (vertical, longitudinal and transverse dispersivity, hydraulic gradient, half-life of the constituent within the aquifer). Other parameters control the simulation characteristics and locations where exposure concentrations are calculated.

HSSM Model

EXAMPLE PROBLEM   As an example, a spill of 1,500 gallons of gasoline over a sandy aquifer was simulated. The area of the release has a radius of 2 m and the duration of the release is 1 day. The water table is at a depth of 5 m and the aquifer has a seepage velocity of 0.413 m/d. The simulation results show that the gasoline first reaches the water table at a time of 2.4 days. The maximum NAPL head of the lens, $h_o = 0.10$ m at a time of 5 days after the release. The lens continues to spread laterally across the capillary fringe over a period of about 5 years, reaching an ultimate radius of 20 m.

Benzene, toluene, and xylene were separately considered as constituents within the gasoline. Benzene is assumed to have a concentration of 8.2 g/L (1.14 percent by mass) in the gasoline. Toluene is assumed to have a concentration of 43.6 g/L (6.07 percent by mass). Xylene is assumed to have a concentration of 71.8 g/L that is 10 percent of the gasoline. Partition coefficients are based on an assumed fraction of organic carbon in the soil and aquifer of $f_{oc} = 0.001$, and with gasoline having a molar concentration of 7 mol/L. With Equation 5.5.14, the soil-water distribution coefficients for benzene, toluene, and xylene are 0.083, 0.30, and 0.83 L/kg, respectively. The corresponding aquifer retardation factors for these chemicals are 1.3, 2.0, and 3.9 for benzene, toluene, and xylene. Equation 5.5.18 gives OIL-water partition coefficients of 312, 1200, and 4240 for benzene, toluene, and xylene, respectively.

HSSM has an option for selecting the effective size of the OIL lens based on its size at the time of the maximum flux from the lens to the aquifer. This size and the corresponding time depend on the partitioning characteristics of the chemical constituent. For this example, these range from an effective radius of 11.4 m for benzene (corresponding to a time of 75 days) to 19.9 m for xylene (corresponding to a time of 420 days). These radii are used to determine the effective size of the Gaussian source in the TSGPLUME model.

Figure 9.6.4 compares the concentrations of benzene, toluene, and xylene at an observation well located 50 m directly downgradient from the center of the release. The peak concentrations of these chemicals are similar (10.4 mg/L at 200 days for benzene, 19.6 mg/L at 380 days for toluene, and 10.9 mg/L at 800 days for xylene), although they were present in far different concentrations

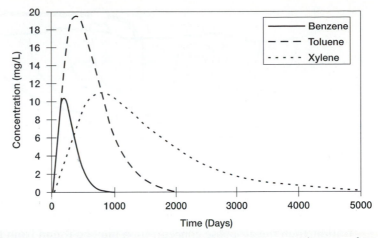

FIGURE 9.6.4   Benzene, toluence and xylene concentrations at a distance of 50 m from the spill

in the gasoline. This behavior is due primarily to the OIL-water partition coefficients. Benzene is leached most rapidly from the OIL lens. Toluene is leached less rapidly than benzene, but because it was present at much higher concentrations in the gasoline, the receptor concentration is higher. Xylene, which is present in gasoline at the highest concentrations, is leached so slowly from the lens that its concentrations are diluted due to mixing with the groundwater and the resulting exposure concentrations are less. In addition, the soil-water partition coefficient for xylene is an order of magnitude larger than that for benzene, so there is more xylene associated with the aquifer soils.

9.7   Soil-Vapor Extraction Systems

Leaking underground storage tanks and spills or other releases of volatile organic liquids can result in contamination of the vadose zone and serve as a long-term source of groundwater contamination as infiltrating rainwater leaches residual hydrocarbons. At one time, remediation of such releases required excavation of contaminated soils and transport to an appropriate landfill. More recently, the technology of soil-venting has emerged as an attractive and cost effective alternative to excavation or pump-and-treat methods for removal of volatile NAPLs in the vadose zone and free product layers at the water table.

The basic idea of soil venting is simple, as shown in Figure 9.7.1. A vacuum or suction pressure is applied at an extraction well that is completed and screened within the vadose zone, and soil gases are pulled to the extraction well carrying contaminants by gaseous advection. The potential recovery rate depends on the vapor phase concentration of the contaminants, the airflow rates that are achieved, and on the airflow path relative to the location of the contaminants within the subsurface (Johnson et al. 1990). The basic principles and limitations are similar to those of pump-and-treat systems for removal of contaminants from the saturated zone.

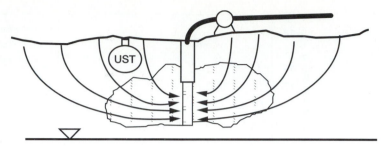

FIGURE 9.7.1   Vacuum Extraction Well

The vapor phase concentration may be estimated using the linear partitioning models described in Section 5.4. If m is the bulk contaminant concentration, then the air phase concentration may be found from Equation 5.4.9.

If the bulk concentration is sufficiently large so that soil venting is considered as a remediation alternative (say greater than 500 mg/kg), then the hydrocarbons will likely be present as a free phase. Raoult's law may then be applied for the vapor phase, giving

$$C_{est} = \sum_i \frac{X_i P_i^v M_i}{RT} \tag{9.7.1}$$

where C_{est} is the estimated contaminant vapor concentration (mg/L), X_i is the mole fraction of component i in liquid-phase residual, P_i^v is the pure component vapor pressure at temperature T (atm), M_i is the molecular weight of component i (mg/mole), R is gas constant (0.0821 L-atm/mole-°K), and T is the absolute temperature (degrees K). Raoult's law says that the vapor pressure of a component of a mixture is equal to the pure component vapor pressure times the component mole fraction. The rest of Equation 9.7.1 is the ideal gas law and Dalton's law of partial pressures.

The airflow rate through the unsaturated zone depends on the hydraulics of soil air, which is discussed in Section 2.4. The continuity equation takes the form of Equation 2.4.10, where compressibility of the soil matrix has been neglected, as have sources and sinks of air volume and the influence of gravity. Furthermore, if the flow is steady and if variations in permeability and viscosity may be neglected, then the continuity equation may be written in cylindrical coordinates as

$$\frac{1}{r}\frac{d}{dr}\left(r\frac{dp^2}{dr}\right) = 0 \tag{9.7.2}$$

so that p^2 satisfies Laplace's equation. Integration with $p = p_w$ at the well ($r = r_w$) and $p = p_{atm}$ at $r = R$ (the radius of influence of the well) gives

$$p(r) = \sqrt{p_w^2 + (p_{atm}^2 - p_w^2)\frac{ln(r/r_w)}{ln(R/r_w)}} \tag{9.7.3}$$

The air discharge per unit length of the well screen, L_w, is calculated from Darcy's law to be

$$\frac{Q_a}{L_w} = \frac{\pi k \overline{k}_{ra}}{\mu_a} \frac{p_{atm}^2 - p_w^2}{p_w \, ln(R/r_w)} \tag{9.7.4}$$

where $\overline{k}_{ra}$ is the average relative permeability of the vadose zone to airflow and μ_a is the air dynamic viscosity. When applying these equations, one must remember that the pressures must be absolute pressures, not gauge pressures. Atmospheric pressure is about 14.7 psia, 101.7 kPa, or 407 inches H_2O. Typical vacuum well pressures range from 20–40 inches H_2O vacuum.

When the suction pressure head does not differ too much from atmospheric pressure, the form of Equation 9.7.4 may be simplified. For these conditions, the following approximation applies

$$\frac{p_{atm}^2 - p_w^2}{p_w} = \frac{(p_{atm} + p_w)(p_{atm} - p_w)}{p_w} \cong 2(p_{atm} - p_w)$$

and Equation 9.7.4 may be approximated as

$$s_a = \frac{Q_a}{2\pi K_{ws} \overline{k}_{ra} L_w} \frac{\mu_a}{\mu_w} \, ln\!\left(\frac{R}{r_w}\right) \tag{9.7.5}$$

In Equation 9.7.5, s_a is the equivalent water drawdown (head) developed by the vacuum system

$$s_a = \frac{p_{atm} - p_w}{\rho_w g} \tag{9.7.6}$$

Equation 9.7.5 looks similar to the Thiem equation for steady groundwater flow to a well in a confined aquifer.

Equations 9.7.3 and 9.7.4 (and thus Equation 9.7.5) are approximate. They are based on the assumption that the flow is steady state and horizontal. In vapor extraction systems, the source of the air is usually directly from the atmosphere at the ground surface so that vertical flow components may be significant. Nevertheless, Equation 9.7.4 or 9.7.5 should provide a reasonable estimate for vapor flow rates.

In Equation 9.7.4 or 9.7.5, k or K_{ws} may be estimated from soil texture or from standard hydraulic testing using methods described in Chapter 3. If a flow test is performed and a number of observation wells are available, then R may be estimated from a plot of p versus r. Typical values range from 30 to 100 ft. The results from Equation 9.7.4 or 9.7.5 are not sensitive to R, and if no other data is available, Johnson et al. (1990) suggest that a value of 40 ft can be used without significant loss of accuracy.

In principle, Equations 9.7.1 and 9.7.4 or 9.7.5 may be used to estimate the pollutant removal rate with vacuum extraction systems. If all of the soil gas passes through the contaminated region, and if equilibrium-partitioning conditions are reached between the soil gas and the residual hydrocarbon phase, then the pollutant removal rate is calculated from

$$\dot{m} = C_{est}Q \tag{9.7.7}$$

Generally, one would like to achieve removal rates greater than 1 kg/d for vapor extraction to be considered as a useful remediation technology. Unfortunately, equilibrium conditions may not be reached for all of the flow paths, and not all of the flow paths come into contact with the pollutant. If a fraction ϕ of the vapor flows through uncontaminated soil, the maximum removal rate is

$$\dot{m} = (1 - \phi)C_{est}Q \tag{9.7.8}$$

Johnson et al. (1990) discusses a number of other limitations and practical applications for design and operation of vapor extraction systems.

9.8   Free Product Recovery of Petroleum Hydrocarbon Liquids

When released to the subsurface in sufficient quantity, petroleum hydrocarbon liquids (gasoline, fuel oils, and other NAPLs that are less dense than water) accumulate and spread as a free-product lens near the water table. In this section we consider recovery of free-product hydrocarbons through use of pumping systems. Free-product recovery is often practiced in conjunction with soil vapor extraction for remediation of petroleum hydrocarbon contaminated sites. There are a number of types of pumping systems available for recovery of floating free product hydrocarbon from an aquifer. When there is only a small amount of free product present, the simplest of these is to place a pump in a well that will skim-off any NAPL which enters the well. These are typically low production rate systems, and when significant quantities of hydrocarbon are present, they are not efficient. Alternative technologies place single- and dual-pump systems in wells to actively pump water and OIL. Water production creates a gradient towards the well that promotes the flow of OIL for recovery. These pumping systems are most useful when there is a significant amount of OIL present and if the porous medium has sufficient permeability. The dual-pump system recovers OIL and water with dissolved constituents through separate pumps. The single-pump system recovers both OIL and water, and requires an API separator at the surface to separate OIL and water for treatment and disposal. Another technology that is especially useful in low-permeability media is vacuum-enhanced free-product recovery. A vacuum is applied to a recovery well, producing airflow. The air-phase gradient creates an OIL gradient that causes the movement of OIL towards the well where the OIL is recovered using a skimmer or other type of pump. Technologies for free-product liquid recovery are discussed in some detail by Charbeneau et al. (1999).

In this section, we present methods for analysis and design of free-product recovery systems using pumping wells and vacuum-enhanced recovery systems. The modeling approach is based upon assignment of effective saturation and relative permeability values to an OIL layer—and basing the performance of the system on the OIL-layer thickness that would be observed in a monitoring well in vertical equilibrium with the formation fluids.

9.8.1   Effective OIL-Layer Saturation and Relative Permeability

The OIL saturation and relative permeability vary with elevation across the thickness of a free-product layer. Figure 9.4.2 shows two examples of the OIL saturation distribution. The saturation is greatest near the top of the layer, and decreases with depth. The relative permeability distribution shows similar characteristics. For modeling purposes the OIL layer is represented by an equivalent layer that has a uniform saturation and relative permeability. These **effective** saturation and relative permeability values must depend on the OIL-layer thickness, for which the monitoring well thickness serves as a useful measure.

If the monitoring well OIL-layer thickness (b_o) is selected as an appropriate measure for the formation OIL-layer thickness, then the effective OIL-layer saturation is calculated from (see Equation 9.4.15)

$$\bar{S}_o = \frac{1}{b_o} \int S_o(z)dz = \frac{D_o}{nb_o} \tag{9.8.1}$$

Equation 9.8.1 appropriately states that the effective layer saturation is equal to the ratio of the volume of free product present to the volume of porous medium void space within the monitoring well OIL thickness. Use of Equation 9.4.18 to predict the formation free-product thickness gives

$$\bar{S}_o = \frac{\beta}{n}\left(1 - \frac{\alpha}{b_o}\right) \; ; \; b_o \geq \alpha \tag{9.8.2}$$

According to this model, there is no free product if $b_o < \alpha$.

The consistent definition of an equivalent relative permeability is given by

$$\bar{k}_{ro} = \frac{1}{b_o} \int k_{ro}(z)dz \tag{9.8.3}$$

In general, neither empirical data nor the analytical forms of models for the NAPL relative permeability function allow Equation 9.8.3 to be evaluated analytically. Charbeneau et al. (1999), however, show that with the Brooks-Corey-Burdine model Equation 9.1.20, a simple model is suggested from a graph of the effective layer relative permeability calculated from Equation 9.8.3 versus the effective layer saturation from Equation 9.8.2. The results from such a plot for 35° API fuel are shown in Figure 9.8.1. The following model gives the curve that is shown on this figure.

$$\bar{k}_{ro} = \bar{S}_o^{\,2} \tag{9.8.4}$$

This model fits the computed data well. The same conclusion is drawn for other density fluids. Charbeneau and Chiang (1995) suggested Equation 9.8.4. With Equation 9.8.2, this model gives

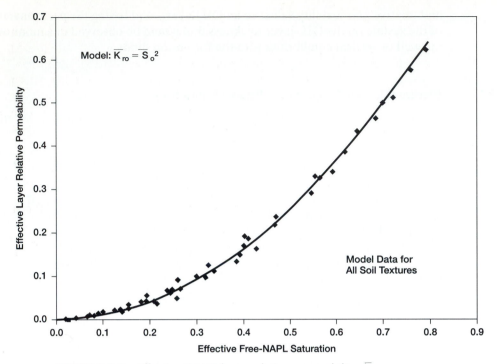

FIGURE 9.8.1   Effective LNAPL layer relative permeability, $\bar{k}_{ro}$

$$\bar{k}_{ro} = \left(\frac{\beta}{n}\right)^2\left(1 - \frac{\alpha}{b_o}\right)^2 \; ; \; b_o \geq \alpha \tag{9.8.5}$$

Equations 9.8.2 and 9.8.5 are used to develop models for design of free-product recovery systems.

9.8.2   OIL Recovery Using Pumping Wells

Single-pump wells produce both water and OIL through a single-pump unit. Compared with dual-pump systems, they are inexpensive and have low maintenance costs. They are simple to operate and may be used to create capture zones to control offsite migration of free product and hydrocarbon dissolved in groundwater. Single-pump wells do require use of a oil/water separation system with increased treatment requirements for produced water.

Dual-pump recovery wells are also able to create capture zones to control offsite migration of free product and dissolved phase hydrocarbons. They have separate production of OIL and water that results in reduced treatment requirements for produced water. They have greater costs and maintenance requirements than single-pump systems, however.

Single-pump and dual-pump systems may be analyzed using the same methods, at least for long-term recovery from the remediation system. The long-term recovery from a remediation well is determined by the hydraulic gradient created towards the well. For both single- and dual-pump systems, this gradient is created through the production of groundwater. The mechanics of flow near the well play a lesser role in long-term recovery. This situation is similar to that found for recovery of groundwater from fully and partially penetrating wells. The water flow towards the well does not differ in direction until it reaches a distance of 1 to 1.5 times the saturated thickness of the aquifer (Bear, 1979). Numerical simulations have confirmed this assessment. Results from long-term simulations (10^+ years) show that the OIL thickness far from the well changes only slowly because of overall free-product recovery. This situation is in contrast with the OIL and water thickness near the well, which fluctuates greatly as the OIL pump is turned on and off. The long-term recovery is independent of these on-and-off fluctuations (Charbeneau and Chiang, 1995).

The drawdown distribution within the OIL layer is created by the same distribution of drawdown in the groundwater layer. This process is shown schematically in Figure 9.8.2. Drawdown is plotted as a function of $ln(r/r_w)$, where r is the radial distance from the recovery well and r_w is the radius of the recovery well. The OIL layer extends out to a radius R_o from the well, and the radius of influence of the well for groundwater flow is R_w. The same radial distribution of drawdown exists in both layers.

Figure 9.8.2 shows that

$$\frac{s_w}{ln(R_w/r_w)} = \frac{s_o}{ln(R_o/r_w)} \qquad \textbf{(9.8.6)}$$

If the drawdown of the water table is small compared with the groundwater saturated thickness, then the Thiem Equation 3.2.3 may be used to predict flow to the well for both the OIL and groundwater layers. With Equations 3.2.3 and 9.8.6, this condition gives

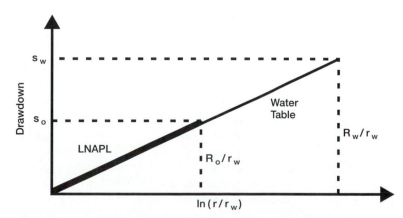

FIGURE 9.8.2 Distribution of drawdown with radial dstance from the well for the oIL and groundwater layers

$$\frac{Q_w}{K_{ws}\bar{k}_{rw}\bar{b}_w} = \frac{Q_o}{K_{os}\bar{k}_{ro}\bar{b}_o} \tag{9.8.7}$$

where $\bar{b}_w$ and $\bar{b}_o$ are the groundwater and OIL layer thicknesses. Charbeneau and Chiang (1995) presented Equation 9.8.7. Assuming that the groundwater relative permeability is one ($\bar{k}_{rw} = 1$), and using Equations 9.1.23 and 9.8.5, the OIL recovery rate may be estimated from

$$Q_o = \left(\frac{\rho_r\beta^2 Q_w}{\mu_r n^2 b_w}\right)\frac{(b_o - \alpha)^2}{b_o} \tag{9.8.8}$$

In Equation 9.8.8, ρ_r and μ_r are the OIL/water density and viscosity ratios, and representative values of α and β are presented in Table 9.4.1. Equation 9.8.8 shows that for small drawdown, the OIL recovery rate is proportional to the water production rate. As the water production rate increases, so does the head gradient towards the well. This process results in a corresponding increase in the OIL recovery rate, independent of the hydraulic conductivity of the formation. Nevertheless, for a given water production rate Q_w, OIL recovery is smaller in fine soils than in coarse soils because β is smaller and α is larger. The OIL recovery rate that is predicted by Equation 9.8.8 is a nonlinear function of the monitoring well OIL-layer thickness. For OIL thickness values much greater than α, however, the relationship between OIL recovery and OIL-layer thickness is nearly linear.

9.8.3   Estimation of Recovery Times for Pumping Well Systems

Recovery times may be estimated by applying the principle of continuity along with the recovery rate equations, such as Equation 9.8.8, to the OIL that is contained within the radius of capture R_c of a recovery well. A simplified picture is shown schematically in Figure 9.8.3. An OIL layer of observation well

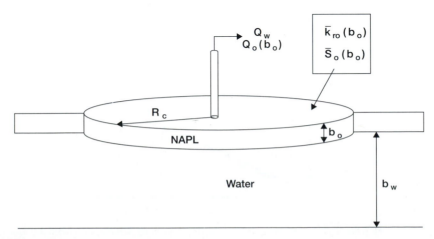

FIGURE 9.8.3   Recovery of LNAPL Layer

thickness b_o is present within the capture radius R_c of the well. As OIL is recovered, the layer thickness decreases as does its effective saturation $(\bar{S}_o)$ and relative permeability $(\bar{k}_{ro})$. These effects are incorporated in Equation 9.8.8.

OIL continuity states that the rate of decrease in free-OIL volume is equal to the OIL recovery rate:

$$-\frac{dV_o}{dt} = -\pi R_c^2 \frac{d(n\bar{S}_o b_o)}{dt} = Q_o \qquad \textbf{(9.8.9)}$$

where V_o is the volume of free-product OIL contained within the radius of recovery. Equations 9.4.18 and 9.8.1 show that

$$n\bar{S}_o b_o = D_o = \beta(b_o - \alpha)$$

Furthermore, if Equation 9.8.8 is used for Q_o, then Equation 9.8.9 may be integrated to give

$$F\left(\frac{b_o(t)}{\alpha}\right) - F\left(\frac{b_o(0)}{\alpha}\right) = \tau \qquad \textbf{(9.8.10)}$$

where the function $F(\ )$ is defined for dummy variable η by

$$F\left(\frac{b_o(\eta)}{\alpha}\right) \equiv \frac{1}{\dfrac{b_o(\eta)}{\alpha} - 1} - ln\left(\frac{b_o(\eta)}{\alpha} - 1\right) \qquad \textbf{(9.8.11)}$$

$$\tau \equiv \left(\frac{\rho_r Q_w}{\mu_r b_w}\right)\left(\frac{\beta}{\pi R_c^2 n^2}\right) t \equiv A_w t \qquad \textbf{(9.8.12)}$$

In Equation 9.8.10, $b_o(0)$ is the initial free product thickness in the observation well and $b_o(t)$ is the free product thickness after time t of recovery. These results were presented by Charbeneau et al. (1999).

9.8.4   Vacuum-Enhanced Recovery Systems

Applying a vacuum to a recovery well can enhance OIL recovery. The vacuum draws air to the well and thus creates an air-pressure gradient towards the well. The air-pressure gradient is transferred directly to the OIL layer, causing OIL flow in response to the head gradient created by the vacuum. This process is completely analogous to OIL flow in response to the gradient created through groundwater recovery. For typical vacuums, the corresponding drawdown created by the vacuum is given by Equation 9.7.5.

Following the same steps leading to Equation 9.8.8, the corresponding equation for OIL recovery caused by the applied vacuum is

$$Q_o = \frac{\mu_a \rho_r Q_a \beta^2}{\mu_0 \bar{k}_{ra} L_w n^2} \frac{(b_o - \alpha)^2}{b_o} \qquad \textbf{(9.8.13)}$$

Application of this equation assumes that OIL is skimmed out of the well at the same rate that it enters from the formation. Equation 9.8.13 has the same form as Equation 9.8.8. In addition, the continuity equation applies for vacuum-enhanced recovery just as it does for the single- or dual-pump systems. Thus, Equation 9.8.10 for time of recovery also applies for vacuum-enhanced recovery, except that

$$A_a = \left(\frac{\mu_a \rho_r Q_a}{\mu_o \bar{k}_{ra} L_w}\right)\left(\frac{\beta}{\pi R_c^2 n^2}\right) \tag{9.8.14}$$

replaces A_w in Equation 9.8.12.

The spreadsheet *LNAPL Recovery.xls* was developed to calculate OIL recovery using pumping wells and vacuum-enhanced systems. This spreadsheet is described in Appendix I.

9.8.5   Example: Design of a Free-Product Recovery System

Problem Statement: The hypothetical problem concerns a tank farm for storage of jet and automotive fuel (estimated that the average NAPL density is 0.82 g/cm³ and viscosity 2 *cp*). Through spills and releases from pipes and surface storage tanks, the subsurface has been impacted by free product. The layout of the hydrocarbon lens is shown in Figure 9.8.4, with the lens having a size of roughly 500 by 400 feet, and an average monitoring well thickness of 3 feet. The formation soil is fine silty sand and may be described as a loamy sand texture. Borehole and pumping tests give an average hydraulic conductivity (K_{ws}) of 0.002 cm/s (5.7 ft/d). The water table is located at a depth of 50 feet and the saturated thickness is 35 feet. The hydraulic gradient is towards the east with a magnitude of 0.005. The site boundary is located 200 feet to the east of the lens.

Data Estimation: The problem statement provides information on soil texture and hydraulic conductivity. Comparison with Table 4.4.1 shows that the

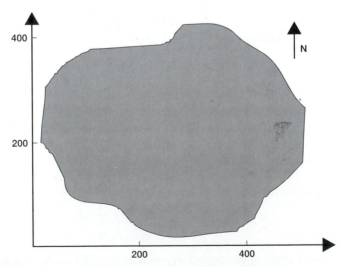

FIGURE 9.8.4 Plan View of Free-Product Hydrocarbon Lens

estimated K_{ws} value is within the expected range for a loamy sand soil. Given that other data are not available, mean soil parameters for a loamy sand soil are selected from Table 4.4.1 (*porosity* = 0.41, S_{wr} = 0.14, Ψ_b = 0.081 m = 0.27 ft, λ = 1.28). The parameters α and β may be found by interpolation from Table 9.4.1. Alternately, the estimated soil parameters may be used with the *LNAPL Distribution.xls* spreadsheet, as described in Appendix H, to calculate the parameters α and β for Equation 9.4.19. The latter approach gives α = 0.21 m = 0.69 ft, β = 0.332.

Free-Product Volume: The parameters α and β, along with the average monitoring well thickness, allow the free-product volume to be estimated using Equation 9.4.18, along with the area of the OIL lens.

$$V_o = D_o \times A_{\text{lens}} = 0.332 \times (3\ \text{ft} - 0.69\ \text{ft}) \times (500\ \text{ft} \times 400\ \text{ft}) = 153{,}000\ \text{ft}^3 = 1{,}147{,}000\ \text{gallons}$$

Technology Selection: For this example, both single-pump wells and vacuum enhanced wells will be evaluated. For further information on technology selection, see Charbeneau et al. (1999).

Evaluation of Single-Pump Wells: Given the formation hydraulic conductivity, we expect that the radius of influence of a pumping well will extend beyond the size of the lens (likely on the order of 1,000 ft). Thus, the number of wells and corresponding radius of capture are determined by limits on recovery times and capabilities for handling produced fluids. A straightforward approach is to select a radius of capture, determine the required number of wells to cover the OIL lens, and then evaluate the performance of the system.

For this example, a radius of capture of R_c = 75 feet is selected (see the following **Discussion**). Potential configurations of wells are determined by drawing circles of radius R_c in a pattern that covers the lens, such as that shown in Figure 9.8.5. This configuration uses 10 wells to cover the area extent of the lens. The discharge per well is determined by limitations on smearing of the hydrocarbon over the formation thickness. If the fraction drawdown is limited to s_w/b_w (same as s_w/H_R) = 0.1, then Equation 3.2.11 gives Γ = 0.19. If we further assume that R_w/r_w = 1000, then Equation 3.2.11 gives

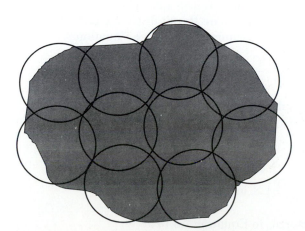

FIGURE 9.8.5 Recovery System with 10 Wells and a Radius of Capture of 75 feet

$$Q_w = \frac{\pi \times 5.7 \text{ ft/d} \times (35 \text{ ft})^2 \times 0.19}{\ln(1000)} = 603 \text{ ft}^3/\text{d} = 3.1 \text{ gpm}$$

This equation represents the water discharge per well.

Equation 9.8.8 may be used to estimate the initial OIL production rate.

$$Q_o = \left(\frac{0.82 \times 0.332^2 \times 603 \text{ ft}^3/\text{d}}{2 \times 0.41^2 \times 35 \text{ ft}}\right) \frac{(3 \text{ ft} - 0.69 \text{ ft})^2}{3 \text{ ft}} = 8.2 \text{ ft}^3/\text{d} = 0.043 \text{ gpm}$$

Thus, with a total liquid production rate per well of 611 ft³/d (=603 + 8.2 ft³/d), approximately 1.3 percent will be OIL and the remaining 98.7 percent groundwater. These represent the initial rates per well during recovery.

To evaluate the time required for free-product remediation, consider the time to recover three-quarters of the free product (leaving a free OIL volume of 38,000 ft³). This value corresponds to a monitoring well thickness $b_o = 1.27$ ft at the end of the recovery period. Using b_o/α values of 4.35 and 1.84 in Equation 9.8.10 gives $\tau = 2.275$. With the given data and Equation 9.8.12, $A_w = 7.9 \times 10^{-4} d^{-1}$. Equation 9.8.12 gives

$$t_{\text{recovery}} = \frac{2.275}{7.9 \times 10^{-4}} = 2{,}900 \text{ d} = 7.9 \text{ years}$$

Achievement of 90 percent recovery corresponds to a final OIL thickness of $b_o = 0.92$ ft, and $b_o/\alpha = 1.33$. For this case, $\tau = 4.8$, and $t_{\text{recovery}} = 6{,}100$ days, or approximately 16.7 years. Alternative designs would consider different capture radii, number of wells, and well locations. The system of single-pump wells will also require a treatment system including an oil/water separator.

Evaluation of vacuum-enhanced systems: The vacuum-enhanced system is useful if surface treatment requirements are to be minimized. As noted in Section 9.7, Johnson *et al.* (1990) suggest that a value of 40 ft can be used for the radius of influence of a vapor extraction well if no other data are available. For this analysis, it is assumed that 32 wells, each with a radius of recovery of 40 ft, will be used. The screen length in the vadose zone is assumed to be 10 ft for each well, and the well pressure (vacuum) is 0.85 atmospheres.

Equation 9.7.5 can be used to estimate the air discharge rate. The equivalent water head developed by the vacuum system is see Equation 9.7.6

$$s_a = \frac{2120 \text{ lb/ft}^2 - 1800 \text{ lb/ft}^2}{1.94 \text{ slugs/ft}^3 \times 32.2 \text{ ft/s}^2} = 5.1 \text{ ft}$$

where a pressure of 1 atmosphere = 2120 pounds per square foot, $\rho_w = 1.94$ slugs/ft³, and $g = 32.2$ ft/sec². The air viscosity is 0.018 cp, and values $k = 0.9$ and $r_w = 0.5$ ft are assumed.

$$Q_a = \frac{2\pi \times 5.7 \text{ ft/d} \times 0.9 \times 10 \text{ ft} \times 5.1 \text{ ft}}{\left(\frac{0.018 cp}{1 cp}\right) \times \ln\left(\frac{40 \text{ ft}}{0.5 \text{ ft}}\right)} = 20{,}800 \text{ ft}^3/\text{d}$$

With this discharge, the factor A_a in Equation 9.8.14 is

$$A_a = \left(\frac{0.018 \text{ cp} \times 0.82 \times 20800 \text{ ft}^3/\text{d}}{2 \text{ cp} \times 0.9 \times 10 \text{ ft}}\right) \times \left(\frac{0.332}{\pi \times (40 \text{ ft})^2 \times 0.41^2}\right) = 0.0067 \text{ d}^{-1}$$

For these conditions, Equation 9.8.12 gives for the time to recover three-quarters of the free product

$$t_{recovery} = \frac{2.275}{0.0067} = 340 \text{ d}$$

The vacuum enhanced system allows for significantly shorter recovery times but requires many more wells. With use of skimmer wells, OIL will be the only liquid produced but an air stream may require treatment before discharge to the atmosphere.

Discussion: This example shows the type of calculations to be performed in the design and analysis of OIL recovery systems. Soil and fluid properties are estimated from field data or tabulations, and site characterization determines the amount of free-product and its areal extent. The primary design variables are the well discharge and the capture radius for single- or dual-pump recovery wells and for vacuum-enhanced wells. For the vacuum-enhanced system, the air discharge is a function of the vacuum applied to the well.

Selection of radius of capture R_c is fundamentally different for single- and dual-pump wells and for vacuum-enhanced wells. For single- and dual-pump wells, the gradient that causes OIL flow is created by water production and extends out to the radius of influence of the well. While this distance may be estimated from field data, it is seldom necessary to do so. The radius of influence of a pumping well usually extends beyond a distance of 1000 ft and should always exceed the radius of capture. Thus, for OIL recovery systems using water production, selection of the radius of capture is directly linked to the number of wells that are used within the area with free product, and limitations on the recovery time and fluid handling capabilities.

For vacuum-enhanced wells, on the other hand, the radius of capture is the same as the radius of influence of the well for airflow, because it is the vacuum that creates the gradient causing OIL flow. The airflow radius of influence is determined by soil properties, and represents a constraint to recovery system design. It is not a variable that may freely be selected. Johnson et al. (1990) discuss methods for estimating the radius of influence of wells used in soil vapor extraction.

Charbeneau et al. (1999) provide more detailed information on selection and design of free-product liquid recovery systems.

Problems

9.1.1. NAPL is present in a monitoring well with the NAPL-water interface located 15 m below the ground surface. If the interfacial tension is 30 dynes/cm and the NAPL density is 0.85 g/cm³, what is the NAPL-water capillary pressure within the formation at the depth of 10 m below the ground surface, assuming vertical equilibrium conditions exist?

9.1.2. An observation well at a contaminated site contains 2 m of hydrocarbon with a density of 0.84 g/cm^3. If the air-hydrocarbon interface is at a depth of 5.2 m and the hydrocarbon-water interface is at a depth of 7.2 m in the well, at what depth would you find the water table if the hydrocarbon were not present?

9.1.3. The figure below shows the plan view from a site characterization investigation of a DNAPL release. The 62 ft and 53 ft contours mark elevations of the base of the aquifer above mean sea level. The DNAPL is a chlorinated hydrocarbon mixture with a density of 1.4 g/cm^3. What total production rate from wells 1 and 2 is required to prevent downgradient migration of the DNAPL if the wells are to pump groundwater at the same rate? The transmissivity of the formation is 50 ft^2/d.

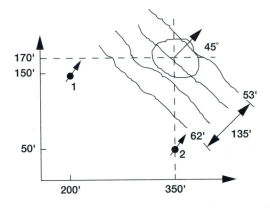

9.1.4. A DNAPL is trapped within a region where the facies (soil texture) show a large variation in the vertical direction, with coarse-grained material above and fine-grained material below. The DNAPL density is 1.4 g/cm^3 and the DNAPL-water interfacial tension is 30 dynes/cm. At a location within fine sand the mean grain diameter is 1 mm. At an adjacent location at 2 cm lower depth the mean grain diameter is 0.1 mm. Estimate the order of magnitude of the equivalent NAPL gradient (dimensionless) to the forces caused by the texture gradation. What is the direction of the gradient?

9.1.5. A soil has a hydraulic conductivity of K_{ws} = 10 ft/d. What is its hydraulic conductivity toward gasoline (μ = 0.29 cp; ρ = 0.68 g/cm^3) and 35 API fuel (μ = 4.0 cp; ρ = 0.85 g/cm^3)? Recall that water has a dynamic viscosity of 1 *cp*.

9.2.1. At a hazardous waste site trichloroethylene is detected in a groundwater sample at a concentration of 30 mg/L. Does this suggest the presence of DNAPL at the site?

9.3.1. DNAPL is pooled on a fractured clay unit to a depth of 25 cm, and soil core samples show that the DNAPL has entered the clay. If σ_{ow} = 30 dynes/cm and ρ_o = 1.3 g/cm^3, what is the size of the larger fracture apertures present?

9.3.2. PERC is trapped in a fractured clay matrix where the largest fracture aperture is 80 μm. Estimate the time period over which PERC will be present within the clay unit as a DNAPL.

9.4.1. A monitoring well in an aquifer has 5 ft of 35 API fuel (μ_o = 4.0 cp; ρ_o = 0.85 g/cm^3) on top of water. The aquifer has the following characteristics: λ = 0.8, n = 0.4, S_{wr} = 0.125, S_{or} = 0, Ψ_{baw} = 0.35 ft. In addition, use S_{orv} = 0.125, S_{ors} = 0.25, and σ_{ow} = σ_{ao} = 30 dynes/cm. What is the formation free product thickness and average saturation for the hydrocarbon layer observed in the monitoring well?

9.4.2. Use the *LNAPL Distribution.xls* spreadsheet to fit a two-segment D_o - b_o curve to Equation 9.4.20 for 38° API fuel with σ_{ao} = σ_{ow} = 25 dynes/cm in loam soil. Use S_{orv} = 0.1, S_{ors} = 0.25. Fit the curve through a monitoring well thickness of 1.5 m, and use the average soil parameters from Table 4.4.1. During an LNAPL recovery project the monitoring well LNAPL thickness is reduced from 1.2 m to 0.6 m over an area of 100 m^2. What volume of LNAPL was produced?

9.4.3. Use the *LNAPL Distribution.xls* spreadsheet to fit a three-segment D_o - b_o curve to Equation 9.4.20 for gasoline (ρ_o = 0.72 g/cm^3, σ_{ao} = 40 dynes/cm, σ_{ow} = 15 dynes/cm) in a sandy loam soil. Assume S_{orv} = S_{ors} = 0.0, and fit the model using van Genuchten parameters through a monitoring well thickness of 1 m. Soil parameters from Table 4.4.1 may be used. Over a 25 m^2 area, how much gasoline must be removed to reduce the monitoring well thickness from 0.25 m to 0.05 m?

9.5.1. A fuel oil storage tank failure results in release of LNAPL to an area of size 100 m^2 surrounded by a retaining berm. The fuel remains ponded with an average depth of 0.25 m for a time period of 1 day, and then infiltrates the soil. The soil is a sandy loam with a water saturation of 0.2, and the fuel oil has a density and viscosity of 0.82 g/cm^3 and 3 cp, respectively. Use the *LNAPL Infiltration.xls* model to determine the time at which the infiltrating oil first reaches the water table at depth 3 m, and the total accumulation of LNAPL at the water table over a 60-day period following the initial release.

9.6.1. An oily waste is applied to a soil during land-farming operations, and tilled into the soil to a depth of 20 cm. Naphthalene is present at a concentration of 17 g/L within the oily waste, and constitutes 2 percent mole fraction. The soil has a *porosity* = 0.40, *bulk density* = 1.6 kg/L, *fraction organic carbon* = 0.02, *volumetric water content* = 0.18, and *oily waste volumetric content* = 0.04. The net infiltration rate is 35 cm/yr. What is the "half-life" of naphthalene within the zone of incorporation considering only the effects of leaching with the infiltrating water?

9.6.2. Estimate the time for loss of naphthalene from the oily waste considered in problem 9.6.1 if volatilization is the only loss mechanism. Assume that the molecular diffusion coefficients in air and liquid are 5(10^{-2}) and 5(10^{-6}) cm^2/s, respectively, and carefully state what other assumptions you are making in your analysis.

9.6.3. Assume the oily waste considered in Problems 9.6.1 and 9.6.2 has been toxic to the microorganisms residing in the soil so that degradation losses are negligible within the zone of original contamination. Estimate the time for loss of naphthalene from the zone of incorporation including both volatile and leaching losses. What fraction of the loss is due to volatilization? Also, estimate the duration over which leachate containing naphthalene will enter an underlying aquifer at a depth of 3 m below the land surface. What is the naphthalene flux rate (mg/d) to the aquifer during this interval if the waste mixture was applied over an area of 200 m²?

9.6.4. A petroleum product containing naphthalene is applied to a land treatment system at an initial volumetric content of 0.03 as immiscible product. Naphthalene is present at an initial concentration of 25 g/L in the petroleum product and constitutes a 3 percent mole fraction. If the petroleum product has an effective half-life of 25 days while naphthalene has a half-life of 45 days, estimate the maximum naphthalene concentration in leachate that may escape from the facility. Assume facility characteristics similar to those of Problem 9.6.1.

9.7.1. A gasoline spill results in vadose zone contamination with a residual OIL saturation of 0.125. The gasoline partitioning characteristics are the same as those presented in Example 5.4.2. Remediation involves soil vapor extraction with suction of 0.25 atmospheres applied to the extraction wells. The screen length for each well is 15 ft, and the soil has a hydraulic conductivity of 15 ft/d. The OIL and water saturation results in an effective air relative permeability of 0.75. The radius of influence and well radius are 40 ft and 0.5 ft, respectively. What is the initial well recovery rate (g/d) for benzene, toluene and xylene if the contaminant flow-fraction ϕ from Equation 9.7.8 is 0.7?

9.8.1. Estimate the recovery rate of 35° API fuel ($\mu_o = 4.0cp$; $\rho_o = 0.85$ g/cm³) from Problem 9.4.1 using a dual pump system with $Q_w = 500$ ft³/d and $K_{ws} = 4$ ft/d. Assume that the thickness of the saturated water layer is 35 ft.

9.8.2. An OIL plume is present beneath a petroleum storage tank farm. The hydrocarbon liquid has an average density and viscosity of 0.8 g/cm³ and 2 cp. Use $\sigma_{oa} = \sigma_{ow} = 25$ dynes/cm and $S_{orv} = S_{ors} = 0.0$. The soil is characterized as loamy sand and the saturated thickness is 10 m. To minimize smearing the maximum well drawdown is 1.5 m. The hydrocarbon plume has an elliptical shape with length 150 m and width 60 m. The average OIL-layer monitoring well thickness is 1.2 m. Design a free project recovery system to recover 75 percent of the recoverable free product over a 3-year time period. Use the van Genuchten model to develop the D_o - b_o relationship. The spreadsheet *LNAPL Recovery.xls* may be used for the calculations.

9.8.3. Repeat problem 9.8.2 using a vacuum-enhanced recovery system with a screen length of 3 m, a well vacuum of 0.85 atmospheres and well radius of influence of 10 m. What is the recovery time to capture 90 percent of the recoverable OIL?

Appendix A
Mathematical Formalisms

Scalar and Vector Fields

Certain mathematical formalisms that are used throughout the textbook are introduced in this appendix for reference. A **field** represents a quantity that is defined over a region of space. The field may also depend upon time. If the quantity is described only by its magnitude, then it specifies a **scalar** field. For example, the pressure p and head h at a point are scalar quantities, and the pressure and head fields are represented by the functions $p(x, y, z, t)$ and $h(x, y, z, t)$.

A *vector* quantity has both magnitude and direction. In this text, a vector is specified by a symbol with a tilde above, such as the velocity vector $\tilde{v}$. The velocity vector field is designated, $\tilde{v}(x, y, z, t)$, or in shorthand notation $\tilde{v}(\tilde{x}, t)$, where the location vector is an argument of the vector-valued function. The direction of a vector is given by the magnitude of its components with respect to a selected coordinate system. We will usually use a Cartesian coordinate system (x, y, z) with **unit** *vectors* $(\hat{i}, \text{etc.})$ pointing in the x, y, and z directions, respectively. Unit vectors have length 1 and are designated by a hat. In terms of the unit vectors, an arbitrary vector $\tilde{a}$ is expressed by

$$\tilde{a} = a_x \hat{i} + a_y \hat{j} + a_z \hat{k} \tag{A.1}$$

The set of scalar quantities (a_x, a_y, a_z) are the components of the vector, specifying the projection of its length in each of the coordinate directions. The magnitude (length) of the vector is calculated from

$$|a| = \sqrt{a_x^2 + a_y^2 + a_z^2} \tag{A.2}$$

Vector Algebra

One may develop algebra for vectors. The product of a scalar and a vector is defined by

$$\alpha \tilde{b} = (\alpha\, b_x)\hat{i} + (\alpha\, b_y)\hat{j} + (\alpha\, b_z)\hat{k} \tag{A.3}$$

for scalar α and vector $\tilde{b}$. The effect of scalar multiplication is simply to change the length of the vector but not its direction. The sum of two vectors $\tilde{a}$ and $\tilde{b}$ is given by

$$\tilde{a} + \tilde{b} = (a_x + b_x)\hat{i} + (a_y + b_y)\hat{j} + (a_z + b_z)\hat{k} \qquad \text{(A.4)}$$

There are two ways to multiply two vectors. The most important for us is the **scalar** or dot product, which is the product of the magnitude of one vector and the magnitude of the projection of the second vector in the direction of the first. Mathematically, this relationship is written

$$\tilde{a} \cdot \tilde{b} = |a|\,|b|\cos(\theta) = a_x b_x + a_y b_y + a_z b_z \qquad \text{(A.5)}$$

where θ is the angle between the two vectors. The scalar product of two vectors is a scalar quantity. In terms of the scalar product, the angle between two vectors is calculated from

$$\theta = \cos^{-1}\left(\frac{a_x b_x + a_y b_y + a_z b_z}{|a||b|}\right) \qquad \text{(A.6)}$$

The second way to define vector multiplication is through the **vector product** or **cross product**, which is designated by $\tilde{a} \times \tilde{b}$. As the name implies, the vector product of two vectors is a vector quantity. The direction of the resultant vector is normal (perpendicular) to the plane containing the two original vectors, and the magnitude of the resultant is

$$|\tilde{a} \times \tilde{b}| = |a|\,|b|\sin(\theta) \qquad \text{(A.7)}$$

where θ is the angle between the two vectors. The vector product of two vectors that are parallel is the zero vector. The components of the cross product may be calculated from the following determinant:

$$\tilde{a} \times \tilde{b} = \begin{vmatrix} \hat{i} & \hat{j} & \hat{k} \\ a_x & a_y & a_z \\ b_x & b_y & b_z \end{vmatrix} = (a_y b_z - a_z b_y)\hat{i} + (a_z b_x - a_x b_z)\hat{j} + (a_x b_y - a_y b_x)\hat{k} \qquad \text{(A.8)}$$

Division of vectors is not formally defined.

Vector Calculus

A calculus of vectors may also be developed as a straightforward extension of single-variable calculus. The concept of the differential is generalized to the "**del operator**", ∇. The definition of the del operator is

$$\nabla \equiv \frac{\partial}{\partial x}\hat{i} + \frac{\partial}{\partial y}\hat{j} + \frac{\partial}{\partial z}\hat{k} \qquad \text{(A.9)}$$

This definition shows that ∇ is a vector quantity, but that by itself it has no meaning (it is an operator waiting to operate on something).

If the del operator is applied to a scalar field, then it defines the **gradient** of that field. For scalar field $c(x, y, z)$, this relationship is written as follows:

$$\nabla c \equiv \text{grad}(c) = \frac{\partial c}{\partial x} \hat{i} + \frac{\partial c}{\partial y} \hat{j} + \frac{\partial c}{\partial z} \hat{k} \qquad \textbf{(A.10)}$$

The gradient of a scalar field is a vector field with direction pointing in the direction of the greatest increase in the scalar magnitude, and the magnitude of the gradient is the rate of increase in c per unit length in that direction.

If $\hat{n}$ is a unit vector, the rate increase in scalar $c(x, y, z)$ in the direction of $\hat{n}$ is calculated from

$$\hat{n} \cdot \nabla c = n_x \frac{\partial c}{\partial x} + n_y \frac{\partial c}{\partial y} + n_z \frac{\partial c}{\partial z} \qquad \textbf{(A.11)}$$

The del operator may also be applied to a vector field to give the **divergence** of the field. For example, with the steady-state velocity field $\tilde{v}(\tilde{x})$, this value is written as follows:

$$\nabla \cdot \tilde{v} \equiv \text{div}(\tilde{v}) = \frac{\partial v_x}{\partial x} + \frac{\partial v_y}{\partial y} + \frac{\partial v_z}{\partial z} \qquad \textbf{(A.12)}$$

The divergence of the vector field is a scalar quantity that represents the local expansion of the field. This value is clearly seen for the velocity field by using the theorem of **Gauss** (the divergence theorem). To write this theorem, we note that we can also associate the vector concept with a surface. The direction of the vector is perpendicular to the surface, and its magnitude is equal to the size of the surface area, such as shown in Figure A.1.a. This representation is especially useful for a surface of unit area, where the unit normal vector is designated $\hat{n}$. Figure A.1.b shows an arbitrary closed surface along with an incremental surface element dA with unit normal $\hat{n}$. The field vector $\tilde{v}(\tilde{x})$ is evaluated on this surface element, and $\tilde{v} \cdot \hat{n} dA$ is the volume flux across (normal to) the surface. With this notation, the Gauss divergence theorem is

$$\iiint_V \nabla \cdot \tilde{v} \, dV = \iint_A \tilde{v} \cdot \hat{n} \, dA \qquad \textbf{(A.13)}$$

which holds for an arbitrary vector field, not just the velocity field. V is the volume enclosed within the closed surface A, shown in Figure A.1.b, and the right side of Equation A.13 is the net volume flux out of the enclosed volume. As the size of the volume shrinks to zero, continuity of the vector field dictates that

$$\nabla \cdot \tilde{v} = \lim_{V \to 0} \frac{1}{V} \iint_A \tilde{v} \cdot \hat{n} \, dA \qquad \textbf{(A.14)}$$

The right side of Equation A.14 is the net volume outflow per unit time per unit volume from the limiting point location of the volume V, which provides a definition of the divergence of the $\tilde{v}(\tilde{x})$-field. It is the local rate of expansion of the field. The same arguments hold for an arbitrary vector field.

An additional vector calculus concept that is used in the text is the divergence of the gradient of a scalar field. If the scalar field is $a(x, y, z)$, this concept gives

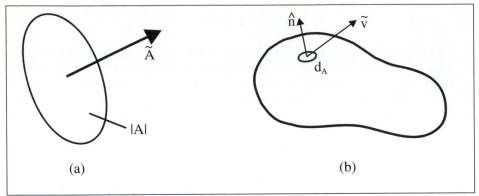

FIGURE A.1   Vector corresponding to an area. (a) Vector area A. (b) Normal to closed surface with field vector ṽ.

$$\nabla \cdot \nabla a \equiv \nabla^2 a = \frac{\partial^2 a}{\partial x^2} + \frac{\partial^2 a}{\partial y^2} + \frac{\partial^2 a}{\partial z^2} \tag{A.15}$$

Equation A.15 defines the **Laplacian** of the scalar field a(x, y, z). The physical meaning of the Laplacian directly follows from that of the divergence of a vector field that is derived from the gradient of a scalar field. In general, it is a measure of the deviation of the local value of a scalar quantity from its average value. Hopf (Hopf, L., *Introduction to the differential equations of physics*, Dover Publications, New York, 1948) shows that

$$\overline{\Phi} - \Phi_o = \frac{L^2}{24} \nabla_o^2 \Phi \tag{A.16}$$

where $\overline{\Phi}$ is the average value of the scalar quantity Φ over a cube with sides of length L centered on point 0, and the Laplacian is evaluated at this point.

Appendix B
Selected Conversions, Parameter Values, and Properties

TABLE B.1 Conversions

Length	1 m = 3.2808 ft = 39.370 in
Area	1 hectare = 10,000 m^2 = 2.47 acres
Volume	1 m^3 = 1000 L = 35.315 ft^3
Discharge	1 m^3/d = 35.315 ft^3/d = 0.1834 gpm
Temperature	°F = 9/5 °C + 32; °C = 5/9 (°F–32); K = °C + 273.16
Pressure	1 atm = 101.3 kPa = 2116.2 lb/ft^2 = 14.7 psi
	1 bar = 10^5 Pa
Hydraulic Conductivity	1 m/d = 3.28 ft/d = $1.16(10^{-3})$ cm/s
Viscosity	1 poise(p) = 1 g/(cm−s) = 0.1 kg/(m−s) = 100 cp = 0.1 Pa-s

TABLE B.2 Properties of Water

Temperature, °C	Density, kg/L	Dynamic Viscosity, Pa-s*	Vapor Pressure, kPa
0	0.9998	$1.781 (10^{-3})$	0.61
5	1.0000	1.518	0.87
10	0.9997	1.307	1.23
15	0.9991	1.139	1.70
20	0.9982	1.002	2.34
25	0.9970	0.890	3.17
30	0.9957	0.798	4.24
40	0.9922	0.653	7.38

*1 Pa equals 0.02089 lb/ft^2.

TABLE B.3   Representative Values of Selected Parameters

Parameter	Value
Density of Water, ρ_w	1.0 g/cm^3 = 1.0 kg/L = 1.94 slugs/ft^3
Dynamic Viscosity of Water, μ_w	0.01 g/cm–s = 0.001 Pa–s = 1 cp = 2(10^{-5}) lbs/ft^2
Acceleration of Gravity, g	9.81 m/s^2 = 981 cm/s^2 = 32.2f t/s^2
Specific Weight of Water, γ_w	9810 N/m^3 = 62.4 lb/ft^3
Latent Heat of Vaporization for Water	2450 J/g
Compressibility of Water	4.5 (10^{-10}) Pa^{-1}
Atmospheric Pressure	101.3 kPa = 1.013(10^6) dyn/cm^2 = 14.7 psia = 407 inches H$_2$O:]
Boltzmann's Constant, κ	1.3805 (10^{-16}) erg/°K
Avogadro's Number, N	6.023(10^{23}) molecules/mol
Gas Constant, R	0.0821 atm–L/mol–°K = 8.21(10^{-5})atm$-^3$/mol–°K

Appendix C
Theis Well Function, W(u)

u x	1.0	2.0	3.0	4.0	5.0	6.0	7.0	8.0	9.0
1.E+00	0.2194	4.89E−2	1.31E−2	3.78E−3	1.15E−3	3.60E−4	1.16E−4	3.76E−5	1.24E−5
1.E−01	1.823	1.223	0.906	0.702	0.560	0.454	0.374	0.311	0.260
1.E−02	4.038	3.355	2.959	2.681	2.468	2.295	2.151	2.027	1.919
1.E−03	6.332	5.639	5.235	4.948	4.726	4.545	4.392	4.259	4.142
1.E−04	8.633	7.940	7.535	7.247	7.024	6.842	6.688	6.555	6.437
1.E−05	10.94	10.24	9.837	9.550	9.326	9.144	8.990	8.856	8.739
1.E−06	13.24	12.55	12.14	11.85	11.63	11.45	11.29	11.16	11.04
1.E−07	15.54	14.85	14.44	14.15	13.93	13.75	13.60	13.46	13.34
1.E−08	17.84	17.15	16.74	16.46	16.23	16.05	15.90	15.76	15.65
1.E−09	20.15	19.45	19.05	18.76	18.54	18.35	18.20	18.07	17.95
1.E−10	22.45	21.76	21.35	21.06	20.84	20.66	20.50	20.37	20.25
1.E−11	24.75	24.06	23.65	23.36	23.14	22.96	22.81	22.67	22.55
1.E−12	27.05	26.36	25.96	25.67	25.44	25.26	25.11	24.97	24.86
1.E−13	29.36	28.66	28.26	27.97	27.75	27.56	27.41	27.28	27.16
1.E−14	31.66	30.97	30.56	30.27	30.05	29.87	29.71	29.58	29.46
1.E−15	33.96	33.27	32.86	32.58	32.35	32.17	32.02	31.88	31.76

Appendix D
Spreadsheet Modules for Calculating Well Functions

The spreadsheet modules are written in the Basic programming language. The modified Bessel functions $I_o(x)$ and $K_o(x)$ use the polynomial approximations presented in Abramowitz and Stegun (1964). The Theis well function (exponential integral) $Wt(u)$ uses the polynomial approximation for $u \leftarrow 1$ and the rational approximation for $u > 1$, both from Abramowitz and Stegun (1964). The Hantush well function $Wh(u,b)$ is evaluated using Equation 3.5.3 and Simpson's rule, as modified from Press et al., (1992).

```
Function Io(x)
        'Io approximation for |x| < 3.75 from Abramowitz and Stegun (1964)
        t = x / 3.75
        t1 = 1 + 3.5156229 * t ^ 2 + 3.0899424 * t ^ 4 + 1.2067492 * t ^ 6
        t2 = 0.2659732 * t ^ 8 + 0.0360768 * t ^ 10 + 0.0045813 * t ^ 12
        Io = t1 + t2
End Function

Function Ko(x)
        'Ko approximation from Abramowitz and Stegun (1964)
                If x < 2 Then
                t = x / 2
                t1 = -Log(t) * Io(x) - 0.57721566 + 0.4227842 * t ^ 2
                t2 = 0.23069756 * t ^ 4 + 0.0348859 * t ^ 6 + 0.00262698 * t ^ 8
                t3 = 0.0001075 * t ^ 10 + 0.0000074 * t ^ 12
                Ko = t1 + t2 + t3
        Else
                t = 2 / x
                t1 = 1.25331414 - 0.07832358 * t + 0.02189568 * t ^ 2
                t2 = -0.01062446 * t ^ 3 + 0.00587872 * t ^ 4 - 0.0025154 * t ^ 5
                t3 = 0.00053208 * t ^ 6
                Ko = x ^ (-0.5) * Exp(-x) * (t1 + t2 + t3)
        End If
End Function
```

```
Function Wt(u)
        'Theis well function (exponential integral) from Abramowitz and Stegun
        (1964)
        If u <= 1 Then
                t1 = -Log(u) - 0.57721566 + 0.99999193 * u - 0.24991055 * u ^ 2
                t2 = 0.05519968 * u ^ 3 - 0.00976004 * u ^ 4 + 0.00107857 * u ^ 5
                Wt = t1 + t2
        Else
                t1 = u ^ 4 + 8.5733287401 * u ^ 3 + 18.059016973 * u ^ 2 +
                8.6347608925 * u + 0.2677737343
                t2 = u ^ 4 + 9.5733223454 * u ^ 3 + 25.6329561486 * u ^ 2 +
                21.0996530827 * u + 3.9584969228
                Wt = Exp(-u) * t1 / (u * t2)
        End If
End Function

Function Wh(u, b)
        'Hantush well function calculated using Simpson's rule as modified from
        Press et al., (1992).
        'Function evaluated using Equation (3.5.3).
        Eps = 0.000001 'Iteration convergence tolerance
        Jmax = 20 'Maximum number of iterations
        ost = -1# * 10 ^ 30
        os = -1 * 10 ^ 30
        'Check limits noting that 2 Ko(b) = W(u,b) + W(b^2/(4u),b)
        v = u 'Set upper limit of integral to v
        If u > b / 2 Then v = b ^ 2 / (4 * u)
        'Main driver loop
        J = 1
        Do While J <= Jmax
                If J = 1 Then
                        st = 0.5 * Exp(-(v + b ^ 2 / (4 * v)))
                Else
                        it = 2 ^ (J - 2)
                        tnm = it
                        del = v / tnm
                        x = del / 2
                        s = 0#
                        I = 1
                        Do While I <= it
                                s = s + Exp(-(x + b ^ 2 / (4 * x))) / x
                                x = x + del
                                I = I + 1
                        Loop
                        st = (st + v * s / tnm) / 2#
                End If
                s = (4# * st - ost) / 3#
                If Abs(s - os) < Eps Then Exit Do
                os = s
                ost = st
                J = J + 1
```

```
        Loop
        If u = v Then
                Wh = 2 * Ko(b) — s
        Else
                Wh = s
        End If
End Function
```

Appendix E

Spreadsheet Module for Calculating Slug Test Well Function

The slug test well function is defined by Equation 3.6.9. While the range of the integral is infinite, for reasonable values of α and β, the integrand is differs from 0 only for u << 1. For this reason, series expansions of the Bessel functions may be used. The following expansions are utilized:

$$J_0(u) = 1 - \frac{u^2}{4}$$

$$J_1(u) = \frac{u}{2} - \frac{u^3}{16}$$

$$Y_0(u) = \frac{2}{\pi}\left(\gamma + ln\left(\frac{u}{2}\right)\right)\left(1 - \frac{u^2}{4}\right) + \frac{u^2}{2\pi}$$

$$Y_1(u) = -\frac{2}{\pi u} + \left(\frac{2\gamma - 1}{2} + ln\left(\frac{u}{2}\right)\right)\frac{u}{\pi} + \left(\frac{5 - 4\gamma}{4} - ln\left(\frac{u}{2}\right)\right)\frac{u^3}{8\pi}$$

With these expansions, the module $I(x, a, b)$ calculates the integrand in Equation 3.6.9, while the module $G(a,b)$ calculates the slug test well function using Simpson's rule (in the limit, one finds that $I(0, a, b) = 0$).

```
Function I(x, a, b)
      'Calculate the integrand for the slug test well function.
      'Use series expansions for Bessel functions Jo, J1, Yo and Y1.
      e = 0.5772156649                  'Euler's gamma constant
      Pi = 3.1415926536                 'Pi
      BF = ((1-a)*x-((2-a)*x^3)/8)^2    'BF is the Bessel function expansion
      g1 = 2* (e + Log(x / 2)) * x + (1-e-Log(x / 2)) * x ^ 3 / 2
      g2 = 4 * a / x-2 * a * ((2 * e-1) / 2 + Log(x / 2)) * x
      g3 = (a / 4) * ((5-4 * e) / 4-Log(x / 2)) * x ^ 3
      BF = BF + ((g1 + g2 + g3) / Pi) ^ 2
      I = Exp(-b * x ^ 2 / a) / (x * BF)
End Function
```

```
Function G(a, b, Eps)
        'Estimate the Slug Test well function
        'Eps = Iteration convergence tolerance (start w/ 0.00001 and modify)
        Jmax = 20           'Maximum number of iterations
        Pi = 3.1415926536
        ost = -1# * 10 ^ 30
        os = -1 * 10 ^ 30
        'Determine the upper limit for the integral
        umax = 10#
        Do While I(umax / 2, a, b) < Eps
                umax = umax / 2
        Loop
        'Main driver loop
        J = 1
        Do While J <= Jmax
                'Simpson's rule implementation after Press et al. (1992)
                If J = 1 Then
                        st = 0.5 * umax * I(umax, a, b)
                Else
                        it = 2 ^ (J-2)
                        tnm = it
                        del = umax / tnm
                        x = del / 2
                        s = 0#
                        K = 1
                        Do While K <= it
                                s = s + I(x, a, b)
                                x = x + del
                                K = K + 1
                        Loop
                        st = (st + umax * s / tnm) / 2#
                End If
                s = (4# * st-ost) / 3#
                If Abs(s-os) < Eps * Abs(os) Then Exit Do
                os = s
                ost = st
                J = J + 1
        Loop
        G = 8 * a * s / Pi ^ 2
End Function
```

Appendix F
Water Balance for Wetting Event

Sections 4.5, 4.6, and 4.7 present models for calculating infiltration, soil water redistribution, and evaporation losses for a rainfall event. This appendix describes a more general model for combining these calculations using a computer spreadsheet that also graphs the water saturation profile. An example showing the computation page and graph is shown in Figure F.1.

The boxes on the left side of the computation page are used for data entry. They include the soil parameters described using the Brooks and Corey (1964) model; infiltration data including the antecedent water saturation, S_a, the rainfall rate, P, and the rainfall duration, t_r; and evaporation data including the potential evaporation rate, e_p, and the effective evaporation profile water saturation, S_{ep}, which is used to calculate desorptivity.

The boxes on the right side of Figure F.1 contain the calculated parameters and computation results. The infiltration water saturation, S_i, is calculated from

$$S_i = S_{fc} + (1 - S_{fc})\left(\frac{q}{K_{ws}}\right)^{1/\varepsilon} \; ; \; q = \text{Min}\left(\frac{K_{ws}}{2}, P\right) \qquad \textbf{(F.1)}$$

Ponding time, t_p, and the cumulative infiltration, $I(t)$, runoff, $R(t)$, and infiltration depth, L_f, are calculated using the methods presented in Section 4.5.

The kinematic model is used to describe the drainage profile so that the water saturation on the upstream side of the wetting front satisfies the following equation:

$$S(z,t) = S_{fc} + (1 - S_{fc})\left(\frac{n(1 - S_{fc})z}{\varepsilon K_{ws}(t - t_r)}\right)^{1/(\varepsilon-1)} \qquad \textbf{(F.2)}$$

The antecedent water saturation is arbitrary so that the general wetting front step equation is used:

$$\frac{dz_f}{dt} = \frac{K_{ws}\left(\dfrac{S - S_{fc}}{1 - S_{fc}}\right)^{\varepsilon} - q_a}{S - S_a} \qquad \textbf{(F.3)}$$

543

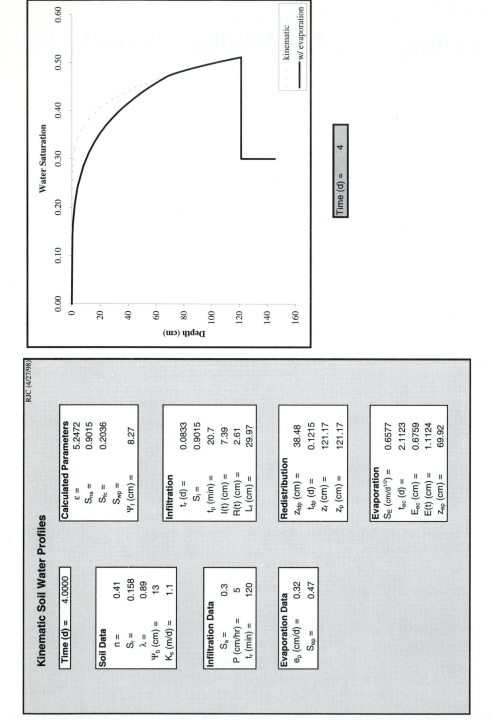

FIGURE F.1 Speadsheet showing the computation page and the soil water content profile.

In Equation F.3, S is the water saturation on the upstream side of the wetting front and is equal to S_i at times before the plateau disappears $(t < t_{dp})$—and is equal to S from Equation F.2 thereafter. q_a is the Darcy flux corresponding to the antecedent water saturation, S_a, which equals the hydraulic conductivity at saturation S_a for the unit gradient model. Equation F.3 is integrated using a modified midpoint method.

The time and depth for the plateau to disappear are found from

$$t_{dp} - t_r = \frac{nL_f}{\dfrac{\varepsilon q}{(S_i - S_{fc})} - \dfrac{q - q_a}{(S_i - S_a)}} \; ; \; z_{fdp} = \frac{L_f}{1 - \left(\dfrac{q - q_a}{\varepsilon q}\right)\left(\dfrac{S_i - S_{fc}}{S_i - S_a}\right)} \quad \text{(F.4)}$$

where q is given in Equation F.1.

The evaporation model follows Section 4.7 and Example 4.7.1 for the evaporation deficit with S_{fc} used in place of S_{ep} for the profile calculation (but not in the calculation of the desorptivity, S_E). z_{ep} is the depth of the evaporation profile.

The example shown in Figure F.1 is for a sandy loam soil with a rainfall rate of 5 cm/hr for 2 hours. Ponding occurs after 20 minutes, and the cumulative rainfall and runoff are 7.39 and 2.61 cm, respectively. At the end of the period of infiltration, the wetting front depth is 30 cm. The water content profile is shown at a time of 4 days, after the beginning of the rainfall event. At this time, the wetting front has reached a depth of 120 cm. The dashed curve shows the kinematic profile without evaporation. The cumulative evaporation is 1.1 cm, and the evaporation profile depth is 70 cm.

This Microsoft Excel spreadsheet titled *Kinematic Profiles.xls* is available for downloading from the world wide web address http://www.prenhall.com.

Appendix G
Error and Complementary Error Functions

The complementary error function appears many times within the study of subsurface fate and transport, and it is of interest to note some of its properties and relationships to other functions. The basic functions that are used in mathematical physics are the error function and its complement. The **Error Function** is defined by

$$\text{erf}(X) = \frac{2}{\sqrt{\pi}} \int_0^x e^{-\zeta^2} d\zeta \tag{G.1}$$

From this definition, it is apparent that

$$\text{erf}(-X) = -\text{erf}(X) \tag{G.2}$$

The **Complementary Error Function** is defined by

$$\text{erfc}(X) = 1 - \text{erf}(X) = \frac{2}{\sqrt{\pi}} \int_x^{\infty} e^{-\zeta^2} d\zeta \tag{G.3}$$

Together, Equations 7.A.2 and 7.A.3 give

$$\text{erfc}(-X) = 1 + \text{erf}(X) \tag{G.4}$$

The error function and its complement are shown in Figure G.1 and are presented in Table G.2. The error function increases from -1 at $X = -\infty$ to 1 at $X = \infty$, while the complementary error function decreases from 2 to zero over this range.

For *large X*, the following asymptotic approximation from Abramowitz and Stegun (1970) is helpful:

$$\text{erfc}(X) \cong \frac{1}{X\sqrt{\pi}} e^{-x_2} \left(1 - \frac{1}{2X^2} + \frac{3}{4X^4} - \cdots \right) \tag{G.5}$$

For *small X*, we have $e^{-X^2} \cong 1$. Using this information in Equation G.1, we have the approximations given by

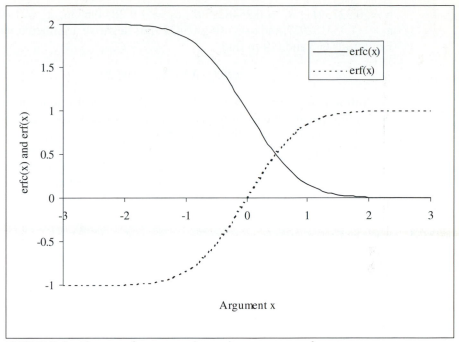

FIGURE G.1   The error function and complementary error function.

$$\mathrm{erf}(X) \cong \frac{2}{\sqrt{\pi}} X \ ; \ \mathrm{erfc}(X) = 1 - \frac{2}{\sqrt{\pi}} X \qquad \textbf{(G.6)}$$

A careful look at Figure G.1 shows that these linear approximations are valid for *small X*.

For computer or spreadsheet calculations, approximations that are valid over the entire range of arguments are necessary. Abramowitz and Stegun (1970) present the following rational approximation for the complementary error function for arguments $0 \leq x \leq \infty$, with $|\varepsilon(x)| \leq 1.5 \times 10^{-7}$:

$$\mathrm{erfc}(x) = (a_1 t + a_2 t^2 + a_3 t^3 + a_4 t^4 + a_5 t^5)e^{-x^2} + \varepsilon(x) \qquad \textbf{(G.7)}$$

where

$$t = \frac{1}{1 + p x}$$

$p = 0.3275911$, $a_1 = 0.254829592$, $a_2 = -0.284496736$, $a_3 = 1.421413741$, $a_4 = -1.453152027$, and $a_5 = 1.061405429$. For negative arguments, we combine Equations G.3 and G.4 to find

$$\text{erfc}(-X) = 2 - \text{erfc}(X) \qquad \textbf{(G.8)}$$

The rational approximation given by Equation G.7 is useful for many applications. For large arguments, however, the asymptotic expansions of Equation G.5 should be used (with only the first term for large arguments).

A module that can be used to calculate the complementary error function for use in spreadsheets is presented in Table G.1. A tabulation of the error function and complementary error function is presented in Table G.2.

TABLE G.1   Module for Calculating the Error Function (Basic Language)

```
Function erfc(x)
    tmp = Abs(x)
    Pi = 3.141592654
    If tmp > 3 Then
        f1 = (1—1 / (2 * tmp ^ 2) + 3 / (4 * tmp ^ 4)—5 / (6 * tmp ^ 6))
        fun = f1 * Exp(-tmp * tmp) / (tmp * Sqr(Pi))
        If tmp = x Then
            erfc = fun
        Else
            erfc = 2—fun
        End If
    Else
        tmp2 = 1 / (1 + 0.3275911 * tmp)
        tmp3 = 0.254829592 * tmp2—0.284496736 * tmp2 ^ 2 + 1.421413741 * tmp2 ^ 3
        tmp4 = −1.453152027 * tmp2 ^ 4 + 1.061405429 * tmp2 ^ 5
        fun = (tmp3 + tmp4) * Exp(-tmp * tmp)
        If tmp = x Then
            erfc = fun
        Else
            erfc = 2—fun
        End If
    End If
End Function
```

TABLE G.2   The Error Functions

X	erf(X)	erfc(X)
0.00	0.000000	1.000000
0.05	0.056372	0.943628
0.10	0.112463	0.887537
0.15	0.167996	0.832044
0.20	0.222703	0.777297
0.25	0.276326	0.723674
0.30	0.328627	0.671373
0.35	0.379382	0.620618
0.40	0.428392	0.571608
0.45	0.475482	0.524518
0.50	0.520500	0.479500
0.55	0.563323	0.436677
0.60	0.603856	0.396144
0.65	0.642029	0.357971
0.70	0.677801	0.322199
0.75	0.711156	0.288844
0.80	0.742101	0.257899
0.85	0.770668	0.229332
0.90	0.796908	0.203092
0.95	0.820891	0.179109
1.00	0.842701	0.157299
1.10	0.880205	0.119795
1.20	0.910314	0.089686
1.30	0.934008	0.065992
1.40	0.952285	0.047715
1.50	0.966105	0.033895
1.60	0.976348	0.023652
1.70	0.983790	0.016210
1.80	0.989091	0.010909
1.90	0.992790	0.007210
2.00	0.995322	0.004678
2.10	0.997021	0.002979
2.20	0.998137	0.001863
2.30	0.998857	0.001143
2.40	0.999311	0.000689
2.50	0.999593	0.000407
2.60	0.999764	0.000236
2.70	0.999866	0.000134
2.80	0.999925	0.000075
2.90	0.999959	0.000041
3.00	0.999978	0.000022

Appendix H
Spreadsheet Calculations of LNAPL Distribution

A spreadsheet has been developed to perform many of the calculations presented in Section 9.4. This Microsoft Excel spreadsheet (workbook), titled *LNAPL Distribution.xls*, is available for downloading from the World Wide Web address *(http://www.prenhall.com/)*. It may be used to calculate many of the significant parameters that are used in design and analysis of *free-product recovery* (FPR) systems, including the effective LNAPL-layer saturation $(\overline{S}_o)$ and effective LNAPL-layer relative permeability $(\overline{k}_{ro})$ as a function of monitoring well LNAPL-layer thickness (b_o). In addition, the parameters α and β from Equation 9.4.20 that are used to calculate LNAPL recovery with the methods presented in Section 9.8 may also be calculated using the spreadsheet. Application and use of this spreadsheet are described in this appendix.

The spreadsheet (workbook) contains four worksheets. The first worksheet is for data entry. Consistent length units should be used (usually meters or feet). An example is shown in Figure H.1 (the computation area of the worksheet is not shown). On the left of the worksheet are the cells for entering monitoring well LNAPL thickness, Brooks and Corey soil characteristic parameters, and fluid parameters. With a color monitor, this section of the worksheet is colored yellow. On the right of the worksheet are the corresponding values for the various parameters that were discussed in Section 9.4. In particular, the values shown in the section colored green include the monitoring well LNAPL thickness, the formation free-product volume, and the calculated effective LNAPL-layer saturation and relative permeability. Data from this section may be copied to the second worksheet for use in calculation of the parameters α and β. If a value of b_o less than the LNAPL entry head is entered on the left, then a value equal to the entry head is used in the calculations and is shown on the right (the LNAPL entry head is labeled $\Delta\Psi$ on the worksheet). For the example shown, the LNAPL entry head is 0.061 meters. Values of b_o less than $\Delta\Psi$, however, are used on the van Genuchten worksheet, because the van Genuchten capillary pressure model has a zero entry head. The van Genuchten parameter values corresponding to the Brooks and Corey parameter values that were entered are also shown to the far right on the worksheet.

RJC (8/14/98)

Brooks-Corey LNAPL Distribution Worksheet

Enter Data in Yellow Region - Use Consistent Length Units

Monitoring Well Thickness

b_o =	1.220	(length)

Soil Characteristic

n =	0.410	porosity
λ =	0.685	pore size dist. Index
Ψ_{baw} =	0.083	displacement pressure head (length)
S_{wr} =	0.159	irreducible water saturation
S_{ors} =	0.050	residual LNAPL saturation (saturated)
S_{orv} =	0.050	residual LNAPL saturation (vadose)
S_{or} =	0.000	resid. LNAPL sat. (rel. perm. calc.)
Z_{orv} =	0.000	elev. vadose zone residual (length)
Z_{ors} =	0.000	elev. saturated zone residual (length)

Fluid Characteristics:

ρ_o =	0.700	LNAPL density (g/cm³)
σ_{aw} =	65.000	air/water surface tension (dyne/cm)
σ_{ow} =	25.000	LNAPL/water surface tension (dyne/cm)
σ_{ao} =	25.000	air/LNAPL surface tension (dyne/cm)

Copy data for Work Chart

b_o =	1.220	monitoring well thickness in computation
D_o =	0.319	formation free-product volume (length)
$\underline{S_o}$ =	0.638	effective LNAPL layer saturation
k_{ro} =	0.339	effective LNAPL layer rel. permeability

ε =	5.920
z_{ao} =	0.366
z_{ow} =	-0.854
z_r =	1.281
Ψ_{bao} =	0.046
Ψ_{bow} =	0.106
$\Delta\Psi$ =	0.061
$z_{ao}+\Psi_{bao}$ =	0.412
$z_{ow}+\Psi_{bow}$ =	-0.748

van Genuchten Parameters

M =	0.471
N =	1.889
α =	7.497
α_{ao} =	13.644
α_{ow} =	5.848

FIGURE H.1 Brooks and Corey Data Entry Worksheet

All workbook cells except for those used in data entry are protected.

The second worksheet, shown in Figure H.2, is used for estimating the (set of) α and β values that are used in Equation 9.4.20. Data (b_o, D_o, $\overline{S}_o$, and $\overline{k}_{ro}$ values) are entered (copied from the first or third worksheet) in the section labeled "Enter data here." Up to 15 data sets may be used. The data must be copied and pasted using the "Edit…Paste Special…Values" sequence. (Alternatively, after the data have been copied from the first or third worksheet, one may press "Alt-E-S-V" and then "Enter.") As data are entered, the graph is automatically updated. One may also enter a brief title, such as "Soil Texture." The second part of the work area is for fitting α and β values to different segments of the data curve. For each segment, initial and final values of b_o and D_o are entered. The corresponding segment is shown as a dashed line on the graph. Up to three segments may be used. The example shown in Figure H.2 has only one segment with an initial (b_o, D_o) point (0.230, 0.000) and final point (2.000, 0.573). The worksheet calculates the corresponding α and β values, where α has the same length units selected for the first worksheet. An example graph with three segments is shown in Figure 9.4.4.

The third worksheet, as shown in Figure H.3, provides more detailed information on the van Genuchten parameters, including the formation product thickness, effective saturation, and relative permeability values calculated using the van Genuchten model for use with the second worksheet.

The fourth worksheet, shown in Figure H.4, shows the LNAPL saturation and relative permeability distribution profiles. The saturation profile is calculated using Equation 9.4.14 for the Brooks and Corey model—and a corresponding equation for the van Genuchten model. The relative permeability distributions are calculated using Equations 9.1.20 and the corresponding van Genuchten model, along with the saturation distributions.

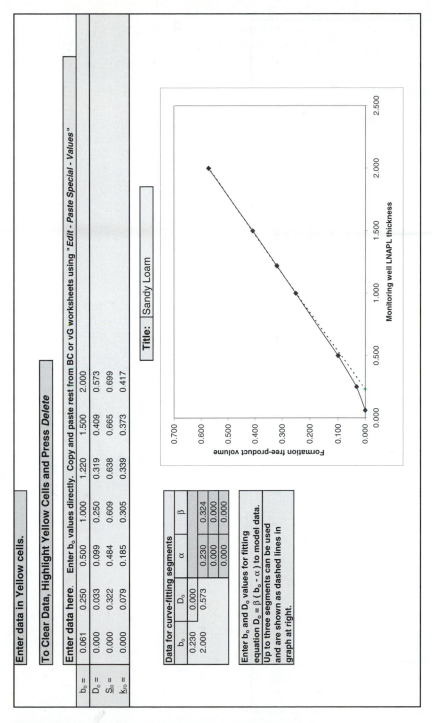

FIGURE H.2 Work Chart Worksheet for Calculating α and β.

Data Sheet for van Genuchten Model of LNAPL Distribution and Permeability

Basic data comes from the Brooks-Corey Worksheet

van Genuchten Parameters				
$M =$	0.471			
$N =$	1.889			
$\alpha =$	7.497			
$\alpha_{ao} =$	13.644			
$\alpha_{ow} =$	5.848			

$b_o =$	0.030	(length)
$S_{wr} =$	0.159	
$S_{ors} =$	0.050	
$S_{orv} =$	0.050	
$Z_{orv} =$	0.000	elev. vadose zone residual (length)
$Z_{ors} =$	0.000	elev. saturated zone residual (length)
$S_m =$	0.000	minimum liquid sat. (rel. perm. calc.)
$z_{ao} =$	0.009	
$z_{ow} =$	-0.021	
$Z_{max} =$	0.032	maximum free-product elevation

Effective LNAPL Saturation and Relative Permeability	
$D_o =$	0.001
$\underline{S}_o =$	0.101
$\underline{k}_{ro} =$	0.043

RJC (8/14/98)

FIGURE H.3 van Genuchten Parameter Worksheet

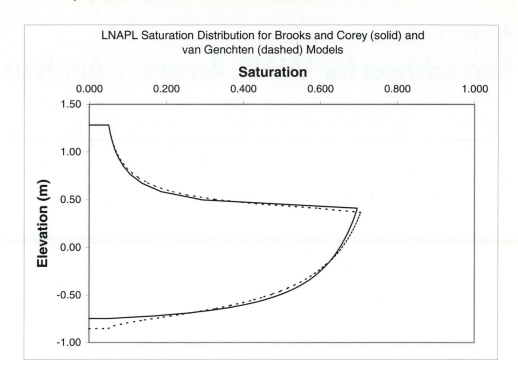

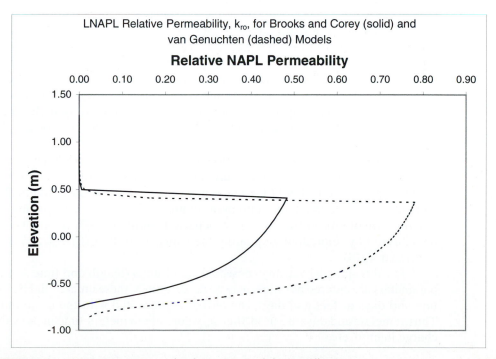

FIGURE H.4   LNAPL Saturation and Relative Permeability Distributions

Appendix I

Spreadsheet for LNAPL Recovery Analysis

A spreadsheet has been developed to perform the LNAPL recovery calculations presented in Section 9.8 for single- and dual-pump recovery systems and for vacuum-enhanced systems. This Microsoft Excel spreadsheet (workbook), titled *LNAPL Recovery.xls*, is available for downloading from the World Wide Web address *(http://www.prenhall.com/)*. It may be used to calculate the free-LNAPL recovery rate, cumulative LNAPL recovery, remaining LNAPL volume, and residual volume as a function of time. Application and use of this spreadsheet are described in this appendix.

The spreadsheet (workbook) consists of a single worksheet. An example is shown in Figure I.1. The data entry cells are located in the upper-left region of the worksheet. The required input data include the fluid (water or air) discharge rate, length of the well screen, fluid (water or air) relative permeability, LNAPL/fluid (water or air) viscosity ratio, porosity, capture radius of the recovery well, and the vadose zone and saturated zone residual LNAPL saturation. The b_o-D_o segment values of α and β are entered, as are the initial and final values of b_o. If a multiple b_o-D_o segment representation is used, the initial time, LNAPL volume recovered, and residual LNAPL volume may be entered for a given segment.

The worksheet output information includes the maximum LNAPL residual volume within the radius of capture if the entire free product is recovered, the initial free-product LNAPL volume, and the final free-product volume. The computation work area table contains a summary of the recovery information as a function of time during the recovery period. Multiple b_o-D_o segments may be analyzed by sequentially updating the values of α, β, $b_o(0)$, $b_o(t_{max})$, $t(0)$, $V_o(0)$ and $V_{or}(0)$.

For this spreadsheet, any consistent set of units (length and time—other parameters are entered as ratios) may be used. Usually, these units would be meters and days or feet and days. The output is presented with the same units. Thus, if meters and days are used, then b_o values are in meters, time in days, discharge in m^3/d, etc.

RJC (8/14/98)

Free-Product Recovery System Analysis

Enter data in Yellow area. Use any consistent units.

Variables:

$Q =$	600	Fluid discharge (water or air)
$L =$	35	Fluid recovery length (aquifer thickness or vacuum well screen length)
$k_{rf} =$	1	Fluid relative permeability (water or air)
$\rho_r =$	0.7	LNAPL/water density ratio
$\propto_{rf} =$	0.3	LNAPL/Fluid (water or air) viscosity ratio
$n =$	0.41	Porosity
$R_c =$	80	Radius of capture
$S_{orv} =$	0.1	Vadose zone residual LNAPL saturation
$S_{ors} =$	0.2	Saturated zone residual LNAPL saturation

Max. resid. LNAPL Vol.
$V_{or}(max) =$ 4548

Product Layer:

$=$	0.7544	Minimum monitoring well LNAPL thickness
$\beta =$	0.324	LNAPL-layer specific yield

Initial Free-LNAPL Vol.
$V_o(0) =$ 21143

$b_o(0) =$	4	Initial LNAPL Thickness
$b_o(t_{max}) =$	1	Final LNAPL Thickness

Final Free-LNAPL Vol.
$V_o(t_{max}) =$ 1600

$t(0) =$	0	Initial time for computation
$V_o(0) =$	0	Initial LNAPL volume recovered
$V_{or}(0) =$	0	Initial residual LNAPL Volume

Well Recovery Coef.
$A_i =$ 0.0048855

Computation Work Area:

Index	thickness b_o	time t	discharge Q_o	recovered V_o	remaining V_o	residual V_{or}	saturation S_o
0	4	0.0	65.78	0	21143	0	0.641
1	3.85	12.0	62.17	767	20166	210	0.635
2	3.7	24.7	58.58	1534	19189	420	0.629
3	3.55	38.2	54.99	2301	18212	631	0.622
4	3.4	52.6	51.42	3068	17235	841	0.615
5	3.25	68.1	47.87	3835	16257	1051	0.607
6	3.1	84.7	44.33	4602	15280	1261	0.598
7	2.95	102.8	40.82	5369	14303	1471	0.588
8	2.8	122.4	37.33	6136	13326	1682	0.577
9	2.65	144.0	33.87	6903	12349	1892	0.565
10	2.5	167.8	30.45	7670	11372	2102	0.552
11	2.35	194.5	27.06	8436	10394	2312	0.537
12	2.2	224.8	23.73	9203	9417	2523	0.519
13	2.05	259.6	20.45	9970	8440	2733	0.499
14	1.9	300.4	17.25	10737	7463	2943	0.476
15	1.75	349.4	14.15	11504	6486	3153	0.450
16	1.6	410.3	11.16	12271	5509	3363	0.418
17	1.45	489.7	8.34	13038	4531	3574	0.379
18	1.3	600.4	5.72	13805	3554	3784	0.332
19	1.15	773.6	3.40	14572	2577	3994	0.272
20	1	1109.5	1.51	15339	1600	4204	0.194

FIGURE I.1 LNAPL Recovery Worksheet.

References

Abramowitz, M. and I. A. Stegun, *Handbook of Mathematical Functions,* Dover, New York, 1965.

Abriola, L. M. and G. F. Pinder, "A Multiphase Approach to the Modeling of Porous Media Contaminated by Organic Compounds, 1 Equation Development," *Water Resourc. Res., 21,* 11–18, 1985.

Adamson, A. W., *Physical Chemistry of Surfaces,* 3rd Ed., Interscience, New York, 1978.

Ahuja, L. R., "Unsaturated Hydraulic Conductivity from Cumulative Inflow Data," *Soil Sci. Soc. Am. Procc.,* 38(5), p. 695–698, 1974.

Ahuja, L. R., J. W. Naney, and R. D. Williams, "Estimating Soil Water Characteristics from Simpler Properties or Limited Data," *Soil Sci. Soc. Am. J., 45*(5), 1100–1105, 1985.

Alexander, M., "Biodegradation: Problems of Molecular Recalcitrance and Microbial Infallibility," *Adv. Appl. Microb., 7,* 35–80, 1965.

Alexander, M., *Introduction to Soil Microbiology*, 2nd ed., John Wiley & Sons, New York, 1977.

Alexander, M. and K. M. Scow, "Kinetics of Biodegradation in Soil," in *Reactions and Movement of Organic Chemicals in Soils*, Sawhney, B. L. and K. Brown (Eds.), SSSA Special Publication Number 22, Soil Science Society of America, Madison, WI, 1989.

American Petroleum Institute, "The Land Treatability of Appendix VIII Constituents Present in Petroleum Industry Wastes," *API Publication 4379,* 1984.

American Society for Testing and Materials, "Test Method for Capillary Moisture Relationships for Coarse and Medium Textured Soils by Porous Plate Apparatus, Method D–2325," *Annual Book of ASTM Standards, Vol. 04. 08,* 302–308, 1992a.

American Society for Testing and Materials, "Test Method for Capillary Moisture Relationships for Fine-Textured Soils by Pressure Membrane Apparatus, Method D–3152," *Annual Book of ASTM Standards, Vol. 04. 08,* 431–436, 1992b.

ASTM (American Society for Testing and Materials), "Standard Guide for Risk-Based Corrective Action Applied at Petroleum-Release Sites," E1739–95, West Conshohocken, PA, 1995.

Amoozegar, A., "A Compact Constant—Head Permeameter for Measuring Saturated Hydraulic Conductivity of the Vadose Zone," *Soil Science Society of America Journal,* 53(5), p. 1356–1361, 1989.

Amoozegar-Fard, A., D. R. Nielsen, and A. W. Warrick, "Soil Solute Concentration Distributions for Spatially Varying Pore Water Velocities and Apparent Diffusion Coefficients," *Soil Sci. Soc. Am. J., 46,* 3–9, 1982.

Amoozegar, A., and A. W. Warrick, "Hydraulic Conductivity of Saturated Soils: Field Methods," *Methods of Soil Analysis, Part 1,* A. Klute (EEd.), American Society of Agronomy, Madison, WI, 735–798, 1986.

Anat, A., Duke, H. R., and A. T. Corey, "Steady Upward Flow from Water Tables," Colorado State University, CO, Hydrology Paper No. 7, June, 1965.

Badon-Ghyben, W., "Notes on the Probable Results of Well Drilling Near Amsterdam (in Dutch)," *Tijdschrift van het Koninklijk Inst. van Ing.,* The Hague, p. 21, 1888.

Baehr, A. L., "Selective Transport of Hydrocarbons in the Unsaturated Zone Due to Aqueous and Vapor Phase Partitioning," *Water Resour. Res., 23*(10), 1926–1938, 1987.

Baehr, A. L. and M. Y. Corapcioglu, "A Compositional Multiphase Model for Groundwater Contamination by Petroleum Products, 2. Numerical Solution," *Water Resour. Res., 23*(1), 201–214, 1987.

Barenblatt, G., Zheltov, I. P., and I. N. Kochina, "Basic Concepts in the Theory of Seepage of Homogeneous Liquids in Fissured Rocks," *Sov. Appl. Mech. (Engl. transl.), 24* (5), 852–864, 1960.

Bear, J., "On the Tensor Form of Dispersion in Porous Media," *J. Geophys. Research, 66*(4), 1185–1197, 1961.

Bear, J. and M. Jacobs, "On the Movement of Water Bodies Injected into Aquifers", *Journal of Hydrology, 3,* 37–57, 1965.

Bear, J., *Dynamics of Fluids in Porous Media,* Elsevier, New York, 1972.

Bear, J., *Hydraulics of Groundwater,* McGraw-Hill, New York, 1979.

Biggar, J. W. and D. R. Nielsen, "Diffusion Effects in Miscible Displacement Occurring in Saturated and Unsaturated Porous Materials," *J. Geophys. Res., 65,* 2885–2895, 1960.

Biggar, J. W. and D. R. Nielsen, "Miscible Displacement, 2, Behavior of Tracers," *Soil Sci. Soc. Am. Proc., 26,* 125–128, 1962.

Biggar, J. W. and D. R. Nielsen, "Spatial Variability of the Leaching Characteristics of a Field Soil," *Water Resour. Res., 12*(1), 78–84, 1976.

Bird, R. B. Stewart, W. E., and E. N. Lightfoot, *Transport Phenomena*, John Wiley and Sons, Inc., Madison, WI, 1960.

Borden, R. C. and P. B. Bedient, "Transport of Dissolved Hydrocarbons Influenced by Oxygen-Limited Biodegradation, 1. Theoretical Development," *Water Resour. Res., 22*(13), 1973–82, 1986.

Borden, R. C. and P. B. Bedient, "In Situ Measurement of Adsorption and Biotransformation at a Hazardous Waste Site," *Water Resources Bulletin, 23*(4), 629–636, 1987.

Borden, R. C., Bedient, P. B., Lee, M. D., Ward, C. H., and J. T. Wilson, "Transport of Dissolved Hydrocarbons Influenced by Oxygen-Limited Biodegradation, 2. Field Applications," *Water Resour. Res., 22*(13), 1983–90, 1986.

Boulton, N. S., "Analysis of Data from Nonequilibrium Pumping Tests Allowing for Delayed Yield from Storage," *Proc. Inst. Civil Engrs.* , 26, 469–482, 1963.

Bouwer, H., "Rapid Field Measurements of Air Entry Value and Hydraulic Conductivity of Soil as Significant Parameters in Flow System Analysis," *Water Resour. Res., 2*, 729–738, 1966.

Bouwer, H., *Groundwater Hydrology*, McGraw-Hill, New York, 1978.

Bouwer, H. and R. D. Jackson, "Determining Soil Properties," in *Drainage For Agriculture*, J. V. Schilfgaarde (Ed.), American Society of Agronomy, Agronomy No. 17, Madison, WI, 1974.

Bouwer, H. and R. C. Rice, "A Slug Test for Determining Hydraulic Conductivity of Unconfined Aquifers with Completely or Partially Penetrating Wells," *Water Resour. Res., 12*(3), 423–428, 1978.

Bradbury, K. R., and M. A. Muldoon, Hydraulic conductivity determinations in unlithified glacial and fluvial materials, *Ground Water and Vadose Zone Monitoring, STP 1053,* D. M. Nielsen and A. I. Johnson (Eds.), American Society for Testing and Materials, Philadelphia, PA, 138–151, 1990.

Brady, N. C., *The Nature and Properties of Soils*, 8th Ed., Macmillan, New York, 1974.

Brakensiek, D. L., "Estimating the Effective Capillary Pressure in the Green and Ampt Infiltration Equation," *Water Resour. Res. 13*(3), 680–682, 1977.

Bredehoeft, J. D. and R. L. Cooley, "Comment on 'A Note on the Meaning of Storage Coefficient' by T. N. Narasimhan and B. Y. Kanehiro," *Water Resour. Res., 19*(6), 1632–1634, 1983.

Bresler, E., NcNeal, B. L., and D. L. Carter, *Saline and Sodic Soils*, Springer-Verlag, Berlin, Germany, 1982.

Bresler, E. and G. Dagan, "Solute Dispersion in Unsaturated Heterogeneous Soil at Field Scale: II. Applications," *Soil Sci. Soc. Am. J., 43*, 467–472, 1979.

Bresler, E. and G. Dagan, "Convective and Pore Scale Dispersive Solute Transport in Unsaturated Heterogeneous Fields," *Water Resour. Res., 17*(6), 1683–1693, 1981.

Brigham, W. E., "Mixing Equations in Short Laboratory Columns," *Soc. Pet. Eng. J., 14,* 91–99, 1974.

Brigham, W. E., P. W. Reed, and J. N. Dew, *Soc. Pet. Eng. J., 1,* 1, 1961.

Brooks, R. H. and A. T. Corey. "Hydraulic Properties of Porous Media," Hydrology Paper 3., Colorado State University, Fort Collins, CO, 1964.

Brooks, R. H. and A. T. Corey, "Properties of Porous Media Affecting Fluid Flow," *J. Irrigation and Drainage Div., A. S. C. E., IR2,* 61–88, 1966.

Brown, D. S. and E. W. Flagg, "Empirical Prediction of Organic Pollutant Sorption in Natural Sediments," *J. Environ. Qual., 10*(3), 382–386, 1981.

Bruce, R. R., and R. J. Luxmoore, "Water Retention: Field Methods," in *Methods of Soil Analysis, Part 1,* A. Klute (EEd.), American Society of Agronomy, Madison, WI, 663–686, 1986.

Brusseau, M. L., "Transport of Reactive Contaminants in Heterogeneous Porous Media," *Reviews of Geophysics, 32*(3), 285–313, 1994.

Brutsaert, W., "Some Methods of Calculating Unsaturated Permeability," *Trans. of the Amer. Soc. of Agricultural Engineers, 10,* 400–404, 1967.

Brutsaert, W. F., E. A. Breitenbach, and D. K. Sunada, "Computer Analysis of Free-Surface Well Flow," *J. Irrig. Drain. Div., Proc. American Society of Civil Engineers, 97,* 405–420, 1971.

Buckingham, E., "Studies on the Movement of Soil Moisture," Bulletin 38, United States Department of Agriculture Bureau of Soils, Washington D. C., 1907.

Burdine, N. T., "Relative Permeability Calculations from Pore-Size Data," *Trans. A. I. M. E. 198,* 71–77, 1953.

Burkholder, H. C. and E. L. J. Rosinger, "A Model for the Transport of Radionuclides and Their Decay Products through Geologic Media," *Nuclear Technology, 49,* 150–158, 1980.

Campbell, G. S., "A Simple Method for Determining Unsaturated Conductivity from Moisture Retention Data," *Soil Sci., 117*(6), 311–314, 1974.

Campbell, G. S., and G. W. Gee, "Water Potential: Miscellaneous Methods," in *Methods of Soil Analysis, Part 1,* A. Klute (Ed.), American Society of Agronomy, Madison, WI, 619–633, 1986.

Carman, P. C., "Fluid Flow through Granular Beds," *Trans. Inst. Chem. Eng. London,* England, *15,* 150–166, 1937.

Carman, P. C., *Flow of Gases through Porous Media,* Butterworths, London, England, 1956.

Carsel, R. F. and R. S. Parrish, "Developing Joint Probability Distributions of Soil Water Retention Characteristics," *Water Resour. Res., 24*(5), 755–769, 1988.

Carslaw, H. W. and J. C. Jaeger, *Conduction of Heat in Solids,* 2nd Ed., Oxford University Press (Clarendon), London, England, 1959.

Cary, J. W., Simmons, C. S., and J. F. McBride, 1989. "Permeability of Air and Immiscible Organic Liquids in Porous Media. " *Water Resources Bulletin,* *25*(6), 1205–1216.

Cassell, D. K., and A. Klute. "Water Potential: Tensiometry," in *Methods of Soil Analysis, Part 1*, Second Edition, A. Klute (Ed.), American Society of Agronomy, Madison WI, p. 563–596, 1986.

Charbeneau, R. J., "Groundwater Contaminant Transport with Adsorption and Ion Exchange Chemistry: Method of Characteristics for the Case without Dispersion," *Water Resourc. Res., 17*(3), 705–713, 1981.

Charbeneau, R. J., "Calculation of Pollutant Removal Rates during Groundwater Restoration with Adsorption and Ion Exchange," *Water Resourc. Res., 18*(4), 1117–1125, 1982.

Charbeneau, R. J., "Kinematic Models for Soil Moisture and Solute Transport," *Water Resour. Res., 20,* 699–706, 1984.

Charbeneau, R. J., "Groundwater Restoration with In Situ Uranium Leach Mining," in *Groundwater Contamination*, National Academy Press, Washington, D. C, 1984.

Charbeneau, R. J., "Multicomponent Exchange and Subsurface Solute Transport: Characteristics, Coherence, and the Riemann Problem," *Water Resour. Res., 24*(1), 57–64, 1988.

Charbeneau, R. J., "Liquid Moisture Redistribution: Hydrologic Simulation and Spatial Variability," in *Unsaturated Flow in Hydrologic Modeling: Theory and Practice,* H. J. Morel-Seytoux, Ed., NATO ASI Series C: Mathematical and Physical Sciences, Vol. 275, 1989.

Charbeneau, R. J. and S. G. Wright, "Hydrologic Site Selection for Mining of Deep Basin Texas Lignite," Center for Research in Water Resources Technical Report CRWR–198, The University of Texas at Austin, December 1983.

Charbeneau, R. J., Weaver, J. W. and V. J. Smith, "Kinematic Modeling of Multiphase Solute Transport in the Vadose Zone," R. S. Kerr Environmental Research Laboratory, United States Environmental Protection Agency, EPA/600/2-89/035, 1989.

Charbeneau R. J., N. Wanakule, C. Y. Chiang, J. P. Nevin and C. L. Klein, "A Two-Layer Model to Simulate Floating Free Product Recovery: Formulation and Applications," The Proceedings of the Petroleum Hydrocarbon and Organic Chemicals in Ground Water. p. 333–346, 1989.

Charbeneau, R. J. and R. G. Asgian, "Simulation of the Transient Soil Water Content Profile for a Homogeneous Bare Soil," *Water Resour. Res. 27*(6), 1271–1279, 1991.

Charbeneau, R. J. and D. E. Daniel, "Contaminant Transport in Unsaturated Flow," Chapter 15 in *Handbook of Hydrology,* D. R. Maidment (Ed.), McGraw-Hill Inc., New York, 1993.

Charbeneau, R. J., Weaver, J. W. and B. K. Lien, "The Hydrocarbon Spill Screening Model (HSSM), Volume 2: Theoretical background and source codes," Robert S. Kerr Environmental Research Laboratory, United States Environmental Protection Agency, EPA/600/R–94/039 b, April 1995.

Charbeneau, R. J. and C. Y. Chiang, "Estimation of Free-Hydrocarbon Recovery from Dual-Pump Systems," *Ground Water, 33*(4), 627–634, 1995.

Charbeneau, R. J., Johns, R. T., Lake, L. W., and M. J. McAdams III, Free Product Recovery of Petroleum Hydrocarbon Liquids, American Petroleum Institute, in press (1999).

Chatzis, I., Morrow, N. R. and H. T. Lim, "Magnitude and Detailed Structure of Residual Oil Saturation," *Soc. Petrol. Engrs. of AIME, 23*(2), 311–326, 1983.

Chiang C. Y., J. P. Nevin, and R. J. Charbeneau, "Optimal Free Hydrocarbon Recovery from a Single Pumping Well," The Proceedings of Petroleum Petroleum Hydrocarbons and Organic Chemicals in Ground Water: Prevention, Detection, and Restoration, Houston, Texas. National Water Well Association, 1990.

Childs, E. C. and N. Collis-George, "The Permeability of Porous Materials," *Proc. Roy. Soc. London Ser. A., 201*, 392–405, 1950

Chow, V. T., Maidment, D. R. and L. W. Mays, *Applied Hydrology,* McGraw-Hill, New York, 1988.

Clapp, R. B. and G. M. Hornberger, "Empirical Equations for Some Soil Hydraulic Properties," *Water Resourc. Res., 14*(4), 601–604, 1978.

Cline, P. V., Delfino, J. J. and P. S. C. Rao, "Partitioning of Aromatic Constituents into Water from Gasoline and Other Complex Solvent Mixtures," *Environmental Science and Technology, 23*, 914–920, 1991.

Clothier, B. E., and I. White, "Measurement of Sorptivity and Soil Water Diffusivity in the Field," *Soil Sci. Soc. Am. J., 45*(2), 241–245, 1981.

Clothier, B. E., and K. R. J. Smettem, "Combining Laboratory and Field Measurements to Define the Hydraulic Properties of Soil," *Soil Science Society of America Journal,* 54(2), p. 299–304, 1990.

Cohen, R. M. and J. W. Mercer, *DNAPL Site Evaluation,* C. K. Smoley/CRC Press, Inc., Boca Raton, FL, 1993.

Coleman, J. D., and A. D. Marsh, "An Investigation of the Pressure Membrane Method for Measuring the Suction Properties of Soil," *Journal of Soil Science, 12,* p. 343–360, 1961.

Cooley, R. L., "A Finite-Difference Method for Unsteady Flow in Variably Saturated Porous Media: Application to a Single Pumping Well," *Water Resour. Res., 7,* 1607–1625, 1971.

Cooper, H. H., Jr., and C. E. Jacob, "A Generalized Graphical Method for Evaluating Formation Constants and Summarizing Well Field History," *Trans. Amer. Geophys. Union, 27,* p. 526–534, 1946.

Cooper, H. H., Jr., Bredehoeft, J. D., and I. S. Papadopulos, "Response of a Finite-Diameter Well to an Instantaneous Charge of Water," *Water Resour. Res., 3*, p. 263–269, 1967.

Corapcioglu, M. Y. and A. L. Baehr, "A Compositional Multiphase Model for Groundwater Contamination by Petroleum Products, 1. Theoretical Considerations," *Water Resour. Res., 23*(1), 191–200, 1987.

Corey, A. T., *Mechanics of Heterogeneous Fluids in Porous Media*, Water Resources Publications, Fort Collins, CO, 1977.

Courant, R., *Differential and Integral Calculus, Volume II,* Wiley-Interscience, 1936.

Crank, J., *Mathematics of Diffusion,* 2nd Ed., Oxford University Press, London, 1975.

Dagan, G. and E. Bresler, "Solute Dispersion in Unsaturated Heterogeneous Soil at Field Scale: I. Theory," *Soil Sci. Soc. Am. J., 43*, 461–467, 1979.

Dagan, G. and E. Bresler, "Unsaturated Flow in Spatially Variable Fields, Parts 1–3," *Water Resour. Res., 19*(2), 413–435, 1983.

Dagan, G., *Flow and Transport in Porous Formations*, Springer-Verlag, Berlin, 1989.

Danckwerts, P. V., "Continuous Flow Systems: Distributions of Residence Times," *Chem. Eng. Sci., 2*(1), 1–13, 1953.

Daniel, D. E., "Permeability Test for Unsaturated Soil," *Geotech. Testing J., 6*(2), 81–86, 1983.

Daniel, D. E., S. J. Trautwein, S. S. Boynton, and D. E. Foreman, "Permeability Testing with Flexible-Wall Permeameters," *Geotechnical Testing Journal, 7*(3), 113–122, 1984.

Davis, S. N. and R. J. M. De Wiest, *Hydrogeology*, John Wiley, New York, 1966.

De Finetti, B., *Theory of Probability*, John Wiley & Sons Ltd., Chichester, 1974. (Also available in Wiley Classics Library, 1990.)

De Glee, G. L., "Over grondwaterstromingen bij wateronttrekking door middel van putten." Thesis. J. Waltman, Delft (The Netherlands), 175 p., 1930.

De Jong, G. de J., "Longitudinal and Transverse Diffusion in Granular Deposits," *Trans. Am. Geophys. Union, 39*, 67–74, 1958.

Demond, A. H. and P. V. Roberts, "Effect of Interfacial Forces on Two-Phase Capillary Pressure-Saturation Relations," *Water Resour. Res., 27*(3), 423–437, 1991.

DeVault, D., "The Theory of Chromatography," *J. Am. Chem. Soc., 65*, 532–540, 1943.

Domenico, P. A. and F. W. Schwartz, *Physical and Chemical Hydrogeology*, John Wiley & Sons, New York, 1990.

Dracos, T., "Theoretical Considerations and Practical Implications on the Infiltration of Hydrocarbons in Aquifers," paper presented at the International Symposium on Groundwater Pollution by Oil Hydrocarbons, *Int. Assoc. of Hydrol.*, Prague, June 5–9, 1978.

Driscoll, F. G., *Groundwater and Wells,* Johnson Division, St. Paul, MN, 1986.

Dullien, F. A. L., *Porous Media, Fluid Transport and Pore Structure*, Academic Press, New York, Second Edition, 1992.

Dupont, R. R. and J. A. Reineman, "Evaluation of Volatilization of Hazardous Constituents at Hazardous Waste Land Treatment Sites," NTIS P B86–233939, 1986.

El-Kadi, A. I., "Applicability of Sharp-Interface Models for NAPL Transport: 1, Infiltration," *Ground Water, 30*, 849–856, 1992.

Elzeftawy, A., and R. S. Mansell, "Hydraulic Conductivity Calculations for Unsaturated Steady State and Transient-State Flow in Sand," *Soil Sci. Soc. Am. Proc., 39*(4), 599–603, 1975.

Everett, L. G., "Soil Pore-Liquid Monitoring," in *Subsurface Migration of Hazardous Wastes,* J. S. Devinny, L. G. Everett, J. C. S. Lu, and R. L. Stollan (Eds.), Van Nostrand Reinhold, New York, 306–336, 1990.

Everett, L. G., McMillion, L. G., and L. A. Eccles, "Suction Lysimeter Operation at Hazardous Waste Sites," in *Ground Water Contamination: Field Methods, ASTM STP 963,* A. G. Collins and A. I. Johnson (Eds.), American Society for Testing and Materials, Philadelphia, 304–327, 1988.

Fair, G. M. and L. P. Hatch, "Fundamental Factors Governing the Streamline Flow of Water through Sand," *J. Amer. Water Works Ass., 25*, 1551–1565, 1933.

Farmer, W. J., M. S. Yang, J. Letey, and W. F. Spencer, "Hexachlorobenzene: Its Vapor Pressure and Vapor Phase Diffusion in Soil," *Soil Sci. Soc. Amer. J., 44*, 676–680, 1980.

Farr, A. M., R. J. Houghtalen, and D. B. McWhorter, "Volume Estimation of Light Nonaqueous Phase Liquids in Porous Media," *Ground Water, Vol. 28*, No. 1, p. 48–56, 1990.

Faust, C. R., "Transport of Immiscible Fluids within and Below the Unsaturated Zone—a Numerical Model," *Water Resourc. Res., 21,* 587–596, 1985.

Ferris, J. G., "Cyclic Fluctuations of Water Levels as a Basis for Determining Aquifer Transmissivity," *Int. Assoc. Sci. Hydrol. Pub., 33*, 148–155, 1951.

Ferris, J. G., Knowles, D. B., Brown, R. H. and R. W. Stallman, "Theory of Aquifer Tests," Geological Survey Water-Supply Paper 1536–E, 1962.

Fetter, C. W., *Applied Hydrogeology*, 3rd Ed., MacMillan, New York, 1994.

Forchheimer, P. "Wasserbewegung dur Bodem, *Z. Ver. Deutsch." Ing., 45*, 1782–1788, 1901.

Freeze, R. A. and J. A. Cherry, *Groundwater*, Prentice-Hall, Englewood Cliffs, 1979.

Freyberg, D. L., "A Natural Gradient Experiment on Solute Transport in a Sand Aquifer, 2. Spatial Movements and the Advection and Dispersion of Nonreactive Tracers," *Water Resourc. Res., 22*(13), 2031–2047, 1986.

Fried, J. J. and M. A. Combarnous, "Dispersion in Porous Media," in *Advances in Hydroscience* edited by V. T. Chow, Vol. 7, 169–282, 1971.

Galya, D. P., "A Horizonatal Plane Source Model for Groundwater Transport," *Ground Water, 25*(6), 1987.

Garder, A. O., Jr., Peaceman, D. W. and A. L. Pozzi, Jr., "Numerical Calculation of Multidimensional Miscible Displacement by the Method of Characteristics," *Soc. Pet. Eng. J., 4*, 26–36, 1964.

Gardner, W. H., "Water Content," in *Methods of Soil Analysis. Part 1*, Klute, A. (Ed.), Monograph 9, American Society of Agronomy, Madison, WI, 1986.

Gardner, W. R., "Calculation of Capillary Conductivity from Pressure Plate Outflow Data," *Soil Sci. Soc. Am. Proc., 20*, 317–320, 1956.

Gardner, W. R., "Some Steady State Solutions of the Unsaturated Moisture Flow Equation with Application to Evaporation from a Water Table," *Soil Sci., 85*, (4), 1958.

Gardner, W. R., "Solution of the Flow Equation for the Drying of Soils and Other Porous Media," *Soil Sci. Soc. Amer. Proc., 23*, 183–187, 1959.

Gardner, W. R., Hillel, D. and Y. Benyamini, "Post-Irrigation Movement of Soil Water 2. Simultaneous Redistribution and Evaporation," *Water Resour. Res., 6*, 1148–1153, 1970.

Gee, G. W. and J. W. Bauder, "Particle-Size Analysis," in *Methods of Soil Analysis, Part I*, A. Klute (Ed.), Monograph 9, American Society of Agronomy, Madison, WI, p. 383–412, 1986.

Gee, G. W., Wierenga, P. J., Andraski, B. J., Young, M. H., Fayer, M. J. and M. L. Rockhold, "Variations in Water Balance and Recharge Potential at Three Western Desert Sites," *Soil Sci. Soc. Am. J., 58*, 63–72, 1994.

Gelhar, L. W. and M. A. Collins, "General Analysis of Longitudinal Dispersion in Nonuniform Flow," *Water Resour. Res., 7*(6), 1511–1521, 1971.

Gelhar, L. W., C. Welty and K. R. Rehfeldt, "A Critical Review of Data on Field-Scale Dispersion in Aquifers," *Water Resour. Res., 28*(7), 1955–1974, 1992.

Gelhar, L. W., *Stochastic Subsurface Hydrology*, Prentice Hall, Englewood Cliffs, NJ, 1993.

Ghodrati, M. and W. A. Jury, 1990. "A Field Study Using Dyes to Characterize Preferential Flow of Water," *Soil Sci. Soc. Am. J. 54*, 1558.

Girinskii, N. K., "Generalization of Some Solutions for Wells to More Complicated Natural Conditions," (in Russian), *Dokl. A. N. United States S. R. No. 3, 54*, 1946.

Glotfelty, D. E. and C. J. Schomburg, "Volatilization of Pesticides from Soil," in *Reactions and Movement of Organic Chemicals in Soils*, Sawhney, B. L. and K. Brown (Eds.), SSSA Special Publication Number 22, Soil Science Society of America, Madison, WI, 1989.

Goode, D. J. and L. F. Konikow, "Modification of a Method-of-Characteristics Solute-Transport Model to Incorporate Decay and Equilibrium-Controlled Sorption or Ion Exchange," United States Geological Survey Water-Resources Investigations Report 89–4030, p. 65, 1989.

Green, W. A. and G. A. Ampt, "Studies on Soil Physics, 1. The Flow of Air and Water through Soils," *J. Agr. Sci., 4*, 1–24, 1911.

Green, R. E., and J. C. Corey, "Calculation of Hydraulic Conductivity: A Further Evaluation of Some Predictive Methods," *Soil Sci. Soc. Am. Proc.,* 35(1), 3–8, 1971.

Green, R. E., Ahuja, L. R., and S. K. Chong, "Hydraulic Conductivity, Diffusivity, and Sorbivity of Unsaturated Soils: Field Methods," in *Methods of Soil Analysis, Part 1, Physical and Mineralogical Methods,* (2nd Ed.), A. Klute (Ed.), ASA Monograph 9, American Society of Ayronom, Madison, WI, p. 799–823, 1986.

Grove, B. and W. A. Beetem, "Porosity and Dispersion Calculation for Fractured Carbonate Aquifer, Using the 2-Well Tracer Method," *Water Resour. Res.,* 7(1), 128–134, 1971.

Gupta, S. C., and W. E. Larson, "Estimating Soil Water Retention Characteristics from Particle Size Distribution, Organic Matter Percent, and Bulk Density," *Water Resour. Res., 15*, 1633–1635, 1979.

Guven, O., R. W. Falta, F. J. Molz, and J. G. Melville, "Analysis and Interpretation of Single-Well Tracer Tests in Stratified Aquifers," *Water Resour. Res., 21*(5), 676–684, 1985.

Hagan, G., *Poggendorff's Annalen. d. Physik u. Chemie (2), 46*, 423, 1839.

Hamaker, J. W., "Diffusion and Volatilization," in *Organic Chemicals in the Soil Environment, Vol. 1*, Goring, C. A. I. and J. W. Hamaker (Eds.), Marcel Dekker, New York, 1972.

Hantush, M. S. "Analysis of Data from Pumping Tests in Leaky Aquifers," *Trans. Amer. Geophys. Union,* 37, 702–714, 1956.

Harbaugh, A. W. and M. G. McDonald, "User's Documentation for MODFLOW−96, an Update to the United States Geological Survey Modular Finite-Difference Groundwater Flow Model," United States Geological Survey, Open-File Report 96–485, 1996.

Harleman, D. R. F., P. F. Mehlhorn, and R. R. Rumer, Jr., "Dispersion-Permeability Correlation in Porous Media," *J. Hydraul. Div., ASCE, HY(2)*, 67–85, 1963.

Harleman, D. R. F. and R. R. Rumer, Jr., *J. Fluid Mech., 16*, 385, 1973.

Harpaz, Y. and J. Bear, "Investigations on Mixing of Waters in Underground Storage Operations," *Intern. Assoc. Scientific Hydrology, 64,* 132–153, 1963.

Harr, M. E., *Groundwater and Seepage*, McGraw-Hill, New York, 1962.

Hauber, W. C., "Prediction of Waterflood Performance for Arbitrary Well Patterns and Mobility Ratios," *J. Petrol. Tech.,* 95–103, 1964.

Hemstreet, T. J., "Graphical Determination of Site-Specific Dilution Attenuation Factors for the Soil to Groundwater Pathway," M. S. Thesis in Engineering, University of Texas at Austin, May 1996.

Herzberg, A., 1901. "Die Wasserversorgung einiger Nordseebader" (The Water Supply on Parts of the North Sea Coast, in German), *J. Gasbeleucht. Wasserversorg., 44*, 815–819, 842–844.

Herzog, B. L., and W. J. Morse, "Comparison of Slug Test Methodologies for Determination of Hydraulic Conductivity in Fine-Grained Sediments," in *Ground Water and Vadose Zone Monitoring, ASTM STP 1053,* D. M. Nielsen and A. I. Johnson (Eds.), American Society for Testing and Materials, Philadelphia, p. 152–164, 1990.

Hillel, D., *Fundamentals of Soil Physics*, Academic Press, San Diego, 1980.

Hillel, D., *Introduction to Soil Physics*, Academic Press, Orlando, 1982.

Hochmuth, D. P. and D. K. Sunada, "Groundwater Model of Two-Phase Immiscible Flow in Coarse Materials," *Ground Water, 23,* 617–626, 1985.

Holder, M., Brown, K. W., Thomas, J. C., Zabcik, D., and H. E. Murray, "Capillary-Wick Unsaturated Zone Pore Water Sampler," *Soil Science Society of America Journal,* 55(5), 1195–1202, 1991.

Horner, D. R., "Pressure Buildup in Wells," *Third World Petrol. Congress, Proc., Sect. II*, E. J. Brill (Ed.), Leiden, Holland, 503–521, 1951.

Horton, R. E., "An Approach Towards a Physical Interpretation of Infiltration Capacity," *Soil Sci. Soc. Am. Proc. 5*, 399–417, 1940.

Howard, P. H., Boethling, R. S., Jarvis, W. F., Meylan, W. M., and E. M. Michalenko, *Handbook of Environmental Degradation Rates*, Lewis Publishers, Inc., Chelsea, MI, 1991.

Hubbert, M. K., "The Theory of Ground Water Motion," *J. Geol., 48*, 785–944, 1940.

Hubbert, M. K., "Entrapment of Petroleum under Hydrodynamic Conditions," *Bull. Am. Assoc. Petroleum Geologists, 37*, 1954–2026, 1954.

Hubbert, M. K., "Darcy's Law and the Field Equations of the Flow of Underground Fluids," *Petroleum Transactions, AIME 207*, 222–239, 1956.

Huisman, L., *Groundwater Recovery*, Macmillan, London, 1972.

Huisman, L. and T. N. Olsthoorn, *Artificial Groundwater Recharge*, Pitman Publishing Inc., Boston, MA, 1983.

Huling, S. G. and J. W. Weaver, "Dense Nonaqueous Phase Liquids," *Ground Water Issue*, United States Environmental Protection Agency, Office of Research and Development, EPA/540/4–91–002, 1991.

Hunt, B., "Dispersive Sources in Uniform Groundwater Flow," *J. Hydraulics Division, ASCE, Vol. 104*, No. HY1, 75–85, 1978.

Hunt, J. R., N. Sitar and K. S. Udell, "Nonaqueous Phase Liquid Transport and Cleanup, 1. Analysis of Mechanisms," *Water Resour. Res, 24*(8), 1247–1258, 1988.

Huyakorn, P. S. and G. F. Pinder, "New Finite Element Technique for the Solution of Two-Phase Flow through Porous Media," *Adv. Water Res., 1*, 285–298, 1978.

Huyakorn, P. S., M. J. Ungs, L. A. Mulkey and E. A. Sudicky, "A Three-Dimensional Analytical Model for Predicting Leachate Migration," *Ground Water, 25*(5), 588–598, 1982.

Hvorslev, M. J., "Time Lag and Soil Permeability in Groundwater Observations," Bulletin No. 36, Waterways Experiment Station, Corps of Engineers, United States Navy, Vicksburg, MS, 1951.

Irmay, S., "Multiple-Well Systems in Layered Soils," in *Groundwater Hydraulics*, Rosenshein and Bennett (Ed.), American Geophysical Union, Water Resources Monograph 9, Washington, D. C., 1984.

Jacob, C. E., "On the Flow of Water in an Elastic Artesian Aquifer," *Trans. Am. Geophys. Un., pt. 2,* 574–586, 1940.

Jacob, C. E., "Drawdown Test to Determine Effective Radius of Artesian Wells," *Trans. Am. Soc. Civil Engrs., 112,* 1047–1070, 1947.

Jacob, C. E., "Flow of Groundwater," in *Engineering Hydraulics* (H. Rouse, Ed.), Chapter 5, 321–386, Wiley, New York, 1950.

Jasper, J. J., "The Surface Tension of Pure Liquid Compounds," *J. Phys. Chem. Ref. Data. v. 1*, No. 4, 1972.

Javandel, I., C. Doughty, and C. F. Tsang, *Groundwater Transport: Handbook of Mathematical Models*, American Geophysical Union, Water Resources Monograph 10, Washington, D. C., 1984.

Javandel, I. and C. F. Tsang, "Capture-Zone Type Curves: a Tool for Aquifer Cleanup," *Ground Water, 24*(5), 616–625, 1986.

Jennings, A. A. and D. J. Kirkner, "Criteria for Selecting Equilibrium or Kinetic Sorption Descriptions in Groundwater Quality Models," in *Frontiers in Hydraulic Engineering*, Shen, H. T. (Ed.), American Society of Civil Engineers, New York, 1983.

Johnson, P. A. and A. L. Babb, "Liquid Diffusion in Non-Electrolytes," *Chem. Revs., 56*, 387–453, 1956.

Johnson, P. C., Stanley, C. C., Kemblowski, M. W., Byers, D. L. and J. D. Colthart, "A Practical Approach to the Design, Operation, and Monitoring of In Situ Soil-Venting Systems," *Ground Water Monitoring Review, 10*(2), 159–178, 1990.

Johnson, P. C., Abranovic, D., Charbeneau, R. J. and T. Hemstreet, "Graphical Approach for Determining Site-Specific Dilution-Attenuation Factors (DAFs)," Health and Environmental Sciences Department, American Petroleum Institute, API Publication Number 4659, February 1998.

Jury, W. A., "Simulation of Solute Transport Using a Transfer Function Model," *Water Resour. Res., 18*(2), 363–368, 1982.

Jury, W. A., "Volatilization from Soil," in *Vadose Zone Modeling of Organic Pollutants*, Hern, S. C. and S. M. Melancon (Eds.), Lewis Publishers, Inc., Chelsea, Michigan, 1986.

Jury, W. A., R. Grover, W. F. Spencer, and W. F. Farmer, "Modeling Vapor Losses of Soil-Incorporated Triallate," *Soil Sci. Soc. Amer. J., 44*, 445–450, 1980.

Jury, W. A., D. Russo, G. Sposito, and H. Elabd, "The Spatial Variability of Water and Solute Transport Properties in Unsaturated Soil: I. Analysis of Property Variation and Spatial Structure with Statistical Models, and II. Scaling Models of Water Transport," *Hilgardia, 55*(4), 1–32 and 33–56, 1987.

Jury, W. A., W. F. Spencer, and W. J. Farmer, "Behavior Assessment Model for Trace Organics in Soil: I. Model Description," *J. Environ. Qual., 12*(4), 558–564, 1983.

Jury, W. A., W. J. Farmer, and W. F. Spencer, Behavior Assessment Model for Trace Organics in Soil. II. Chemical Classification Parameter Sensitivity," *J. Environ. Qual. 13*: 567–572, 1984.

Jury, W. A. and G. Sposito, "Field Calibration and Validation of Solute Transport Models for the Unsaturated Zone," *Soil Sci. Soc. Am. J., 49*(6), 1331–1341, 1985.

Jury, W. A., G. Sposito, and R. E. White, "A Transfer Function Model of Solute Transport through Soil, 1. Fundamental Concepts," *Water Resour. Res., 22*(2), 243–247, 1986.

Jury, W. A., L. H. Stolzy, and P. Shouse, "A Field Test of the Transfer Function Model for Predicting Solute Transport," *Water Resour. Res., 19*(2), 369–375, 1982.

Jury, W. A., W. R. Gardner, and W. H. Gardner, *Soil Physics*, 5th Ed., Wiley, New York, 1991.

Kalurrachhi, J. and J. C. Parker, "An Efficient Finite Element Method for Modeling Multiphase Flow in Porous Media," *Water Resourc. Res., 25*, 43–54, 1989.

Kaluarachhi, J., J. C. Parker, and R. J. Lenhard, "A Numerical Model for Areal Migration of Water and Light Hydrocarbon in Unconfined Aquifers," *Adv. Water Res., 13*, 29–40, 1990.

Karickhoff, S. W., "Sorption Kinetics of Hydrophobic Pollutants in Natural Sediments," *Contaminants and Sediments, Vol. 2* (Ed. R. A. Baker), Ann Arbor Science Publishers, Inc., Ann Arbor, MI, 193–205, 1980.

Karickhoff, S. W., "Semi-Empirical Estimation of Sorption of Hydrophobic Pollutants on Natural Sediments and Soils," *Chemosphere, 10*(8), 833–846, 1981.

Karickhoff, S. W., "Organic Pollutant Sorption in Aquatic Systems," *J. Hydr. Engrg., ASCE, 110*(6), 707–735, 1984.

Karickhoff, S. W., D. S. Brown, and T. A. Scott, "Sorption of Hydrophobic Pollutants on Natural Sediments," *Water Research, 13*, 241–248, 1979.

Kemblowski, M. W. and C. Y. Chiang, "Hydrocarbon Thickness Fluctuations in Monitoring Wells," *Ground Water, 28*, 244–252, 1990.

Kemper, W. D. and J. C. van Schaik, "Diffusion of Salt in Clay-Water Systems," *Soil Sci. Soc. Am. Proc., 30*(5), 535–540, 1966.

Kinniburgh, D. G., "General Purpose Adsorption Isotherms," *Environ. Sci. Technol., 20*(9), 895–904, 1986.

Klute, A., "Water Retention: Laboratory Methods," in *Methods of Soil Analysis, Part 1,* A. Klute (Ed.), American Society of Agronomy, Madison, WI, 635–686, 1986.

Klute, A., and C. Dirksen, "Hydraulic Conductivity and Diffusivity: Laboratory Methods," in *Methods of Soil Analysis, Part 1,* A. Klute (Ed.), American Society of Agronomy, Madison, WI, 687–734, 1986.

Knisel, W. G. "CREAMS: A Field-Scale Model for Chemicals, Runoff, and Erosion from Agricultural Manangement Systems," United States Department of Agriculture, Conservation Research Report No. 26, 1980.

Konikow, L. F. and J. D. Bredehoeft, "Computer Model of 2-Dimensional Solute Transport and Dispersion in Groundwater," in *USGS Techniques of Water Resources Investigations, Book 7, Chap. C2*, United States Geological Survey, Washington, DC, 1978.

Konikow, L. F., Granato, G. E. and G. Z. Hornberger, "User's Guide to Revised Method-of-Characteristics Solute-Transport Model (MOC-Version 3. 1)," United States Geological Survey Water-Resources Investigations Report 94–4115, 1994.

Konikow, L. F., Goode, D. J. and G. Z. Hornberger, "A Three-Dimensional Method-of-Characteristics Solute-Transport Model (MOC3D)," United States Geological Survey, Water-Resources Investigations Report 96–4267, 1996.

Kostiakov, A. N., "On the Dynamics of the Coefficient of Water Percolation in Soils and on the Necessity of Studying it from a Dynamic Point of View for Purposes of Amelioration," *Trans. Com. Int. Soc. Soil Sci., 6th Moscow, A*, 17–21, 1932.

Kroszynski, U. I. and G. Dagan, "Well Pumping in Unconfined Aquifers: The Influence of the Unsaturated Zone," *Water Resour. Res., 11*, 479–490, 1975.

Kruseman, G. P. and N. A. de Ridder, *Analysis and Evaluation of Pumping Test Data*, 2nd Edition, Intern. Inst. Land Reclamation and Improvement, Publication 47, Wageningen, The Netherlands, 1991.

Kueper, B. H. and D. B. McWhorter, "The Behavior of Dense, Nonaqueous Phase Liquids in Fractured Clay and Rock," *Ground Water*, 29(5), 716–728, 1991.

Kuhn, E. P. and J. M. Suflita, "Dehalogenation of Pesticides by Anaerobic Microorganisms in Soils and Groundwater—A Review," in *Reactions and Movement of Organic Chemicals in Soils*, Sawhney, B. L. and K. Brown (Eds.), SSSA Special Publication Number 22, Soil Science Society of America, Madison, WI, 1989.

Kung, K-J. S., "Preferential Flow in a Sandy Vadose Zone: 1. Field Observation," *Geoderma,* 46: 51–58, 1990.

Kuppusamy, T., J. Sheng, J. C. Parker, and R. J. Lenhard, "Finite-Element Analysis of Multiphase Immiscible Flow through Soils," *Water Resourc. Res., 23,* 625–631, 1987.

Landau, L. D. and E. M. Lifshitz, *Fluid Mechanics*, Pergamon Press, Oxford, 1959.

Lehr, J. H., "How Much Ground Water Have We Really Polluted?" (editorial), *Ground Water Monitoring Rev.* (Winter), 4–5, 1982.

Lenhard, R. J. and J. C. Parker. 1990. "Estimation of Free Hydrocarbon Volume from Fluid Levels in Monitoring Wells. " Ground Water. Vol. 28, No. 1, p. 57–67.

Lenhard, R. J., Parker, J. C. and S. Mishra, "On the Correspondence Between Brooks-Corey and Van Genuchten Models," *Journal of Irrigation and Drainage Engineering, 15*(4), 744–751, 1989.

Letey, J. and W. J. Farmer, "Movement of Pesticides in Soil," in *Pesticides in Soil and Water*, W. D. Guenzi (Ed.), Soil Science Society of America, Madison, WI, 1974.

Levenspiel, O., *Chemical Reaction Engineering*, Chapter 9, John Wiley, New York, 1972.

Leverett, M. C., "Capillary Behavior in Porous Media," *Trans. A. I. M. E. 142*, 341–358, 1941.

Levich, V. G., *Physicochemical Hydrodynamics*, Prentice-Hall, Englewood Cliffs, N. J., 1962.

Lindstrom, F. T., R. Haque, V. H. Freed, and L. Boersma, "Theory on the Movement of Some Herbicides in Soils: Linear Diffusion and Convection of Chemicals in Soils," *J. Environ. Sci. Tech., 1*(7), 561–565, 1967.

List, E. J. and N. H. Brooks, "Lateral Dispersion in Saturated Porous Media," *Journ. Geophys. Res., 72*, 2531–2541, 1967.

Loehr, R. C. and R. J. Charbeneau, "Understanding the Fate and Transport of Contaminants in the Unsaturated Zone of Soil," Groundwater Quality Protection Pre-Conference Workshop Proceedings, Water Pollution Control Federation 61st Annual Conference, Dallas, TX, October 1988.

Lohman, S. W., "Ground-Water Hydraulics," Geological Survey Professional Paper 708, United States Gov. Printing Office, Washington, D. C., 1972.

Lyman W. J., W. F. Reehl and D. H. Rosenblatt, *Handbook of Chemical Property Estimation Methods Environmental Behavior of Organic Compounds*, McGraw-Hill, New York, 1982.

Mackay, D. M., P. V. Roberts, and J. A. Cherry, "Transport of Organic Contaminants in Groundwater," *Environ. Sci. Technol., 19*(5), 384–392, 1985.

Mackay, D. M., Freyberg, D. L., Roberts, P. V., and J. A. Cherry, "A Natural Gadient Experiment on Solute Transport in a Sand Aquifer, 1. Approach and Overview of Plume Movement," *Water Resour. Res., 22*(13), 2017–2029, 1986.

Marle, C. M., *Multiphase Flow in Porous Media*, Gulf Publishing Company, Houston, 1981.

Marshall, T. J., "A Relation between Permeability and Size Distribution of Pores," *J. Soil Sci., 9*, 1–8, 1958.

Marsily, G. de, *Quantitative Hydrogeology. Groundwater Hydrology for Engineers*, Academic Press, Orlando, 1986.

Maxwell, J. B., *Data Book on Hydrocarbons*, van Nostrand Company. Princeton, NJ, 1950.

McCord, J. T., D. B. Stephens and J. L. Wilson, "Toward Validating State-Dependent Macroscopic Anisotropy in Unsaturated Media: Field Experiments and Modeling Considerations," *J. Contaminant Hydrology, 7*, 145, 1991.

McDonald, J. M. and A. W. Harbaugh, "A Modular Three-Dimensional Finite-Difference Groundwater Flow Model," *Techniques of Water Resources Investigations of the United States Geological Survey, Book 6*, 586 p, 1988.

Mein, R. G. and C. L. Larson, "Modeling Infiltration Dring a Steady Rain," *Water Resour. Res., 9*, 384–394, 1973.

Meinzer, O. E., "Outline of Ground-Water Hydrology," United States Geological Survey Water-Supply Paper 494, 1923.

Meinzer, O. E. (Ed.), *Hydrology*, Dover, New York, 1942.

Mercer, J. W. and R. M. Cohen, "A Review of Immiscible Fluids in the Subsurface: Properties, Models, Characterization, and Remediation," *J. Contaminant Hydrology, 6*, 107–163, 1990.

Mercer, J. W., D. C. Skipp, and D. Griffin, "Basic Pump-and-Treat Ground-Water Remediation Technology", R. S. Kerr Environmental Research Laboratory, Ada, OK, United States Environmental Protection Agency, EPA/600/8–90/003, March 1990.

Messing, I., "Estimation of the Saturated Hydraulic Conductivity in Clay Soils from Soil Moisture Retention Data," *Soil Sci. Soc. Am. J., 53*(3), 665–668, 1989.

Meyer, P. D., Rockhold, M. L., Nichols, W. E. and G. W. Gee, "Hydrologic Evaluation Methodology for Estimating Water Movement through the Unsaturated Zone at Commercial Low-Level Radioactive Waste Disposal Sites," United States Nuclear Regulatory Commission, Washington DC, NUREG/CR–6346, 1996.

Miller, C. A., and P. Neogi, *Interfacial Phenomena*, Surfactant Science Series. v. 17. Marcel Dekker, Inc, 1985.

Miller, C. T., Poirier-McNeill, M. M., and A. S. Mayer, "Dissolution of Trapped Nonaqueous Phase Liquids: Mass Transfer Characteristics," *Water Resour. Res., 26*(11), 2783–2796, 1990.

Miller, R. J., J. W. Biggar, and D. R. Nielsen, "Chloride Displacement in Panoche Clay Loam in Relation to Water Movement and Distribution," *Water Resour. Res., 1*(1), 63–73, 1965.

Millington, R. J., "Gas Diffusion in Porous Media," *Science,* 130, 100–102, 1959.

Millington, R. J. and J. M. Quirk, "Permeability of Porous Solids," *Trans. Faraday Soc., 57,* 1200–1207, 1961.

Milne-Thomson, L. M., *Jacobian Elliptic Function Tables*, Dover Publications Inc., New York, 1950.

Monad, J., *Recherches sur la Croissance des Cultures Bacteriennes*, Herman & Cie, Paris, 1942.

Moore, R. E., "Water Conduction from Shallow Water Tables," *Hilgardia 12*, 383–426, 1939.

Moore, W. J., *Physical Chemistry*, 4th Ed., Prentice-Hall, Inc., Englewood Cliffs, N. J., 1972.

Morel-Seytoux, H. J., "Analytical-Numerical Method in Waterflooding Predictions," *Soc. Petrol. Engin. J.,* 247–258, 1965.

Morel-Seytoux, H. J., "Unit Mobility Ratio Displacement Calculations for Pattern Floods in Homogeneous Medium," *Soc. Petrol. Engin. J.* , 217–227, 1966.

Morel-Seytoux, H. J., "Two-Phase Flows in Porous Media," in *Advances in Hydroscience, Vol. 9*, V. T. Chow, Ed., Academic Press, New York, 1973.

Morel-Seytoux, H. J. and D. Khanji, "Derivation of an Equation of Infiltration," *Water Resour. Res. 10*, 795–800, 1974.

Morel-Seytoux, H. J., "Some Recent Developments in Physically Based Rainfall-Runoff Modeling," in *Frontiers in Hydrology*, Water Resources Publications, Littleton, Colorado, 1984.

Morel-Seytoux, H. J., "Conjunctive Use of Surface and Ground Waters," in *Artificial Recharge of Groundwater,* T. Asano, Ed., Butterworth Publishers, Boston, MA, 1985.

Morel-Seytoux, H. J., "Chapter 3 Multiphase Flows in Porous Media," in *Developments in Hydraulic Engineering—4*, P. Novak (Ed.), Elsevier Applied Science, New York, 1987.

Morrow, N. R., Chatzis, I. and J. J. Taber, "Entrapment and Mobilization of Residual Oil in Bead Packs," *SPE Reservior Engineering, 3*(3), 927–934, 1988.

Mualem, Y., "A New Model for Predicting the Hydraulic Conductivity of Unsaturated Porous Media," *Water Resour. Res., 12,* 513–522, 1976.

Mualem, Y., "A Catalog of Hydraulic Properties of Unsaturated Soils," Israel Institute of Technology, Haifa, Israel, 1976a.

Mualem, Y., "Hydraulic Conductivity of Unsaturated Soils: Prediction and Formulas," *Methods of Soil Analysis, Part 1,* A. Klute (Ed.), American Society of Agronomy, Madison, WI, 799–823, 1986.

Mull, R., "The Migration of Oil-Products in the Subsoil with Regard to Groundwater Pollution by Oil," in Proceedings of the *Fifth International Conference on Advances in Water Pollution Research*, Vol. 2, edited by S. A. Jenkins, p. HA 7(A)/1–8, Pergammon, New York, 1971.

Mull, R., "Calculations and Experimental Investigations of the Migration of Oil Products in Natural Soils," paper presented at the International Symposium on Groundwater Pollution by Oil Hydrocarbons, *Int. Assoc. Hydrol.*, Prague, June 5–9, 1978.

Muskat, M. and M. W. Meres, *Physics 7*, 346, 1936.

Muskat, M., Wyckoff, R. D., Botset, H. G., and M. W. Meres, "Flow of Gas Liquid Mixtures through Sands," *Trans. A. I. M. E. Petrol. 123*, 69–96, 1937.

Muskat, M., *The Flow of Homogeneous Fluids Through Porsous Media*, J. W. Edwards, Inc., Ann Arbor, MI, 1946.

National Research Council, *Health Effects of Exposure to Low Levels of Ionizing Radiation*, National Academy Press, Washington, D. C., 1990.

National Research Council, *Opportunities in the Hydrologic Sciences*, P. S. Eagleson, (Ed.), National Academy Press, Washington, D. C., 1992.

National Research Council, *In Situ Bioremediation, When Does it Work?*, National Academy Press, Washington, D. C., 1993.

National Research Council, *Alternatives for Ground Water Cleanup*, National Academy Press, Washington, D. C., 1994.

National Research Council, *Groundwater and Soil Cleanup: Improving Management of Persistent Contaminats,* National Academy Press, Washington, D. C., 1999.

Narasimhan, T. N. and Y. Kanehiro, "A Note on the Meaning of Storage Coefficient," *Water Resour. Res., 16*(2), 423–429, 1980.

Narasimhan, T. N., "Reply," *Water Resour. Res., 19*(6), 1636–1640, 1983.

NCRP, *Natural Background Radiation in the United States*, Report 45, National Council on Radiation Protection and Measurements, Washington, D. C., 1975.

Nelson, W. R., "Evaluating the Environmental Consequences of Groundwater Contamination: Parts 1 and 2," *Water Resour. Res., 14*(3), 409–428, 1978.

Neuman, S. P., "Theory of Flow in Unconfined Aquifers Considering Delayed Response of the Water Table," *Water Resour. Res., 8*, 1031–1045, 1972.

Neuman, S. P., "Analysis of Pumping Test Data from Anisotropic Unconfined Aquifers Considering Delayed Gravity Response," *Water Resour. Res., 11*, 329–342, 1975.

Neuman, S. P., "Wetting Front Pressure Head in the Infiltration Model of Green and Ampt," *Water Resour. Res., 12* (3), 564–566, 1976.

Neuman, S. P., "A Eulerian-Lagrangian Numerical Scheme for the Dispersion-Convection Equation Using Conjugate Space-Time Grids," *J. Computational Physics, 41*, 270–294, 1981.

Neuman, S. P., "Computer Prediction of Subsurface Radionuclide Transport—An Adaptive Numerical Method," United States Nuclear Regulatory Commission, NUREG/CR–3076, 1983.

Nielsen, D. R., and J. W. Biggar, "Measuring Capillary Conductivity," *Soil Science, 92,* 192–193, 1961.

Nielsen, D. R., R. D. Jackson, J. W. Cary, and D. D. Evans, *Soil Water*, American Society of Agronomy, Madison, WI, 1972.

Nielsen, D. R., J. W. Biggar, and K. T. Erh, "Spatial Variability of Field-Measured Soil-Water Properties," *Hilgardia, 42*(7), 215–259, 1973.

Nielsen, D. R., van Genuchten, M. Th., and J. W. Biggar, "Water Flow and Solute Transport Processes in the Unsaturated Zone," *Water Resour. Res., 22*(9), 89–108S, 1986.

Nofziger, D. L., and J. R. Williams, "Interactive Simulation of the Fate of Hazardous Chemicals During Land Treatment of Oily Wastes: RITZ User's Guide," R. S. Kerr Environmental Research Laboratory, Ada, Oklahoma, United States Environmental Protection Agency, EPA/600/8–88–001, January 1988.

Ogata, A. and R. B. Banks, "A Solution of the Differential Equation of Longitudinal Dispersion in Porous Media," *United States Geol. Survey, Prof. Paper no. 411–A*, 1961.

Ogata, A., "Theory of Dispersion in a Granular Medium," *United States Geol. Surv. Prof. Paper 411–I*, 34p., 1970.

Olson, R. S., and D. E. Daniel, "Measurement of the Hydraulic Conductivity of Fine-Grained Soils," *Permeability and Groundwater Contaminant Transport, STP 746*, T. F. Zimmie and C. O. Riggs (eds.), Am. Soc. for Testing and Materials, Philadelphia, PA, 18–64, 1981.

Osborne, M. and J. Sykes, "Numerical Modeling of Immiscible Organic Transport at the Hyde Park Landfill," *Water Resourc. Res., 22,* 25–33, 1986.

Paetzold, R. F., de los Santos, A., and B. A. Matzkanin, "Pulsed Nuclear Magnetic Resonance Instrument for Soil-Water Content Measurement: Sensor Configurations," *Soil Science Society of America Journal, 51*(2), 287–290, 1987.

Pankow, J. F. and J. A. Cherry, *Dense Chlorinated Solvents*, Waterloo Press, Portland, Oregon, 1996.

Papadopulos, I. S. and H. H. Cooper, Jr., "Drawdown in a Well of Large Diameter," *Water Resour. Res., 3*, p. 241–244, 1967.

Papadopulos, S. S., Bredehoeft, J. D. and H. H. Cooper, Jr., "On the Analysis of 'Slug Test' Data," *Water Resour. Res., 9*, p. 1087–1089, 1973.

Parizek, R. R., and B. E. Lane, "Soil Water Sampling Using Pan and Deep Pressure-Vacuum Lysimeters," *Journal of Hydrology,* 11, 1–21, 1970.

Parker, B. L., Gillham, R. W. and J. A. Cherry, "Diffusive Disappearance of Immiscible-Phase Organic Liquids in Fractured Geologic Media," *Ground Water, 32*(5), 805–820, 1994.

Parker, J. C. and M. Th. van Genuchten, "Flux-Averaged and Volume-Averaged Concentrations in Continuum Approaches to Solute Transport," *Water Resour. Res., 20*(7), 866–872, 1984.

Parker, J. C. and A. J. Valocchi, "Constrains on the Validity of Equilibrium and First-Order Kinetic Transport Models in Structured Soils," *Water Resour. Res., 22(3)*, 399–408, 1986.

Parker, J. C., Lenhard, R. J., and T. Kuppusamy, "A Parametric Model for Constitutive Properties Governing Multiphase Flow in Porous Media," *Water Resour. Res., 23*, 618–624, 1987.

Parker, J. C. and R. J. Lenhard, "Vertical Integration of Three-Phase Flow Equations for Analysis of Light Hydrocarbon Plume Movement," *Transport in Porous Media, 5*, 187–206, 1989.

Parker, J. C., J. J. Kaluarachchi, V. J. Kremesec, and E. L. Hockman, "Modeling Free-Product Recovery at Hydrocarbon Spill Sites," in the Proceedings of Petroleum Hydrocarbons and Organic Chemicals in Ground Water: Prevention, Detection, and Restoration, Houston, TX, National Water Well Association, November, 1990.

Parlange, J. Y., I. Lisle, R. D. Braddock, and R. E. Smith, "The Three-Parameter Infiltration Equation," *Soil Sci. 133*, 337–341, 1982.

Parlange, J. Y., Vauclin, M., Haverkamp, R., and I. Lisle, "Note: The Rlation Between Desorptivity and Soil-Water Diffusivity," *Soil Sci., 139*, 458–461, 1985.

Parlange, J. Y. and R. Haverkamp, "Infiltration and Ponding Time," in *Unsaturated Flow in Hydrologic Modeling: Theory and Practice*, H. J. Morel-Seytoux, Ed., NATO ASI Series C: Mathematical and Physical Sciences, Vol. 275, 1989.

Peaceman, D. W., *Fundamentals of Numerical Reservoir Simulation*, Elsevier Scientific, Amsterdam, 1977.

Peaceman, D. W., "Interpretation of Well-Block Pressures in Numberical Reservoir Simulation" *Soc. Pet. Eng. J.*, 183–194, 1978.

Penman, H. L., "Natural Evaporation from Open Water, Bare Soil and Grass," *Proc. R. Soc. London Ser. A, 193*, 120–146, 1948.

Perkins, T. K. and O. C. Johnston, "A Review of Diffusion and Dispersion in Porous Media," *J. Soc. Petrol. Eng., 3*, 70–83, 1963.

Perroux, K. M., P. A. C. Raats, and D. E. Smiles, "Wetting Moisture Characteristic Curves Derived from Constant-Rate Infiltration into Thin Samples," *Soil Sci. Soc. Am. J., 46*(2), 231–234, 1982.

Perroux, K. M., and I. White, "Designs for Disc Permeameters," *Soil Sc. Am. J., 52*(5), 1205–1215, 1988.

Perry, R. H., and C. H. Chilton, *Chemical Engineering Handbook*, Fifth edition, McGraw-Hill, New York, 1973.

Pfannkuch, H. O., "Contribution à l'Étude des Déplacements de Fluides Miscibles dans un Milieu Poreux," *Revue de l'Institut Français du Petrole, 18,* 215–270, 1963.

Philip, J. R., "Numerical Solution of Equations of the Diffusion Type with Diffusivity Concentration Dependent," *Trans. Faraday Soc., 51,* 885–892, 1955.

Philip, J. R., "The Theory of Infiltration: 1. The Infiltration Euation and its Solution," *Soil Sci. 83,* 345–357, 1957a.

Philip, J. R., "The Theory of Infiltration: 4. Sorptivity and Algebraic Infiltration Equations," *Soil Sci. 84,* 257–264, 1957b.

Philip, J. R., "Evaporation, Moisture, and Heat Fields in the Soil," *J. Meteorol. 14,* 354–366, 1957c.

Philip, J. R., "Numerical Solution of Equations of the Diffusion Type with Diffusivity Concentration-Dependent II.," *Austr. J. of Physics 10,* 29–42, 1957d.

Philip, J. R., "Theory of Infiltration," in *Advances in Hydroscience, Vol. 5,* V. T. Chow, Ed., Academic Press, New York, 215–305, 1969.

Philip, J. R., "Approximate Analysis of the Borehole Permeameter in Unsaturated Soil," *Water Resour. Res., 21*(7), 1025–1033, 1985.

Phillips, F. M., Mattick, J. L., Duval, T. A., Elmore, D. and P. W. Kubik, "Chlorine 36 and Tritium from Nuclear Weapons Fallout as Tracers for Long-Term Liquid and Vapor Movement in Desert Soils," *Water Resour. Res., 24*(11), 1877–1891, 1988.

Pickens, J. F., R. E. Jackson, K. J. Inch, and W. F. Merritt, "Measurement of Distribution Coefficients Using a Radial Injection Dual-Tracer Test," *Water Resour. Res., 17*(3), 529–544, 1981.

Pickens, J. F. and G. E. Grisak, "Scale-Dependent Dispersion in a Stratified Granular Aquifer," *Water Resour. Res, 17*(4), 1191–1211, 1981.

Pinder G. F. and L. M. Abriola, "On the Simulation of Nonaqueous Phase Organic Compounds in the Subsurface," *Water Resourc. Res., 22,* 109S–119S, 1986.

Plehm, S. W., "Underground Tankage, the Liabilities of Leaks," Petroleum Marketing Education Foundation, Alexandria, VA, 1985.

Poiseuille, J. L. M., *Comptes Rendus, 11,* 961 and 1041; and *12,* 112, 1840.

Pollock, D. W., "User's Guide for MODPATH/MODPATH-PLOT, Version 3: A Particle Tracking Post-Processing Package for MODFLOW," the United States Geological Survey finite-difference ground-water flow model, United States Geological Survey Open-File Report 94–464, 234, 1994.

Powers, S. E., Abriola, L. M., and W. J. Weber, Jr., "An Experimental Investigation of Nonaqueous Phase Liquid Dissolution in Saturated Subsurface Systems: Steady State Mass Transfer Rates," *Water Resourc. Res., 28*(10), 2691–2705, 1992.

Prats, M., "The Breakthrough Sweep Efficiency of the Staggered Line Drive," *Petrol. Trans. AIME, 207,* 361–362, 1956.

Prats, M., W. R. Strickler, and C. S. Matthews, "Single-Fluid Five-Spot Floods in Dipping Reservoirs," *Petroleum Trans., AIME, 204,* 160–174, 1955.

Press, W. H., Teukolsky, S. A., Vetterling, W. T. and B. P. Flannery, *Numerical Recipes in Fortran, The Art of Scientific Computing,* Cambridge University Press, Cambridge, 1992.

Prickett, T. A., and C. G. Lonnquist, *Selective Digital Computer Technique for Groundwater Resource Evaluation, Bulletin 55,* Illinois State Water Survey, Urbana, 1971.

Pye, V. I. and J. Kelley, "The Extent of Groundwater Contamination in the United States," in *Groundwater Contamination,* National Academy Press, Washington, D. C, 1984.

Rainwater, K. A., W. R. Wise, and R. J. Charbeneau, "Parameter Estimation through Groundwater Tracer Tests," *Water Resour. Res., 23*(10), 1901–1910, 1987.

Raisbeck, J. M. and M. F. Mohtadi, "The Environmental Impacts of Oil Spills on Land in the Arctic Rgions," *Water Air Soil Pollut., 3,* 195–208, 1974.

Rao, P. S. C. and J. M. Davidson, "Estimation of Pesticide Retention and Transformation Parameters Required in Nonpoint Source Pollution Models," in *Environmental Impact of Nonpoint Source Pollution,* Overcash, M. R. and J. M. Davidson (Eds.), Ann Arbor Science, Ann Arbor, MI, 1980.

Rao, P. S. C., A. G. Hornsby, and R. E. Jessup, "Indices for Ranking the Potential for Pesticide Contamination of Groundwater. " *Proc. Soil Crop Sci. Soc.* FL. 44: 1–8, 1985.

Rawlins, S. L., and G. S. Campbell, "Water Potential: Thermocouple Psychrometry," in *Methods of Soil Analysis, Part 1,* A. Klute (Ed.), American Society of Agronomy, Madison, WI, 597–618, 1986.

Rawls, W. J. and D. L. Brakensiek, "Estimating Soil Water Retention from Soil Properties," *J. Irrig. and Drain. Div., ASCE, 108*(IR2), 166–177, 1982.

Rawls, W. J. and D. L. Brakensiek, "Prediction of Soil Water Properties for Hydrologic Modeling," in *Proceedings of Symposium on Watershed Management,* p. 293–299, American Society of Civil Engineers, New York, 1985.

Rawls, W. J. and D. L. Brakensiek, "Estimation of Soil Water Retention and Hydraulic Properties," in *Unsaturated Flow in Hydrologic Modeling: Theory and Practice,* H. J. Morel-Seytoux (Ed.), NATO ASI Series C: Mathematical and Physical Sciences, Vol. 275, Kluwer Academic Publishers, Dordrecht, 1989.

Rawls, W. J., Ahuha, L. R., Bradensiek, D. L. and A. Shirmohammadi, "Infiltration and Soil Water Movement," in *Handbook of Hydrology,* Maidment, D. R. (Ed.), McGraw-Hill, Inc., New York, 1993.

Reible, D. D., Illangasekare, T. H., Doshi, D. V. and M. E. Malhiet, "Infiltration of Immiscible Contaminants in the Unsaturated Zone," *Ground Water, 28*, 685–692, 1990.

Remson, I., Hornberger, G. M., and F. J. Molz, *Numerical Methods in Subsurface Hydrology*, Wiley (Interscience), New York, 1971.

Reynolds, W. D., D. E. Elrick and B. E. Clothier, "The Constant Head Well Permeameter: Effort of Unsaturated Flow," *Soil Sci., 139*(2), 172–180, 1985.

Reynolds, W. D., and D. E. Elrick, "A Laboratory and Numerical Assessment of the Guelph Permeameter Method," *Soil Science,* Vol. 144, p. 282–299, 1987.

Reynolds, W. D., and D. E. Elrick, "Determination of Hydraulic Conductivity Using a Tension Infiltrometer," *Soil Science Society of American Journal, 55*(3), p. 633–639, 1991.

Rhee, H., B. F. Bodin and N. R. Amundson, "A Study of the Shock Layer in Equilibrium Exchange Systems," *Chem. Eng. Sci., 26*, 1571–1580, 1971.

Richards, L. A., "Capillary Conduction of Liquids through Porous Medium," *Physics 1*, 318–333, 1931.

Richards, L. A., "A Pressure-Membrane Extraction Apparatus for Soil Suction," *Soil Science, 51*, p. 377–386, 1941.

Richards, S. J., and L. V. Weeks, "Capillary Conductivity Values from Moisture Yield and Tension Measurements in Soil Columns," *Soil Sci. Soc. Am. Proc., 17*, 206–209, 1953.

Ritzema, H. P., (Editor-in-Chief), *Drainage Principles and Applications*, International Institute for Land Reclamation and Improvement, ILRI Publication 16, Second Edition, Wageningen, The Netherlands, 1994.

Roberts, P. V., Goltz, M. N., and D. A. Mackay, "A Natural Gradient Experiment on Solute Transport in a Sand Aquifer, 3. Retardation Estimates and Mass Balances for Organic Solutes," *Water Resourc. Res., 22*(13), 2047–2059, 1986.

Rorabaugh, M. I., "Graphical and Theoretical Analysis of Step-Drawdown Test of Artesian Wells," *Trans. Am. Soc. Civil Engrs., 79*, 362, 1953.

Rouse, H. and S. Ince, *History of Hydraulics*, Dover Publications, Inc., New York, 1957.

Rubin, J., "Numerical Method for Analyzing Hysteresis-Affected, Post-Infiltration Redistribution of Soil Moisture," *Soil Sci. Soc. Am. Proc., 31*, 13–20, 1967.

Rubin, J., and R. V. James, "Dispersion-Affected Transport of Reacting Solutes in Saturated Porous Media: Galerkin Method Applied to Equilibrium-Controlled Exchange in Unidirectional Steady Water Flow," *Water Resour. Res., 9*(5), 1332–1356, 1973.

Russell, T. F. and M. F. Wheeler, "Finite Element and Finite Difference Methods for Continuous Flows in Porous Media," *The Mathematics of Reservoir Simulation*, R. E. Ewing (Ed.), SIAM, Chapter II, p. 35–106, 1983.

Russo, D. and E. Bresler, "Soil Hydraulic Properties as Stochastic Processes: I. An Analysis of Field Spatial Variability," *Soil Sci. Soc. Am. J., 45*, 682–687, 1981.

Saffman, P. G., "A Theory of Dispersion in Porous Media," *J. Fluid Mech., 6*(6), 321–349, 1959.

Sallam, A., W. A. Jury, and J. Letey, "Measurement of Gas Diffusion Coefficient Under Relatively Low Air-Filled Porosity," *Soil Sci. Soc. Amer. J. 48* , 3–6, 1984.

Sauty, J-P., "An Analysis of Hydrodispersive Transfer in Aquifers," *Water Resour. Res., 16*(1), 145–158.

Scanlon, B. R., "Evaluation of Moisture Flux from Chloride Data in Desert Soils," *J. Hydrology, 128*, 137–156, 1991.

Scheidegger, A. E., "Statistical Hydrodynamics in Porous Media," *J. Appl. Phys., 52*, 994–1001, 1954.

Scheidegger, A. E., "General Theory of Dispersion in Porous Media," *J. Geophys. Research, 66*(10), 3273–3278, 1961.

Scheidegger, A. E., *The Physics of Flow through Porous Media,* University of Toronto Press, Toronto, 1974.

Schiegg, H. O., "Consideration on Water, Oil, and Air in Porous Media," *Water Science and Technology, 17*, 467–476, 1985.

Schroeder, P. R., B. M. McEnroe, R. L. Peyton and J. W. Sjostrom, "The Hydrologic Evaluation of Landfill Performance (HELP) Model," Vol. 4, Documentation for Version 2, IA #DW 21931425–01–3, United States Army Engineer Waterways Experiment Station, October 1989.

Schwarzenbach, R. P. and J. Westall, "Transport of Nonpolar Organic Compounds from Surface Water to Groundwater: Laboratory Sorption Studies," *Environ. Sci. Technol., 15*(11), 1360–1367, 1981.

Schwille, F., "Petroleum Contamination of the Subsoil—A Hydrological Problem," in *The Joint Problems of the Oil and Water Industries*, P. Hepple (Ed.), The Institute of Petroleum, London, 1967.

Scow, K. M., "Rate of Biodegradation," in *Chemical Property Estimation Methods*, Lyman, W. J., W. F. Reehl, and D. H. Rosenblatt (Eds.), McGraw Hill, New York, 1982.

Seidell, A., *Solubilities*, 1, 4th Ed., American Chemical Society, D. van Nostrand Co., Princeton, N. J., 1958.

Shearer, R. C., J. Letey, W. J. Farmer, and A. Klute, "Lindane Diffusion in Soil," *Soil Sci. Soc. Amer. Proc., 37*, 189–194, 1973.

Short, T. E., "Movement of Contaminants from Oily Wastes During Land Treatment," in *Soils Contaminated by Petroleum: Environmental and Public Health Effects*, Calabrese, E. J. and P. T. Kostecki (Eds.), Wiley, New York, 317–330, 1988.

Simmons, C. S., "A Stochastic-Convective Transport Representation of Dispersion in One-Dimensional Porous Media Systems," *Water Resour. Res., 18*, 1193–1214, 1982.

Sisson, J. B., Ferguson, A. H., and M. Th. van Genuchten, "Simple Methods for Predicting Drainage from Field Plots," *Soil Sci. Soc. Amer. J., 44*, 1147–1152, 1980.

Smith, R. E., "Approximate Soil Water Movement by Kinematic Characteristics," *Soil Sci. Soc. Amer. J., 47*, 3–8, 1983.

Smith, R. E. and J. Y. Parlange, "A Parameter-Efficient Hydrologic Infiltration Model," *Water Resour. Res. 14*, 533–538, 1978.

Smith, V. J. and R. J. Charbeneau, "Probabilistic Soil Contamination Exposure Assessment Procedures," *J. Environ. Eng., 116*(6), 1143–1163, 1990.

Smoller, J., *Shock Waves and Reaction-Diffusion Equations*, Springer-Verlag, New York, 1983.

Soil Conservation Service, *SCS National Engineering Handbook, Section 4, Hydrology*, United States Department of Agriculture, Washington, D. C., 1972.

Sokol, D., "Position and Fluctuations of Water Level in Wells Perforated in More than One Aquifer," *J. Geophysical Research, 68*(4), 1079–1080, 1963.

Solley, W. B., Pierce, R. R. and H. A. Perlman, "Estimated Use of Water in the United States in 1995," United States Geological Survey Circular *1,200*, p. 71, 1998.

Spencer, W. F., W. J. Farmer, and M. M. Cliath, "Pesticide Volatilization," *Residue Rev., 49*, 1–47, 1973.

Spencer, W. F., W. J. Farmer, and W. A. Jury, "Review: Behavior of Organic Chemicals at Soil, Air, Water Interfaces as Related to Predicting the Transport and Volatilization of Organic Pollutants," *Environ. Toxicol. Chem., 1*, 17–26, 1982.

Sposito, G. and W. A. Jury, "The Lifetime Probability Density Function for Solute Movement in the Subsurface Zone," *J. Hydrology, 102*, 503–518, 1988.

Sposito, G., R. E. White, P. R. Darrah, and W. A. Jury, "A Transfer Function Model of Solute Transport through Soil, 3. The Convection-Dispersion Equation," *Water Resour. Res., 22*(2), 255–262, 1986.

Squillance, P. J., Zogorski, J. S., Wilber, W. G. and C. V. Price, "Preliminary Assessment of the Occurrence and Possible Sources of MTBE in Groundwater in the United States, 1993–1994," *Environmental Science and Technology, 30*(5), 1721–1730, 1996.

Stannard, D. I., "Tensiometers—Theory, Construction, and Use," *Ground Water and Vadose Zone Monitoring, ASTM STP 1053,* D. M. Neilsen and A. I. Johnson (Eds.), American Society for Testing and Materials, Philadelphia, PA, p. 34–51, 1990.

Steenhuis, T. S., W. Staubitz, M. S. Andreini, J. Surface, T. Richard, R. Paulsen, N. B. Pickering, J. R. Hagerman, and L. D. Geohring, "Preferential Movement of Pesticides and Tracers in Agricultural Soils," *J. Irrig. and Drain., 116*, 50, 1990.

Strack, O. D. L., *Groundwater Mechanics,* Prentice Hall, Englewood Cliffs, N. J., 1989.

Stephens, D. B., and S. P. Neuman, "Vadose Zone Permeability Tests: Summary," *J. Hydr. Div., ASCE, 108*(5), 623–639, 1982.

Stephens, Daniel B. & Associates, Inc., "Review of Methods to Estimate Moisture Infiltration, Recharge and Contaminant Migration Rates in the Vadose Zone for Site Risk Assessment," American Petroleum Institute, Washington DC, (Draft), September 1995.

Sudicky, E. A., "A Natural Gradient Experiment on Solute Transport in a Sand Aquifer: Spatial Variability of Hydraulic Conductivity and its Role in the Dispersion Process," *Water Resour. Res., 22*, 2069–2082, 1986.

Texas Department of Water Resources, "Water for Texas, A Comprehensive Plan for the Future," Volume 1, Austin, 1984.

Theis, C. V., "The Relationship Between the Lowering of the Piezometric Surface and the Rate and Duration of Discharge of a Well Using Groundwater Storage," *Trans. Amer. Geophys. Union, 2*, p. 519–524, 1935.

Terzaghi, K., "Die Berechnung der Durchlassigkeitsziffer des Tones aus dem Verlauf der hydrodynamischen Spannungserscheinungen," *Sitz. Akad. Wiss.* Wien, Austria, *132*, 125–138, 1923.

Thibault, D. H., Sheppard, M. I. and P. A. Smith, "A Critical Compilation and Review of Default Soil Solid/Liquid Partition Coefficients, K_d, for Use in Environment Assessments," Atomic Energy of Canada Limited Research Company, AECL–10125, Pinawa, Manitoba, ROE 1L0, March 1990.

Thibodeaux, L. J. and S. T. Wang, "Landfarming of Petroleum Wastes— Modeling the Air Emission Problem," *Environmental Progress, 1*(1), 42–46, 1982.

Thiem, G., "Hydrologische Methoden", J. M. Gebhardt, Leipziq, 1906.

Thomas, R. G., "Volatilization from Soil," in *Chemical Property Estimation Methods*, Lyman, W. J., W. F. Reehl, and D. H. Rosenblatt (Eds.), McGraw Hill, New York, 1982.

Todd, D. K., *Ground Water Hydrology*, Wiley, New York, 1959.

Topp, G. C. and E. E. Miller, "Hysteresis Moisture Characteristics and Hydraulic Conductivities for Glass-Bead Media," *Soil Sci. Soc. Am. Proc., 30*, 156–162, 1966.

Topp, G. C., A. Klute, and D. B. Peters, "Comparison of Water Content— Pressure Head Data Obtained by Equilibrium, Steady State, and Unsteady State Methods," *Soil Sci. Soc. Am. Proc., 31*(3), 312–314, 1967.

Topp, G. C., Davis, J. L. and A. P. Annan, "Electromagnetic Determination of Soil Water Content: Measurements in Coaxial Transmission Lines," *Water Resourc. Res., 16*, 574–582, 1980.

Topp, G. C. and J. L. Davis, "Measurement of Soil Water Using Time-Domain Reflectometry (TDR): Field Evaluation," *Soil Sci. Soc. Am. J., 49*, 19–24, 1985.

Toth, J., "A Theory of Groundwater Motion in Small Drainage Basins in Central Alberta," *J. Geophys. Res., 67*, 4375–4387, 1962.

Toth, J., "A Theoretical Analysis of Groundwater Flow in Small Drainage Basins," *J. Geophys. Res., 68*, 4795–4812, 1963.

Travis, C. C. and E. L. Etnier, "A Survey of Sorption Relationships for Reactive Solutes in Soil," *J. Environ. Qual., 10*(1), 8–17, 1981.

Trescott, P. C., G. F. Pinder, and S. P. Larson, "Finite-Difference Model for Aquifer Simulation in Two Dimensions with Results of Numerical Experiments," Book 7, Chapter C1, United States Geological Survey, 1976.

UNESCO, Union of Soviet Socialist Republics National Committee for the International Hydrologic Decade, "World Water Balance and Water Resources of the Earth," English translation, *Studies and Reports in Hydrology, 25*, UNESCO, Paris, 1978.

United States Environmental Protection Agency, "Waste Disposal Practices and Their Environmental Effects on Ground Water," *Report to Congress,* Washington, D. C., 512, 1977.

United States Environmental Protection Agency, "Planning Workshops to Develop Recommendations for a Ground Water Protection Strategy," appendixes, Washington, D. C., 171, 1980.

United States Environmental Protection Agency, "Part II Underground Storage Tanks; Technical Requirements and State Programs; Final Rule," *Federal Register, Vol. 53*, No. 185, 1988.

United States Environmental Protection Agency, "Limiting Values of Radionuclide Intake and Air Concentration and Dose Conversion Factors for Inhalation, Submersion, and Ingestion," Federal Guidance Report No. 11, Office of Radiation Programs, EPA–520/1–88–020, Washington, D. C., 1988.

United States Environmental Protection Agency, "Evaluation of Ground-Water Extraction Remedies, Volume 1," Summary Report, Office of Emergency and Remedial Response, Washington, D. C., September, EPA/540/2–89/054, 1989.

United States Environmental Protection Agency, "Evaluation of Ground-Water Extraction Remedies: Phase II, Volume 1, Summary Report," Office of Emergency and Remedial Response, Washington, D. C., Publication 9355.4–05, February 1992.

United States Environmental Protection Agency, "Cleaning Up the Nation's Waste Sites: Markets and Technology Trends," Office of Solid Waste and Emergency Response, Washington, D. C., EPA 542–R–92–102, 1993.

United States Environmental Protection Agency, "How to Effectively Recover Free-Products for Leaking Underground Storage Tank Sites: A Guide for State Regulators," draft report, Office of Underground Storage Tanks, Washington, D. C., August 1995.

United States Environmental Protection Agency, Soil Screening Guidance: Technical Background Document," Washington, D. C., EPA 540/R–95/128, 1996.

United States Salinity Laboratory, "Diagnosis and Improvement of Saline and Alkali Soils," *United States D. A. Agr. Handbook 60*, 1954.

Valentine, R. L. and J. L. Schnoor, "Biotransformation," in *Vadose Zone Modeling of Organic Pollutants*, Hern, S. C. and S. M. Melancon (Eds.), Lewis Publishers, Inc., Chelsea, MI, 1986.

Valocchi, A. J., "Validity of the Local Equilibrium Assumption for Modeling Sorbing Solute Transport Through Homogeneous Soils," *Water Resour. Res., 21*(6), 808–820, 1985.

Van Bavel, C. H. M., N. Underwood, and R. W. Swanson, "Soil Moisture Measurement by Neutron Moderation," *Soil Sci., 73*, 91–104, 1956.

Van Dam, J., "The Migration of Hydrocarbons in a Water-Bearing Stratum," in *The Joint Problems of the Oil and Water Industries*, P. Hepple (Ed.), The Institute of Petroleum, London, 1967.

Van de Pol, R. M., P. J. Wierenga, and D. R. Nielsen, "Solute Movement in a Field Soil," *Soil Sci. Soc. Am. J., 41*(1), 10–13, 1977.

Van Genuchten, M. T. and P. J. Wierenga, "Mass Transfer Studies in Sorbing Porous Media: Analytical Solutions," *Soil Sci. Soc. Am. J., 40*, 473–479, 1976.

Van Genuchten, M. T. and P. J. Wierenga, "Mass Transfer Studies in Sorbing Porous Media: Experimental Evaluation with Tritium," *Soil Sci. Soc. Am. J., 41*, 778–785, 1977.

Van Genuchten, M. T., "A Closed-Form Equation for Predicting the Hydraulic Conductivity of Unsaturated Soil," *Soil Sci. Soc. Am. J.,* Vol. *44*, 892–898, 1980.

Van Genuchten, M. Th., "Convective-Dispersive Transport of Solutes Involved in Sequential First-Order Decay Reactions," *Computers & Geosciences, 11*(2), 129–147, 1985.

Van Genuchten, M. Th., Leij, F. J. and S. R. Yates, "The RETC Code for Quantifying the Hydraulic Functions of Unsaturated Soils," United States Environmental Protection Agency, EPA/600/2–91/065, 1991.

Vauclin, M., "Flow of Water and Air: Theoretical and Experimental Aspects," in *Unsaturated Flow in Hydrologic Modeling: Theory and Practice*, H. J. Morel-Seytoux, Ed., NATO ASI Series C: Mathematical and Physical Sciences, Vol. 275, 1989.

Walton, W. C., Selected analytical methods for well and aquifer evaluation, Bulletin 4-9, Illinois State Water Survey, Urbana, 81 pp., 1962.

Walton, W. C., *Groundwater Resource Evaluation*, McGraw-Hill, New York, 1970.

Walton, W. C., *Groundwater Pumping Tests—Design and Analysis*, Lewis Publishers, Chelsa, MI, 1987.

Ward, J. C., "Turbulent Flow in Porous Media," *Proc. Amer. Soc. Civ. Engrs, 90* (HY5), 1–12, 1964.

Weaver, J. W., "A Kinematic/Dynamic Model of Nonaqueous Phase Liquid Transport in the Vadose Zone," Ph. D. Dissertation, Department of Civil Engineering, The University of Texas at Austin, 1988.

Weaver, J. W. and R. J. Charbeneau, "Hydrocarbon Spill Exposure Assessment Modeling," in *Proceedings of the Petroleum Hydrocarbons and Organic Chemicals in Ground Water: Prevention, Detection and Restoration Conference*, National Water Well Association and the American Petroleum Institute, Houston, TX, October 31–November 2, 1990.

Weaver, J. W., Charbeneau, R. J. and B. K. Lien, "A Screening Model for Nonaqueous Phase Liquid Transport in the Vadose Zone Using Green-Ampt and Kinematic Wave Theory," *Water Resourc. Res., 30*(1), 93–105, 1994.

Weaver, J. W., Charbeneau, R. J., Tauxe, J. D., Lien, B. K., and J. B. Provost, "The Hydrocarbon Spill Screening Model (HSSM)," Volume 1: User's Guide, United States Environmental Protection Agency, EPA/600/R–94/039 a, April 1994.

Weeks, I. V., and S. J. Richards, "Soil-Water Properties Computed from Transient Flow Data," *Soil Sci. Soc. Am. Proc., 31,* 721–735, 1967.

Wetherold, R. G. and W. D. Balfour, "Volatile Emissions from Land Treatment Systems," in Loehr, R. C. and J. F. Malina, Jr., (Eds.), *Land Treatment, A Hazardous Waste Management Alternative*, Water Resources Symposium Number Thirteen, Center for Research in Water Resources, The University of Texas at Austin, Austin, TX, 1986.

White, F. M., *Fluid Mechanics*, Second Edition, McGraw-Hill Book Company, New York, 1986.

White, R. E., J. S. Dyson, R. A. Haigh, W. A. Jury, and G. Sposito, "A Transfer Function Model of Solute Transport through Soil, 2. Illustrative Applications," *Water Resour. Res., 22*(2), 248–254, 1986.

Whitham, G. B., *Linear and Nonlinear Waves*, John Wiley & Sons, New York, 1974.

Wierenga, P. J., "Solute Distribution Profiles Computed with Steady-State and Transient Water Movement Models," *Soil Sci. Soc. Am. J., 41*, 1050–1055, 1977.

Wilkerson, M., D. Kim, and M. Nodell, "The Pesticide Groundwater Prevention Act: Setting Specific Numerical Values. " California Department of Food and Agriculture Rep. September 1984. State of California, Dep. of Food and Agriculture, Sacramento, CA, 1984.

Wilson, J. L. and P. J. Miller, "Two-Dimensional Plume in Uniform Ground-Water Flow," *J. Hydraulics Division, ASCE, Vol. 104*, No. HY4, 503–514, 1978.

Wilson, J. L. and S. H. Conrad, "Is Physical Displacement of Residual Hydrocarbons a Realistic Possibility in Aquifer Restoration?", in Proceeding of *Petroleum Hydrocarbons and Organic Chemicals in Ground Water*, NWWA, Houston, TX, p. 274–298, 1984.

Wilson, J. L., Conrad, S. H., Mason, W. R., Peplinski, W., and E. Hagan, "Laboratory Investigation of Residual Liquid Organics from Spills, Leaks, and the Disposal of Hazardous Wastes in Groundwater." R. S. Kerr Environmental Research Laboratory, United States Environmental Protection Agency, Ada, Oklahoma, EPA/600/6–90/004, April 1990.

Wise, W. R. and R. J. Charbeneau, "In Situ Estimation of Transport Parameters: A Field Demonstration," *Ground Water*, *32*(3), 420–430, 1994.

Wyckoff, R. D. and H. G. Botset, "The Flow of Gas-Liquid Mixture through Unconsolidated Sands," *Physics 7*, 325–345, 1936.

Wyllie, M. R. J. and G. H. F. Gardner, "The Generalized Kozeny-Carman Equation," *World Oil, 146*, 210–228, 1958.

Zheng, C., "MT3D, A Modular Three-Dimensional Transport Model for Simulation of Advection, Dispersion, and Chemical Reactions of Contaminants in Groundwater Systems," S. S. Papadopulos & Associates, Inc., Rockville, Maryland, 1990.

Index